# DNA and Biotechnology

**Companion Web site**

Ancillary materials are available online at:
www.elsevierdirect.com/companions/9780120489305

# DNA and Biotechnology

## Third Edition

Molly Fitzgerald-Hayes and
Frieda Reichsman

AMSTERDAM • BOSTON • HEIDELBERG • LONDON • NEW YORK • OXFORD • PARIS
SAN DIEGO • SAN FRANCISCO • SINGAPORE • SYDNEY • TOKYO

Academic Press is an imprint of Elsevier

Academic Press is an imprint of Elsevier
30 Corporate Drive, Suite 400, Burlington, MA 01803, USA
525 B Street, Suite 1900, San Diego, California 92101-4495, USA
84 Theobald's Road, London WC1X 8RR, UK

**Library of Congress Cataloging-in-Publication Data**
Fitzgerald-Hayes, Molly.
   DNA and biotechnology : the awesome skill / Molly Fitzgerald-Hayes, Frieda Reichsman.
      p. cm.
   ISBN 978-0-12-048930-5 (hard cover : alk. paper)
  1. DNA. 2. Biotechnology—Research. I. Reichsman, Frieda. II. Title.
   QP624.F58 2010
   660.6'5–dc22

                                                          2009021293

**British Library Cataloguing-in-Publication Data**
A catalogue record for this book is available from the British Library.

ISBN: 978-0-12-048930-5

For information on all Academic Press publications
visit our Web site at www.elsevierdirect.com

# Contents

# Contents

## Acknowledgments

### Very Special Thank You to Our Families

The authors extend their deepest appreciation to their families for their love, patience, and constant support during the years we worked on this book. We could not have done it without you.

### Thank you to UMass Biochemistry and Molecular Biology (BMB) Students

The authors thank the BMB undergraduate majors at the University of Massachusetts at Amherst who made special contributions to the book: Stephanie Murphy, who wrote about RNA interference (RNAi); Ervin Pejos, who wrote about carbon-based nanotechnology in medicine; Anne Osborn, who conducted updated information about human genetic diseases in the post-human genome era. Thank you also to Katerina Papa, Nana Biney, and Chris Broderick.

### Thank You to the Professionals

The authors wish to express their deep appreciation to the scholars who made significant contributions to the book in their areas of scientific expertise: Sharon Thoma, Robin Wesselschmidt, Bernice Schachter, Ron Michaelis, and Alan H. McGowan.

### Thank You to Anonymous Chapter Reviewers

The authors also wish to thank the anonymous reviewers who took the time and had the patience to provide the helpful comments that we used to improve the book.

### The Buck Stops Here

The authors are responsible for the contents of this book, including any mistakes. Others who contributed to drafts of the book are not responsible for the final product after the text was revised—this was our task. In the end, the authors take full responsibility for getting it right, and if not, our most sincere apologies are extended—as well as the hope that you will contact us with corrections at dnabiotechnology3@gmail.com

We hope you enjoy reading *DNA and Biotechnology*!

Molly Fitzgerald-Hayes and Frieda Reichsman

## ABOUT THE AUTHORS

As friends and colleagues for many years, we've taught university non-science majors, retirees, clubs, Elderhostel classes, and various community groups about DNA, genes, and the human genome. We've enjoyed giving hands-on DNA workshops for K-12 teachers and middle and high school students, and we have even taught people how to isolate their own DNA from the cheek cells inside their mouths (safe, painless, cheap, and easy!).

The invitation to write the new edition of Alcamo's book brought a surge of anticipation and excitement, and years of work! We hope that the book reflects Ed Alcamo's clear and fluent writing style about science. To bring the 3rd edition as up to date as is possible (for a printed book), we added several new chapters covering advances in gene therapy, stem cells, drug design and development, bioinformatics, and animal and plant biotechnology.

## ABOUT OUR TARGET AUDIENCE

We wrote *DNA and Biotechnology* for a wide audience that includes college and high school students as well as laypeople with varying backgrounds in science. Our goal for the book is to help people to better understand how genes control cell function, how persistent and ingenious scientists tracked down the means of control, and how we as a society now use this knowledge to explore further, to heal, and to attempt to improve life. We tried to make this book very readable without oversimplifying the picture of a living cell, full of the thousands of molecular machines made from protein and RNA parts, reading the DNA genome, and performing the functions that make life possible.

## ABOUT LEARNING FEATURES IN THE BOOK

The 3rd edition includes features designed to make learning about DNA much easier:

**Chapter Outline:** Each chapter starts with a convenient outline that gives you a succinct overview of how the topics and subtopics interrelate.

**Hot Topic Box:** Every chapter draws in readers with a recent, attention-grabbing news headline, brief story, and explanation of its relevance to the scientific information in the chapter.

**Looking Ahead:** This section presents broad learning objectives that orient readers to fundamental goals for understanding and communicating about the chapter topics.

**Special Topic Boxes:** These sections emphasize people and scientific discoveries that have a special connection or relevance to chapter topics.

**Boldface Terms:** New terms are introduced using boldface type. All boldface terms are defined in the Glossary at the end of the book.

**Summary Statements:** Short summary statements punctuate each chapter, helping readers to identify important points and orient themselves when reviewing the information.

**Summary:** The summary at the end of each chapter brings together the key points and relates them to each other more immediately than the full chapter treatment allows.

**Review Questions:** Ten broad-based review questions provide an opportunity for readers to test their recall and comprehension of the information in the chapter.

**Additional Reading:** Recommendations include both current and historically relevant sources (books, newspapers, magazine and research articles, and web sites) that help readers delve further into topics of particular interest.

**Glossary:** A glossary at the end of the book defines the boldface terms introduced in each chapter.

## ABOUT USING THE BOOK

### Organization of Chapters (3rd edition)

Chapter 1: The Roots of DNA Research
Chapter 2: The DNA Double Helix
Chapter 3: DNA in Action
Chapter 4: Tools of the DNA Trade
Chapter 5: Working with DNA

The 16 chapters in *DNA and Biotechnology* are arranged in three groups that give teachers the flexibility to select chapters based on the scientific background of the students in the class. The first five chapters (Chapters 1–5), form a core of content that is essential for understanding the rest of the book. This core includes the basic structure and functions of DNA, RNA, and proteins in cells (Chapters 1 and 2), explains how gene expression controls cell function (Chapter 3), and describes the recombinant DNA cloning technologies that fundamentally changed DNA research (Chapters 4 and 5).

Building on the foundation of the first five chapters, Chapters 6–10 provide an opportunity for readers who are curious about the role of DNA and genes in modern research. Automated DNA sequence analysis has become routine, and hundreds of genome sequences have been analyzed, in addition to the entire human genome (Chapter 6). The deluge of primary sequences has fed the emerging bioinformatics field, which uses information technology to store, explore, and annotate DNA, RNA, and protein sequences (Chapter 7). DNA technology has enabled us to seek out and identify specific DNA sequences in many contexts, with applications in fields such as criminal forensics (fingerprinting) and medical diagnostics (Chapter 8). Molecular genetics research explains how multiple genome mutations can cause a cell to lose growth control and turn into a cancer cell (Chapter 9). The chromosome locations of genes and mutations that cause many genetic diseases such as cystic fibrosis, sickle cell anemia, muscular dystrophy, Huntington's disease, and many more, have been identified (Chapter 10). These chapters tie together the basic functions of genes and proteins in cells (Chapters 1–5) with the advances in DNA based technologies in the research lab, many of which harness the same molecules. The biological mechanisms employed by the cell to replicate DNA and make RNA have led to the development of the most important techniques used in molecular biology and genetics.

Chapters 11–16 focus on several specialized applications of DNA biotechnology. For example, finding a mutant gene that causes a genetic disease opens the door to the possibility of a gene therapy treatment (Chapter 11). The science of human embryonic stem cells is described, as are the very exciting iPS (induced pluripotent stem) cells, which are derived from adult human skin cells but look and act like embryonic stem cells (Chapter 12). New genetic strategies for designing drugs and the development of nanocarriers that deliver drugs directly into cells are just two examples of new areas of pharmaceutical research (Chapter 13). Advances include transgenic animals and plants genetically engineered to produce antibiotics, drugs, and hormones (Chapters 14 and 15). DNA research shows that individual human genomes are almost identical in DNA sequence and that all people have exactly the same genes. What does this mean about our understanding of "race"? (Chapter 16).

# The Roots of DNA Research

## My Genome, Myself: Seeking Clues in DNA

The New York Times, November 17, 2007

By Amy Harmon

The exploration of the human genome has long been relegated to elite scientists in research laboratories. But that is about to change. An infant industry is capitalizing on the plunging cost of genetic testing technology to offer any individual unprecedented—and unmediated—entree to their own DNA.

For as little as $1,000 and a saliva sample, customers will be able to learn what is known so far about how the billions of bits in their biological code shape who they are. Three companies have already announced plans to market such services, one yesterday.

Offered the chance to be among the early testers, I agreed, but not without reservations. What if I learned I was likely to die young? Or that I might have passed on a rogue gene to my daughter? And more pragmatically, what if an insurance company or an employer used such information against me in the future?

But three weeks later, I was already somewhat addicted to the daily communion with my genes. (Recurring note to self: was this addiction genetic?)

[To read on, go to http://tinyurl.com/2zuqsh.]

From the preceding article, we can see that we are at the start of an era of personalized genomes. How long will it be before, alongside tongue depressors, cotton balls, and blood pressure cuffs, a plastic card with your DNA chip becomes a routine part of a visit to the doctor's office? The article represents the type of advances that 150 years of research has brought us, research that started with the studies discussed in this chapter.

We will begin with the work of a monk in the 1850s whose different ideas and meticulous methods of investigation yielded a foundation from which the next two generations of scientists could grow, toward an understanding of heredity at the molecular level. And grow they did, developing the roots of what we now know to be DNA science. As you will see, researchers often had to struggle against preconceived notions of what could (and could not) be the biological material that transferred characteristics from one generation to the next. It took persistence to establish DNA, a material with only four components, as a carrier of information, when proteins, which have 20 components, were much more familiar and well understood. (In fact we can wonder about what preconceived notions the scientists and students of the future will discover concerning our generation). Here we will retrace the steps of these persistent, open-minded scientists who paved our way to DNA.

## LOOKING AHEAD

DNA technology has its foundations in genetics, the science of heredity. It is appropriate, therefore, to open this book by exploring the insights and experiments that led scientists to recognize DNA as the hereditary substance. When you have completed the chapter, you should be able to do the following:

- Understand the differences between prokaryotic and eukaryotic cells.
- Recognize how the experiments of Gregor Mendel focused attention on cellular factors as the basis for inheritance.

- Understand the circumstances under which Mendel's experiments were verified and how Sutton related Mendel's "factors" to cellular units called chromosomes.
- Show how Morgan related eye color in fruit flies to chromosomes.
- Appreciate the origin of the term "gene" and describe how the gene concept emerged.
- Recount Miescher's work on nuclei, and conceptualize how Feulgen and Mirsky contributed to the insight that genes are composed of DNA.
- Understand the significance of Griffith's experiments in bacterial transformation, and conceptualize how the transforming principle was identified as DNA.
- Explain the seminal experiments of Hershey and Chase, and describe why their results pointed to DNA as the substance controlling protein and nucleic acid synthesis.
- Increase your vocabulary of terms relating to DNA technology.

## INTRODUCTION

In past centuries, it was customary to explain **inheritance** by saying, "it's in the blood." People believed that children received blood from their parents and that a union of bloods led to the blending they saw in one's characteristics. Such expressions as "blood relations," "blood will tell," and "bloodlines" reflect this belief.

However, by the 1850s, scientists were questioning the blood theory of inheritance. They could see quite clearly that semen contained no blood, and it was apparent that blood was not being transferred to the offspring. But if blood was not the hereditary substance, then what was?

It was a long road to understanding that DNA mediates inheritance. By the end of the 1800s, the blood basis of heredity was challenged and eventually discarded. In its place, scientists developed an interest in nucleic acid molecules organized into functional units called **genes**. Scientists guessed that genes control heredity by specifying the production of proteins. But even the gene basis of heredity was hard to believe because the amount of nucleic acid in the cell seemed insignificant.

The gene basis for heredity has become one of the foundation principles of biology. In the pages ahead, we will explore the development of the gene theory and note how interest grew in DNA as the substance of the gene. Long before scientists could apply the fruits of DNA research to modern technology, they had to learn what DNA was all about. "What purpose," they asked, "does DNA serve in a living cell?"

---

### Box 1.1   Cell Geography Sets the Stage

The term "cell," in general, refers to a small room or compartment. The smallest compartment of an organism that is considered to be alive is a cell, and so it is regarded as the fundamental unit of life. Cells are the natural environment for all the processes discussed in this book. Getting the lay of the land, then, is important in understanding DNA and how it works.

There are two kinds of biological cells: **prokaryotic** and **eukaryotic**.

Prokaryotic cells (Figure 1.1A) contain one continuous space in which cellular materials are organized, but not separated by membranes. A cell wall surrounds all prokaryotic cells. Prokaryotes are usually single-celled organisms and include both bacteria and archaea. The archaea live in extreme environments (for example, boiling hot thermal vents, freezing cold arctic waters, and oil wells).

In contrast, eukaryotic cells (Figure 1.1B) contain subcellular compartments. A membrane surrounds each compartment, and there are several different types of compartments, called **organelles**. Eukaryotic cells are usually 10- to 100-fold larger than prokaryotic cells (though there are a few exceptionally large bacteria that defy this rule). Eukaryotes include both single-celled organisms (the majority) and multicellular organisms. Of the eukaryotes, only plant cells are surrounded by a cell wall, and its composition is very different than a prokaryotic cell wall.

In eukaryotes, each type of cellular compartment is specialized. The **nucleus** is home to the vast majority of the DNA, where is it complexed with proteins. The endoplasmic reticulum and the Golgi complex compartments are involved in protein synthesis and trafficking (that is, sending proteins to their correct destinations). Mitochondria possess a membrane specialized for energy production. Chloroplasts are centers for photosynthesis.

A prokaryote's DNA is tightly coiled with proteins in a region called the **nucleoid**. The nucleoid is not a compartment; it is just the DNA and proteins compacted together. In prokaryotes, some membrane-associated functions, such as energy production, are accomplished by the plasma membrane.

The presence or absence of subcellular compartments leads to differences in how cellular events take place. Compartments allow for increased complexity and regulation. For example, in prokaryotes, RNA synthesis and protein synthesis take place in the same compartment, so while RNA is being made from a DNA template, it can also be read nearly simultaneously to synthesize a protein. In eukaryotes, RNA synthesis occurs in the nucleus, and protein synthesis occurs in the cytoplasm. The separation or uncoupling of these processes into compartments allows

Box 1.1 Continued

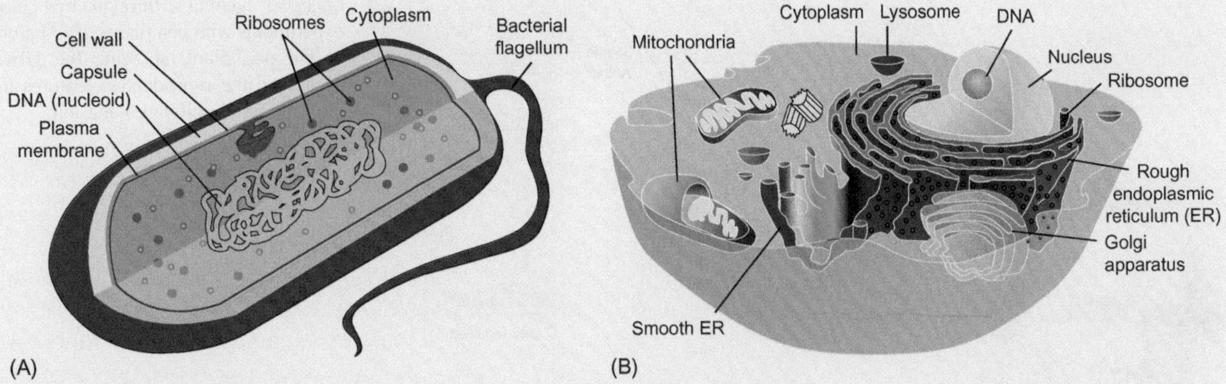

**FIGURE 1.1** (A) A generalized prokaryotic cell. (B) A generalized eukaryotic cell.

for intermediate steps to occur between them. For example, splicing of RNA in different patterns allows several different proteins to be produced from a single gene. This is one of the contributions to the complexity of eukaryotes as compared to prokaryotes.

Features common to prokaryotic and eukaryotic cells include **ribosomes**, the molecular machines that synthesize proteins (although the exact makeup of the prokaryotic and eukaryotic ribosomes is different), the plasma membrane, and the watery interior environment, the **cytoplasm**.

## DEVELOPING A THEORY OF INHERITANCE

### When Counting Counts: Mendel's Approach Yields the Basis of Modern Gene Theory

In the mid-1800s (around the same time that scientists started questioning the blood theory of inheritance), a relatively obscure Austrian monk named Gregor Mendel (pictured in Figure 1.2) was conducting experiments to reveal the statistical pattern of inheritance. Mendel's great contribution to science was the discovery of a predictable mechanism by which inherited characteristics move from parents to offspring. His work with plants laid the groundwork for intensive studies in genetics, a science that would blossom in the early part of the twentieth century.

Mendel lived in a region that relied heavily on agriculture, so it was not uncommon for educated individuals to have an interest in animal and plant breeding. Mendel had studied plant science at the University of Vienna, and he continued his interest in plants at the monastery at Brno (now a part of the Czech Republic). He began a series of experiments to learn more about the breeding patterns of pea plants. Peas were well suited for his work because they were easy to cultivate. Moreover, they had a short growing season, they could be fertilized artificially, and they resisted interference by foreign pollen.

Other important features of pea plants were their easily distinguished **traits**. Mendel observed, for example, that his garden had some pea plants with wrinkled

seeds and others with smooth seeds; some had green pods, and others had yellow pods; some had white flowers, and others had red flowers. Figure 1.3 shows this diversity. The more Mendel pondered the source of variations, the more his curiosity was aroused. He set out to determine how the variations originated and how the traits were passed to the next generation.

A key ingredient in Mendel's success was the plant he used to track inherited traits. The short generation time, obvious characteristics, and ease of breeding of pea plants provided a fertile ground (so to speak) for his observations and experiments to flourish.

Mendel studied pea plants by crossing plants having a certain characteristic with others having a contrasting characteristic. He then studied how traits were expressed in the offspring plants. Mendel found, for example, that by breeding selected tall plants to selected short plants, he could obtain plants that were exclusively tall. The trait for shortness had apparently disappeared. But when he bred the tall plants from this first generation among themselves, some short plants reappeared in the next generation among the tall plants. These results were unexpected and perplexing.

Mendel's forté was mathematics. He carefully counted the plants displaying a particular characteristic and the plants having the contrasting characteristic (for example, tall plants and short plants); he discovered

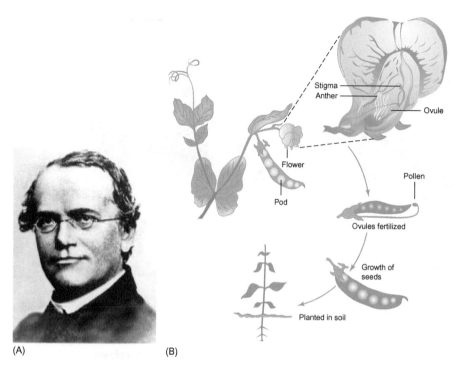

**FIGURE 1.2** Mendel and his pea plants. (A) Gregor Mendel (1822–1884), the Austrian monk who established the principles of genetics through meticulous experiments with pea plants. (B) Anatomy of the pea plant, showing the growth cycle and the reproductive features that make artificial pollination feasible.

(A)          (B)

similar ratios of traits among the offspring. He noted, for example, that crossing the first generation's tall plants among themselves always seemed to yield three tall plants for every short plant, as Figure 1.4 shows. (By that time, the monks in the monastery were noticing that peas had become a fairly regular item on the dinner table.)

Many scientists of the 1850s believed that a single factor controlled a trait, but Mendel, reasoning that one of the factors was obtained from the male and one from the female, began with the assumption that each trait was controlled by two factors (although the nature of the factor was unknown). He guessed that the factors express themselves in the offspring, but that one is **dominant** over the other. For example, the factor for tall plants dominates over the factor for short plants (it suppresses the short-plant factor, which is said to be **recessive**). The factors are then passed on to the next generation. Today we know Mendel's factors as **genes**.

From his work, Mendel developed a theory of inheritance completely at odds with the blood basis of heredity. Mendel's results implied that sperm and egg cells, not blood cells, carry the factors of inheritance. Moreover, Mendel surmised that the factors are discrete units, not some vague, mysterious elements of the blood. Aware of the unconventional nature of his suppositions, Mendel avoided controversy by keeping his suppositions largely to himself.

Mendel's theory came to be known as the theory of transmissible factors. Although it was revolutionary for the times, Mendel did not stop there. For many years he investigated how one factor in the pair dominates the other factor and how a pair of factors separates during transmission to the next generation. He experimented up to the early 1860s and published his results in 1866 in the *Proceedings of the Society of Natural Sciences in Brno*. Mendel included a detailed analysis of his theories in the publication, and he communicated his findings to other scientists of the times through a series of letters. In retrospect, Mendel's observations are regarded as one of the great insights in science and the beginning of the discipline of **genetics**.

Mendel's assumptions were different from those of other scientists studying the same topic, so he was led to interpret his observations differently, developing the concept of transmissible factors we now know to be genes.

Unfortunately, scientists of his time paid little attention to Mendel's work or its implications. One probable reason is that they had little understanding of biological chemistry. Another is that they failed to appreciate the significance of the cellular nucleus, the chromosomes, or the process of fertilization. Also, during the late 1800s, biologists were largely immersed in studying the theory of evolution, first promulgated in 1859 in Charles Darwin's epic work *On the Origin of Species*. Research on inheritance and breeding was placed on the proverbial back burner as the biological, social, and economic implications of the theory of evolution continued to capture the attention and imagination of scientists and laypeople. Not until the year 1900 would interest in genetics once again come to the forefront of science.

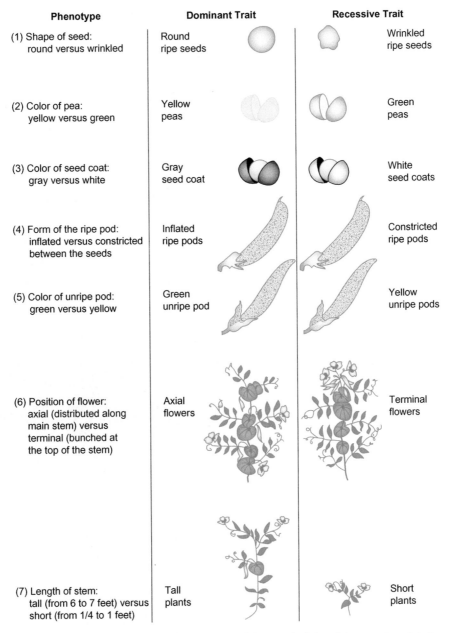

| Phenotype | Dominant Trait | | Recessive Trait |
|---|---|---|---|
| (1) Shape of seed: round versus wrinkled | Round ripe seeds | | Wrinkled ripe seeds |
| (2) Color of pea: yellow versus green | Yellow peas | | Green peas |
| (3) Color of seed coat: gray versus white | Gray seed coat | | White seed coats |
| (4) Form of the ripe pod: inflated versus constricted between the seeds | Inflated ripe pods | | Constricted ripe pods |
| (5) Color of unripe pod: green versus yellow | Green unripe pod | | Yellow unripe pods |
| (6) Position of flower: axial (distributed along main stem) versus terminal (bunched at the top of the stem) | Axial flowers | | Terminal flowers |
| (7) Length of stem: tall (from 6 to 7 feet) versus short (from 1/4 to 1 feet) | Tall plants | | Short plants |

**FIGURE 1.3**  The traits of pea plants studied by Mendel. The dominant allele is on the left and the recessive allele is on the right. A description of the phenotypes associated with specific alleles (traits) is provided.

In the spring of 1900, three European botanists, working independently of each other, repeated and verified Mendel's work. Each botanist cited Mendel's article in his research, and each awakened the scientific community to the work of the pioneering monk. It was not so unusual that all three should be aware of Mendel's work, but it was remarkable that the rediscovery of his theories was made almost simultaneously by three investigators; indeed, the happenstance remains one of the unusual coincidences of scientific history. Within weeks, a wave of enthusiasm for inheritance research sprang up. The discoveries made by

Mendel had been forgotten for almost 40 years. Now they would change scientific thinking forever.

## Morgan's Fruit Fly Experiments Reveal That Mendel's Factors Are on Chromosomes

During the first years of the twentieth century, Mendel's experiments were carefully studied, and the belief emerged that Mendel's factors were related to parts of the cell called **chromosomes**. Chromosomes (literally "colored bodies") are threadlike strands of chemical material located in the cell nucleus. The threads

Parents = P generation

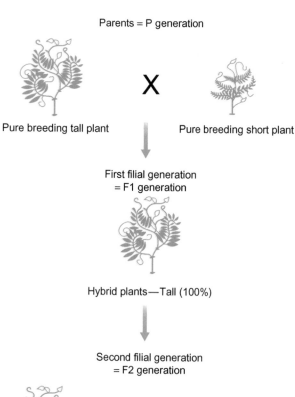

Pure breeding tall plant                     X                     Pure breeding short plant

First filial generation
= F1 generation

Hybrid plants—Tall (100%)

Second filial generation
= F2 generation

Tall plants 651(75.87%)                                    Short plants 207(24.13%)

**FIGURE 1.4** Mendel's experiments with tall and short pea plants. Mendel bred purebred tall plants to purebred short plants (these constitute the P generation). He discovered that all the offspring plants were tall in the first filial (F1) generation. He then bred the tall plants of the F1 generation among themselves and found that short plants appeared as well as tall plants in the F2 generation. His meticulous calculations revealed that about 75% of the plants in the F2 generation were tall, and 25% were short. This 75% to 25% ratio was equivalent to 3:1. This led to his assumption that two "factors" for height exist in pea plants and suggested that one factor dominates over the other.

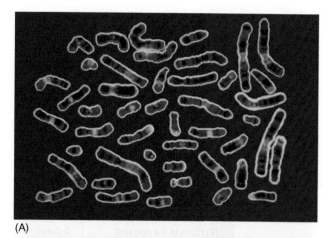

(A)

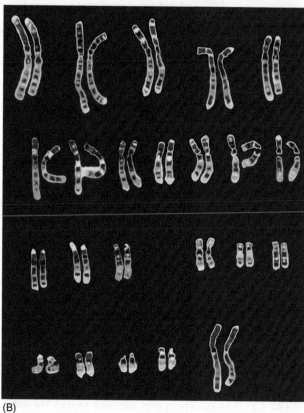

(B)

**FIGURE 1.5** Color-enhanced human chromosomes seen under a microscope after separation from the nucleus. Chromosomes assume these compact shapes during cell division. With the notable exceptions of reproductive cells and red blood cells, 46 chromosomes (23 pairs) are present in each human cell. (A) Each individual chromosome is outlined in white. (B) In a karyotype, chromosomes photographed under the microscope are cut and pasted into an orderly display from largest to smallest, except for the sex chromosomes, which are placed at the end. Images and text: Copyright © 2009 by Photo Researchers, Inc. All rights reserved.

of each chromosome consolidate and become clearly visible under the microscope (Figure 1.5) when a cell is dividing. With few exceptions, all human body cells have 46 chromosomes, and the 46 chromosomes are organized in 23 pairs. (Red blood cells have no chromosomes, and sperm and egg cells have only 23 chromosomes.) It is now known that chromosomes contain the DNA that carries the cell's genetic message.

Among the leaders in chromosome research at the turn of the century was the American biologist W. H. Sutton. In 1902, Sutton wrote that certain of Mendel's rules of inheritance could be explained if Mendel's factors were located on or in the chromosomes. Mendel had written, for instance, that inheritance factors occur in pairs, one member of the pair received from each

parent. By 1900, cell biologists had established that chromosomes also occur in pairs, one chromosome derived from each parent. Moreover, Mendel theorized that during the production of sperm and egg cells,

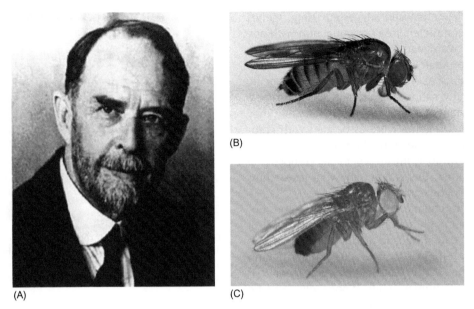

**FIGURE 1.6**  Morgan and his fruit flies. (A) Thomas Hunt Morgan, whose experiments revealed that colorless white eyes in fruit flies are based on the presence of a single chromosome. His experiments related an inherited characteristic to a chromosome. (B) A fruit fly with red eyes. (C) A fruit fly with white eyes.

the paired factors separate and move as units to each cell. Studies in cell biology showed that chromosomes behave similarly during cell reproduction. Sutton pointed out that chromosomes could be the hypothetical inheritance factors Mendel thought responsible for heredity. Perhaps, he suggested, chromosomes and inheritance factors were identical.

To demonstrate the validity of the chromosomal theory of inheritance, scientists had to relate at least one trait to a cell's chromosome. But in the early 1900s, the members of a chromosome pair could not be distinguished from each other visually. Thus, it was impossible to relate a single trait to a single chromosome by sight alone.

The problem was resolved in 1910 by Thomas Hunt Morgan of Columbia University (pictured in Figure 1.6). Morgan used the fruit fly *Drosophila melanogaster* in his work. By careful observations, he determined that one of the four pairs of chromosomes in the fruit fly determines its sex. This chromosome pair, he discovered, also determines colorless white eyes. Through an exhaustive series of genetic crosses and statistical analyses, Morgan determined that the male fruit fly inherits only one chromosome for sex determination. Thus, it must also inherit only one chromosome for white eye color. Therefore, white eye color must depend on a single chromosome. By providing statistical evidence for the relationship between sex and eye color in *Drosophila*, Morgan placed the chromosomal theory of inheritance on a firm footing and enhanced the role of the chromosome as the possible vehicle of inheritance.

Morgan, like Mendel, made careful counts and analyzed them to arrive at firm conclusions about inheritance. By doing so, he was able to provide ample evidence supporting Sutton's explanation that inheritance factors resided on chromosomes.

The next question was whether the whole chromosome or a part of a chromosome is responsible for an inherited trait. Writing in 1903, Sutton proposed that merely a part of a chromosome is the basis for a trait because not enough chromosomes are possible to account for all an individual's traits. Sutton suggested that "the chromosome may be divisible into smaller entities." Most other scientists agreed, and before long, the concept of the gene as the "smaller entity" gained prominence.

## Factors Become Genes, and DNA Is Discovered

In the early 1900s, geneticists began using the terms "inheritance unit" and "genetic particle" to describe the factors occurring on the chromosomes of Mendel's pea plants. By the 1920s, however, these terms had been discarded, and at the suggestion of the Scandinavian scientist Willard Johannsen, geneticists agreed to use the word "gene" instead. ("Gene" is derived from the Greek *gennan*, meaning "to produce.") The term was originally used as part of Darwin's word "pangenesis" to describe the theory that the whole body (including every atom and unit) "produces" itself over and over.

In a 1910 article, Johannsen suggested using "gene" because it was completely free of connection with any hypothesis. The word was and continues to be a less cumbersome term than "inheritance unit."

Scientists of the 1920s viewed the gene as a specific and separate entity sitting on the cell's chromosomes, but not strictly a part of a chromosome. Fortunately, they also reasoned that if genes were associated with chromosomes, a first step in learning the chemical nature of the gene would be to learn the chemical composition of the chromosomes; this is precisely what researchers attempted to do in the early 1900s.

One possible chemical component of chromosomes was a seemingly unique organic compound of the cell nucleus called nucleic acid. Nucleic acid was first described in 1869 by the Swiss researcher Johann Friedrich Miescher. With great difficulty, Miescher separated nuclei from human white blood cells, and he searched for evidence of protein within these nuclei. Instead of protein, however, he found a substance unlike any class of chemicals then known. Miescher named the new substance "nuclein" (relating to its source). When he identified phosphorus in nuclein, he postulated that the substance was a storehouse for phosphorus in the cell.

As Miescher continued his study of nuclein, he located it in yeast cells and kidney, liver, and testicular cells. He also made the notable observation that nuclein was abundant in sperm cells obtained from a salmon. Some years later, chemists led by Phoebus Levene (Chapter 2) used this information in their studies and determined the components of nuclein. They gave it the more descriptive and technical name **deoxyribonucleic acid (DNA)**. Coincidentally, Levene was born in 1869, the year of Miescher's first report of nuclein.

> Looking for the material that makes up chromosomes, Miescher discovered in the nucleus an abundant, initially mysterious substance he dubbed nuclein. After further chemical characterization by others, it was named deoxyribonucleic acid (DNA).

By the 1920s, it was clear that chromosomes had a role in heredity, and the isolation of DNA from cell nuclei made this organic substance a good candidate for the hereditary substance. Interest in DNA was further strengthened by a 1924 discovery attributed to the German biochemist Robert Feulgen, who observed that a dye (now called Feulgen stain) turns bright purple when it reacts with DNA. (Figure 1.7 shows an example of this staining characteristic.) The dye could be used to locate cellular DNA and to determine the

**FIGURE 1.7** Stained chromosomes. A photomicrograph of a plant cell stained with Feulgen stain to highlight the chromosomes of the cell nucleus. In this view, the chromosomes have replicated and are in the process of separating into two newly forming cells.

**TABLE 1.1** A comparison of the DNA content in the tissue cells and sperm cells of various animals

| Organism | Tissue cells | Sperm cells |
|----------|-------------|-------------|
| Cow | 6.6 | 3.3 |
| Human | 6.4 | 3.2 |
| Chicken | 2.6 | 1.3 |
| Frog | 15.0 | 7.5 |

concentration of DNA at various times in the cell's life cycle. Isolating DNA was a difficult chore at that time, but Feulgen's dye technique allowed DNA research to leap ahead without the burden of complex chemical isolations.

Another observation in that period helped forge the link between DNA and heredity. In the late 1920s, Alfred Mirsky and his coworkers at New York's Rockefeller Institute reported that, with only two exceptions, all cells of an organism have virtually the same amount of DNA in their nuclei. The two exceptions are reproductive sperm and egg cells. These cells contain precisely half the amount of DNA found in nonreproductive cells such as muscle cells. Table 1.1 presents these data. Mirsky's observation correlated with the theory that sex cells are the vehicle for bringing half the genetic information from each parent to the offspring.

The Rockefeller group led by Mirsky also experimented with the **zygote**, the cell resulting from the union of sperm and egg cells. The researchers found that the zygote contains the same amount of DNA as other cells of the body. This observation reinforced the

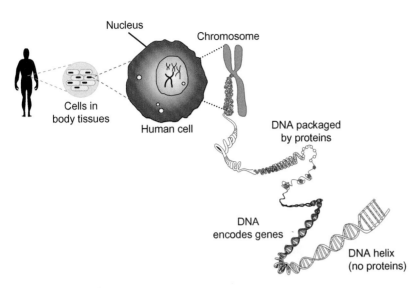

**FIGURE 1.8**   Relationship between DNA, genes, chromosomes, and cells in the human body.

Nucleus

Chromosome

Cells in body tissues

Human cell

DNA packaged by proteins

DNA encodes genes

DNA helix (no proteins)

notion that the DNA from two parents comes together during fertilization. In effect, it also verified the work of Mendel performed more than 60 years before because Mendel proposed that genetic factors separate when reproductive cells form, then come together in the offspring. The visual observations and chemical analyses of DNA led to the conclusion that the DNA was reduced by half during formation of the reproductive cells, then reconstituted in the offspring zygote. Figure 1.8 shows the relationship of DNA to other aspects of the body.

> The relative amount of DNA in the nuclei of resting, dividing, and reproductive cells made DNA a good contender for the substance of chromosomes, but direct evidence that DNA was the hereditary material was lacking.

## RELATING DNA TO HEREDITY

By the late 1920s, two concepts were evolving: genes are involved in heredity, and genes are composed of DNA. However, most evidence continued to be indirect and based on guesswork from laboratory deductions. Scientists needed direct evidence linking DNA to the development of observable traits. Two series of studies would provide that evidence. The first series was performed by Griffith, Alloway, Avery, and their colleagues. It drew a connection between DNA and the appearance of traits in bacteria. The second, performed by Hershey and Chase, linked DNA to the synthesis of protein. Both were landmark achievements in DNA technology; both connected enough dots to show that DNA is the substance of heredity.

## A "Transforming Factor" Changes the Inherited Characteristics of Bacteria

In 1928, a British medical officer named Frederick Griffith reported some puzzling results in his work with bacteria. Griffith was performing experiments with *Streptococcus pneumoniae*, a cause of bacterial pneumonia. The bacterium, commonly known as the pneumococcus, occurs in two strains. One strain is designated S because it forms smooth colonies when growing on bacteriological medium; the other strain is designated R because it forms rough colonies. (Both forms are shown in Figure 1.9.) It is well known that S strain pneumococci are lethal to mice (as well as to humans), whereas R strain pneumococci are harmless.

Griffith's initial experiments produced the expected results. He confirmed that the S strain pneumococci were deadly to mice and that the R strain bacteria were harmless. Then he performed some variations of these experiments. Griffith found that he could inject debris from dead S strain bacteria into mice without harming the animals. As before, these results were not surprising because the S strain bacteria were dead. However, what happened next was remarkable.

Griffith mixed a sample of live R strain pneumococci (harmless) together with debris from dead S strain pneumococci (also harmless because no living cells were present). Then he injected the mixture into the mice. By all expectations, the mice should have lived (both the S strain debris and the R strain bacteria were harmless). But pneumonia developed in the mice and they died, as Figure 1.10 displays. Why did the animals die? Griffith's answer came when he performed an autopsy on the animals: their lungs were full of live S strain pneumococci, the deadly strain. Apparently, the

 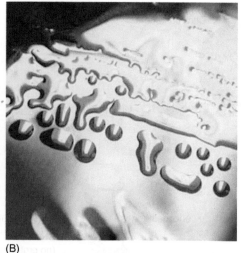

(A)  (B)

FIGURE 1.9 A macroscopic view of colonies of the two types of bacteria used by Griffith in his experiments in transformation. The bacteria are growing on the surface of a nutrient medium in a Petri dish and can be distinguished by their appearance (among other means). (A) The harmless R (rough) strain colonies of *Streptococcus pneumoniae*. (B) The disease-causing S (smooth) strain colonies.

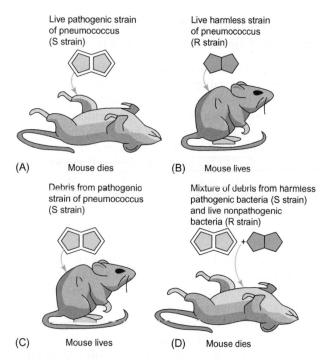

FIGURE 1.10 The transformation experiments performed by Griffith. (A) When live pathogenic pneumococci (S strain) were injected into mice, the animals died. (B) When live, harmless pneumococci (R strain) were injected, the animals remained healthy. (C) When debris from heat-killed S strain bacteria were injected into animals, they lived and were healthy. All these results were as anticipated. (D) However, when live, harmless bacteria (R strain) were mixed with cell debris of heat-killed S strain bacteria and the mixture was injected into animals, the animals died. On autopsy, Griffith found live, pathogenic bacteria (S strain); some of the live R strain bacteria had been transformed to live S strain bacteria.

harmless R strain bacteria had changed into deadly S strain bacteria, which killed the animals.

Griffith's "biochemical magic" appeared to work each time he performed the experiment. Moreover,

other researchers were able to confirm his findings shortly thereafter. Griffith postulated that something in the S strain chemical debris (a protein, he believed) was entering the R strain bacteria and "transforming" them by changing their biochemistry. Unfortunately, he was unable to identify the **transforming substance**. Nor would he live to see the significance of his work. Griffith died during the German air attacks on London in 1941.

Mixing living, harmless bacteria with dead, formerly disease-causing bacteria led to a "transformed" strain of living bacteria that caused disease. The "factor" that transformed the bacteria from harmless to harmful remained unknown.

Griffith's work did not go unnoticed by microbial geneticists. In 1933, James Lionel Alloway and his group at Rockefeller Institute successfully purified cell debris from S strain pneumococci and produced a **cell-free extract**. Then they used the extract to transform R strain bacteria to S strain bacteria. Moreover, they performed the transformation in test tubes, a procedure freeing them from the rigors of working with animals. Alloway noted that the transforming principle could be extracted from the mixture with alcohol and that the chemical appeared as "a thick, stringy precipitate … which slowly settled out on standing." From the description, modern biochemists recognize the transforming principle as DNA. At the time, however, Alloway was inclined to believe the substance was made of protein.

In 1935, Oswald T. Avery (Figure 1.11) and his associates, Colin MacLeod and Maclyn McCarty, began a series of exhaustive chemical analyses to identify the transforming substance. Beginning with crude extracts

**FIGURE 1.11** Oswald Avery, the Rockefeller Institute researcher who led the effort to identify DNA as the transforming principle. Avery's success focused attention on DNA as the chemical material of heredity.

from bacteria, they used a process of elimination to determine the nature of the chemical substance by finding out what it was not. As the months and years unfolded, they dismissed proteins, fats, and carbohydrates as possible candidates for the transforming substance. It appeared that the only possible thing left was nucleic acid. Finally, the search narrowed down to one nucleic acid: deoxyribonucleic acid (DNA). In a seminal article published in 1944, Avery and his colleagues presented evidence that DNA is the transforming principle (and opened the way to modern DNA technology). The researchers were not bold enough to claim that DNA was the hereditary material, but the implication was clear: DNA was apparently able to transform bacteria so dramatically that a harmless strain changed to a deadly strain.

Most scientists were blind to Avery's discovery and were reluctant to accept DNA as the hereditary substance. The majority of geneticists of the 1940s were not trained in biochemistry, and Avery's experiments were difficult for some to repeat. In addition, many scientists were hard pressed to believe that results obtained from bacteria could be applied to more complex organisms such as humans. Moreover, the results were published in the *Journal of Experimental Medicine,* a publication that the geneticists and bacteriologists of that period usually did not read; and the preoccupation with World War II had restricted the dissemination and flow of scientific knowledge while limiting funds for scientific research. Therefore, the impact of Avery's finding was lost.

Nevertheless, Avery's group continued their efforts to prove that DNA was more than an inert chemical

of the cell nucleus. With great difficulty they isolated a quantity of DNase from beef pancreas cells. DNase is a biological catalyst, an **enzyme** that destroys DNA but has no effect on other molecules. The investigators mixed their transforming substance with DNase and noted that the mixture lost its ability to transform bacteria. Still, many were not convinced.

Opposition was also voiced from the "protein supporters." These scientists believed that protein was the genetic material because protein appeared to have the necessary complexity to encode all the biochemical information in the cell's nucleus. Proteins are chains of 20 different and relatively simple organic molecules called amino acids (Chapter 2). A protein can be made of 10 amino acids or 1000 amino acids or 10,000 amino acids. The crucial element of any protein is the sequence of amino acids—that is, which amino acid follows which in the chain. This sequence of amino acids is as important to proteins as is the sequence of letters in a word ("ton" has a much different meaning than "not"). Protein supporters were of the opinion that the amino acid sequence in one protein serves as a model for constructing a new protein. Their outlook would be dealt a severe blow by the 1952 experiments of Hershey and Chase.

> Despite experimentation that eliminated all other known candidates as the transforming factor, there was difficulty in accepting the idea that DNA could be the agent of transformation and hence biological inheritance.

## Using Viruses, Hershey and Chase Establish DNA as the Agent of Inheritance

Alfred Hershey and Martha Chase (Figure 1.12) worked at the Cold Spring Harbor Laboratory in New York. The pair studied bacteria and the viruses that multiply within those bacteria (Figure 1.13). In 1952, scientists knew that certain viruses use bacteria as chemical factories for producing new viruses; however, the actual mechanism was uncertain. Biochemists were also aware that bacterial viruses are composed of a core of DNA enshrouded in a protein coat. What they did not know was whether the nucleic acid or the protein (or both) directs replication of the virus. Hershey and Chase would answer that question and in so doing, they would establish the essential role played by DNA in cellular biochemistry and inheritance.

Hershey and Chase made use of the observation that viral DNA contains phosphorus (P) but no sulfur (S). By contrast, the outer protein coat of the virus has sulfur (S) but no phosphorus (P). In their first experiments,

(A)  (B)

**FIGURE 1.12** (A) Alfred Hershey and (B) Martha Chase performed the key experiments demonstrating that DNA is responsible for directing the synthesis of viral proteins in a host cell.

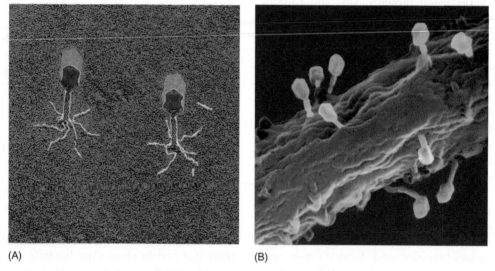

(A)  (B)

**FIGURE 1.13** Bacterial viruses. (A) Colored transmission electron microscopic view of viruses that bind to bacterial cells and replicate within them, thereby producing new viral particles. Viruses such as these are called bacteriophages ("bacteria-eaters"). Viruses consist of little more than a fragment of nucleic acid such as DNA (red) enclosed in a protein coat (orange). These viruses are using their tails as syringes to inject their viral DNA into an *E. coli* cell (blue). (B) Several viruses attacking *E. coli*. The viruses attach by their "tails" to the surface of the bacterium. The protein coat remains outside the bacterium while the DNA enters the bacterium to direct the replication of viruses. Images and text: Copyright © 2009 by Photo Researchers, Inc. All rights reserved.

Hershey and Chase cultivated viruses with the radioactive forms of phosphorus ($^{32}$P) and sulfur ($^{35}$S). They successfully prepared viruses whose nucleic acid was radioactive with $^{32}$P and whose protein was radioactive with $^{35}$S.

Now came the seminal experiments. Hershey and Chase mixed the radioactive viruses with a population of bacteria. Then they waited just long enough for viral replication to begin. At this point, they used an ordinary household blender (now a museum piece) to

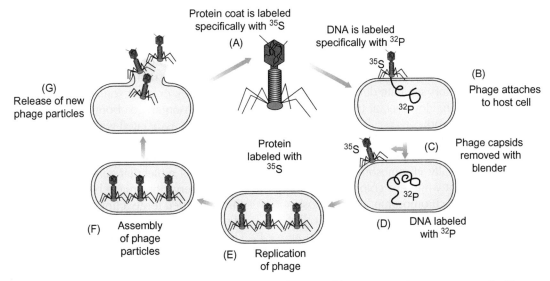

**FIGURE 1.14** The Hershey-Chase experiment with viruses (bacteriophages) and bacteria. (A) Viruses are prepared with two radioactive labels, one in the coat protein (³⁵S) and one in the DNA (³²P). (B) The viruses are combined with host bacterial cells and given an opportunity to interact. (C) Then the mixture is agitated in a blender, and the empty viral protein coats are removed, and are found to contain the radioactive ³⁵S. (D) The DNA carrying the radioactive ³²P is found within the host cell. (E) The phage replicate and (F) Assemble into new phage. (G) The new phage are then released from the cell. The results indicate that the phage DNA directs the synthesis of both the DNA and the protein added to make the new phage.

shear away any viruses and debris clinging to the bacterial surface.

Then the analysis began. Hershey and Chase tested the bacteria and surrounding fluid to find out where the radioactivity was. This would enable them to develop a biochemical glimpse of viral replication. After experimentally bursting the bacteria, the researchers found most of the ³²P within the contents of the bacterial cytoplasm. This finding indicated that viral DNA was entering the bacteria. Then they discovered that the ³⁵S was largely in the sheared-away remains of the viruses and in the surrounding fluid. This observation indicated that the protein part of the viruses was remaining outside the bacteria. The results led Hershey and Chase to the inescapable conclusion that viral DNA enters the bacterium, whereas the viral protein remains outside. (Figure 1.14 shows the experiment.) Thus, DNA was the sole element responsible for viral replication. Protein had no role in the process.

Certain experiments stand out as turning points in scientific history, and the experiments performed by Hershey and Chase are one such turning point. In retrospect, we can see how their results had substantial impact on the thinking of that era. Hershey and Chase clarified the important aspect of viral replication that nucleic acid goes inside the cell, whereas the protein coat remains outside. In broader terms, their results strengthened the place of DNA in cellular biochemistry. Bacterial viruses, it should be remembered, are composed solely of nucleic acid and protein, and the Hershey-Chase experiments

reinforced the concept that DNA, and not protein, dictates the synthesis of both nucleic acid and protein.

> It was becoming clear that all the biochemical information for the synthesis of both nucleic acid and protein is stored in the DNA. Avery's experiments performed eight years earlier had linked DNA to the genetic material. The work of Hershey and Chase considerably strengthened that link.

Scientists are usually reluctant to run to a window and shout their discoveries to the world, and Hershey and Chase were no exceptions. Instead, they wrote soberly in a 1952 scientific journal, "This protein probably has no function in the growth of intracellular phage. The DNA has some function." Their caution was shared by other biochemists who were equally hesitant to apply processes learned from viral replication to human cell functions. Indeed, late in 1952 James D. Watson, the co-discoverer of DNA's structure (Chapter 2), read a telling letter from Hershey to a scientific group at Oxford, England. Said Hershey, "My own guess is that DNA will not prove to be a unique determiner of genetic specificity." Scientific history has proven otherwise—considerably otherwise.

## SUMMARY

The roots of DNA research can be traced back to the innovative experiments of Gregor Mendel conducted in the mid-1800s. Mendel postulated that an inherited trait is controlled not by blood, but by "factors" obtained from the parents. His work with pea plants implied that one factor is obtained from each parent and that a particular factor may dominate the other. It also appeared that the factors separate during transmission to the next generation. However, at the time, little attention was paid to the work.

At the beginning of the 1900s, Mendel's experiments were repeated and verified, and scientists postulated that his "factors" were really chromosomes. Morgan's work of 1910 showed that white eye color in fruit flies is determined by a single chromosome, and he postulated that a single chromosome determines a trait. But because individual chromosomes could not explain all traits, molecular geneticists came to believe that entities on the chromosome called "genes" were the hereditary factors. Evidence presented by Miescher, Feulgen, and Mirsky indicated that chromosomes are composed of deoxyribonucleic acid, or DNA.

Experiments performed by other biologists and chemists strengthened the link between genes and DNA. In 1928, for example, Griffith showed that the characteristics of certain bacteria could be changed (i.e., the bacteria could be transformed) if they were mixed with debris from another type of bacterium. Alloway and his group found they could purify the debris to increase its potential to transform bacteria, and Avery and his colleagues discovered that the transforming material in the debris was DNA.

The experiments of Hershey and Chase provided the final proof for DNA's involvement in heredity. Hershey and Chase performed experiments with bacteria and the viruses that replicate within them. Bacterial viruses are composed primarily of protein and DNA. The experimental results showed that DNA alone directs the replication of new viruses—that is, DNA contains the biochemical information for the synthesis of both the viral protein and the viral DNA. The results obtained by Hershey and Chase confirmed that DNA is the hereditary material and stimulated additional interest in the study of molecular genetics.

## REVIEW

This chapter has recounted the process by which DNA was identified as the substance of heredity. To test your comprehension of the chapter's major ideas, answer the following review questions:

1. What was Mendel's great contribution to science, and how did his work lay the groundwork for studies in genetics?

2. Draw a picture or diagram that contains all of the following, and label each: cell, chromosomes, gene, DNA, nucleus, and cytoplasm. Is the cell eukaryotic or prokaryotic? Label it and state why. Learn to draw such a picture without looking at the textbook or other source of information.

3. How did Morgan go about demonstrating that a trait can be associated with a chromosome?

4. How was a gene viewed in the 1920s?

5. Describe the contributions of Johann Miescher to the study of molecular genetics.

6. What observations made by Feulgen and Mirsky helped forge a link between DNA and heredity?

7. Describe the experiments performed by Frederick Griffith, and explain why they were significant to the development of genetics.

8. Identify the accomplishment of Avery and his collaborators, and indicate why their work did not receive acclaim.

9. Explain the experiments performed by Hershey and Chase, indicating what the "tools" of the experiment were and what the results were.

10. Summarize how the Hershey-Chase experiment was a deciding factor in linking DNA to protein and DNA synthesis.

## ADDITIONAL READING

Cummings, M.C., 2008. Human Heredity, Principles and Issues, eighth ed. Brooks Cole, Belmont, CA.

Dubos, R., 1976. The Professor, The Institute, and DNA. Rockefeller University Press, New York.

Griffith, F., 1928. The significance of pneumococcal types. J. Hyg. 27, 113–115.

Lewis, R., 2007. Human Genetics, eighth ed. WC Brown, McGraw-Hill, Dubuque, IA.

Olby, R., 1974. The Path to the Double Helix. University of Washington Press, Seattle, WA.

## WEB SITES

Cells Alive! Interactive bacterial, animal, and plant cell diagrams. www.cellsalive.com/cells/3dcell.htm

DNA from the Beginning: Classical Genetics. Dolan DNA Learning Center (2002) Cold Spring Harbor Laboratory. www.dnaftb.org

The Inner Life of a Cell (animation). Viel, A., Lue, R.A., Liebler, J., 2007. Howard Hughes Medical Institute. multimedia.mcb. harvard.edu

# The DNA Double Helix

**Nanomaterials: Golden Handshake**

*Nature: News and Views,* January 30, 2008

By John C. Crocker

Three-dimensional nanoparticle arrays are likely to be the foundation of future optical and electronic materials. A promising way to assemble them is through the transient pairings of complementary DNA strands.

One of the staple concepts of nanotechnology is that of "growing" useful materials or devices by coaxing a random mixture of microscopic parts to assemble spontaneously into a desired structure. Versatile self-assembly schemes have been demonstrated that use DNA as the primary building material.... Two research teams have built on the successes with DNA to aid the self-assembly of gold nanoparticles. Their technique should also work for other varieties of technologically exciting nanoparticles.

Progress in achieving the directed self-assembly of nanoparticles had been elusive, owing to one potentially daunting requirement: selective adhesion. Each microscopic part must be engineered so that it sticks only to the others it should abut in the desired final structure...

This is where DNA comes into its own. Particles carrying complementary strands of DNA selectively adhere to each other when the strands "hybridize" to form the familiar DNA double helix. *The final architecture is thus determined not by chemistry or charge, but by the lengths and nucleotide sequences of the DNA strands.* That promises a versatile assembly scheme that might be used with particles of nearly any material to fabricate nanocomposites or "metamaterials" with unusual electronic and optical properties. The applications of such materials might include high-efficiency solar panels and lasers, super-resolution microscopes—and even coatings to render objects invisible.

More than 50 years after the discovery of the DNA double helix, our knowledge of the structure of DNA continues to pay off in ways that even Watson and Crick could not have dreamed of in 1953. Now, almost a decade into the twenty-first century, scientists worldwide are using the unique properties of DNA structure, including its ability to store information in the double helix, for new and amazing applications in science and biomedical research. DNA is being used to assemble individual atoms into designer molecules using **nanotechnology**. Scientists are learning how to assemble molecules with some amazing properties, from fibers that are hundreds of times as strong as steel yet weigh one-sixth as much, to nanofactories that can assemble nearly anything, from a new iPod to the clothes you wear, starting with individual atoms.

*Nanomaterials: Golden Handshake* describes how the structure of DNA is already playing a novel role in the development of these futuristic materials. In this chapter, we'll explore how the structure of DNA was determined, a puzzle that was solved only by interpreting and integrating data from several different scientific fields. This historic achievement was accomplished by two scientists who juggled the scientific puzzle pieces in their minds (and with cardboard cutouts), but who did not actually perform a single hands-on experiment with DNA.

## LOOKING AHEAD

Determining the structure of DNA was one of the major scientific achievements of the twentieth century. Knowing the structure of DNA gave scientists insight into how heredity works and made the revolution in molecular biology and DNA technology possible. Moreover, the structure of DNA had an enormous impact on our understanding of gene function and DNA replication in cells. On completing this chapter, you should be able to do the following:

- Recognize the three fundamental building blocks of nucleotides used to assemble DNA.
- Describe how sugar and phosphate groups link to each other to form the "backbone" of the DNA molecule.
- Summarize Erwin Chargaff's findings, and indicate why they were important in solving the puzzle of DNA structure.
- Discuss the different contributions of Franklin, Wilkins, Watson, and Crick in determining the structure of DNA.
- Explain what is meant by semiconservative DNA replication.
- Describe the important functional characteristics of the DNA polymerase enzymes involved in duplicating genome DNA.
- Use your newly acquired DNA vocabulary to read with understanding about DNA-related topics online (google "DNA"), and expand your confidence in learning about DNA.

## INTRODUCTION

By the 1950s, it was becoming increasingly clear to the scientific community that the deoxyribonucleic acid (DNA) molecule is the basis of genetic heredity. It is hard to believe these days, but at the time, very little was known about DNA structure. Scientists realized that they needed to know the molecular structure of DNA because the structure of the DNA molecule

might shed light on the hereditary process; also, understanding DNA structure might explain how the molecule duplicates during cell reproduction. The processes of genetic heredity and cellular reproduction are among the most fundamental and important events in biology, and the quest for this knowledge started a race to figure out what a DNA molecule actually looks like.

During the 1940s, top scientists worldwide were studying the chemical characteristics of DNA, work that was a critical step to prepare for understanding the three-dimensional structures of biological macromolecules like proteins and DNA. In fact, world-famous scientist Linus Pauling, then a professor at the California Institute of Technology, used x-ray crystallography to solve the structures of large proteins in a series of cutting-edge papers published in 1951. At that time, the race was on among scientists to determine the structure of DNA based on what was known about its chemical and physical characteristics. The race included researchers James Watson and Francis Crick, who determined the structure of DNA and in so doing not only gained international fame but also opened a door to the molecular investigation of heredity. As this chapter will show, the work of Watson and Crick was the jumping-off point for the science behind DNA and biotechnology.

## THE STRUCTURE OF DNA

Establishing the structure of DNA was one of the major achievements of the twentieth century. Not only did it yield myriad practical benefits, but it also gave scientists the philosophical pride of understanding how heredity works. Biology has many bedrock principles—for example, the cellular basis of living things, the germ theory of disease, and the process of evolution are three—and the chemical basis of heredity is another. Unlocking the secret of DNA was the key to understanding this principle.

### DNA is Constructed from Nucleotide Units

Although the structure of DNA was unknown in the 1940s, the basic chemical components of DNA had been studied for two decades. In the 1920s, Phoebus A. T. Levene determined the chemistry of nucleic acids. Working with his colleagues at Rockefeller Institute in New York City, Levene studied two types of nucleic acid—ribonucleic acid (RNA) and deoxyribonucleic acid (DNA)—isolated from yeast cells and animal thymus tissue. Levene's analyses revealed three fundamental components in both types of nucleic acids: (1) a five-carbon sugar, which could be either ribose (in RNA) or deoxyribose (in DNA); (2) phosphate, a chemical group derived

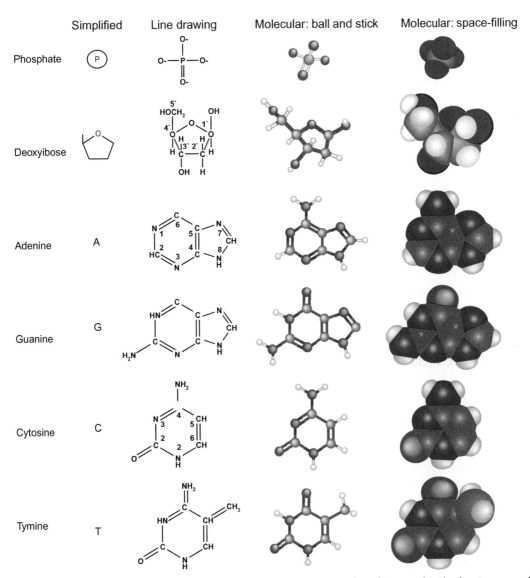

**FIGURE 2.1**  Research revealed the fundamental components of nucleic acids. Research in the 1940s by Phoebus Levene and colleagues revealed the three fundamental components of both types of nucleic acids: phosphate, that is, a chemical group derived from phosphoric acid molecules, a five-carbon sugar, which could be either ribose (in RNA, not shown) or deoxyribose (in DNA), and four different compounds containing nitrogen and having the chemical properties of bases (A, G, C, and T).

from phosphoric acid molecules; and (3) four different compounds containing nitrogen (Figure 2.1).

Because of their nitrogen content and basic qualities, the four nitrogenous compounds are simply referred to as **bases**. In DNA, the four most common bases are **adenine (A), thymine (T), guanine (G),** and **cytosine (C)**. RNA, the other important nucleic acid in cells, contains the A, G, and C bases, but a base called **uracil (U)** replaces **thymine (T)**. The adenine (A) and guanine (G) bases are double-ring molecules called **purines**, whereas the cytosine (C), thymine (T), and uracil (U) bases are single-ring molecules called **pyrimidines**.

Levene concluded that DNA is composed of three essential components that form units, which are in turn strung together to form a long DNA chain. In contemporary biochemical terms, the units are called **nucleotides.** In DNA, each nucleotide consists of a deoxyribose sugar attached to a phosphate group and to

Units called nucleotides are the basic building blocks of DNA and RNA. A nucleotide consists of a base, a sugar, and a phosphate group. The identity of the base is the only feature that distinguishes one DNA nucleotide from another, or one RNA nucleotide from another.

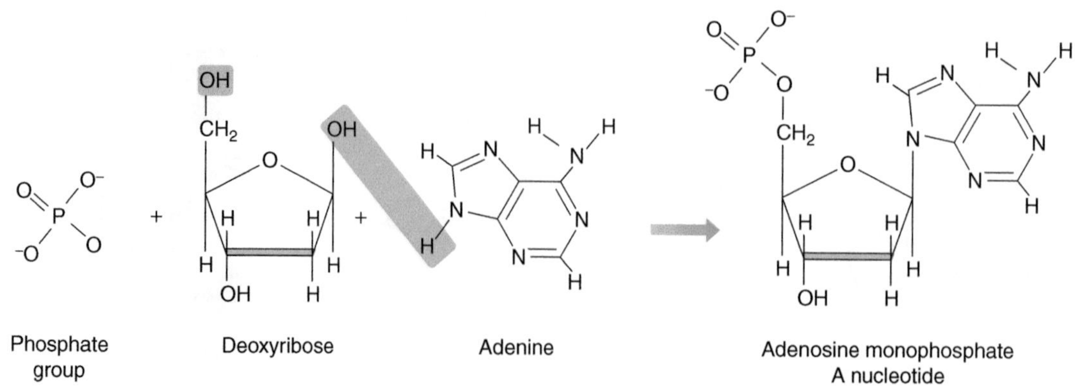

FIGURE 2.2   Synthesis of the DNA building block unit: a nucleotide. A nucleotide is composed of a phosphate group, a deoxyribose molecule, and an adenine (A) molecule. The shaded –OH and –H groups are removed during the synthesis of the nucleotide. This nucleotide is called adenosine monophosphate.

a base (Figure 2.2). Each of the four nucleotides differs from the other three only in its base component.

## DNA Nomenclature Helps to Understand DNA Function

Chemists have developed a nomenclature system to identify the molecular structures of many thousands of chemical compounds, including the components of DNA. The casual student of DNA can perhaps skim over these details, but beware that many of the modern terms used routinely to discuss DNA refer to a specific feature of the DNA molecule that is important in a practical sense. For example, the two ends of a DNA strand are not identical. In laboratory experiments, proteins that respond differently to each DNA end are used routinely, so understanding the terminology that refers to each end can be critical for success. Thus, students who plan to pursue further studies in DNA technology must become familiar with fundamental DNA facts and terminology.

Using standard chemical nomenclature, numbers are assigned to the ring atoms of the base and sugar (Figure 2.3). When specifying where a chemical group is attached to a molecule, it is customary to refer to the number assigned to that specific carbon atom (Figure 2.3). The carbon atoms in deoxyribose are numbered 1′ to 5′ (pronounced "one-prime" and "five-prime") with the numbering system starting to the right of the oxygen atom and proceeding clockwise around the molecule. The phosphate group attached to the 5′ carbon atom of the deoxyribose establishes the 5′ end of the DNA strand. The functionally important –OH (hydroxyl group) is attached at the 3′ carbon atom of the deoxyribose molecule. This 3′ –OH group is required for the addition of nucleotide units during DNA synthesis (as discussed

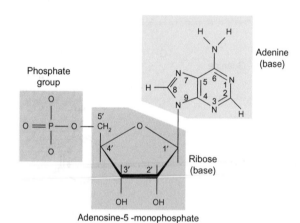

FIGURE 2.3   Numbering system in a nucleotide unit of DNA. The ribose molecule's carbons are numbered starting to the right of the oxygen atom and the numbering proceeds clockwise. A prime (′) is placed next to each number of the sugar, whereas the base component is numbered without primes, thereby distinguishing the atoms of the sugar from those of the base of a nucleotide. When constructing a DNA strand, another nucleotide unit will attach at the 3′ end.

later in this chapter). The 3′ and 5′ carbons, as we'll see, are important markers for distinguishing the chemical direction (polarity) of a DNA strand (Figure 2.4). The bases are attached to the sugar through the 1′ carbon. Note that the atom numbers in the bases do not use "prime" and thus are distinguished from the carbons in deoxyribose.

## The Two Ends of a Single DNA Strand are Chemically Distinct

Because the 3′ end of the DNA strand contains a hydroxyl group (–OH) and the 5′ end of the DNA has a phosphate group (P), the DNA strand has **polarity** (Figure 2.4). In the DNA helix, the two DNA strands

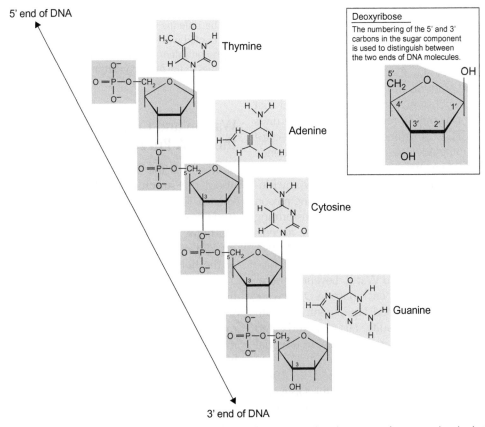

**FIGURE 2.4** Nucleotide units are connected to form a single strand of DNA. The phosphate group forms a molecular bridge between the 5′ carbon atom of one nucleotide and the 3′ carbon of the next nucleotide. This type of chemical bridge is called a phosphodiester chemical bond, and it links together the sugars and phosphates into the backbone of the DNA molecule. A single strand of DNA has two different ends: the 3′ end of the DNA molecule, which has an –OH group that is available for adding more nucleotides onto the DNA strand, and the 5′ end, which has a phosphate group. DNA strands grow at the 3′ end, but not at the 5′ end.

are oriented in opposite directions and are said to be **antiparallel**. The chemical structures at the ends of DNA strands in both single-stranded and double-stranded DNA enable the DNA ends to engage in important processes in the cell. The chemical characteristics of the ends of the DNA strands are particularly important when the DNA is manipulated in the lab. For example, many enzymes can act only on the 3′ or 5′ end of the DNA strand.

> The atom numbers of nucleotides allow us to easily and unambiguously refer to important landmarks in the DNA molecule. A case in point is the two ends of a DNA strand, referred to as 5′ and 3′, which have different properties that are vital to understanding how DNA works in a cell.

Special multiprotein enzymes in the nucleus of the cell synthesize DNA by linking nucleotides to each other; the 3′ carbon of one nucleotide is joined to the phosphate group of a second nucleotide. Note that because the "prime" notation follows the 3, we must be referring to carbon 3 on the sugar, not the base. As Figure 2.4 shows, the phosphate group essentially forms a bridge connecting the two deoxyribose molecules. To continue building the DNA molecule, linkages to other nucleotides are forged in the same way, building up a strand of tens or thousands (or millions, or any number) of nucleotides linked to one another in the DNA molecule. By agreement among scientists, the sequence of the bases in a DNA molecule is read in the **5′ to 3′ direction**, starting at the 5′ end of the DNA strand and reading toward the 3′ end.

> DNA molecules are synthesized in the 5′ to 3′ direction. By convention, nucleotide sequences are written and read from 5′ to 3′ as well.

## Levene and Chargaff Provided Chemical Clues to the Structure of DNA

Levene's studies in the 1920s indicated that all four nitrogenous bases, A, G, C, and T, were present in virtually the same amounts in DNA. Later proven false, at the time this conclusion encouraged the belief that DNA is simply a polymer of repeating nucleotide units (e.g., TGACTGACTGAC; thymine-guanine-adenine-cytosine-thymine-guanine-adenine-etc.). Without sequence variation in a repeating chain, it was difficult to see how DNA could provide sufficient diversity to carry any biochemical or hereditary information. This was one reason that Avery's identification of DNA as the transforming substance in bacteria did not receive immediate acceptance (see Chapter 1). At the time it appeared that DNA was nothing more than a structural component of chromosomes. However, after World War II, Levene's chemical analyses of the amounts of each base in DNA were repeated with more sophisticated equipment and with very different experimental results. His tests indicated that the four nitrogenous bases were present in unequal amounts in DNA.

Erwin Chargaff's experiments, reported in 1949, indicated that regardless of the source of the DNA, the amounts of adenine (A) and thymine (T) were similar to each other, and the same was true for the amounts of cytosine (C) and guanine (G) (Table 2.1). It appeared that for every adenine molecule there was a thymine molecule (and vice versa), and for every cytosine molecule there was a guanine molecule (and vice versa). These observations suggested that DNA is not a simple repeating polymer. Moreover, if the amounts of each base vary in an organism's chromosomes, perhaps DNA might have the properties necessary to code for information. And if different organisms have different amounts of bases in their DNA, maybe the bases have something to do with difference in the organisms. Years would pass before the significance of these observations would be fully understood.

The differing amounts of the bases in various organisms, coupled with the fact that of A and T were always present in equivalent amounts, as were G and C, were important clues to the function and structure of DNA.

## Watson and Crick Set Their Sights on Solving the DNA Puzzle

Students of the twenty-first century are often taught about the structure of DNA as if it has always been known. They learn about DNA's components, and they study its double-stranded spiral form known as the double helix. But in reality, in the early 1950s, biochemists did not know about either the number of strands or the helical arrangement of DNA, nor did they have a clear understanding of DNA functions. Many scientists thought that DNA functioned only as a structural support for the chromosomes apparent in the visible microscope. Although the components of DNA and their relative amounts were known, the spatial arrangement of the components remained a mystery. Inspired by the need to know, scientists began an unofficial race to determine the molecular structure of DNA.

In early 1950, against this historical backdrop and in this political environment, a young American graduate student named James D. Watson arrived at Cambridge University in London to work with prominent biochemist Francis H. C. Crick. In pursuit of their goal to solve the three-dimensional structure of the DNA molecule, Watson and Crick would do no laboratory bench work—surprisingly, their greatest contribution to science was the result of the amazing ability to imagine DNA structures in three dimensions. In addition to scrutinizing the research results of many other scientists, Watson and Crick tested homemade, two-dimensional puzzle pieces (paper cutouts of sugars, bases, phosphates) in various combinations to figure out which pieces fit together. Eventually they were

**TABLE 2.1** The base compositions of DNA from various species, as determined by Erwin Chargaff

| Species | A | T | G | C |
|---|---|---|---|---|
| *Homo sapiens* | 31.0 | 31.5 | 19.1 | 18.4 |
| *Drosophila melanogaster* | 27.3 | 27.6 | 22.5 | 22.5 |
| *Zea mays* | 25.6 | 25.3 | 24.5 | 24.6 |
| *Neurospora crassa* | 23.0 | 23.3 | 27.1 | 26.6 |
| *Escherichia coli* | 24.6 | 24.3 | 25.5 | 25.6 |
| *Bacillus subtilis* | 28.4 | 29.0 | 21.0 | 21.6 |

*Note that the percents of adenine and thymine are consistently similar, as are the percents of cytosine and guanine.*

able to propose a three-dimensional model representing the DNA structure.

In the early 1950s, scientists used a technique called **x-ray diffraction** (Figure 2.5) to determine the molecular structures of various compounds including proteins. In this process, crystallized molecules are rotated and bombarded with x-rays. The atoms in the crystal deflect (or "diffract") the x-ray beams, which hit a photographic plate and make a pattern. The diffraction pattern gives a large number of clues to the three-dimensional position of the atoms in the crystal. It requires sophisticated mathematics to interpret the diffraction pattern and arrive at a molecular structure. At the simplest level, the diffracted x-rays can be

compared to ripples in a lake created by tossing a rock into the water. (The ripples give an idea of the size and shape of the rock.) It is important to remember that even today solving the three-dimensional molecular structure of biomolecules by x-ray diffraction is a complex task that requires computers and specialized software. Imagine how difficult it was to use this technology over 50 years ago.

> X-ray diffraction studies are key experiments in constructing models of large molecules such as proteins and DNA.

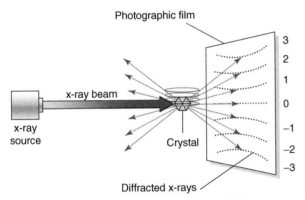

**FIGURE 2.5** How crystallography and x-ray diffraction reveal structure. In x-ray diffraction, the x-ray beam is focused on a crystal painstakingly made from a pure solution of the molecule(s) of interest. The uniform structures inside the crystal diffract the x-rays onto photographic film, creating patterns of spots that are characteristic of certain molecular structures.

## Franklin's X-ray Diffraction "Images" Played a Crucial Role in Visualizing DNA Structure

By now, students routinely explore 3D models of proteins and DNA rendered with amazing accuracy and in a user-friendly format by computers. But in their time, Watson and Crick were engaged in a much different type of model building to figure out the structure of DNA (Figure 2.7). They had gathered as much information as they could from available experimental results and proceeded to try to integrate the information into an accurate theoretical model of DNA structure. They worked with molecular components represented by cardboard cutouts, sticks, and wire, cut to size according to experimentally derived dimensions. Watson and Crick tried to fit everything together, just like solving a three-dimensional jigsaw puzzle.

---

### Box 2.1    Linus Pauling, Expert Protein Scientist, Enters the DNA Race

Linus Pauling was already a world-renowned scientist at the California Institute of Technology when he entered the race to solve the structure of DNA. Pauling was expert at protein crystallography and x-ray diffraction, and he solved the structures of several large proteins in a series of cutting-edge papers published in 1951. He brilliantly characterized the now well-known alpha helical protein structure (Figure 2.6A). Pauling was also experienced with DNA and had proposed a DNA replication mechanism (with Max Delbruck). He was familiar with Oswald Avery's extraordinary work in 1944 (Chapter 1), indicating that nucleic acids could transmit genetic information, and again ahead of his time, Pauling proposed in 1948 that genes consist of mutually complementary molecules. However, like many scientists, Pauling still believed that the key to inheritance would be found in the amazing variety of protein structures and not in the apparently repetitive DNA polymer. Pauling actually spoke in person with Erwin Chargaff in 1947. Chargaff told him about his observations on DNA

bases, but for some reason Pauling, who found Chargaff's personality disagreeable, did not heed this important clue to DNA structure. Pauling and his colleagues continued to build DNA models with three DNA strands wrapped around the axis of the DNA helix, with the bases exposed on the outside of the structure (Figure 2.6B).

How did a double Nobel Laureate scientist like Linus Pauling make such an error? Pauling was at a deficit because he did not have access to the beautiful x-ray diffraction patterns of DNA generated by Rosalind Franklin and Maurice Wilkins at King's College in London. Pauling wrote to Wilkins and asked to see Franklin's x-ray pictures of DNA. When Pauling was denied, he wrote to Wilkins' superior and was again denied. After the end of World War II, Pauling was active in a group of concerned scholars that warned the public about nuclear war, a touchy subject at the time. Pauling was awarded the Presidential Medal for Merit by President Truman and given a certificate of appreciation from the War Department.

Box 2.1    (Continued)

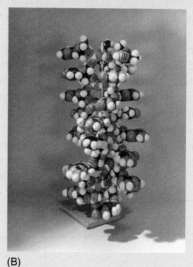

(A)                                                                                          (B)

**FIGURE 2.6**   Linus Pauling with proposed models of protein and DNA. (A) Linus Pauling used x-ray crystallography to determine the structures of many large proteins in a series of groundbreaking papers published in 1951, describing the structure of the alpha helix protein motif (OSU Special Connections: Linus Pauling). (B) Pauling and Corey's incorrect, triple-helical DNA structure, one of the most famous mistakes in twentieth-century science.

But many viewed his opposition to atomic bombs and atmospheric nuclear testing as unpatriotic and even subversive. In 1952, Pauling's application for a passport to travel to England was denied because it "would not be in the best interests of the United States." Linus Pauling stayed home and missed a symposium held in honor of his achievements solving protein structures. He also missed a chance to see Franklin's x-ray diffraction pictures of DNA. Eventually Pauling's passport application was approved, much too late for the symposium, but he finally visited England in the summer of 1952. For some reason, Pauling did not visit King's College or ask to see Franklin's DNA data.

In 1953 Linus Pauling and his collaborator, Robert Corey, published a paper called "A Proposed Structure for the Nucleic Acids" in *The Proceedings of the National Academy of Sciences*, which argued for a triple-helix DNA structure. This paper would turn out to be one of the most famous incorrect theories in twentieth-century science.

At the time Watson arrived in England, Rosalind Franklin (Figure 2.8) and Maurice Wilkins (Figure 2.9) were also working on the problem of DNA structure at King's College in London. Franklin was an expert at x-ray diffraction, but she was new to working with DNA. Wilkins was new at x-ray diffraction, but he had extensive experience with DNA. Unfortunately, Wilkins and Franklin became scientific adversaries. Rosalind Franklin used her extensive knowledge and experimental skills to capture the best diffraction patterns of DNA ever made. The infamous pattern called "Photograph 51" (Figure 2.8B) clearly revealed (to those experienced in interpretation of diffraction patterns) not only that the DNA molecule was a double helix containing two DNA strands but also the previously unknown dimensions of the DNA molecule: the diameter, the distance per turn of the helix, and the interval between repeating units of the helix (Figure 2.10). In his book *The Double Helix* (W.W. Norton & Co., 1980, p. 98)

written many years later, Watson described his view of the photograph as follows:

> The instant I saw the picture my mouth fell open and my pulse began to race.... the black cross of reflections which dominated the picture could arise only from a helical structure... mere inspection of the X-ray picture gave several of the vital helical parameters.

These important dimensions turned out to be essential for building the correct DNA model structure. Franklin's data were scheduled to be published in 1953. In 1952, Watson met with Wilkins and during discussions was shown some of Franklin's data, including Photograph 51, without her knowledge. When Watson and Crick returned to model building, they eventually went on to construct their final (and correct) DNA model (Figure 2.7).

Photograph 51 contained critical data, including that the diameter of the DNA molecule was 2.0

**FIGURE 2.7**   Watson and Crick with their model of the DNA double-helix in 1953 and 50 years later. (A) Watson (left) and Crick (right) figured out the structure of DNA by putting together all the pieces of a three dimensional puzzle. (B) The discoverers of DNA structure pose for a recreation of the original picture, this time with a DNA model in 2003.

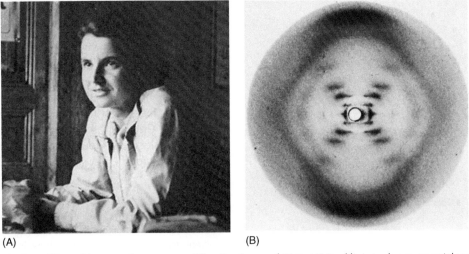

**FIGURE 2.8**   Rosalind Franklin and her most famous x-ray diffraction image of DNA. (A) Franklin's work was essential to Watson and Crick in their discovery of the structure of DNA. (B) This x-ray diffraction pattern of DNA, made by Rosalind Franklin and called "Photograph 51", was the source of some essential information about the structure of DNA.

nanometers (nm, a billionth of a meter). At this time, many models, including the early Watson and Crick models and the triple helix published by Linus Pauling (see Box 2.1 and Figure 2.6B), had the backbone of the molecule located in the center, with the bases radiating outward. Franklin strongly disagreed with this on the basis of her x-ray diffraction data, insisting that the backbone must be on the outside of the molecule. At this point, Watson and Crick realized that if the bases were arranged to point inward (as Franklin had maintained), the width of DNA would start to approach the 2.0-nm diameter observed in Photograph 51.

Franklin's diffraction measurements also showed a 0.34-nm distance between successive nucleotides on the strand, with 3.4 nm per turn of the DNA helix.

> Rosalind Franklin's x-ray diffraction studies revealed critical details about DNA structure that Watson and Crick were able to put together in constructing their model of DNA.

Noting that 3.4 is exactly ten times 0.34, the model was built with ten nucleotides per turn of the helix (Figure 2.10). The data suggested that a DNA molecule is composed of two nucleotide chains wound like a spiral staircase around a hypothetical central axis. The deoxyribose-phosphate combinations form the backbones of both chains, and the nitrogenous bases point inward. Watson and Crick's DNA model was getting closer, but it was not yet right. How were the bases oriented in the center so they would fit together perfectly?

**FIGURE 2.9** Maurice Wilkins. Wilkins shared the Nobel Prize with Watson and Crick for solving the structure of DNA.

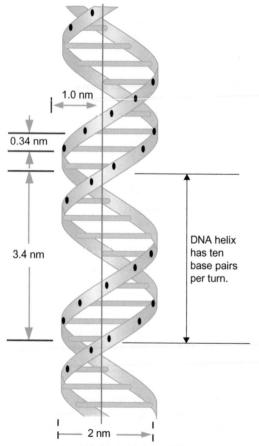

1.0 nm

0.34 nm

3.4 nm

DNA helix has ten base pairs per turn.

2 nm

**FIGURE 2.10** DNA helix dimensions derived from x-ray diffraction. Franklin's x-ray diffraction data revealed a 0.34-nm distance between successive nucleotides on the helix (between adjacent base pairs), 3.4 nm per turn of the helix, and a 2.0-nm diameter.

## Watson and Crick Integrate Available Data into a DNA Model

Watson tested his cardboard cutouts to try to approximate the 2.0-nm diameter of DNA with different combinations of base pairs. He finally had a model with each base lined up to pair with itself, A-A, C-C, and so on. Jerry Donohue, a former student of Linus Pauling, pointed out to Watson that his model had the bases in their **enol** forms, as suggested by the textbooks at the time. But Donohue knew of unpublished research suggesting that the DNA bases exist in the **keto** form in cells. Once Watson changed the structures of the bases in the model to their keto forms, he immediately realized that with these changes the adenine and thymine bases formed an A-T base pair that fit the 2-nm diameter of DNA. The guanine and cytosine (G-C) base pair did the same. Watson and Crick knew that the proposed A-T base pair would satisfy Chargaff's requirement that in any DNA molecule, the amounts of adenine and thymine are identical. Watson and Crick could envision that for every adenine base on one DNA strand there must be a thymine base on the other DNA strand. Similarly, because DNA has equal amounts of guanine and cytosine, there must be a guanine base in the DNA for every cytosine base. As a result, Watson and Crick's final DNA model contains pairs of bases, A-T and G-C, which each fit perfectly into the internal diameter of the DNA helix. When the DNA model was complete, Watson and Crick had solved the universal structure of the DNA molecule: a double-stranded DNA helix containing intertwined DNA strands (Figure 2.11).

Watson and Crick certainly won the race to solve the mystery of DNA structure, but their accomplishment incorporated data from Franklin, Wilkins, Pauling, and many other scientists. This story emphasizes how interactions among scientists and the free exchange of information play critically important roles in promoting scientific breakthroughs.

The two strands of DNA form a helix with two unequally sized grooves, a narrow one called the **minor groove** and a wide one called the **major groove**. The two grooves wind continuously around the entire length of the DNA double helix (Figure 2.12). Many different proteins in the cell bind to the major groove in the DNA helix; which proteins bind to which DNA helix often depends on the specific DNA base sequence in the major groove. This type of DNA-protein binding regulates when and how genes are turned on and off (expressed) in the cell.

Given the base sequence—that is, the order of the bases—along one DNA strand, the base sequence along the complementary strand is automatically

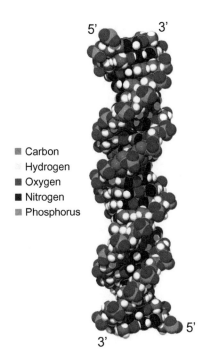

■ Carbon
   Hydrogen
■ Oxygen
■ Nitrogen
■ Phosphorus

**FIGURE 2.11** Double-stranded DNA. The atoms of this space-filled model of double-stranded DNA are colored by element. The two DNA strands of a double helix are always antiparallel; they are arranged in opposite directions. Both ends of each strand are labeled to indicate the directionality of the strands.

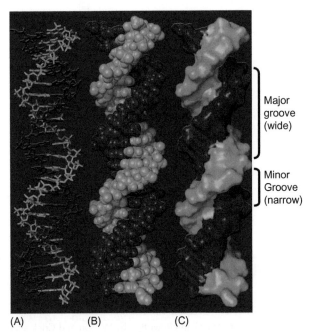

Major groove (wide)

Minor Groove (narrow)

(A)      (B)      (C)

**FIGURE 2.12** Three representations of a double-stranded DNA helix. (A) "Stick" models of DNA show the bonds that underlie the structure of DNA. (B) Spacefill DNA models show each atom and the approximate space it occupies. (C) A surface representation shows off the dimensions of the grooves along the DNA. The DNA strands coil around each other in a manner that creates two different-sized grooves along the molecule: a wide major groove, and a narrow minor groove, as shown.

determined using simple base pairing rules; adenine base pairs with thymine (A-T) and cytosine base pairs with guanine (C-G). The DNA helix structure has an intrinsic duplication mechanism: the base sequence in one strand, the template, guides the synthesis of the bases in a second, complementary strand. We will see how this important process works in cells later in the chapter.

Watson and Crick's paper announcing the structure of DNA opened with the lines "We wish to suggest a structure for the salt of deoxyribose nucleic acid..." (Figure 2.13) and was greeted enthusiastically by the scientific community; the proposed DNA model was elegant in its simplicity as illustrated by the DNA helix sketch drawn by Crick's wife Odile and included in the famous paper (Figure 2.14). The proposed structure of DNA made it easy to see how DNA could provide hereditary information. Biochemists saw that the nitrogenous bases, occurring in highly variable sequences, could provide a code of heredity. The sequence was not boring and repeated (TGAC...TGAC...TGAC) as Levene's work had suggested. Rather, the DNA base sequence was variable (TGGACTTGCCTAAGCGATA...), with the ability to encode a specific sequence of amino acids in a protein chain.

> The model proposed by Watson and Crick included the basic elements needed for the hereditary material: variation in base sequence and base pairing that suggested a mechanism for duplication, and hence inheritance, of DNA.

## Data from Many Laboratories Contributed to Solving the DNA Structure Puzzle

Watson and Crick were not experts in any of the scientific areas they used in constructing their DNA model. Franklin on the other hand was a superb crystallographer, Chargaff understood base relationships thoroughly, and numerous other scientists provided information on the chemistry of DNA. Watson and Crick achieved their goal because they were able to see the big picture and took risks in building models that they could not yet establish as correct but that fit all the available data. They took what they needed from several disciplines and used it to compose something greater than its parts. In effect, they saw the proverbial "forest for the trees."

More than 50 years later, controversy continues to swirl around the relationships between Watson, Crick, Franklin, and Wilkins. It is clear that Franklin and Wilkins disliked each other and that Franklin was excluded from some scientific exchanges with Wilkins. Questions remain about who influenced whom, how

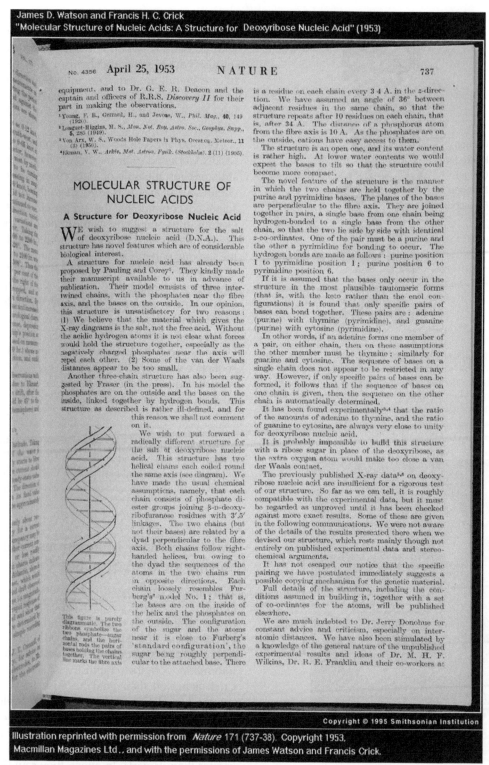

**FIGURE 2.13**  The publication of the DNA double-helix structure. This classic, one-page *Nature* article written by Watson and Crick accurately described the three-dimensional molecular structure of DNA for the first time. The paper included a single figure, an original sketch of a novel DNA double-helix structure, rendered by artist Mrs. Francis Crick.

Franklin's data came to be shared with Watson, and whether adequate credit was given to her. Although the questions will probably never be resolved, the controversy gives us a glimpse of how these famous scientists went about uncovering the truths waiting to be discovered in nature. Readers who wish to learn more about the process and controversy can consult the resources listed at the end of this chapter.

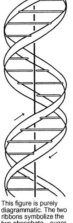

This figure is purely diagrammatic. The two ribbons symbolize the two phosphate—sugar chains, and the horizontal rods the pairs of bases holding the chains together. The vertical line marks the fibre axis.

**FIGURE 2.14** Dr. Francis H. C. Crick and wife Odile. In 2003, Odile and Dr. Francis H. C. Crick attended a dinner in La Jolla, California, honoring the 50th anniversary of the discovery of the structure of DNA by Dr. Crick and James D. Watson. Mrs. Crick's original sketch of a DNA double-helix structure appeared in the famous paper in the journal *Nature* in 1953 (*right*).

The contributions to science made by Watson, Crick, Franklin, and Wilkins were original and carried the impact sufficient to merit a place in the annals of scientific fame; the DNA double helix became the charter molecule of molecular biology. In 1962, Watson, Crick, and Wilkins were awarded the Nobel Prize in physiology or medicine. Unfortunately, Franklin died of cancer in 1958 and because the Nobel committee does not cite individuals posthumously, Franklin did not share in the award. However, Franklin's contributions were mentioned in the *Nature* paper and are now universally acknowledged. Recently the Royal Society (a distinguished academy of the sciences in the United Kingdom) established an annual award in Franklin's name, and the University of Health Sciences in the United States changed its name to Rosalind Franklin University in 2004.

In 1953, a single-page paper in *Nature* proposed a simple, elegant double-helix model to explain the molecular structure of DNA. Nine years later, the Nobel Prize was awarded to Watson, Crick, and Wilkins.

After the *Nature* article was published in 1953, evidence favoring the double-helix structure of DNA proposed by Watson and Crick accumulated rapidly. Few biologists doubted the accuracy of the molecular model, and most were intrigued with the implications of

the double helix for coding genetic information within the DNA structure. Scientists also recognized that the double-helix architecture of paired bases on two strands would accommodate the biochemical requirements for DNA replication. Still no one yet understood how the DNA double helix in a chromosome could pass genetic information on to the next generation. It was quite apparent that solving the molecular structure of DNA had not answered all the important questions about DNA, but in fact solving this scientific puzzle was the amazing beginning of the science of molecular biology. With time, this information gave rise to the fruits we know of as DNA science and biotechnology.

## DNA REPLICATION

### The Structure of DNA Leaves an Open Question

The paired bases (A-T and G-C) in the DNA double helix accommodate the biochemical requirements for reproducing or duplicating the DNA structure and information. Indeed this critically important point was not lost on Watson and Crick, who wrote one of the greatest understatements of all time in their DNA helix paper:

> It has not escaped our notice that the specific base pairing we have postulated immediately suggests a possible copying mechanism for the genetic material.

This is one of the most appealing features of the Watson–Crick DNA model; the structure of the DNA double helix provides insight into how the DNA molecule duplicates. In cells, this process is called DNA replication. The copying and distribution of DNA is a key process in cells with major implications for the mechanisms that operate to transmit genetic information from one generation to the next. From the DNA structure it is easy to see that one strand of the double helix has a base sequence that unambiguously determines the base sequence of the opposite complementary strand (Figure 2.15). Any sequence of A, G, C, or T bases can reside on one DNA strand, but the second strand must contain the complementary base sequence. For example, if the base sequence in one strand is G-T-A-C-C-A-T..., the base sequence of the partner strand must be C-A-T-G-G-T-A.... For the DNA replication process to occur, the DNA double helix must unwind and the DNA strands must "unzip" and separate. This allows both single strands of DNA to become accessible to the enzymes that use each strand as a template to create a new, complementary DNA strand (Figure 2.15).

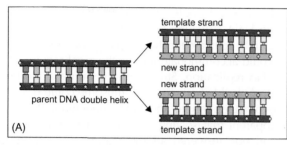
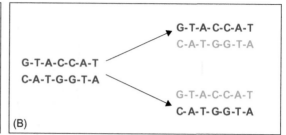

**FIGURE 2.15** Overview of DNA replication. (A) The two DNA strands (*red backbones*) in the parent DNA double helix are both used as templates by DNA polymerase and are copied into new DNA strands (*gold backbones*). (B) Any sequence of A, G, C, or T bases can reside on the DNA template strand (*red*), but after replication, the new DNA strand (*blue*) must have the complementary base sequence.

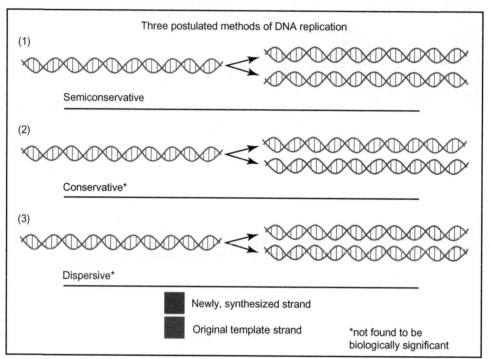

**FIGURE 2.16** Originally proposed mechanisms of DNA replication. The proposed DNA double-helix structure did not immediately explain the complete mechanism for how the DNA helix replicates to produce two DNA helix molecules. Three possible mechanisms of DNA replication were proposed: (1) each parent strand combines with the complementary new strand to reform a new double helix (semiconservative), (2) the parent strands reunite with each other, leaving the newly synthesized strands to form the second double helix (conservative), or (3) the products of DNA replication contain alternating regions of conservative replication (dispersive). Meselson and Stahl designed an elegant experiment to determine the correct answer.

## Meselson and Stahl Answer the Question

The DNA double-helix model left another important question unanswered, because the structure of DNA by itself did not automatically predict a specific mechanism used by the cell to copy (replicate) the DNA helix. Scientists suggested three possible ways that DNA might be replicated in the cell, called the dispersive, conservative, and semiconservative modes of DNA replication (Figure 2.16).

The structure of DNA suggested base pairing as a general mechanism for DNA replication. However, the process by which each strand of the double helix was copied remained to be discovered.

The definitive experiment to address this question was published in 1957 by Matthew Meselson and Franklin W. Stahl of the California Institute of Technology, who figured out a way to experimentally distinguish among the three possible modes of DNA replication. The Meselson and Stahl DNA replication experiment is one of the best demonstrations of how a key scientific theory or hypothesis can be conclusively resolved by a simple, well-designed, and definitive experiment (Figure 2.17).

To start this experiment, Meselson and Stahl grew bacteria in liquid medium containing a "heavy isotope" of nitrogen called $^{15}$N. At each round of bacterial cell division, as the DNA was replicated, the heavy $^{15}$N isotope became incorporated into the bacterial

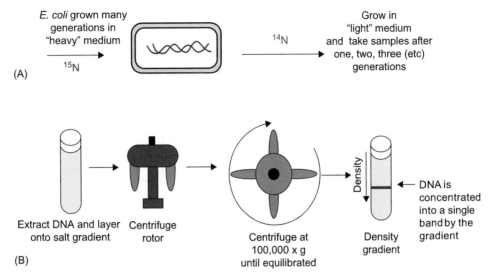

**FIGURE 2.17**   Overview of the Meselson and Stahl experiment. (A) Bacteria are grown in $^{15}$N-containing medium until virtually all of the nitrogen in DNA is $^{15}$N. Then they are grown in $^{14}$N-containing medium. Samples of DNA are isolated at each stage. (B) The DNA samples are centrifuged in a salt density gradient. During centrifugation, the DNA will move through the salt gradient until the density of the DNA and salt solution are equal. The more $^{15}$N is in the DNA, the denser it is, and the further it will move down the gradient.

chromosome DNA, instead of the normal isotope, $^{14}$N. As a result, the DNA in this population of bacteria was heavier than normal DNA because the heavy $^{15}$N atoms have replaced the lighter $^{14}$N atoms normally found in the nitrogenous bases. These two types of DNA, "heavy" (H) and "light" (L), have different densities and can be physically separated by density gradient centrifugation methods (Figure 2.17).

First, Meselson and Stahl grew the bacteria in medium containing the heavy isotope long enough that virtually all DNA helices in the bacteria would contain heavy isotopes in both DNA strands (H/H). This was the starting point of the experiment (Figure 2.18A and B). After removing aliquots (samples) at this stage of the experiment, the bacteria growing in $^{15}$N medium were then transferred to fresh growth medium containing only the light $^{14}$N isotope and were allowed to undergo a single round of bacterial cell division (during which the DNA in the cells replicates once and only once). Aliquots of the growing bacteria were removed for analysis. Then, following more rounds of cell division, additional aliquots were removed. The DNA was isolated from cells in the aliquots for density gradient analysis.

When they analyzed the aliquots (Figure 2.18B), Meselson and Stahl confirmed that the DNA first synthesized by the growing bacteria was heavy (H/H) because it contained only the heavy isotope $^{15}$N. The second set of samples, from when the bacteria reproduced once taken in the light isotope ($^{14}$N growth medium) a different form of DNA was found

with a lighter density. Most interesting, the scientists noted that the newly made DNA was not as light as the DNA containing only $^{14}$N (L/L). Rather, the new DNA species had an intermediate density falling in between the all-heavy DNA (H/H) (containing only $^{15}$N) and the all-light DNA (L/L) (Figure 2.18). This eliminated the possibility that the new DNA strands form an entirely new DNA molecule (L/L) and the parent DNA strands forming a second molecule (H/H).

After another round of bacterial reproduction takes place, in addition to some intermediate-density (L/H) DNA, the density analysis revealed that the bacteria now formed light-density DNA (L/L) containing only $^{14}$N (Figure 2.18B). The light-density (L/L) DNA represented the molecule formed from the $^{14}$N (L) parent strand of the L/H DNA and a newly synthesized DNA strand using the $^{14}$N in the culture medium. This result indicated decisively that a replicated double helix contains a parent DNA strand and a new DNA strand. Neither the conservative nor the dispersive mechanism (Figure 2.16) could explain these results, whereas the semiconservative mechanism fit the data perfectly.

The DNA replication mechanism suggested by the Watson–Crick DNA helix model and confirmed by Meselson and Stahl was termed **semiconservative** replication because within each new DNA double helix, one of the two DNA strands ("semi") is from the original double helix ("conserved"). Semiconservative DNA replication occurs in all cells just before they undergo cell division (mitosis). Each "parent" strand of DNA acts

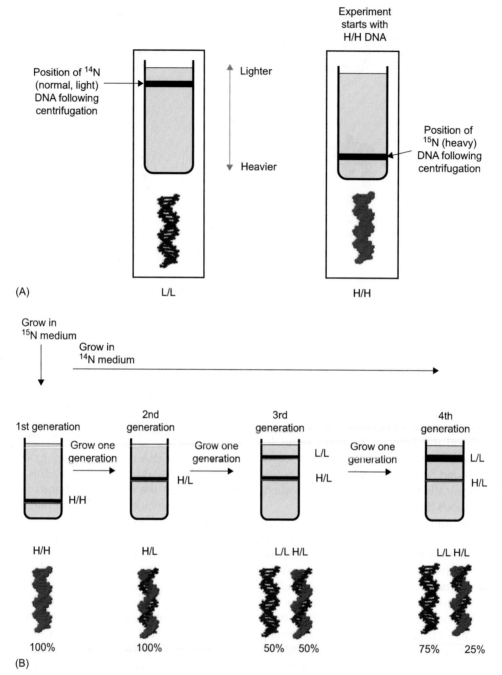

Position of $^{14}N$ (normal, light) DNA following centrifugation

Lighter

Heavier

Experiment starts with H/H DNA

Position of $^{15}N$ (heavy) DNA following centrifugation

(A)    L/L    H/H

Grow in $^{15}N$ medium

Grow in $^{14}N$ medium

1st generation

Grow one generation

2nd generation

Grow one generation

3rd generation

Grow one generation

4th generation

H/H

H/L

L/L

H/L

L/L

H/L

H/H

H/L

L/L H/L

L/L H/L

100%    100%    50%    50%    75%    25%

(B)

**FIGURE 2.18** Interpreting the results of the Meselson and Stahl experiment. (A) Initial comparison in density gradient centrifugation of DNA from normal and experimental growth conditions. Bacteria were grown in medium with the normal "light" isotope of nitrogen ($^{14}N$), which was incorporated into the bacterial chromosome DNA as the cells divided and the DNA replicated. Their DNA was compared with that of bacteria grown in "heavy," $^{15}N$-containing medium. Density gradient centrifugation was used to physically separate and identify the different DNA species made in the cells. $^{15}N$ DNA (H/H) is denser than normal DNA because the heavy $^{15}N$ atoms replace the lighter $^{14}N$ atoms in both strands of DNA. As a result, it migrates farther during centrifugation. (Lighter indicates lower density, heavier indicates higher density.) (B) The H/H bacteria were grown for one generation at a time with $^{14}N$, the light isotope of nitrogen. DNA was extracted from sample aliquots after each generation and centrifuged for comparison of density. With each generation, the density of the DNA decreased with successive generations. The percentage beneath each generation indicates the relative amount of DNA in each band.

as a template for the synthesis of a new DNA strand. In effect, the parent strand dictates that a nucleotide containing adenine (A) must be placed opposite one containing thymine (T) (and vice versa); it dictates that a nucleotide containing guanine (G) will be positioned opposite one containing cytosine (C) (and vice versa). These "base-pairing rules" are the key to the secret of how the DNA helix passes on genetic information.

Meselson and Stahl designed and executed a clever and efficient experiment to discriminate between three possible means of DNA replication. Their results indicated that each strand of the parent DNA helix remains intact and becomes paired with an entirely new strand, in a process called semiconservative replication.

**FIGURE 2.19**  Father and son Nobel laureates. Arthur Kornberg *(left)* with his son Roger *(right)* in 2006. Jamie Kripke Photography.

## DNA Polymerases Copy Complementary DNA Strands from DNA Templates

When DNA replication occurs, a **DNA polymerase** enzyme reads the template DNA base sequence and adds the complementary nucleotide bases to the new, growing DNA strand one by one. In the 1950s, Arthur Kornberg (Figure 2.19) discovered a DNA polymerase enzyme in *E. coli* bacteria and showed that DNA

polymerase could synthesize DNA outside a cell using building block components provided in a test tube. This was an awesome accomplishment. In the presence of Roger, his then 12-year-old son, Arthur Kornberg received the 1959 Nobel Prize in physiology or medicine for his groundbreaking work on DNA synthesis. Kornberg's research characterized replication, the process of DNA copied into duplicate DNA strands when cells divide. This universal process in all cells is essential to understanding how genetic information is transferred from parent cells to progeny cells. Arthur returned to Stockholm again in 2006, but this time the Nobel Prize in chemistry went to his son, for Roger's cutting-edge work showing how the genetic information in DNA is copied into a messenger RNA (mRNA) strand using the process of **transcription** (see Chapter 3). The cell absolutely requires transcription for life. If transcription stops, the organism dies quickly. This is the cause of death from eating certain mushrooms, which contain a lethal toxin that quickly blocks transcription in the cells.

As it turned out, Arthur Kornberg's DNA polymerase is not the principal enzyme used to replicate the genome DNA in the cell. We now know Kornberg's enzyme as ***E. coli* DNA polymerase I**, one of three DNA polymerases that replicate chromosome DNA in *E. coli* cells (Table 2.2). All organisms contain DNA polymerases, which range from small **monomer** proteins to enormous **multiprotein** complexes. Smaller DNA polymerase enzymes are often involved in repairing damaged DNA (see Chapter 9), whereas larger multiprotein polymerase complexes participate in the replication of genome DNA. Although some aspects of DNA replication vary among prokaryotic and eukaryotic cells and viruses, the chemical mechanism and the protein machinery involved in DNA replication are highly conserved.

**TABLE 2.2**  Summary of properties of three of the five DNA polymerase enzymes in *E. coli*

|  | *E. coli* DNA polymerases | | |
|---|---|---|---|
|  | **I** | **II** | **III** |
| *E. coli* gene for the polymerase subunit | ***polA*** | ***polB*** | ***polC*** |
| Number of subunits | 1 | >4 | >10 |
| Proofreading (3′ to 5′) exonuclease activity | yes | yes | yes |
| 5′ to 3′ exonuclease activity | yes | no | no |
| Polymerization rate (nucleotides added per second) | 16–20 | 5–10 | 250–1000 |
| Processivity (nucleotdes added before dissociation) | low (3–200) | high (10,000) | very high (500,000) |

*Source of table: http://www.mun.ca/biochem/courses/3107/Topics/DNA_polymerases.html.*

## The Replication Fork Allows Both DNA Strands to be Synthesized in the 5′ to 3′ Direction

The relatively simple idea of replicating the DNA helix by separating the DNA strands and copying the template to make complementary DNA strands is only a starting point for understanding the process of DNA replication. Along with matching the incoming nucleotides to the template strand, the polymerase forges a bond between the existing strand and the new nucleotide. The energy for this reaction comes from the removal of two phosphates from the incoming nucleotide (Figure 2.20). In fact, DNA polymerases can only add nucleotides to the 3′ end of a DNA strand.

Some details of the replication process are key in making extremely high-fidelity (almost error-free) DNA copies. The overall rate of errors in eukaryotic DNA replication is only about one in a billion bases. Part of the reason for this is that DNA polymerases are good at correcting their own errors using proofreading and repair activities. DNA polymerases can add onto an existing RNA **primer**, a short strand of RNA that is base paired to the template DNA strand at the point on the helix where DNA replication is to start. The primer has a

hydroxyl group (–OH) group at its 3′ end providing a site for the DNA polymerase to add the next nucleotide (Figure 2.20B). The RNA primer is removed later by an additional enzyme, and replaced with DNA bases.

As a consequence of the requirement for a primer with a 3′ –OH group, DNA polymerase enzymes can copy a DNA template in only one direction, adding new nucleotide units on to the 3′ end, and synthesizing new DNA strands in the 5′ to 3′ direction only. If the polymerase worked on the 5′ end of the DNA instead, the high-energy bond would be on the growing strand, and not on the incoming nucleotide. That would not allow for correction of mistakes, because if an incorrect nucleotide were removed, there would no longer be a high-energy bond available for adding a nucleotide (the triphosphate moiety is at the opposite end of the incoming nucleotide). Thus, the directionality of DNA strand synthesis contributes to fidelity of replication.

> The initiation of DNA replication requires an RNA primer as a starting point to provide a 3′ –OH for DNA polymerase to extend with additional nucleotides into a new DNA strand. Replication proceeds only in the 5′ to 3′ direction.

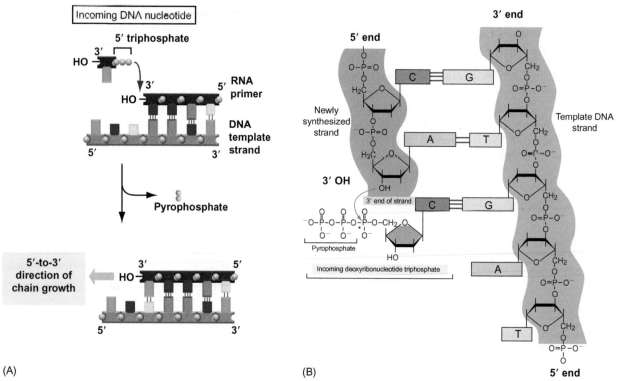

(A)

(B)

**FIGURE 2.20** Incoming nucleotides are added to the 3′ OH during replication. (A) To start to copy DNA, the DNA polymerase enzyme needs a primer with an available 3′ OH. Thus the DNA strand grows in the 5′ to 3′ direction only. The two phosphates on the end of the incoming nucleotide are removed as pyrophosphate providing energy for the formation of a bond between the two nucleotides. (B). The chemical structure of the DNA backbone is shown as a C nucleotide is added onto the 3′ end of the DNA, base-pairing with a G nucleotide. The 3′ OH plays a role in breaking the bond between the phosphates of the incoming deoxyribonucleotide triphosphase (dNTP).

During DNA replication, many enzymes and proteins work together at the site of replication, called the **replication fork**, as it works its way along the length of the DNA molecule. A very simplified view of the replication fork (Figure 2.21A) highlights a problem posed by the fact that DNA polymerases synthesize DNA strands in the 5′ to 3′ direction—since the two strands of a DNA double-helix molecule are antiparallel to each other, only one strand can be synthesized, it would seem, as the fork advances. The cell solves this problem using two DNA polymerase enzymes, one enzyme working continuously in the 3′ to 5′ direction synthesizing the **leading strand**, and the other enzyme performing DNA synthesis in a fragmentary fashion along the **lagging strand**.

A more complex model (Figure 2.21B) shows that the replication fork is the site of a complicated choreography. A multi-protein polymerase complex includes both polymerases, clamp proteins that help keep the polymerases on the DNA, and a **helicase** enzyme that separates the two parent strands of DNA as the replication fork moves forward. On the leading strand template, the DNA polymerase complex performs **continuous replication** moving toward the advancing replication fork. However, the parent template for the lagging strand forms a large loop that feeds into the polymerase complex. On the lagging strand, a polymerase performs **discontinuous replication** and a **primase** enzyme repeatedly lays down short RNA primers to provide a 3′ –OH for adding nucleotides (on the leading strand, a primer is needed only where replication starts). The primers are subsequently removed, and DNA polymerase fills in the resulting gaps. The resulting DNA strands, called **Okazaki fragments**, are joined to each other by a DNA ligase enzyme to complete lagging strand synthesis. The large loop made for the lagging DNA strand allows for coordination of the activities of the DNA polymerases at the constantly moving replication fork, placing the two actively replicating regions of DNA adjacent to each other.

> The replication fork is the site of DNA synthesis on both strands of a DNA double helix. DNA polymerase enzymes at the fork add nucleotides to each growing strand of DNA in the 5′ to 3′ direction. One strand therefore is synthesized continuously; the other is synthesized as small fragments that are subsequently joined together.

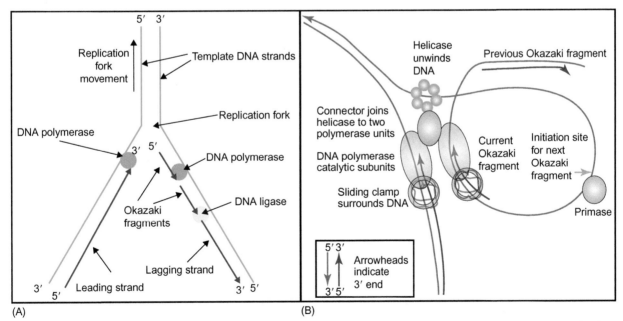

**FIGURE 2.21** DNA replication occurs at the replication fork. (A) A simplified schematic diagram shows the two parent strands of DNA opened for replication, forming the replication fork. A DNA polymerase operates on each strand, synthesizing a new DNA strand moving in the 5′ to 3′ direction. One polymerase creates one continuous new strand, the leading strand. On the opposite strand, a discontinuous series of Okazaki fragments is created on the lagging strand. The resulting Okazaki fragments are sealed into a continuous DNA strand by a DNA ligase enzyme. (B) A more detailed diagram of the replication fork shows the loop that forms enabling a polymerase dimer to copy both strands as a unit. Also shown are sliding clamp proteins that help the polymerases to stay on the DNA, a helicase enzyme that separates the two parent strands, and a primase enzyme that lays down the RNA primers to initiate synthesis of the Okazaki fragments on the lagging strand.

## Box 2.2    DNA Replication Is a High-Fidelity Process That Makes Nearly Error-Free DNA Copies

The accuracy of DNA synthesis is very important for correctly transmitting genetic information over many generations and for avoiding mutations that can initiate and promote diseases such as cancer. Studies on *E. coli* chromosome replication show that the error rate of the bacterial replication machinery *in vivo* (in the cell) is very low, making only 1 error in every 10 million to 100 million bases copied into DNA. It is impressive that in a growing bacterium, the DNA replication enzyme complex moves like a locomotive along the DNA helix at the astonishing rate of 500 nucleotides (bases) per second, simultaneously reading the DNA template and incorporating the complementary nucleotides into the growing strand. DNA replication in eukaryotic cells is even more accurate, in part because of safeguards that operate before and after DNA replication in the cell cycle (see Chapter 9 on cell fate). Believe it or not, DNA polymerases not only *proofread* their work to find errors, but they also have the ability to actually *correct the replication mistakes* made in the genome DNA before the enzyme moves farther down the DNA template.

The DNA polymerase proofreading activity can detect and repair replication mistakes quickly. The DNA polymerase enzyme contains a 5' to 3' exonuclease activity and a 3' to 5' exonuclease activity, which are both involved in maintaining DNA accuracy and integrity. *E. coli's* DNA polymerase I can be cleaved by certain enzymes into two smaller proteins. The larger of these two, called the Klenow fragment (Figure 2.22), has been studied closely and retains both the ability to synthesize DNA and the 3' to 5' exonuclease activity. The three-dimensional structure of the Klenow fragment is shaped approximately like a right hand, which helps people to better visualize how a polymerase functions (Figure 2.23).

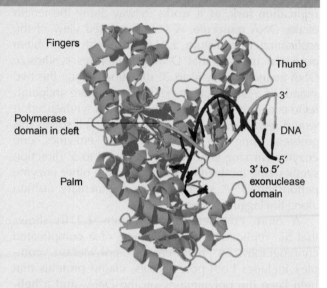

**FIGURE 2.23**    DNA polymerase has a "hand" in DNA replication and proofreading. The Klenow fragment DNA polymerase can be visualized as a partially open hand with thumb, fingers, and palm domains. The DNA helix is cradled in the hand of the protein, with the palm domain making a plate at the bottom of the cleft formed by the thumb and fingers. The polymerase activity takes place in the cleft at the base of the fingers and the thumb. The 3' to 5' exonuclease site is adjacent to the palm domain, and contains an incorrectly added base at the 3' end of the DNA strand being synthesized (purple). This is the site that removes an incorrectly paired nucleotide. The template DNA strand (beige) was truncated in order to crystallize the complex for x-ray diffraction.

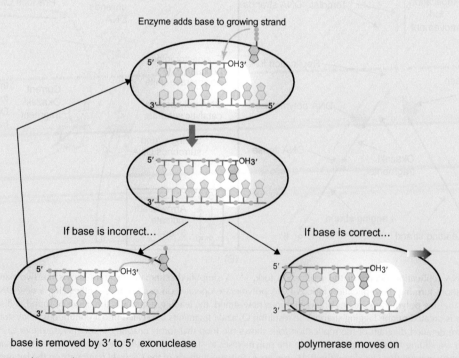

**FIGURE 2.22**    DNA polymerase has a 3' to 5' proofreading activity that corrects mistakes. When DNA polymerase adds an incorrect base to the the 3' end of the growing DNA strand, the 3' to 5' exonuclease activity removes the incorrect base and the DNA polymerase then continues synthesis by adding the correct base.

## Replication Origins in DNA Control the Start of DNA Synthesis

It is essential that DNA synthesis be closely regulated to ensure that the cell does not replicate the genome DNA more than once per cell cycle. Creating additional copies of the genome could have disastrous effects on gene expression resulting in the death of the organism. An important step in controlling DNA replication (Chapter 9) occurs when the cell "decides" to initiate DNA synthesis. Replication of the bacterial chromosome (Figure 2.24) begins at a specific region of DNA called the **origin of replication (ori)**. A collection of proteins controls the use of the ori DNA. To initiate replication of a DNA helix, proteins separate the DNA strands in the ori region, which contains a high percentage of A-T base pairs.

(A-T base pairs form only two hydrogen bonds, compared to G-C pairs which form three, so it is easier to separate "A-T rich" DNA strands.) Once DNA replication has been initiated, two replication forks proceed in opposite directions away from the point of origin in the DNA.

In humans, DNA replication is particularly impressive because of the massive size of the genome compared to bacteria. Every time a human cell divides, the billions of DNA bases distributed over 46 chromosomes must be quickly and accurately replicated to produce enough chromosomes for correct cell division. Eukaryotic chromosomes have multiple replication origins distributed along each DNA genome (Figure 2.25). The cells must control the rate at which DNA replication initiates at each origin on each chromosome. The goal is to restrict duplication of the cell genome to only

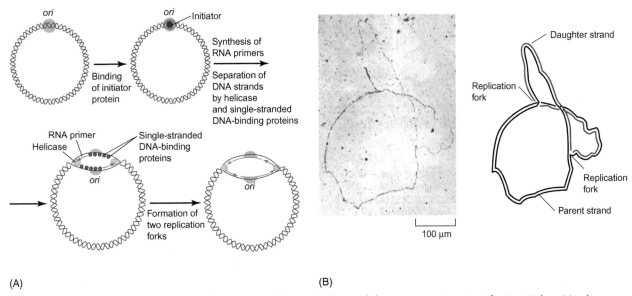

(A)                                                                 (B)

**FIGURE 2.24**  *E. coli* chromosome replication. (A) Replication of the circular bacterial chromosome starts at a specific site (ori, the origin of replication). Two replication forks are formed by separating the strands of the double helix, and they proceed in opposite directions around the circular molecule. (B) The electron micrograph on the left shows the circular *E. coli* chromosome during the process of replication, diagrammed at the right.

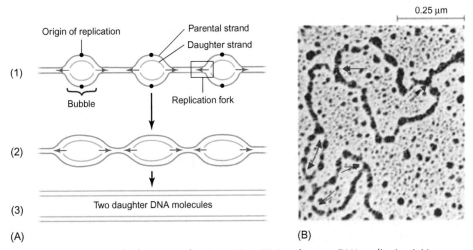

(A)                                                                 (B)

**FIGURE 2.25**  Eukaryotic genomes have multiple DNA replication origins. (A) In eukaryotes DNA replication initiates at many sites along the very long chromosome DNA molecule, making many small replication bubbles (1); bidirectional DNA replication increases sizes of bubbles (2), and eventually the bubbles resolve into two DNA helices (two daughter DNA molecules) (3). (B) Electron micrograph shows three replication bubbles along chromosome DNA in Chinese hamster cells. (Red arrows show the direction of movement of the replication forks in each bubble.)

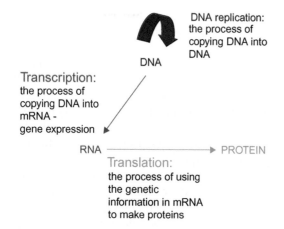

**FIGURE 2.26** Keeping perspective: How is DNA replication related to gene expression? Replication insures the perpetuation of genetic information in an organism as it grows and develops, and between generations of the organism. In contrast, gene expression involves making use of the stored genetic information, via the processes of transcription and translation.

once per cell cycle. The timing and temporal order of initiation at the hundreds of DNA replication origins in the genome are regulated by specialized replication control proteins (see the discussion in Chapter 9 on cell fate [includes cell cycle] and S phase [DNA synthesis]).

> The cell must regulate the timing of genome replication during each cell cycle. Proteins bind to the replication origin(s) in the genome DNA to control the initiation of DNA synthesis.

Frequently during our exploration of cell function it is important to touch base and recognize how each specialized process, such as DNA replication, fits into the big picture of what is going on inside the cell (Figure 2.26). Accurate chromosome duplication and transmission are key processes that must be completed correctly and at the correct time during the cell's growth cycle, or the cell will not survive.

## SUMMARY

Scientists wanted to solve the structure of DNA because they knew that the DNA molecule would provide important clues to the function of DNA and reveal information about how DNA replicates every time the cell divides. In the 1920s, Phoebus Levene and his colleagues identified the three basic components of DNA: a five-carbon sugar named deoxyribose (ribose in RNA); phosphate groups; and the four nitrogenous bases,

adenine (A), thymine (T), guanine (G), and cytosine (C). These components make up the nucleotides, the building blocks of a DNA strand. Studies by Chargaff indicated that the amounts of the DNA bases vary in different organisms, suggesting that organisms may be different because their DNAs are different. Regardless of the source of the DNA, however, Chargaff found that the amount of adenine (A) always equals the amount of thymine (T) and the amount of guanine (G) always equals the amount of cytosine (C). These and other experimental data made essential contributions to solving the puzzle of DNA structure, which was published by Watson and Crick in 1953. Using the x-ray diffraction photographs of DNA obtained by Franklin and Wilkins and incorporating chemical data on DNA, Watson and Crick proposed that DNA is a double-stranded helical molecule, the DNA double helix, with each adenine (A) paired to a thymine (T) and each guanine (G) paired to a cytosine (C).

The arrangement of bases paired between the two DNA strands of the double helix suggested a possible replication mechanism in which each DNA strand carries a base sequence that intrinsically encodes its complementary strand. In 1957, Meselson and Stahl showed that DNA replication proceeds using a semiconservative mechanism where one of the parental DNA strands is conserved in each of the newly formed DNA double helices. In each case, the parental DNA strand serves as a template for the synthesis of a brand-new, complementary strand of DNA, while also becoming paired with the new strand. DNA polymerase is the type of enzyme that copies DNA into DNA when the cell divides. During DNA replication, the DNA polymerase enzyme reads the template base sequence and adds the correct complementary nucleotide base to the growing DNA strand. As a consequence of the replication mechanism, DNA polymerase copies the DNA template in only one direction; each new DNA strand is synthesized in the 5' to 3' direction, with nucleotide units added on to the 3' end. In the cell, DNA replication starts at specific sites on the chromosome DNA called origins of replication. Specialized proteins associated with the origin DNA play critically important roles in initiating DNA replication and in restricting the occurrence of DNA replication to once per cell cycle.

## REVIEW

This chapter describes the development of thought and experiments that led up to the historic discovery of the double-helix structure of DNA and beyond it to an understanding of the process of DNA replication.

To assess your comprehension of these topic areas, answer the following review questions:

1. According to Levene's research, what are the three basic components of the nucleotide units in DNA?
2. Describe the method by which nucleotide units are linked to one another in DNA.
3. What two significant observations were made by Erwin Chargaff's group, and how were they important to establishing the structure of DNA?
4. Explain how Rosalind Franklin's x-ray diffraction studies helped to determine the structure of DNA.
5. Given the base sequence of one strand in a DNA helix, explain how you can deduce the base sequence of the opposite DNA strand.
6. In the race to solve the structure of the DNA helix, what were the key features of the DNA model proposed by the famous scientist Linus Pauling?
7. Describe the structure of DNA referred to as a "replication fork," and explain how it is related to the process of DNA replication.
8. Determine what the results of the Meselson and Stahl experiments would have been if the mechanism of replication were actually (a) conservative and (b) dispersive (see Figure 2.16) instead of semiconservative.
9. Describe the activities of some of the proteins required for DNA replication in addition to DNA polymerase.
10. Explain the role of replication origins in bacterial and eukaryotic cells.

## ADDITIONAL READING

Check, W., 1994. DNA helix turns 40. ASM News. ASM Press, Washington, DC.

Clayton, J., Dennis, C. (Eds.), 2003. 50 Years of DNA. Palgrave Macmillan, Basingstoke, UK. Foreword by Campbell, P.

Flannery, M., 1997. The many sides of DNA. Am. Biol. Teach. 59, 54–60.

Hall, S., February 1990. James Watson and the search for the Holy Grail. Smithsonian.

Hoffman, R., August 1993. Unnatural acts. Discover.

Holmes, F.L., 2001. Meselson, Stahl, and the replication of DNA: A History of "The Most Beautiful Experiment in Biology." Yale University Press.

Jaroff, L., March 15 1993. Happy birthday, double helix. Time.

Johnson, S., Mertens, T.R., 1989. An interview with Nobel Laureate Maurice Wilkins. Am. Biol. Teach. 51, 151–157.

Meselson, M., Stahl, F.W., 1958. The replication of DNA in *Escherichia coli*. Proc. Natl. Acad. Sci. 44, 671–682 PMID 16590258.

Pauling, L., Corey, R., 1953. A proposed structure for the nucleic acids. Proc. Natl. Acad. Sci.

Sayre, A., 1975. Rosalind Franklin and DNA. W. W. Norton and Co., New York.

Watson, J., 1968. The Double Helix. Athenum Press, New York.

Watson, J.D., Crick, F.C.H., 1953. Molecular structure of nucleic acids; a structure for Deoxyribose Nucleic Acid. Nature 171, 737–738.

## WEB SITES

Force of Nature: The Life of Linus Pauling, by Tom Hager. http://lpi.oregonstate.edu/ss03/triplehelix.html

Linus Pauling and the Race for DNA: A Documentary History. Helen and Linus Pauling Papers. http://osulibrary.orst.edu/specialcollections/coll/pauling/dna/

Rosalind Franklin: Light on a Dark Lady, by Anne Piper. http://www.physics.ucla.edu/~cwp/articles/franklin/piper.html

Meselson–Stahl Experiment. http://en.wikipedia.org/wiki/Meselson-Stahl_experiment

DNA Structure: An Interactive Animated Nonlinear Tutorial, by Eric Martz. http://www.umass.edu/molvis/tutorials/dna

Exploring DNA, by Andrew Carter. http://www.umass.edu/molvis/bme3d/materials/jtat_080510/exploringdna/contents/contents.htm

DNA Interactive, from the Dolan DNA Learning Center, Cold Spring Harbor Laboratory. http://www.dnai.org

# DNA in Action

---

**Genetic Tweak Produces Mighty Mouse to Outrun Rivals**

*The Guardian,* November 2, 2007
  By Ian Sample
  Scientists have created a real-life Speedy Gonzales by genetically engineering a mouse which can easily outrun its natural cousins.

When let loose on a treadmill in the laboratory, the mouse ran for up to six hours without stopping, covering many kilometres before finally taking a rest. Normal mice gave up after covering just 200 metres at the same speed.

The souped-up rodent consumes 60% more food than other mice, but remains fitter and leaner. Surprisingly, the species also lives longer and is able to breed until a later age.

With the goal of investigating exercise effects on cancer, scientists generated a mouse with an extra copy of the PEPCK-C gene, placing it under the direction of a special DNA control region. This control region causes the mouse's muscle cells to make high levels of PEPCK-C proteins. The PEPCK-C protein functions in an important biochemical pathway that mitochondria use to obtain energy for the cell via the citric acid cycle. (Mitochondria are the organelles in eukaryotic cells that produce energy for the cell's functions.) PEPCK-C genes are usually most active in liver cells, making high levels of PEPCK-C protein, but little if any of this protein is normally made in muscle cells. The high level of PEPCK-C protein in the mouse's muscle cells appears to have transformed these mice into "super mice." Interestingly, the muscle cells from the super mice make many more mitochondria than normal muscle cells, and their muscle cells produce less lactate, the chemical that causes a burning sensation when produced in muscles during exercise.

But beware! Before you head out to look for a PEPCK-C protein supplement in the store, understand that the results in mice probably have limited application to human health. For one thing, the PEPCK-C gene may not function the same way in humans as it does in mice. In addition, the super mouse experiments show that increasing the amount of protein produced by a gene can have unanticipated effects. This experiment highlights the importance of controlling gene expression, the subject of this chapter.

## LOOKING AHEAD

This chapter explores the succession of observations and experiments that led scientists to make many critical connections between deoxyribonucleic acid (DNA) and protein synthesis. We will describe the mechanisms by which the genetic information in DNA is converted into biochemical action through the encoded gene products, mostly proteins. On completing this chapter, you should be able to do the following:

- Understand how scientists made the connection between chromosomes and biochemical activities taking place in the cell.
- Follow the reasoning that led scientists to understand the genetic code in DNA, and helped them relate the code to the specific amino acid sequence in a protein.
- Recognize the roles played by ribonucleic acid (RNA) in transmitting the genetic information in DNA into the amino acid sequence in a protein.
- Explain why the genetic code Is considered to be degenerate.
- Describe the role of translation in the synthesis of proteins in the cell.
- Understand the need for gene control in cells.
- Summarize the mechanisms behind some positive and negative gene expression controls.

## INTRODUCTION

In the beginning of the 20th century, scientists significantly advanced human understanding when they realized that patterns of heredity could be explained by the activity of chromosomes. That concept led not only to the modern science of genetics but also to great advances in medicine, biotechnology, agriculture, and many other fields. Genetics has profoundly influenced how we think about ourselves and our world because it removed the mystery from heredity and made the biological nature of life seem more logical and approachable.

The great discoveries concerning DNA and chromosomes also raised a question that would influence biological thinking for more than half a century: What exactly is the biochemical connection between chromosomes and the hereditary traits visible in organisms? Put simply, how do chromosomes (Figure 3.1) and genes work?

Modern biologists can partly answer that question. We know, for example, that the genetic information in chromosomes is put into action by directing the production of molecules, mainly proteins that both form the structures and carry out biochemical actions in our cells, tissues, and organs. Hence, all parts of human anatomy reflect the activity of our chromosomes. Moreover, the entire chemistry of our bodily activities (whether moving or thinking or digesting) is governed by biological catalysts called **enzymes**, and most enzymes are proteins. Thus, the structure and chemistry of our bodies revolve around proteins. The genes, carried on chromosomes, specify the proteins (Figure 3.2).

The fundamental ideas that relate chromosomes to genes and proteins were not arrived at in a sudden eureka moment but were developed slowly by a

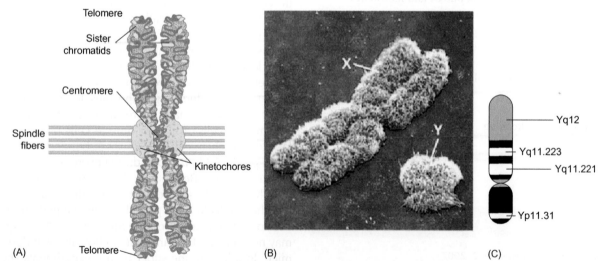

**FIGURE 3.1** Eukaryotic chromosomes. (A) A duplicated, condensed eukaryotic chromosome during mitosis. In the duplicated state, each chromosome of the pair is called a sister chromatid. Each sister chromatid contains a single, linear, double-stranded DNA molecule extending from telomere to telomere through the centromere. During mitosis, the chromosome is attached to the mitotic spindle fibers via the centromere. The DNA helix is packaged by histone proteins (not shown) into chromosomes. (B) Electron micrograph of condensed X and Y chromosomes that have been extracted from a cell. (C) A map of the Y chromosome showing the long and short arms and a view of the Y chromosome banding pattern (Yq12, Yq11.223, Yq11.221, Yq11.31).

succession of scientists working over many decades. This process began with the understanding that deoxyribonucleic acid (DNA) is a key substance of chromosomes (Chapter 1), and continued with the discovery that the DNA structure has the capacity to encode genetic information (Chapter 2). This chapter focuses on the seminal experiments that related chromosomal DNA and genes to the production of proteins in the cell. We will consider the mechanisms by which the biochemical information in DNA is accessed and then put into action in the cell, mostly in the form of proteins. Proteins are linear strings of amino acids, whose sequences are encoded in DNA and interpreted through RNA.

The DNA double helix itself is designed to encode, duplicate, and store genetic information. The chapter title, "DNA in Action," refers to the biological mechanisms that permit DNA to direct the deeds of the cell via the synthesis and activity of specific proteins and RNA molecules. By considering the information in DNA as it is put into action, we are really studying how genes work; in doing so we are answering one of the great questions in biology: How do our chromosomes express our hereditary traits?

> Much of the information contained in our chromosomes is accessed and executed through the synthesis of proteins, which are the end product of most of our hereditary instructions.

## FUNDAMENTAL SCIENCE CONNECTS DNA AND TRAITS

### The First Link between Inheritance and Enzymes

In 1902, a prominent British physician named Archibald Garrod (Figure 3.3) wrote that certain diseases seemed to occur time and again in selected families. Garrod studied the disease alkaptonuria in detail. People with this disease expel urine that rapidly turns black when exposed to air. The color changes because the urine contains alkapton (the chemical name is homogentisic acid), which darkens on exposure to oxygen. In normal individuals, alkapton is broken down in the body, but the alkaptonuria disease prevents this process, and alkapton is excreted in the urine. Garrod

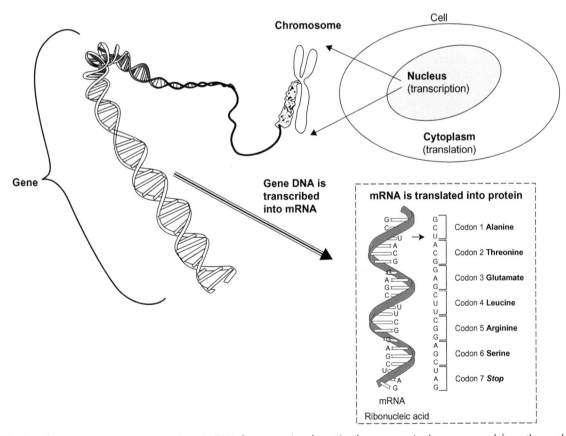

**FIGURE 3.2**  Chromosomes carry genes written in DNA language. A eukaryotic chromosome is shown removed from the nucleus and unpackaged to indicate the linear double-stranded DNA molecule that carries the gene. Eukaryotic genomes have a large amount of noncoding DNA. In eukaryotes, genes encoded along the DNA are transcribed into mRNA in the nucleus, and the RNA is processed before carrying the genetic code for a new protein into the cytoplasm for translation.

**FIGURE 3.3** Garrod suggested a link between inheritance and enzymes. Archibald Garrod's observations of diseases that occurred over and over again in the same families led him to postulate connections between disease, inherited factors, and enzymes—all relatively new concepts in medicine at the time.

studied several generations of different families and became convinced that alkaptonuria, and some other metabolic disorders, were each controlled by a single inherited factor. Mendel's work was in the process of being rediscovered in Garrod's time (Chapter 1), and it was acceptable to think of inherited traits in terms of Mendelian "markers" (or genes, as they would later be called). Garrod concluded that certain diseases and disorders could be inherited much as flower color was inherited in pea plants. He further postulated, quite remarkably for his time, that a genetic disease is caused by a change in an ancestor's genetic material, and the defect is passed along within the family.

Although Garrod lacked a background in biochemistry, he suggested that the cells in people with alkaptonuria could not break down alkapton because they lacked the necessary biochemical enzyme. Thus, the patient with alkaptonuria could have an enzyme defect, and if that defect was inherited, it followed that a gene defect could cause an inherited enzyme deficiency. The concept of an "enzyme" was also relatively new in the early 1900s, and relating a chemical reaction to an enzyme put one at the forefront of science. Other scientists accepted Garrod's insight, and the concept gradually developed that genes have something to do with enzyme production. This concept would be greatly strengthened by the work of Beadle and Tatum, but the world would wait 40 years for their work to be performed.

George Beadle                      Edward Tatum

**FIGURE 3.4** Beadle and Tatum linked genes and enzymes. George Beadle and Edward Tatum postulated the one gene–one enzyme theory that related gene activity to protein synthesis.

In the early 1940s, while Avery and his colleagues were attempting to identify the role of DNA as the transforming principle in bacteria (Chapter 1), other researchers were trying to clarify the functions of chromosomes. Among the leaders in chromosome research were George Beadle and Edward Tatum at Stanford University (Figure 3.4). The Beadle and Tatum experiments were innovative and imaginative, especially the now famous "one gene–one enzyme" experiment. Beadle and Tatum wanted to find out what would happen to the biochemistry of an organism if they mutated an organism's chromosome DNA. For these experiments they selected the common bread mold *Neurospora crassa* (Figure 3.5). At the time,

In the early twentieth century, Garrod noted that some diseases that "run" in families were inherited in patterns similar to ordinary traits. He also recognized for the first time that a defect in an enzyme could cause an inherited disease.

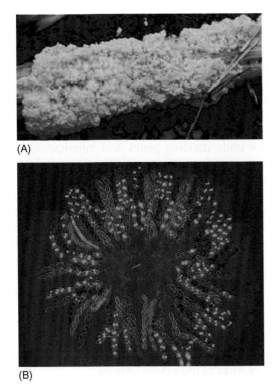

(A)

(B)

**FIGURE 3.5** *Neurospora crassa*. (A) *Neurospora crassa* is a reddish bread mold. (B) Fluorescence microscopy of neurospora cells expressing histone proteins with bright green fluorescent markers. Histones are proteins that package DNA into chromosomes in the nucleus.

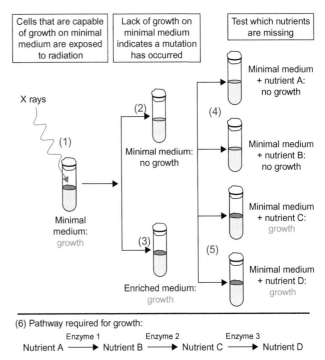

**FIGURE 3.6** Beadle-Tatum: the one gene-one enzyme experiment. (1) *Neurospora* cells were exposed to x-rays to cause mutations. Then the culture was split and transferred into (2) minimal medium lacking certain nutrients and (3) enriched medium containing all nutrients. The cells did not grow on the minimal medium, or on (4) minimal medium with nutrients A or B added. (5) Cells grew only on minimal medium containing added nutrients C and D. (6) The positions of nutrients A-D in the metabolic pathway indicates that the cells must be missing enzyme 2 from the pathway, leading to the conclusion that the mutation damaged enzyme 2 in pathway.

*Neurospora* cells were among the few types of cells that could be grown in the laboratory on a defined growth medium where the exact nature and amount of each nutrient is controlled (for example, the medium might contain precisely 10g of glucose per liter). Beadle and Tatum knew that the ability to control the growth conditions of the cells by changing the composition of the medium was essential for the experiment. They needed what geneticists now call a **genetic screen**, a way to find rare *Neurospora* cells that could not grow without certain nutrients supplied in the medium.

The genetic screen used in the Beadle and Tatum experiment is most easily understood by following the protocol they used, shown in Figure 3.6. First a culture of *Neurospora* cells was exposed to x-rays to cause **mutations** in the genome DNA sequences. Then the culture was split and half was transferred into "minimal medium," that is, a growth medium lacking certain specific nutrients (called nutrients A, B, C, and D in Figure 3.6). The other half of the cell culture was transferred into "enriched medium" containing all the nutrients. The scientists observed that the irradiated cells did not grow in the minimal medium or in minimal medium with nutrient A or B added. Importantly,

the cells did grow on minimal medium with nutrient C or D added. Beadle and Tatum concluded that because the cells grew only when nutrients C or D were added to the medium, it followed that the cells must be missing enzyme 2 in the biochemical pathway (Figure 3.6). This meant that in these rare cells, the x-rays most likely damaged the DNA that encodes enzyme 2. This established the first experimental connection between a specific site in the DNA genome and a specific protein product.

Beadle and Tatum used haploid *Neurospora* cells in their experiment; each nucleus contained only one copy of each chromosome (and hence one copy of each gene). When haploid cells are treated with a **mutagen**, which causes mutations in DNA, a change in phenotype can be observed when one copy of the gene is mutated, because haploid cells do not carry a second copy of any gene. Diploid cells, which each contain two copies of every chromosome (and hence, two copies of each gene), decrease the odds for success because the **mutagenesis** treatment often must disable both copies of the target gene for the phenotype of the cell to be visibly affected by the mutation. Most mutations are silent or recessive in diploid cells because the single remaining normal gene is sufficient to do the work in the absence of the second gene product. Less often the mutant gene has a dominant phenotype, which means that the mutant gene function overrides the normal gene function and the diploid cells are visibly affected by the mutation (see Chapter 10).

Through their experiments, Beadle and Tatum noted that mutant *Neurospora* cells passed the mutation to the next generation, and each mutation could be explained as damage to a specific cellular catalyst (a specific **enzyme** protein). Most enzymes are protein molecules that carry out virtually all the biochemical reactions in a cell (a few enzymes are RNA molecules or protein-RNA complexes). The key feature of enzymes is that they speed up biological reactions, so that an entire biochemical reaction occurs in a fraction of the time needed to complete the reaction without enzyme catalysis. Additionally, the chemical structure of the enzyme does not change as a result of the reaction, so the same enzyme will catalyze the same reaction repeatedly. Beadle and Tatum's work indicated that a mutation in a cell's DNA, is correlated with the inability of the cell to produce an important growth substance. This missing substance could be traced to a defect in a specific enzyme that catalyzes the reaction for synthesis of the substance.

Two major concepts came from Beadle and Tatum's work:

- Mutations can be traced to the hereditary constituents in the cell (genes).
- DNA in genes has something to do with the cellular production of key enzymes.

Beadle and Tatum came up with the idea that each gene in a cell influences the production of a cellular enzyme: the "**one gene–one enzyme hypothesis**." They proposed that altering one enzyme would in turn disrupt at least one event in a series of chemical reactions in one or more biochemical pathways in the cell.

George Beadle and Edward Tatum found that when the chromosomes in the *Neurospora* cells were damaged, the resulting cellular defects were passed on to the next generation of cells. In addition, they theorized that each chromosome defect interferes with the production of one enzyme.

In the 1940s, biochemists agreed that enzymes are specific types of proteins and that enzymes are catalysts that drive the synthesis of all of an organism's structural parts (e.g., proteins, carbohydrates, lipids, nucleic acids, blood components, hormones, antibodies, hair fiber, muscle proteins, and on and on). It stood to reason that if genes affect the production of enzymes and if enzymes are proteins, then genes must affect protein production. This is a fundamental tenet of molecular biology. However, even though the "one gene–one enzyme" theory is critically important for understanding genes and inheritance, it is an oversimplification.

From a historical standpoint, it is easy to understand why scientific interest in DNA grew during the 1940s. Avery's group was in the process of identifying DNA as the substance of bacterial transformation, Hershey and Chase were working to show that DNA was the determining molecule in viral replication (Chapter 1), and Beadle and Tatum were focusing attention on genes for their ability to transmit mutations between generations and for the connection between genes and the synthesis of proteins. The implications of DNA in cellular functions captured the imaginations of biologists, and deservedly so because the function of DNA stands at the very roots of molecular biology.

## Proteins are Linear Sequences of Amino Acids

**Proteins** do much of the work and comprise many of the structures in all cells, be they microbial, plant, animal, or other cells. Chemically, proteins are composed of linear chains of building-block units called **amino acids** (Figure 3.7(A). Cells produce protein chains of different lengths that are composed of different sequences (order) of the 20 amino acids found in biological organisms (Figure 3.7B). The sequence of the human genome has revealed about 20,000 different human genes, but scientists predict that there are more than 100,000 different combinations of the 20 amino acids that make up human proteins (see Chapter 6). It is essential to understand that one protein is distinguished from another protein by the sequence of the amino acids in the protein. In protein production (**protein synthesis**), specific amino acid sequences encoded in the DNA are linked together into proteins. Think for a moment how many different English words have been composed using a 26-letter alphabet. For proteins, the alphabet is not composed of 26 letters but of 20 amino acids, and the variety of possible combinations of amino acids leads to a tremendous variety of different proteins.

The amino acids in a protein are linked together by peptide bonds made between adjacent amino acids (Figure 3.8). The amino acid sequence of each protein is unique to that specific protein. For example, insulin is a protein made by the cells in your pancreas to regulate

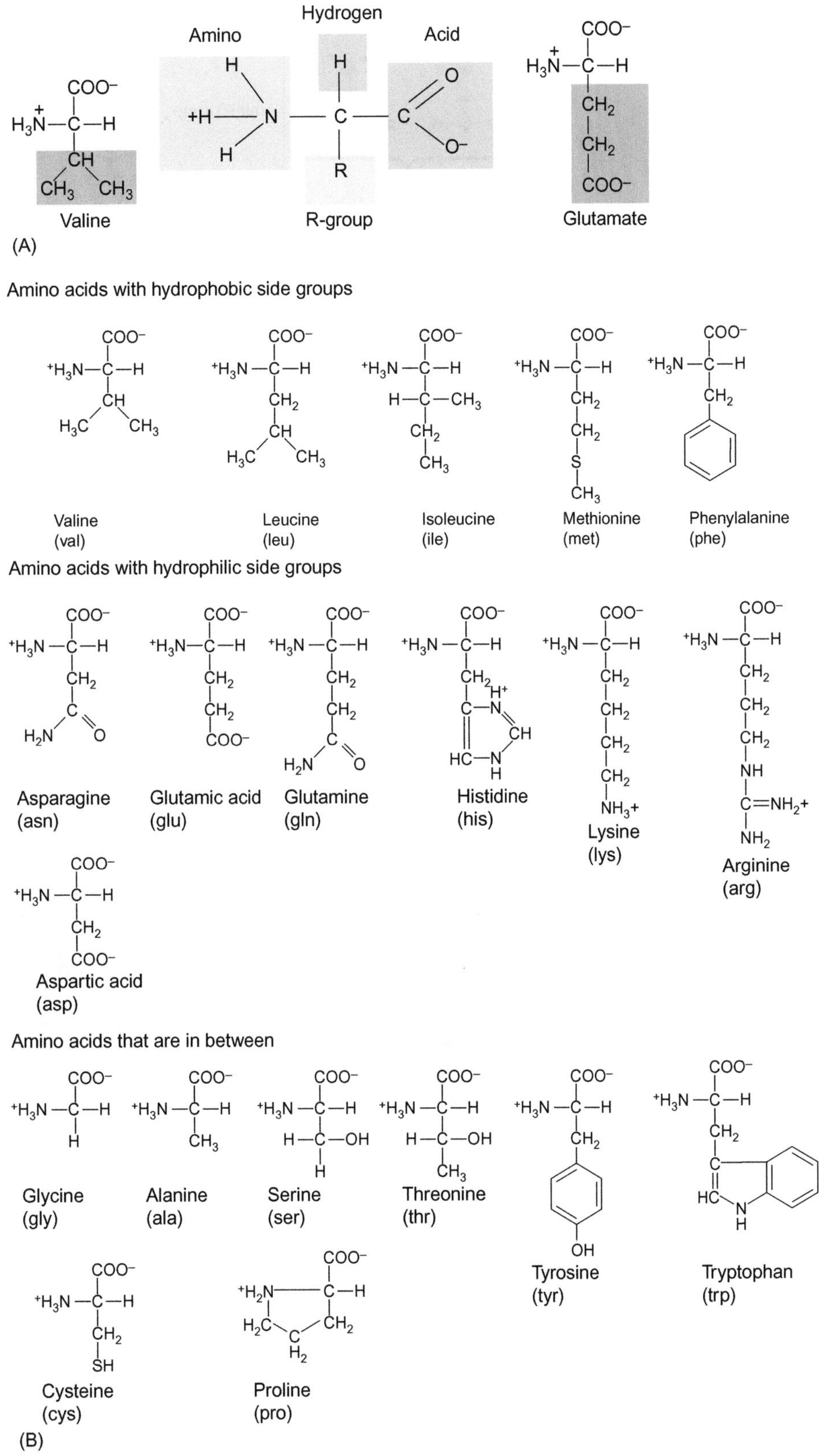

**FIGURE 3.7** Amino acids are the building blocks of proteins. (A) Each amino acid has four chemical groups: amino, hydrogen, carboxyl, and R group, attached to a central or "α" carbon atom. Only the R group differs for each amino acid and confers specific chemical and physical properties of polarity, size, and charge. The R groups on the valine and glutamate amino acids are indicated. (B) Proteins in living cells contain combinations of the 20 different amino acids shown. Each of the 20 amino acids has a different R group, also called a side chain, that confers unique properties to each amino acid as indicated.

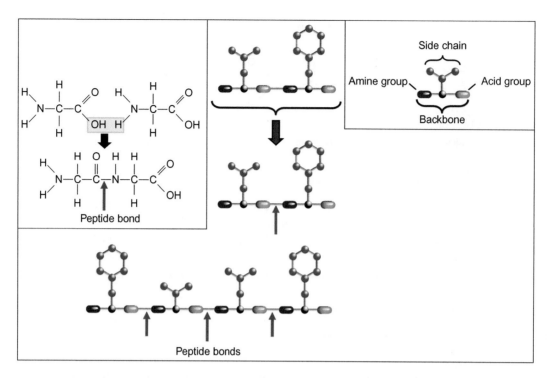

**FIGURE 3.8** Proteins are made by linking amino acids together with peptide bonds. During protein synthesis, the amino portion of one amino acid and the acid portion of another amino acid join together in a dehydration synthesis reaction to form a peptide bond. A dipeptide and a water molecule result from the reaction. Additions at the carboxyl (acid) end of this dipeptide can lengthen the protein. Protein chains are linear, never branched, and nearly all fold into a specific shape to function.

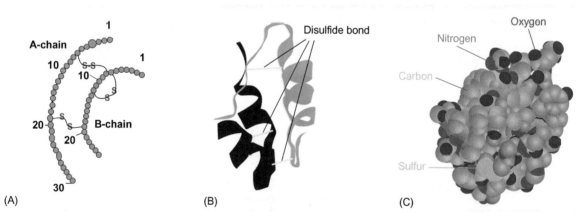

**FIGURE 3.9** Views of the insulin protein. (A) The insulin protein is made in the pancreas as a single protein chain that is connected internally by two disulfide bonds (yellow) which form between its cysteine amino acid side chains. Each amino acid of insulin is represented here as a small circle. Insulin is subsequently cleaved into three protein chains and active insulin contains only the A-chain and B-chain, connected to each other by two disulfide bonds. (B) A ribbon drawing of the cleaved insulin protein chain illustrates the three-dimensional folds of the protein into three helices, three turns, and a loop. The ribbon traces the path of the protein's "backbone" in space. (C) The overall shape of insulin is shown in a space-filling model. All atoms are shown as solid spheres, except hydrogen atoms, which are omitted for clarity. The sizes of the remaining atoms are inflated slightly to make up for the missing hydrogen. The atoms are colored by element as shown.

the amount of glucose sugar in the bloodstream. All the insulin proteins made by a human pancreas have the exact same order (sequence) of amino acids in the protein chain; they all fold into identical three-dimensional shapes and participate in the same biochemical processes and reactions in the body (Figure 3.9).

The importance of the specific sequence of amino acids to the function of each protein was highlighted by the pioneering work of Vernon M. Ingram and his group at Cambridge University in 1956. Ingram discovered that a single genetic mutation in the β globin protein causes sickle cell disease (see Chapter 10).

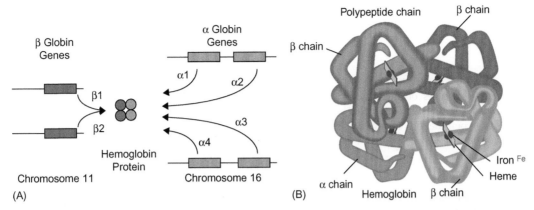

FIGURE 3.10  α and β globin proteins make hemoglobin. (A) In humans, the gene for the β subunit of hemoglobin is located on chromosome 11. Because humans have two copies of each chromosome, there are two β globin genes in the human genome. Both copies are used to synthesize β globin protein. The gene for the α subunit appears twice on chromosome 16 in humans. All four copies of the gene contribute to producing the α hemoglobin protein chain. Overall, equal numbers of α and β globin proteins are produced, and two of each type combine to form hemoglobin. (B) The hemoglobin is shown as a space-fill model. Heme is a small organic molecule that binds in a pocket on each subunit of the protein, enabling the hemoglobin to carry oxygen.

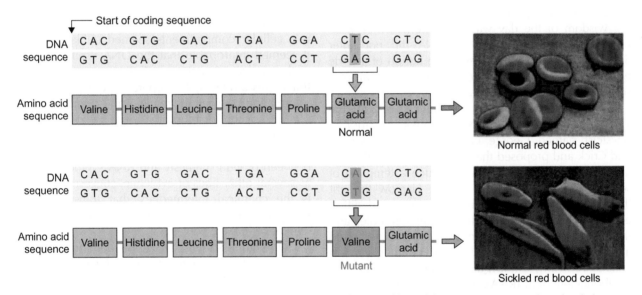

FIGURE 3.11   Mutation in beta globin protein causes sickle-shaped cells. The normal beta globin DNA sequence is aligned with the amino acids in the protein *(top)*. The mutation in the DNA causes a change in the amino acid sequence of the beta globin protein that promotes the formation of long fibers of hemoglobin proteins. The hemoglobin fibers distort the normally disk-shaped red blood cells into sickle shapes, and as a consequence they get stuck in the small blood vessels and capillaries, causing severe disease symptoms.

The globin proteins are part of a large multigene family that includes the α and β globin proteins (Figure 3.10), myoglobin, and the plant leghemoglobins; all are proteins that bind to and carry oxygen. Hemoglobin contains two α globin proteins and two β globin proteins arranged in a complex that responds to the amount of oxygen in the environment by changing its shape, which affects the ability of hemoglobin to bind to oxygen (see Figure 3.10).

Ingram's experiments indicated that sickle cell disease is caused by one incorrect amino acid out of the 146 amino acids in the beta globin protein (Figure 3.11). The replacement of the charged (highly polar) glutamic acid with a nonpolar valine residue causes the mutant hemoglobin complexes to form long fibers that physically distort the cells into a crescent or sickle shape. The sickle-shaped red blood cells cannot move freely through the smaller blood vessels and capillaries. This interferes with oxygen transport to the cells, resulting in painful symptoms and, sometimes, early death. At the time Ingram began his work, sickle cell disease was known to be inherited, but Ingram's research took science a step further by showing that a gene defective in its encoding of a single amino acid causes a devastating disease.

> Genes encode the sequences of amino acids in protein chains. Ingram's research pointed to the possibility that a devastating disease could be the result of single incorrect amino acid in an entire protein.

## Scientists Link Genes with Amino Acid Sequence of a Protein

One of the first to suggest a cellular coding mechanism for proteins was the German physicist Erwin Schrödinger. In his 1944 book *What Is Life?*, Schrödinger suggested that the cellular chromosomes might contain a "code-script." He proposed that variations of atoms in the chromosome could produce a Morse code–like packet of information. The prevailing wisdom of the day was that chromosomes were composed of protein, and so DNA did not enter Schrödinger's thinking; however, the notion of a code embedded in the arrangement of atoms in chromosomes was an important step.

Shortly after the 1953 announcement of the Watson-Crick model of DNA (Chapter 2), another physicist named George Gamow wrote to Watson and Crick and proposed that the arrangement of bases along the double helix might produce a template of sorts, a series of chemical "holes." Gamow thought the holes could have different structures, and different amino acids could then fit into different holes and link up to form a protein. In this model, DNA acted as a direct, physical template for building protein chains. Ironically, Gamow also became very interested in ribonucleic acid (RNA), and in 1954 he organized the first "RNA Club" to investigate the role of RNA in protein synthesis. Members wore a tie with a sinuous lime-green curl of nucleic acid flanked by boxy yellow outlines of purines and pyrimidines on a black background.

> Gamow's model correlated the order of DNA bases (the DNA sequence) with the order of the amino acids in a protein, but it required physical contact between the DNA and protein.

As the 1950s progressed, evidence continued to accumulate that supported DNA as the material of heredity and the idea that the genetic message in DNA

is expressed through proteins. Many scientists also supported the concept that some sort of genetic code exists in DNA and that the code actually specifies the sequence of amino acids in a protein. Eventually the questions about the genetic code were answered, but only after considerable experimentation and educated guesswork provided a much better grasp of the biochemical realities of living things, starting with the components inside cells.

To understand the process of protein synthesis, biochemists had to know whether the chemical information in DNA passes directly from the nucleus to the site of protein synthesis or whether some intermediary substance is required to transfer the information from DNA to protein. Recall that because prokaryotes do not have nuclei, transcription and translation take place in the same cellular compartment in bacterial cells. However, in eukaryotic cells, protein synthesis takes place in the cytoplasm, whereas the chromosomes are confined to the nucleus (mitochondria each contain a DNA genome). In eukaryotic cells, the direct passage of information between the two separate cellular compartments seemed unlikely—that is, it seemed improbable that the DNA itself moved out of the nucleus. A better possibility, scientists began to think, was a relay system operating to transfer genetic information between the nucleus and cytoplasm. In the 1940s, biochemists found that cells and organs undergoing protein synthesis made unusually large amounts of RNA, a chemical cousin of DNA. This correlation between protein synthesis and RNA synthesis, as well as several other lines of evidence, suggested that RNA is the chemical intermediary that transfers genetic information from the DNA in the nucleus to the protein made in the cytoplasm.

Although RNA and DNA are both nucleic acids with similar molecular structures (Figure 3.12), the molecules differ in three important ways:

- The sugar in the backbone of RNA is ribose; in DNA it is deoxyribose.
- RNA has the base uracil, whereas DNA has thymine.
- An RNA molecule is usually single-stranded, and base-pairs with itself; DNA usually has two DNA strands that are base-paired to each other (double-stranded).

Despite the chemical differences between DNA and RNA, it is now clear that base pairing is an essential feature of both the DNA and RNA molecules. DNA base pairing establishes the double-stranded DNA helix, whereas RNA is typically single-stranded, but contains many regions where the RNA strand base-pairs with itself, creating loops and hairpin structures (Figure 3.13).

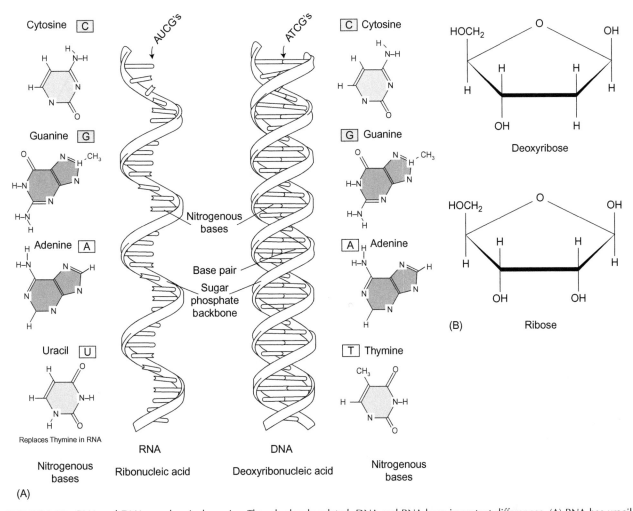

**FIGURE 3.12** RNA and DNA are chemical cousins. Though closely related, DNA and RNA have important differences. (A) RNA has uracil (U) instead of thymine (T) nucleotides; DNA is typically double stranded in the cell, whereas RNA is single stranded but usually base pairs with itself. (B) Ribose, the sugar in RNA, has a hydroxyl (–OH) group on the 2′ carbon, whereas deoxyribose in DNA has a hydrogen at the 2′ carbon. This makes RNA more reactive and less stable than DNA.

Each DNA base sequence is a template for RNA synthesis, much as DNA acts as a template for the synthesis of new DNA during replication (Chapter 2). Biochemists proposed that if the synthesis of RNA from DNA occurs in the nucleus, then the RNA copy could travel to the cytoplasm, where it would convey the amino acid sequence information for the synthesis of a protein. Indeed, radioactive isotope tracer experiments in eukaryotic cells showed that RNA molecules are made in the nucleus and then move from the nucleus to the cytoplasm.

Important evidence supporting the role of RNA in information transfer in the cell came from experiments with viruses. Bacteriophages are viruses that infect bacterial cells (but do not attack eukaryotic cells). Researchers found that when bacteria are infected by bacteriophages, the DNA genomes enter the cells

(Figure 3.14). The bacteria synthesize the bacteriophage RNA before they begin synthesizing protein. Therefore, RNA appeared to be the intermediary compound between DNA and protein. Moreover, infections with the tobacco mosaic virus (TMV), which has no DNA—its small genome is entirely RNA (Figure 3.15)—causes infected tobacco leaf cells to produce viral proteins. Scientists reasoned that the TMV RNA genome carries the information needed to synthesize the viral proteins without requiring DNA. The functional links between DNA, RNA, and protein began to emerge.

Mounting evidence suggested that RNA molecules transfer genetic instructions from DNA to the protein synthesis machinery of the cell.

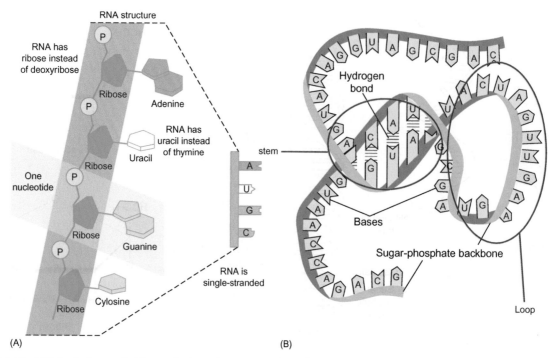

**FIGURE 3.13** RNA is single stranded but easily forms base pairs with itself. (A) The four nucleotide bases form the backbone of an RNA molecule. (B) Diagrammatic representation of a longer RNA molecule forming a stem-loop structure.

## FRANCIS CRICK STARTS TO UNRAVEL THE GENETIC CODE

At the start of the 1960s, the evidence that DNA carries genetic information was very compelling. By then, biochemists knew about the DNA double helix; they had learned that cells linked amino acids together to make protein chains, and the concept that genetic information "flowed" in the cell from DNA to RNA to protein was widespread. Still, fundamental questions remained: How is the genetic message in DNA transmitted to RNA, and how is the RNA carrier involved in protein synthesis? More specifically, how does the order of bases in the DNA encode the order of amino acids in the corresponding protein?

Answers to these questions came from a series of elegant experiments reported in 1961 by Francis Crick and his colleagues. They reasoned that the simplest genetic code in DNA probably consists of a series of blocks of chemical information, with each block having a linear relationship with one amino acid in the protein. In addition, they thought that a block of three DNA bases would be sufficient to universally specify an amino acid. To test these ideas, Crick's group performed experiments on viruses containing DNA genomes. They altered the DNA sequence of the viral genome, then infected bacterial cells and asked what would happen to the virus, the viral protein products, and the cells.

The experiments were simple yet ingenious. Crick's group altered the viral DNA using acridine compounds, which inserts or deletes a single base pair in a DNA strand. The Crick group sought to find out whether deleting one DNA base would change the amino acids in the protein synthesized. For example, let's start with the following DNA sequence:

AGGCATGCAATG (parent DNA)

After acridine treatment, this sequence might have been changed, for example, to

AGGATGCAATG (mutant DNA)

In this case, the fourth base (C) was deleted.

If the genetic code were read in blocks of three DNA bases (Figure 3.16A), the first block of the preceding sequence would remain unchanged, but the succeeding blocks of three would be different:

AGG CAT GCA TTG (parent DNA)

AGG ATG CAT TG (mutant DNA)

The result of this mutation is demonstrated because the first amino acid would appear in its proper position in the mutant protein, but the subsequent amino acids

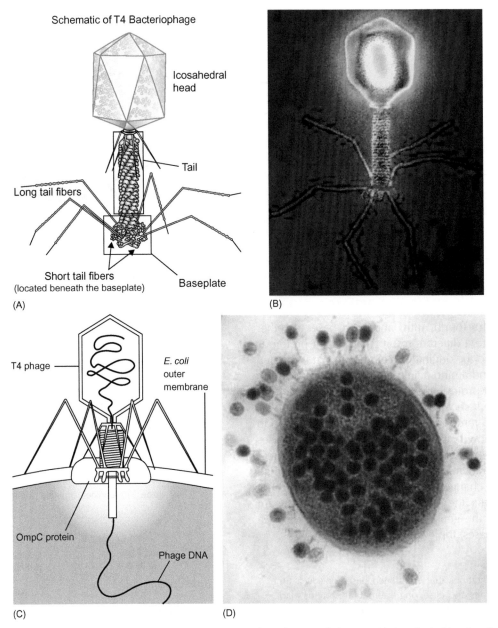

Schematic of T4 Bacteriophage

Icosahedral
head

Tail

Long tail fibers

Short tail fibers
(located beneath the baseplate)

Baseplate

(A)

(B)

T4 phage

E. coli
outer
membrane

OmpC protein

Phage DNA

(C)

(D)

**FIGURE 3.14** Bacteriophage T4 infects bacterial cells. (A) Diagram of T4 phage, including capsid (icosahedral head), tail proteins, baseplate, and tail fibers. (B) Scanning electron micrograph of a T4 bacteriophage virus. (C) The T4 DNA is in the phage head and is injected through the cell membrane and into the bacterial cell. (D) An electron micrograph of an infected bacterial cell producing many T4 bacteriophages offspring.

in the mutant protein would change due to the new blocks of three nucleotides generated by the mutation. Indeed, in Crick's experiments, the first amino acid occurred in its correct position in the protein, but all succeeding amino acids were different.

Next, Crick and his colleagues deleted two DNA bases, predicting that the first block would remain the same, but the subsequent blocks would change yet again:

AGG CAT GCA TTG → (delete bases 4 and 5, C and A)
→ AGG TGC ATT G

As you can see, deleting two bases at a fixed point changes the three-base blocks after the deletion. In the protein, the amino acid encoded by AGG occurs in the correct position, but all succeeding amino acids are different because all the following blocks (e.g., TGC and ATT) are different.

Finally, the key experiment was performed. Crick and his colleagues removed three bases to observe the effect on the amino acids in the protein:

AGG CAT GCA TTG → (delete C, A, T)
→ AGG GCA TTG

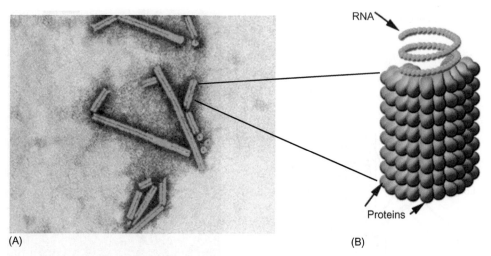

**FIGURE 3.15** Tobacco mosaic viruses have RNA genomes. TMV research provided evidence that RNA has a central role in protein synthesis. (A) Electron micrograph of tobacco mosaic viruses (TMV). (B) TMV has a thin, rodlike shape, defined by a coat made up of repeated protein units that assemble around the single-stranded RNA genome.

Deletion of the fourth, fifth, and sixth bases together (CAT) has quite a different result. This mutant protein is more similar to the original protein because it is missing only the second amino acid, and the other amino acids in the mutant protein are correct. This is a key experiment because it shows that the reading of the code was restored when three bases were removed. Crick's group went on to show that this was the case whenever bases were added or removed from DNA in multiples of three. Adding or removing other numbers of bases resulted in wholesale changes of the amino acid sequence of the protein after the site of the mutation. Crick's team deduced that the genetic code is read in sets of three consecutive bases, triplets, later termed **codons**.

> By testing mutations that eliminated bases one at a time and in multiples and observing the mutant proteins made, Crick's team established that a nucleotide block of three bases corresponds to a single amino acid.

## The Genetic Code is Cracked Using Synthetic RNA

Additional elegant experiments in 1962, reported by Marshall Nirenberg with Heinrich Matthaei, and independently by Har Gobind Khorana, were designed to determine the genetic codons for all 20 amino acids. To determine which mRNA codons call for which amino acids, the biochemists combined *in vitro* (outside the cell) different individual synthetic RNA molecules with enzymes, amino acids, and other essential compounds to see what protein would be made in the test tube (Figure 3.16B). They found that the synthetic RNA molecule consisting of only uracil residues (U-U-U-U-U-U-U-U-U) produced a protein consisting of just

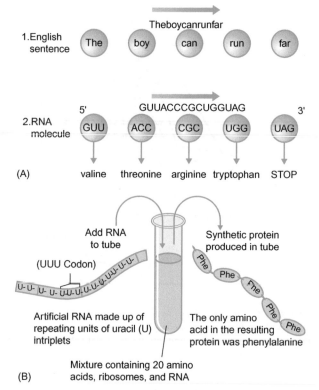

**FIGURE 3.16** Codons deduced for all 20 amino acids. (A) The message in (1) is written in three-letter English words reading from left to right. Each three-letter word has meaning in the message. In (2), we have a biochemical message written in three-letter RNA bases reading from left to right. Each three-letter "word" (or codon) specifies an amino acid in the protein with an AUG start codon (not shown) and at the far right is a stop signal. (B) The experiment that broke the genetic code for the amino acid phenylalanine. Researchers synthesized a strand of RNA containing only uracil (U-U-U-U-U-U-U-U…) and placed it into a test tube with 20 different amino acids and all the materials needed to synthesize protein. The system generated a protein consisting only of phenylalanines. Thus, from these results scientists knew that the codon UUU specifies the amino acid phenylalanine.

Second base in codon

| | | U | C | A | G | |
|---|---|---|---|---|---|---|
| | | Phe | Ser | Tyr | Cys | U |
| | U | Phe | Ser | Tyr | Cys | C |
| | | Leu | Ser | STOP | STOP | A |
| | | Leu | Ser | STOP | Trp | G |
| | | Leu | Pro | His | Arg | U |
| | C | Leu | Pro | His | Arg | C |
| | | Leu | Pro | Gln | Arg | A |
| | | Leu | Pro | Gln | Arg | G |
| | | Ile | Thr | Asn | Ser | U |
| | A | Ile | Thr | Asn | Ser | C |
| | | Ile | Thr | Lys | Arg | A |
| | | Met | Thr | Lys | Arg | G |
| | | Val | Ala | Asp | Gly | U |
| | G | Val | Ala | Asp | Gly | C |
| | | Val | Ala | Glu | Gly | A |
| | | Val | Ala | Glu | Gly | G |

First base in codon (left axis) — Third base in codon (right axis)

**FIGURE 3.17** Genetic code table. This chart shows how the three-letter mRNA codons correspond to specific amino acids. For example, to look up the amino acid for the codon AGU, the first base in the codon is used to select a row in the far left column of the table (row A). The second base is used to select a column from the top of the table (column G). From within the box at the intersection of the selected row and the column, use the third codon base (U) in the final column of the table to select the amino acid. The amino acid is identified as serine (Ser). The AGU codon specifies serine in the protein.

the amino acid, phenylalanine. From this result, scientists surmised that the RNA three-letter code word, or codon, for phenylalanine is U-U-U. Working backward, scientists reasoned that the DNA codon for phenylalanine is A-A-A. Similar experiments rapidly led to the identification of RNA codons for all 20 amino acids and showed that many amino acids are encoded by more than one codon (Figure 3.17). The codons UAG, UGA, and UAA are translation "stop" signals. For their work in determining the chemical nature of the genetic code, Nirenberg and Khorana were awarded the 1968 Nobel Prize in physiology or medicine.

In the ensuing years, biochemists demonstrated that the **genetic code** is virtually universal: the same three-base codons specify the same amino acids regardless of whether the codons reside in a bacterium, bee, buttercup, or bear. The essential differences among species of organisms is not the nature of the DNA bases but the order in which the bases occur in the DNA molecules or, in the case of proteins, the order of the amino acids.

The three-letter genetic code is universal and is used to specify proteins in all life on earth. This seemingly simple system allows for all the consistency and all the biological diversity on our planet!

DNA sequences are copied into RNA sequences, and the order of the bases in the RNA transcript specifies the sequence of the amino acids in the protein

product. These events reflect the core element of the so-called **central dogma** of molecular biology, which encompasses the major gene expression pathways in living cells (see Figure 3.18).

The work of many scientists over much of the twentieth century elucidated the relationship between chromosomes and proteins, as well as the genetic code that enables DNA to specify the proteins made by a given cell. The chemically similar polymer, RNA, acts as an intermediary and carries an imprint of the genetic information to be processed into a protein. Next, we will explore how the information travels from DNA to RNA and is then used to synthesize proteins—the process of gene expression.

## GENE EXPRESSION

Genomes vary in size from just under 500 genes in the smallest known genome to more than 45,000 genes in the largest genome known in 2009, that of the black cottonwood tree. No matter the size of the genome, the genes in the genome are not all "turned on," or **expressed**, at the same time. Genes are turned on and off as needed, by signals from the external environment as well as signals from within the organism. In multicellular organisms, the pattern of gene expression determines cell type. For example, a neuron is different from a skin cell by virtue of the genes that have been expressed in its development and that are expressed as it performs its specialized functions.

The control of gene expression is a complex and intricate business. It involves adjusting the level as well as the timing of gene activation and silencing (repression) in different cells. Control sequences in both DNA and RNA are "read" by regulatory proteins and RNA molecules that act in response to internal and external signals and turn genes on and off. The primary level of control over gene expression is at the level of **transcription**, a word coined by Francis Crick in 1956 to refer to the process of copying DNA sequences into RNA. The product of the transcription of a gene is a single-stranded RNA molecule.

Differences in cell structure, genomic structure, and complexity between prokaryotes and eukaryotes dictate that they exhibit important differences in gene expression pathways. Eukaryotic cells have nuclei, and eukaryotic genes usually contain intron and exon sequences. These differences have a fundamental impact on gene expression.

Life demands that organisms respond to changing conditions. At the molecular level, these responses are often made via changes in gene expression. Differences in cell structure and complexity lead to major differences in how gene expression occurs in prokaryotic and eukaryotic cells.

The term "**central dogma** of molecular biology" was coined in 1958 by Francis Crick, one of the discoverers of the DNA double helix. The central dogma, often stated as "DNA makes RNA makes protein," for many years was the only known pathway describing gene (DNA) expression. Now, many, many experiments later, we have learned that there are many exceptions to the central dogma; some organisms have RNA chromosomes, and some copy RNA *into* DNA. This is how the HIV virus causes AIDS; once HIV enters a human cell, the HIV RNA genome is copied into DNA by HIV's highly specialized **reverse transcriptase** enzyme. The DNA copied from the RNA genome becomes inserted (integrated) into the host cell DNA genome, and the HIV DNA becomes a permanent part of the human cell's chromosome. It became routine for scientists to discover new exceptions to the central dogma rules! As a result of such discoveries, articles with titles such as "The Death of the Central Dogma" and "Dog Eat Dogma" proclaim the central dogma to be useless, and some say even harmful. They argue, why are we still teaching and modeling research on something as outmoded as this concept? And isn't science supposed to be the antithesis of dogma, which can be thought of as an authoritative proclamation to be generally accepted without proof?

Author Horace Freeland Judson asked Francis Crick about the meaning of "central dogma." Crick explained, "My mind was that a dogma was an idea for which there was *no reasonable evidence*. You see?!" And Crick gave a roar of delight. "I just didn't *know* what dogma *meant*. And I could just as well have called it the 'Central Hypothesis'…. which is what

I meant to say. Dogma was just a catch phrase." So actually, Crick's use of the word must be taken in the context revealed through his own words. In 1970, Crick, criticized in light of reports that claimed to refute the central dogma, wrote a paper in the journal *Nature* that clarified his meaning, and helped people to better visualize the possible paths used to transfer genetic information between DNA, RNA, and protein (Figure 3.18). The central dogma rules out some of the possible pathways (removes arrows), and suggests that other gene expression pathways only occur under special circumstances. Still unchanged, however, is the idea that once an organism or cell has expressed its genetic information at the protein level, the genetic information in the protein cannot be sent in reverse into RNA or DNA.

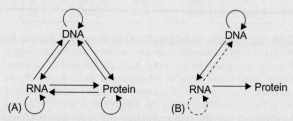

**FIGURE 3.18**  The meaning of the "central dogma" of molecular biology. (A) The many theoretical paths for genetic information to pass between DNA, RNA, and protein. (B) The central dogma rules out some of the possible pathways (removes arrows) entirely and suggests that other gene expression pathways only occur under special circumstances or in certain organisms (dotted lines).

## Several Types of RNA are Transcribed from DNA

The major gene expression pathway in cells begins with **transcription**, copying DNA into RNA (Figure 3.19). Cells transcribe several types of RNAs, which fall into to two broad categories: RNA that encodes proteins (**messenger RNA, mRNA**) and RNA that does not encode proteins (**noncoding RNA, ncRNA**). Two ncRNAs play direct roles in protein synthesis. **Transfer RNA (tRNA)** molecules ferry amino acids to the protein synthesizing machinery, the ribosomes, and **ribosomal RNA (rRNA)** provides major functional and structural aspects of ribosomes (Figure 3.20). In addition, a variety of other ncRNAs perform essential functions including the chemical modification of other RNA molecules, RNA splicing, and the regulation of gene expression. Because RNA is single stranded, it will form stable base-paired structures if its bases are unimpeded from interaction with each other. In particular, ncRNAs form three-dimensional RNA structures that carry out important cell functions. Although

mRNAs sometimes form base-paired stems and loop structures, the main function of mRNAs is to carry sequence information to the ribosomes.

## mRNA Encodes Protein

The RNA sequence of an mRNA carries the information that will be decoded into a chain of amino acids during protein synthesis on ribosomes in the cell. In prokaryotes, mRNAs often contain the coding regions for several proteins, one after another, and are immediately used to synthesize proteins, even while the mRNA is still being transcribed from the DNA gene. In eukaryotes, mRNA is synthesized in the nucleus as a long precursor RNA strand that is processed extensively before it leaves the nucleus to be translated into protein in the cytoplasm.

## Ribosomal RNA

rRNA provides essential structural and enzymatic functions in the protein synthesis machine, the ribosome

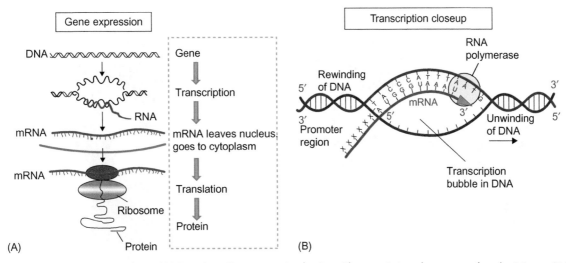

FIGURE 3.19   Gene expression pathway. (A) Overview: Gene expression begins with transcription, the process of synthesizing an RNA copy from a sequence of DNA. Noncoding RNAs, such as tRNA and rRNA, fold into functional molecules (not shown); mRNA, which encodes proteins, is used as a template by a ribosome to synthesize a protein chain from amino acid building blocks. In eukaryotes, the mRNA is transported out of the nucleus and into the cytoplasm where translation occurs. (B) Transcription: Strands of DNA separate as the RNA polymerase enzyme copies the DNA bases into an mRNA strand. The sequence of the RNA is determined by copying and base-pairing to the DNA template.

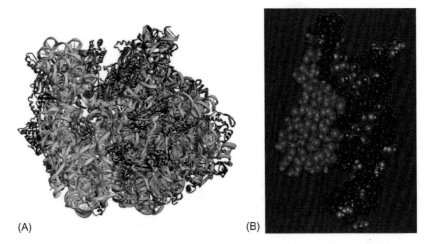

FIGURE 3.20   Ribosome and ribosomal RNA structure. (A) Three-dimensional structure of a bacterial ribosome. Proteins (grey, blue) and rRNAs (purple) come together to make the protein synthesis machinery shown here. Prokaryotic ribosomes are made of more than 50 proteins and three rRNAs. (B) Close-up of a small part of the ribosome: three-dimensional structure of an rRNA fragment and ribosomal protein. This small fragment of the 16S rRNA (blue backbone, gray bases) illustrates how its folding pattern allows it to bind closely with a ribosomal protein (magenta).

(Figure 3.20). The critical functions of rRNA are reflected in the fact that rRNA genes are the most highly conserved sequences (least varied) in evolution. Ribosomes in both prokaryotes and eukaryotes are formed from two subunits, each of which consists of many proteins and RNAs. The rRNA components provide the enzyme activity that catalyzes the creation of the peptide bond between amino acids during protein synthesis, and they interact with mRNA, enabling it to bind to the ribosome. Prokaryotic ribosomes are

smaller than eukaryotic ribosomes and contain a total of three rRNAs, whereas eukaryotic ribosomes contain four rRNAs.

## Transfer RNA

tRNAs (Figure 3.21) are adaptor molecules that deliver the amino acid corresponding to a specific mRNA codon while the mRNA is associated with the ribosome. During protein synthesis, the bases in the tRNA

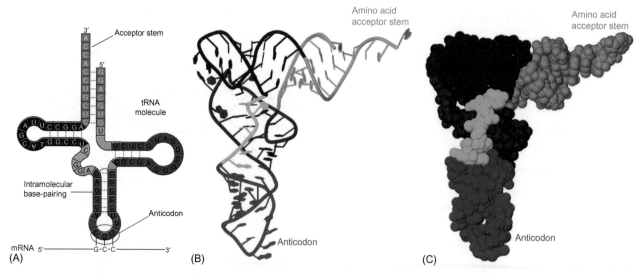

**FIGURE 3.21** Transfer RNA structure. (A) The characteristic base-pairing pattern or secondary structure of all tRNAs resembles a cloverleaf. (B) The three-dimensional folding of tRNA molecules. At the 3′ (upper right) end of the molecule is the amino acid attachment site. At the lower portion of the molecule contains the tRNA anticodon, shown base paired to the mRNA codon. (C) A space-filling model of the tRNA molecule shown in the same orientation as (B).

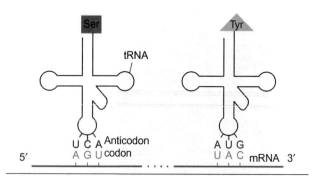

**FIGURE 3.22** Role of tRNAs in translation. An amino acid is attached to the acceptor stem of the appropriate tRNA determined by the anticodon sequence. Once inside the ribosome, the anticodon in the tRNA base-pairs with the codon in the mRNA being translated.

**anticodon** pair with complementary bases in the codon of the mRNA strand (Figure 3.22) bringing the correct amino acid into position to be added to the growing protein chain in the ribosome. These codon-anticodon interactions translate the genetic code in the DNA into an amino acid sequence in the protein.

## Other Non-coding RNA

Many types of ncRNAs perform a wide variety of functions in eukaryotes. These include micro RNAs (miRNA) and small interfering RNAs (siRNA), which are important in controlling gene expression, small nuclear RNAs (snRNAs) that are involved in processing mRNA, and small nucleolar RNAs (snoRNAs) that help to modify the structures of certain bases in rRNA

and tRNA molecules. Prokaryotes have fewer kinds of ncRNAs, some of which modulate gene expression.

> The three most well known types of RNA—mRNA, tRNA, and rRNA—are involved directly in protein synthesis. Various other noncoding RNAs have important roles in the control of gene expression and chemical modification of other RNA molecules.

## RNA POLYMERASES COPY DNA INTO RNA

**RNA polymerase** (RNAP) enzymes synthesize all the different types of RNA transcripts made in both eukaryotic and prokaryotic cells. These enzymes are large multiprotein complexes that range in size from 100 kDa (kilo Daltons) in T7 bacteriophage, to 400 kDa in bacteria (Figure 3.23) and 500 kDa in eukaryotes. Like the DNA polymerases that replicate DNA, RNA polymerases catalyze the addition of single nucleotides to a growing chain in the 5′ to 3′ direction, but RNAPs make RNA copies that are composed of *ribo*nucleotides instead of *deoxy*ribonucleotides, which are found in DNA.

The *E. coli* RNA polymerase contains a core set of protein subunits and one variable protein subunit, called the sigma factor. All the different forms of the sigma protein fit into the core enzyme and enable it to start transcription, but each sigma factor allows the polymerase to transcribe different sets of genes. The bacterial RNA polymerase enzyme copies DNA into

RNA at about 50 bases per second and makes a mistake in only about 1 in 10,000 bases added. A proofreading mechanism is built into the enzyme complex, allowing it to detect and remove an incorrectly added nucleotide.

In eukaryotes, transcription is necessarily more complex because of the larger number of genes to be expressed, and the increased gene regulation required to control the expression of thousands of genes in hundreds of different types of cells. This complexity is reflected in the three different RNA polymerase enzymes that transcribe different types of RNA molecules in eukaryotic cells. These large enzyme complexes each have a core of protein subunits, five common additional protein subunits, and a variable number of specific subunits. Different combinations of conserved and unique subunits allow the eukaryotic RNAP enzymes to transcribe different sets of genes. For example, RNAPI transcribes most rRNA genes whereas RNAPII transcribes most protein genes (Table 3.1).

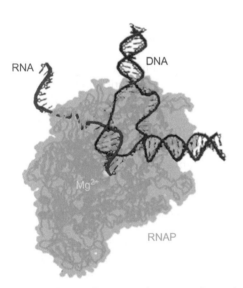

**FIGURE 3.23** A bacterial RNA polymerase. The polymerase enzyme has channels for the entry of the template and nontemplate strands of the DNA helix, and the emergence of the mRNA.

The specific types of RNA transcribed by the three different eukaryotic RNA polymerases were elucidated using an inhibitor of transcription called α-amanitin, a toxin derived from deadly mushrooms that blocks the function of some eukaryotic RNA polymerases; each year more than a hundred people are fatally poisoned by eating mushrooms containing α-amanitin, which blocks transcription. However, α-amanitin is useful in the laboratory to distinguish between the actions of the three polymerases. Each enzyme responds differently when α-amanitin is added to the cell. The toxin binds tightly to RNA polymerase II, which immediately blocks mRNA transcription by that enzyme. Higher concentrations of α-amanitin are required to inhibit RNA polymerase III function, whereas RNA polymerase I activity is unaffected by the α-amanitin toxin. Using the differential sensitivity of RNAP enzymes to α-amanitin, scientists developed assays to reveal the functional connections between the three RNA polymerase enzymes and the many types of RNA gene products.

**RNA polymerase I** (RNAP I) synthesizes the eukaryotic rRNAs, 5.8S, 18S, and 28S but not the 5S rRNA. The rRNA genes exist in tandem repeats on eukaryotic chromosomes, and are rapidly transcribed by many RNAP I enzymes, as shown in electron micrograph images of the rDNA genes that display dramatic arrays of progressively longer rRNA transcripts that form a characteristic Christmas tree structure (Figure 3.24). The image also demonstrates that many RNA polymerases can transcribe a gene one following right after another. This high level of rRNA production allows the cell to respond rapidly to the need for new proteins by making the numerous ribosomes necessary to do the protein synthesis work. **RNA polymerase II (RNAP II)** synthesizes mRNA from all genes encoding proteins, and also makes many ncRNAs. **RNA polymerase III** transcribes the transfer RNAs, the 5S rRNA needed for ribosomes and some of the snRNAs. Because RNAP II transcribes more protein coding genes than RNAP I or III, we will focus on transcription by RNAP II in eukaryotic cells.

**TABLE 3.1** Eukaryotic RNA Polymerases

| Type | Location | Cellular RNAs transcribed | α-amanitin |
|------|----------|---------------------------|------------|
| I | Nucleolus | 18S, 5.8S, and 28S rRNA | Insensitive |
| II | Nucleus | mRNA, most snRNA, miRNA, siRNA, snoRNA | Inhibited by low concentrations |
| III | Nucleus | tRNA and 5S rRNA, some snRNA and other ncRNAs | Inhibited by high concentrations |

*From www.ncbi.nlm.nih.gov/books/bv.fcgi?rid = stryer.table.3982. Biochemistry, 5th ed., 2002, by Jeremy M. Berg, John L. Tymoczko, Lubert Stryer, and Neil D. Clarke, WH Freeman & Co, New York. Further adapted by FR with information from Molecular Biology of the Cell, Alberts, 2008, Garland Science, New York.*

RNA polymerases are large, multisubunit protein enzymes that synthesize RNA molecules from DNA templates. Prokaryotes have a single RNA polymerase to transcribe genes, whereas eukaryotic cells have three RNAPs.

## Genes Contain "Start" and "Stop" Signals

Genes are not actively transcribed into RNA at all times; it is critical for cells to respond to signals that indicate which genes are required for the needs of the organism at a given time. For example, in single-celled organisms, these signals often indicate what kind of food source is available. Multicellular organisms must respond to the environment as well as to signals from cells and tissues. Shutting down transcription of a gene eliminates the need for long-term storage of the RNA and protein gene products in the cell. Although some 'housekeeping' genes are expressed for the life span

A family of identical genes encode ribosomal RNA

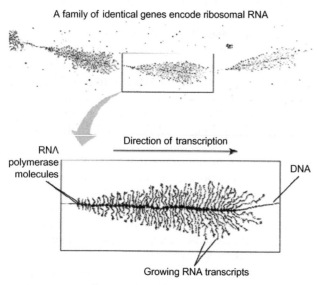

FIGURE 3.24   RNA polymerase I transcribes rRNA. Eukaryotic genes for rRNAs are present in hundreds of identical copies distributed in tandem on the chromosome. When cells are growing and need large amounts of protein synthesis, the genes are transcribed by many RNA polymerases one after the other. The top panel shows three complete genes and part of a fourth. The enlargement of one of the genes shows the gradually increasing lengths of rRNA transcripts that are each emerging from an enzyme on the DNA.

of the organism, most genes are transcribed only for a specified period of time and then are turned off when the gene product is no longer needed. In addition, in some cell types in multicellular organisms, many genes are never turned on because they are used only in other cell types.

Genes can be regulated at several steps along the gene expression pathway, including the start signal for an RNA polymerase to begin transcription of DNA into RNA, which is encoded in DNA located **upstream** of (preceding) the gene's coding region, called a **promoter**. RNA polymerase and various proteins bind to this DNA region and work together to initiate transcription. The promoter DNA sequence is highly conserved in prokaryotic genes and includes a TATA sequence at −10 (the bases are counted backward from the first base that is transcribed into RNA at +1) called **the TATA box** and another conserved sequence located at about −35 (Figure 3.25). Promoters function to attract RNA polymerase to the DNA at the start of a gene, which is required for transcription to begin. A promoter **consensus sequence** is created by comparing many promoter sequences and choosing the bases that appear most frequently at each position in the sequence (see Figure 3.25). Genes with promoters that closely resemble the promoter consensus sequence are transcribed more frequently. Also encoded in prokaryotic genes are a series of bases that serve as a transcription **termination signal**. Termination signals in the mRNA transcript form a stem-loop structure that releases the mRNA from the template DNA. Termination can also involve the action of additional proteins that bind to the termination sequence and promote release.

Eukaryotic RNAPII promoters are varied in DNA sequence, but typically encompass large regions of DNA located in front of the coding region of the gene. Some RNAPII promoters contain TATA sequences that are usually located within 50 bases upstream of the transcription start site. Most eukaryotic protein-coding genes also contain additional control sequences called **enhancers** that are located at large distances from the gene. RNAPII transcription terminates by a different mechanism than in prokaryotic cells.

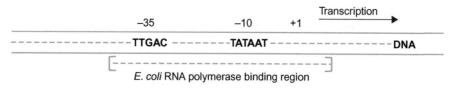

FIGURE 3.25   Prokaryotic promoter recognition. The promoter represents where RNA polymerase binds to DNA to initiate transcription of a gene. The consensus sequences for the two regions of prokaryotic promoters are shown. The ideal distance between the two regions is 17 bp the exact positioning varies from promoter to promoter. Transcription starts at nucleotide number 1. Note that no promoter is identical to the consensus sequences.

Promoters are DNA sequences that indicate where transcription should start. In addition, the promoter influences which genes are expressed in the cell.

Promoter selection by RNAP enzymes is a major mechanism used to control gene expression in both prokaryotic and eukaryotic cells. Exercising control at this initial point in the gene expression pathway has the advantage of conserving cellular energy and resources. The first level of transcriptional control is **promoter strength**, measured by the rate of successful **transcription initiation** in the absence of any additional regulation. Weak promoters support only occasional transcription, whereas strong promoters initiate more frequent transcription events.

In prokaryotes, different sigma factor subunits enable RNA polymerase to recognize and bind tightly to specific DNA promoter regions to initiate the transcription of selected genes. Each sigma subunit enables RNAP to transcribe sets of genes that share similar promoters. Various sigma factors are synthesized in response to changes in environmental conditions. *E. coli*, which lives in the sheltered environment of the gut, has eight different sigma factors that enable the cells to alter gene expression quickly when faced with conditions that cause stress or changes in metabolism. Many bacterial species live under less controlled conditions and have far greater numbers of sigma factors to respond to a wider variety of stresses. Once transcription has started, the sigma factor leaves the RNAP complex and RNAP works its way along the DNA, separating the strands and catalyzing the formation of the messenger RNA (mRNA) copied from the DNA template strand (Figure 3.26). Most prokaryotic genes are arranged in groups called **operons** that are under the control of a single promoter. In operons, the RNAP creates an mRNA copy that encodes multiple genes, called a **polycistronic transcript**.

Each of the three eukaryotic RNA polymerases interacts with a different set of **transcription factor proteins** that enable the RNA polymerase to transcribe the correct genes (Figure 3.27). In addition to general transcription factors, there are many tissue-specific transcription factors that bind to the enhancer sequences sometimes located thousands of bases away from the promoter. This observation was puzzling to scientists until it was established that the DNA helix forms a loop that brings the enhancer DNA region and the bound proteins into the proximity of RNAP II, which allows the formation of a huge complex containing RNAPII and all the proteins required for transcription to start (Figure 3.28). A protein called **mediator** coordinates all the activators with RNAP II and the transcription factors.

Access to the DNA template is further complicated in eukaryotic cells because the chromosome DNA in the nucleus is in the form of **chromatin**, DNA and proteins that form nucleosomes every few hundred base pairs (Figure 3.29) (see Chapter 12). Chromatin remodeling machines precede RNA polymerase II on the DNA helix, removing the nucleosomes from the DNA template preceding transcription. Once all of the component parts are assembled, RNAP II consists of more than 100 proteins. As the transcript emerges from RNAP II, specialized proteins and RNA-protein particles bind to the transcript. These complexes serve a variety of purposes in the processing and transport of the eukaryotic RNA transcript.

The initiation of eukaryotic transcription is more complex than that of prokaryotes. In addition to having promoter regions at the start of genes, eukaryotic genes have enhancer regions that are often located at a great distance from the genes they regulate. Many proteins must assemble into a large molecular machine before transcription can start.

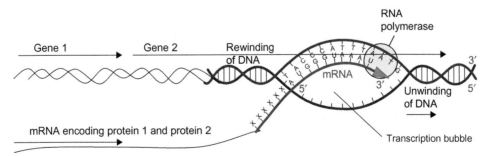

**FIGURE 3.26** Prokaryotic transcription. After the sigma factor dissociates from RNA polymerase, transcription proceeds. A region of DNA opened inside the polymerase is called the transcription bubble. The polymerase is shown working its way through the second of two adjacent genes in the DNA; both will be encoded in one mRNA transcript.

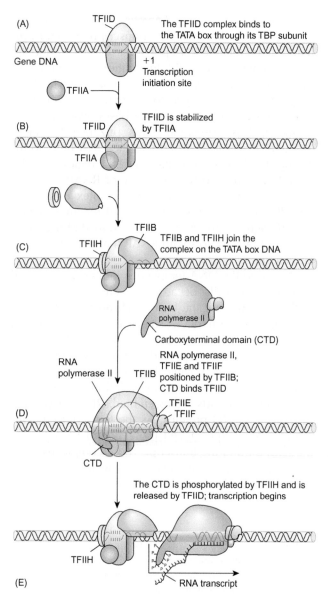

**FIGURE 3.27** Eukaryotic RNA polymerase II and general transcription factors. All eukaryotic polymerases require a set of general transcription factors to initiate transcription. (A) For RNAP II to initiate transcription, transcription factor TFIID binds to the TATA box, where it severely distorts the path of the DNA helix, probably helping to make the region stand out in a very large genome. (B) TFIIB binds to the -35 region of the promoter and helps to position RNAP II correctly at the start site. Transcription factor TFIIH unwinds the double helix and phosphorylates (transfers phosphate groups) to the "tail" region of the RNAP II enzyme, which triggers the RNA polymerase to begin transcription of the gene. TFIIE and TFIIF help to attract the other transcription factors and stabilize the complex.

## Eukaryotic mRNA Transcripts Require Processing

A nucleus gives eukaryotic cells an advantage by increasing the number of steps in the flow of genetic information where the cell can control the fate of gene products. This is particularly true for RNA transcripts that require extensive **posttranscriptional processing** before the mature mRNAs can be transported into the cytoplasm and translated.

Eukaryotic mRNA transcripts are synthesized as **precursor mRNAs** that carry the necessary genetic information to make a protein, but the precursor mRNAs must undergo three major processing events before export from the nucleus:

- A methylguanosine cap structure is added to the 5' end of the mRNA.
- The precursor RNA is spliced to remove introns and join exons together.
- A poly-A tail is added to the 3' end of the precursor RNA.

A modified guanine nucleotide cap is added to the 5' end of the RNA transcript as soon as the RNA transcript emerges from the polymerase complex (Figure 3.30). Following cleavage to remove one phosphate from the 5' end of the RNA, the guanyl transferase enzyme adds a guanosine to the RNA transcript using a reverse, 5' to 5' bond (as opposed to the usual 5' to 3' linkage), and the guanine cap is methylated. A cap-binding complex binds to the completed 5' cap, which prevents the precursor mRNA from being nonspecifically degraded in the nucleus. During protein synthesis, the 5' cap binds to the small ribosome subunit. Although other types of eukaryotic RNAs are chemically modified, the methyl G cap is added only to RNA molecules synthesized by RNAP II.

Most eukaryotic genes are **interrupted genes**: the precursor mRNA transcript produced by RNAP II contains introns and exons that must be **spliced** to create the mature mRNA (Figure 3.31). The concept of interrupted genes was first demonstrated in 1977 by Richard Roberts at Cold Spring Harbor Laboratory and by Philip Sharp at Massachusetts Institute of Technology, who shared the 1993 Nobel Prize in physiology or medicine for their discovery of RNA splicing. Precursor mRNAs contain exons that carry the expressed genetic information in the gene. In the precursor RNAs, the exons are separated by introns that vary in length from 100 to 100,000 bases in length. The introns are the intervening sequences in the RNA precursor, which must be removed by splicing before the RNA can leave the nucleus. After the introns in the precursor RNA molecules are removed, the exons are pasted together in a precise splicing process that preserves the reading frame from one exon to the next, rarely making a mistake.

The multicomponent **spliceosome** is composed of small nuclear ribonucleic particles (snRNPs), which are, in turn, made up of small nuclear RNAs (snRNAs) and proteins. The snRNA components perform essential

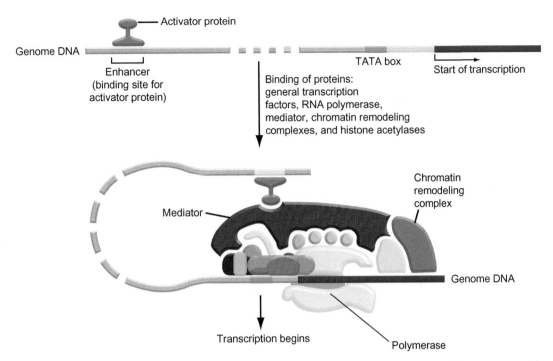

**FIGURE 3.28**   RNA polymerase II transcription initiation requires enhancers. Regulatory proteins bind to enhancer regions of the DNA and make contact with RNAP through a mediator protein, causing the DNA to form loops involving large regions of genome DNA (1 to 20 kilo bp). The gene's promoters and enhancers create a gene regulatory region that can be much larger than the gene itself, with many enhancers (both negative and positive) modifying recruitment of RNA polymerase, and allowing very fine tuning of gene expression. In addition, chromatin remodeling proteins precede the complex to unwrap the DNA nucleosomes.

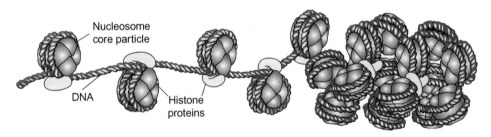

**FIGURE 3.29**   Nucleosomes package eukaryotic DNA. In eukaryotes, DNA is wrapped around histone proteins (blue and yellow), into structures called nucleosomes that occur about every 200 base pairs. The nucleosomes are compacted together into higher orders of chromatin structures that are responsible for the dense packing of eukaryotic chromosomes during mitosis (cell division). The chromosomes together with these proteins are called chromatin.

aspects of the splicing process, including the critically important selection of the exact sites for cutting and ligating the RNA and are involved in the catalytic activity of the spliceosome. Splice sites are represented as DNA and RNA consensus sequences that base pair to the snRNAs as part of the process of splice-site selection. However, much more is involved in choosing the splice site than just the splice-site consensus sequences. During transcription, components of the spliceosome are carried on the RNP II complex and transfer to the precursor mRNA to mark the location of the splice site and guides the choice of the next splice site as the RNA is transcribed. In a typical pre-mRNA splicing reaction, snRNPs bind to the consensus sequence sites in the precursor RNA and then come together to form a spliceosome complex (Figure 3.32). During the splicing reactions, the intron forms a lariat shape. Of course, in genes with many introns, this reaction must take place many times to remove the introns from the final mRNA. The lariat structures are removed and subsequently degraded.

snRNPs assist in the essential processing of eukaryotic mRNAs. Several mechanisms work together to guarantee the extraordinary accuracy of splicing.

## Alternative RNA Splicing Generates Complexity

Alternative splicing is a critically important mechanism controlling the types of proteins synthesized in eukaryotic cells. The actual number of human genes thought to be about 20,000, yet the human body expresses more than 100,000 different proteins in its 60–100 trillion cells. In order to "stretch" the 20,000 genes to more than 100,000 proteins, the expression of at least some genes must allow for the production of more than one protein product. Alternative splicing can allow the cell to select which exons in each transcript will be removed or retained by the splicing process (Figure 3.33). An exon might be retained in some mRNAs if the exon encodes a special protein function, or contains an alternative 3' cleavage and poly A addition site (discussed in the next section). In addition, the cell can select certain intron sequences to be retained in the mature mRNA.

The variation in the protein products from a single gene often plays a role in the tissue-specificity of the protein produced; one version of the protein may be produced in skeletal muscle, for example, whereas a different form of the protein is required in heart muscle cells. Scientists have found many examples of complicated but beneficial alternative splicing patterns at work in eukaryotic cells. The existence of alternative splicing has challenged the traditional concept of a gene and the interpretation of the one gene–one enzyme hypothesis. The alternative RNA

**FIGURE 3.30** 5' cap added to eukaryotic mRNAs. RNAP II adds a methylated guanosine cap to the 5' end of the emerging mRNA transcript. Following this, a cap binding complex (CBC) protects the mRNA from degradation, helps in transport out of the nucleus, and assists in binding to the ribosome.

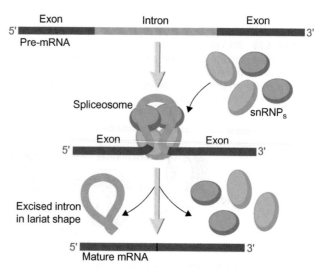

**FIGURE 3.32** snRNPs form spliceosomes. (A) The spliceosome is formed by snRNP particles consisting of proteins and small nuclear RNAs. The spliceosome forms on the pre-mRNA, and the reaction requires base pairing between the snRNA components and the pre-mRNA. The snRNAs catalyze RNA splicing and intron removal via the formation of a lariat structure.

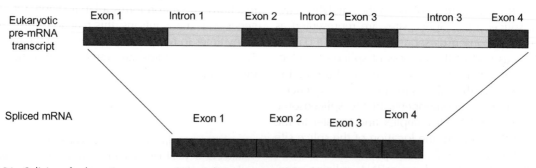

**FIGURE 3.31** Splicing of eukaryotic precursor mRNA. Splicing consists of the precise removal of intron sequences and the ligation of exons together to make a much shorter transcript containing an intact protein coding region.

splicing option adds greatly to the genetic flexibility of the eukaryotic genomes by offering additional ways to fine-tune gene expression. However, coupled with the ambiguity of splicing consensus sites, it also contributes greatly to the difficulty of predicting protein sequences from genomic data (see Chapter 7).

## Poly-A Tail Is Added When Transcription Terminates

The act of adding a poly-A tail to the mRNA transcript is tightly coupled with the termination of eukaryotic transcription. Signals encoded in the DNA trigger an enzyme complex associated with RNAP II to cleave the mRNA, which effectively terminates transcription of that gene. The poly-A polymerase associated with the RNAP II complex then adds many adenine (A) nucleotides, one after another, to the 3′ end of the RNA. This poly-A tail is not encoded in the DNA; it is attached to the RNA after transcription of the gene is complete. The AAUAAA sequence, which is located about 20 bases upstream from the 3′ end of the transcript, is required for cleavage of the RNA molecule and the addition of the poly-A tail (polyadenylation). The poly-A tail is essential for the function of RNAPII transcripts. The poly-A tail interacts with several poly-A binding proteins that function to protect the completed RNA transcript.

On completion of the processing of eukaryotic precursor mRNA into mature mRNA, the capped, spliced, poly-adenylated mRNA (Figure 3.34) is exported through pores in the nuclear membrane into the cytoplasm, where translation takes place. The transport of mRNA is regulated, so that only correctly and fully processed mRNAs are exported out of the nucleus.

> Eukaryotic mRNAs are capped, spliced, and polyadenylated before leaving the nucleus. RNA splicing provides the opportunity for creating multiple protein products from a single gene.

In the cytoplasm, the cap binding complex is replaced by a protein factor that interacts with the proteins bound to the poly-A tail on the same RNA transcript. This creates a circular RNA molecule that can be translated by many ribosomes at once, which greatly speeds the process of making new proteins (Figure 3.35).

## Prokaryotes Couple Transcription and Translation

The RNAs made in prokaryotic cells typically have no introns and lack a 5′ cap and a poly-A tail; prokaryotic mRNAs are usually identical to the original RNA transcript copied from the DNA. Because of the absence

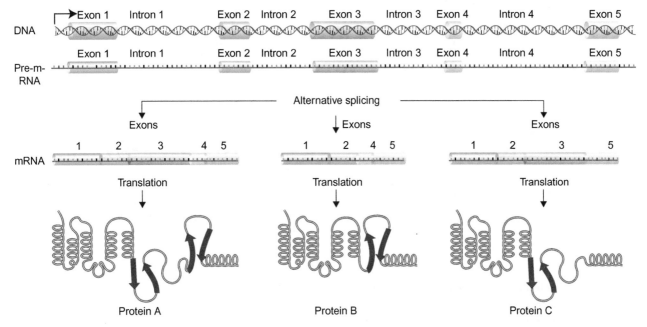

**FIGURE 3.33** Alternative RNA splicing increases genetic flexibility. Alternate splicing patters produce three different proteins from one gene. Different exons can be selected for inclusion or removal by the spliceosome. Proteins A, B, and C differ because of the different exons included in the alternatively spliced mRNAs.

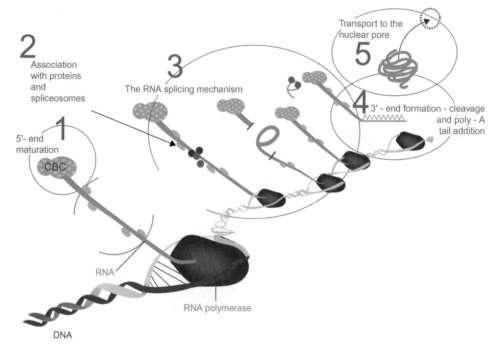

**FIGURE 3.34** Summary of mRNA processing in eukaryotes. The processes of mRNA maturation are shown in the order they occur during transcription: (1) Shortly after RNA polymerase initiates transcription, the emerging 5′ end of the transcript is capped, and CBP binds to the cap. (2) As the transcript continues to emerge, it collects proteins that mark exon/intron boundaries, and snRNPs bind to the pre-mRNA and form spliceosomes. (3) Spliceosomes read consensus sequences and other signals for splicing and remove the introns. (4) In response to a specific sequence in the DNA, the 3′ end of the pre-mRNA is cleaved and the poly-A tail is added. (5) The mature mRNA transcript travels to the cytoplasm through the nuclear pores.

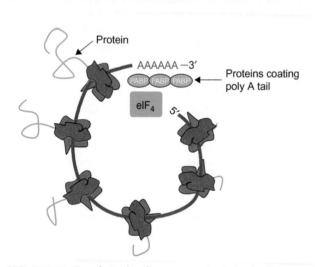

**FIGURE 3.35** Translation by ribosomes on circularized mRNA transcripts. In eukaryotes, an elongation factor (eIF$_4$) replaces the cap-binding complex and engages with the proteins that coat the poly-A tail, circularizing the transcript while it is translated. Many ribosomes can translate a single transcript at the same time.

of nuclei in bacterial cells, transcription and translation can be coupled together (Figure 3.36) leading to one continuous, economical process where genes are copied into RNA and the RNA is translated into protein nearly simultaneously. As we have seen, the process in eukaryotes is quite different, involving processing of

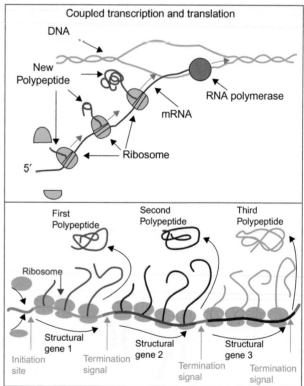

**FIGURE 3.36** Bacterial cells couple transcription and translation together. In prokaryotes, ribosomes start translation as mRNA transcripts emerge from RNA polymerase. Multiple ribosomes translate one mRNA simultaneously. Polycistronic mRNA transcripts enable groups of genes arranged in operons to be efficiently transcribed and translated.

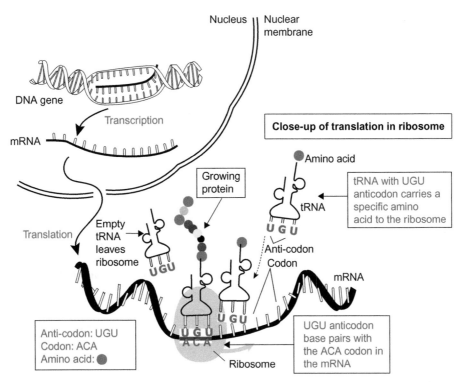

Nucleus | Nuclear membrane

DNA gene

Transcription

mRNA

**Close-up of translation in ribosome**

Amino acid

tRNA

U G U

Anti-codon

Codon

tRNA with UGU anticodon carries a specific amino acid to the ribosome

Growing protein

Translation

Empty tRNA leaves ribosome

U G U

mRNA

Anti-codon: UGU
Codon: ACA
Amino acid: ●

U G U

U G U
A C A

Ribosome

UGU anticodon base pairs with the ACA codon in the mRNA

**FIGURE 3.37** Overview of translation on a ribosome. Ribosomes engage with mRNA near the transcript's 5′ end, forming three binding sites for tRNAs when fully assembled. When a tRNA anticodon base pairs to the codon in the first site, the amino acid it carries is added to the growing amino acid chain at the heart of the ribosome. "Empty" tRNAs exit as the ribosome moves along the transcript.

mRNA and the complete physical separation of transcription and translation.

## PROTEIN SYNTHESIS REQUIRES MRNA AND RIBOSOMES

Translation is the process whereby the genetic code embedded in an mRNA molecule is decoded to create the chain of amino acids that constitutes a specific protein (Figure 3.37). The genetic code, read in groups of three nucleotides at a time, is universal in all life on earth with few exceptions (which are chiefly found in the DNA of mitochondria). The three-nucleotide codon specifies the same amino acid in an *E. coli* cell as it does in a human cell. This conservation of the genetic code is reflected in the very similar protein synthesis machinery and mechanisms in prokaryotes and eukaryotes, and it is used to great advantage in bioinformatics databases that allow scientists to find similar genes and proteins in different organisms to explore their functions.

The matching of codons to amino acids is not direct—that is, an amino acid does not bind to a codon in order to be added to the protein chain. All 20 amino acids are ferried to the site of protein synthesis by tRNA molecules, which contain three-base

anticodons that bind to the codons in the mRNA. The opposite end of the tRNA from the anticodon contains a site where an amino acid is attached to the tRNA by an **aminoacyl tRNA synthetase** enzyme. There are 20 such enzymes in the cell, one for each amino acid. As the genetic code table shows (see Figure 3.17), there is more than one codon that specifies most of the 20 amino acids. Because each tRNA molecule has a different anticodon, there must be multiple tRNAs carrying the same amino acid in the cell. Each enzyme must recognize and attach its amino acid to more than one tRNA, but not to an incorrect tRNA. To accomplish this, the enzyme binds closely to its cognate tRNAs, and contacts several different sites in the tRNA in addition to the anticodon to ensure that it is a correct (Figure 3.38). The result is a tRNA bonded to an amino acid, which is called a "charged" tRNA.

Translation in prokaryotes and eukaryotes is similar. In the process of translation, tRNAs carry amino acids to a binding site in the ribosome where the anticodon in tRNA is matched by base pairing to a specific mRNA codon.

In both prokaryotes and eukaryotes, ribosomes are the cellular machines of protein synthesis. They

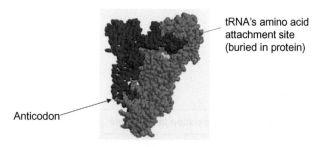

**FIGURE 3.38** Aminoacyl tRNA synthetases charge tRNAs. Every tRNA engages with an enzyme that attaches the correct amino acid to the tRNA. Cells have 20 aminoacyl tRNA synthetases, one for each amino acid. Because there are more than 20 tRNAs in every cell, most of these enzymes must recognize multiple tRNAs. It is clear from examining the complex formed by the glutamine synthetase and its partner, a glutamate tRNA, that many points of contact are examined by the enzyme in order to ensure that the tRNA it operates on is one of the correct ones for that amino acid.

consist of two subunits that come together to form a ribosome only in the presence of mRNA, a special initiator tRNA, and initiation factor proteins. Ribosomes differ in size and exact composition in eukaryotes and prokaryotes but are very similar to each other (Figure 3.39). Fully assembled ribosomes contain one binding site for mRNA and three binding sites for tRNAs (Figure 3.40). The tRNA binding sites are named for their functions in protein synthesis. The aminoacyl (A) site is where tRNAs bearing an amino acid enter the ribosome (except the first tRNA; discussed later). In the peptidyl (P) site, a tRNA carries the *peptide* being synthesized. Finally, the 'empty' tRNA leaves the ribosome via the exit (E) site.

Protein synthesis begins with the assembly of the ribosome and a tRNA onto the mRNA transcript, with the help of protein **initiation factors** (Figure 3.41A).

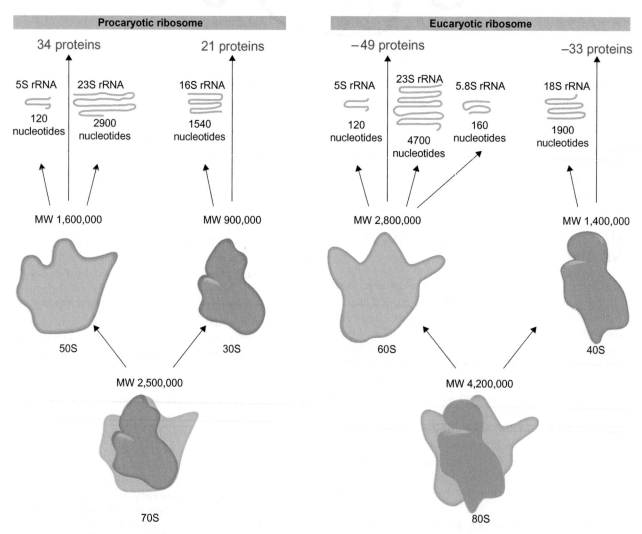

**FIGURE 3.39** Prokaryotic and eukaryotic ribosomes. Fully assembled prokaryotic and eukaryotic ribosomes are extremely large complexes with molecular weights in the millions of daltons. In both cases, dozens of proteins come together with rRNAs to form two subunits. The ribosomal proteins, which make up about one-third of the ribosome structure, mainly seem to support and stabilize an rRNA core, which has the peptidyl transferase enzyme activity, that forges the peptide bond between amino acids to make the protein.

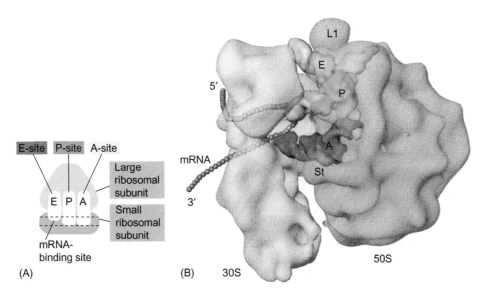

**FIGURE 3.40** Binding sites in ribosomes. Three sites for tRNAs are formed when the ribosomal subunits come together, along with a channel for the mRNA to travel through. These are shown (A) schematically and (B) as a molecular model, determined by x-ray crystallography, of the prokaryotic ribosome with bound tRNAs. The tRNAs are bound in the A, P, and E sites. The anticodons of the A and P site tRNAs are in contact with their respective codons on the mRNA transcript, shown as a string of gold balls that mark the path of the mRNA transcript through the ribosome. The growing polypeptide is not shown.

The initiation factors in eukaryotes and prokaryotes are different, but they accomplish the same goal: uniting the mRNA, ribosomal subunits, and the initiator tRNA. The codon signifying the start of translation is AUG, which codes for the amino acid methionine (met). Therefore, all proteins initially start with methionine as the first amino acid, but the methionine is often removed later.

The first step in initiation is the binding of the initiator tRNA, which always carries the amino acid methionine (in prokaryotes it is chemically modified to formylmethionine), to the P site of the small ribosomal subunit with the help of two initiation factors. Next, the mRNA binds to the complex with the assistance of additional initiation factors. In eukaryotes, two initiation factors bind to the 5′ methylguanosine cap of mRNA, which enables it to bind to the small ribosomal subunit. The small subunit then moves along the mRNA looking for the AUG start codon. Additional ribosomes can engage the transcript after the first ribosome moves on. Prokaryotic mRNAs contain a **Shine-Dalgarno** sequence just upstream from the start codon. With the help of prokaryotic initiation factors, this RNA sequence base pairs to a sequence in the rRNA present in the small subunit of the ribosome. This arrangement places the AUG start codon in the correct position to permit the large ribosomal subunit to join the complex. This results in the formation of a complete ribosome containing the initiator tRNA base paired through its anticodon to the mRNA AUG start codon. Energy is provided for translation

initiation by the hydrolysis of high-energy chemical bonds in two molecules of GTP.

With the initiator tRNA in the P site, the ribosome engaged at the start codon is ready to receive a charged tRNA in the A site (Figure 3.41B). Each successive tRNA is accompanied to the site by an **elongation factor**: EF-Tu (prokaryotes) or EF-1 (eukaryotes). These protein factors help to ensure that the codon-anticodon base pairing is correct. If it is, GTP is hydrolyzed and the tRNA is shifted slightly into an ideal position for peptide bond formation to occur. The peptidyl transferase activity of the ribosome, supplied entirely by RNA, forges the peptide bond between the two amino acids, transferring the resulting dipeptide to the tRNA in the A site. At this point, another elongation factor bound to GTP binds to the ribosome, and GTP is hydrolyzed to provide energy for the ribosome to move three nucleotides along the mRNA. This moves the A and P site tRNAs into the P and E sites, respectively, and the tRNA in the E site dissociates from the ribosome. The tRNA in the P site is holding the dipeptide and is ready for the cycle to repeat, resulting in sequential additions of amino acids to the growing polypeptide chain. As the protein chain emerges from the ribosome, it begins the process of folding. Some proteins can fold fully as they exit the ribosome; others require proteins called **chaperones**, which help newly synthesized proteins to fold into proper 3D structures. Elongation by the ribosome continues until a stop codon is encountered in the mRNA sequence.

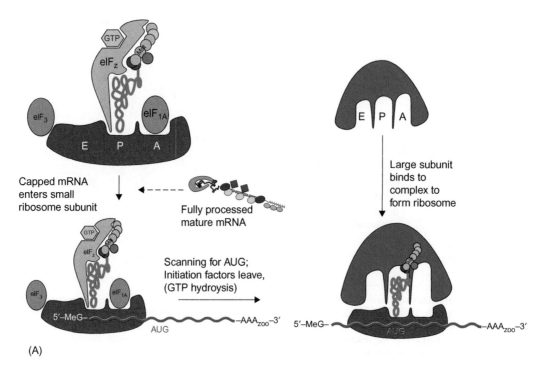

(A)

**FIGURE 3.41** Eukaryotic translation initiation, elongation, and termination on ribosome. (A) Translation initiation requires several initiation factors, only a few of which are shown here. The factors assist in the assembly of the initiator tRNA with the small subunit, followed by the binding of mRNA, which is recognized both by its cap and poly-A tail. This complex scans for the first AUG in the transcript (prokaryotes use a different mechanism to find the first AUG). When the AUG codon is engaged, the large subunit joins to form the complete ribosome complex. This process costs the cell two high-energy molecules in the form of GTP. (B) Elongation of the protein chain involves a repeated cycle of events as the ribosome moves along the mRNA. The tRNAs landing in the A site provide the new amino acids to add; the existing polypeptide is added on to the newly arrived amino acid, and then the ribosome translocates one codon further, moving the tRNA holding the polypeptide to the P site. Additional protein elongation factors assist in the process, and GTP hydrolysis again provides energy. (C) Stop codons terminate protein synthesis. A protein release factor that mimics the shape of a tRNA binds to the stop codon in the A site, and the protein chain is released. The ribosomal subunits dissociate and can re-form the ribosome on the beginning of another, or even the same, transcript.

Three codons in the genetic code function as "stop" signals for translation. Proteins that mimic the shape and properties of a tRNA molecule interact with the stop codons in the A site (Figure 3.40). These molecular mimics, called **release factors**, do not carry amino acids, but they do trigger the activity of the peptidyl transferase. The enzyme adds a water molecule to the polypeptide in place of an amino acid so that instead of being transferred to a waiting amino acid, the new protein is released and leaves the ribosome. The ribosomal subunits separate from each other and from the mRNA, ready to bind to the initiator components and begin translation again.

> Translation takes place in three phases: initiation, repeated cycles of elongation, and termination. In addition to mRNAs and ribosomes, each phase of translation requires protein factors and energy derived from GTP hydrolysis.

To become functional, the newly made protein must finish folding correctly. Proteins that fold incorrectly are usually degraded; if this process is faulty, serious disease can result. In some cases, proteins join together as subunits to form complexes or molecular machines. Many proteins must bind to small helper molecules to function; an example is hemoglobin, which must contain heme in order to bind and carry oxygen. In eukaryotes, many proteins also undergo **posttranslational modifications** in order to achieve full functionality. There are numerous types of modifications to proteins: formation of disulfide bonds, covalent attachment of carbohydrates or lipids, the addition of small organic groups such as phosphates, methyl and acetyl groups, and the removal of part of the polypeptide chain are just a few of the possibilities.

## Gene Expression Responds to Changes In Environment and Development

Precise control of gene expression is an essential feature of the life of an organism. Genes are turned on or off in response to signals from the outside environment and from inside the cell as conditions change. The

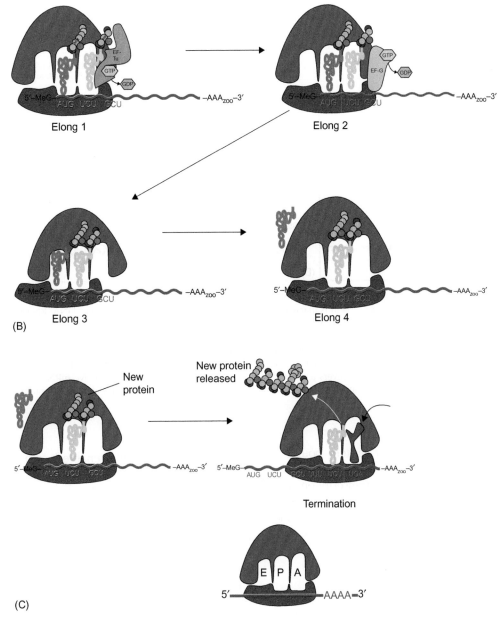

(B)

(C)

**FIGURE 3.41** Continued

mechanisms used by cells to control or regulate gene expression differ greatly in prokaryotic and eukaryotic cells, in part because prokaryotic cells largely control gene expression at the level of transcription initiation, whereas eukaryotic cells not only regulate transcription initiation but also impose controls at many other points along the gene expression pathway.

## Prokaryotic Gene Control: Under the Influence of an Operon

Consider how dramatically and rapidly the growth conditions can change for a bacterium living in the

human gut. The bacterium must adapt constantly and the ability to rapidly change its gene activity helps it to adapt and survive. Many bacterial genes are organized into **operons**, small groups of related **structural genes** that are transcribed as a unit under the control of a promoter and an additional regulatory sequence called an **operator** (Figure 3.42A). Bacterial operons typically encode enzymes that participate in a particular biochemical pathway. Seventy-five different operons controlling 250 genes have been identified in the *E. coli* genome.

At any given time, most prokaryotic genes are inhibited from transcription by **repressor proteins** that bind to operator sequences. The operator is near or overlaps

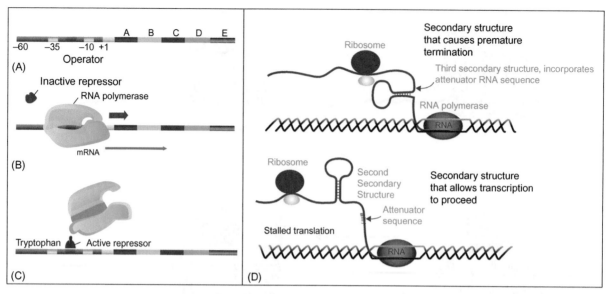

**FIGURE 3.42** Trp operon control. (A) The trp operon consists of the 5 genes for the biosynthesis of tryptophan, plus the control region. (B) When tryptophan levels are low, the repressor protein is inactive and unable to bind to the operator region of DNA; RNA polymerase binds to the promoter and transcription proceeds. (C) As tryptophan levels rise, tryptophan can bind to a site on the repressor, changing the repressor conformation so that it binds to the operator DNA. This prevents RNA polymerase from binding to the promoter, shutting off transcription. (D) Two stem-loop structures can form in the mRNA transcript of the trp operon. Which one forms is dependent on the speed of the ribosome on the mRNA—a faster ribosome *(upper panel)* covers a specific part of the transcript, leading to a structure that causes premature transcription termination. A slow ribosome *(lower panel)* covers a different area, and the mRNA forms the alternative mRNA structure that allows transcription of the trp operon to proceed. The speed of the ribosome is, in turn, dependent on the availability of tryptophan, because it is translating a peptide that contains tryptophan.

with the promoter, so a repressor protein bound to an operator physically blocks the RNA polymerase's access to the promoter, preventing transcription from occurring. We will look at two well-studied bacterial operons that demonstrate how repressors and other regulatory proteins respond to signals to regulate gene expression.

> Proteins mediate repression and activation of bacterial genes in response to signals from inside and outside the cell. Prokaryotic genes are often organized into operons that can be controlled as a unit.

The genes of the pathway for biosynthesis of the amino acid tryptophan are in the bacterial *trp* operon. When tryptophan is plentiful in the environment, the trp operon is silent (not transcribed) because of a repressor protein that binds to the trp operator DNA, preventing RNA polymerase from binding to the promoter (Figure 3.42). However, when tryptophan is not readily available, the trp operon is actively transcribed, and the enzymes needed for the biosynthesis of tryptophan are expressed (Figure 3.42C).

How does the bacterial cell know when to make tryptophan? The answer is that tryptophan itself binds to the repressor protein to enable it to bind the operator site and block transcription. Without tryptophan, the repressor is unable to bind to the operator, and transcription proceeds. This negative feedback loop allows the cell to respond to an abundance of tryptophan by ceasing to make the amino acid.

The trp operon is also an excellent example of a means of further adjusting the rate of gene expression, called **attenuation** (Figure 3.42D). The mechanism of attenuation depends on the coupling of transcription and translation in prokaryotes: while prokaryotic mRNA is being transcribed, ribosomes bind and work their way along the transcript, synthesizing the encoded protein(s). The trp operon contains a "leader" sequence upstream of the genes in the operon. Part of the leader RNA, which encodes a short polypeptide, can base pair with itself to form two different stem-loop RNA structures. Depending on how quickly the ribosome can translate the leader polypeptide, one of two different stem-loop structures forms in the mRNA located between the ribosome and RNA polymerase.

One of these structures allows the RNA polymerase, which is ahead of the ribosome, to continue transcription; the other stem-loop structure causes the RNA polymerase to release the DNA and transcription stops. The key is the position of the ribosome along the mRNA, and whether or not it physically covers parts of

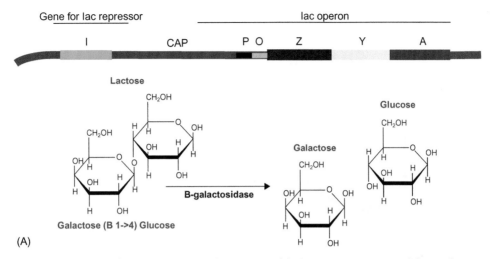

**FIGURE 3.43** *lac* operon structure and function. (A) *(upper)* The structure of the lactose operon DNA and the regulatory gene that precedes it. The lac operon contains a promoter (P), an operator (O), and three genes (Z, Y, A) required for metabolism of lactose. Upstream from the operon, the lac repressor gene (I) is encoded. Also upstream is a region that binds the catabolite activator protein (CAP). *(lower)* The biochemical reaction to breakdown lactose into glucose and galactose, which is catalyzed by the enzyme beta galactosidase (encoded by gene Z). (B) *(upper)* In the absence of allolactose, the repressor binds and blocks transcription of the lac operon. *(lower)* When lactose is available, the allolactose inducer molecule binds to the repressor protein, which inactivates the repressor and removes it from the operator, allowing a low level of transcription of the genes for metabolism of lactose. (C) A summary of the lac operon under three different environmental conditions.

the mRNA required for each of these RNA structures to form. If the ribosome can translate the leader peptide quickly, it covers a section further along the transcript, and the mRNA forms a structure that triggers RNA polymerase to fall off the operon DNA before it transcribes the trp biosynthesis genes. If the ribosome is too slow, the alternative stem-loop structure forms, which allows RNA polymerase to proceed along the DNA and transcribe the operon. So the question becomes, what controls the speed of the ribosome in translating the leader sequence?

If you think it should have something to do with how much tryptophan is present, you are correct. The leader peptide encodes two tryptophan amino acids. Without a ready supply of tryptophan, the ribosome stalls as it waits for the amino acid, and it covers a specific part of the mRNA transcript. As a result, the mRNA forms the stem-loop structure that allows RNA polymerase to proceed with transcription. In contrast, when enough tryptophan is present to make the leader peptide, the ribosome moves quickly through the region of mRNA, uncovering the mRNA that forms the structure that causes RNA polymerase to fall off the mRNA.

> The *trp* operon exemplifies the typical default repression of many prokaryotic operons and the activation triggered in response to a cellular need. In addition, attenuation control of the operon allows for more delicate regulation.

The **lac operon** of *E. coli* has been studied since the 1950s and exemplifies how genes can be regulated in response to more than one environmental circumstance. *E. coli*'s preferred source of energy is glucose, but when necessary, the bacterium can also metabolize other sugars, such as lactose, the sugar found in milk. The *lac* operon contains the genes for breaking lactose down into its constituent chemical parts (glucose and galactose) (Figure 3.43A). We'll consider three conditions that the lac operon should respond to appropriately: (1) lactose is available but there is little or no glucose available; (2) only glucose is available as a food source; (3) glucose and lactose are both available. Under condition 1, the lac genes should be highly expressed, and under condition 2, the lac operon should be silent (turned off). Under condition 3, a small amount of lac operon expression is useful.

The presence or absence of lactose plays an important, but incomplete, role in lac operon expression (Figure 3.43B). When lactose is absent, the lactose repressor protein (product of the *lacI* repressor gene, not part of the lac operon) binds to the operator region and covers part of the promoter, preventing RNA polymerase from binding to the promoter site and thereby blocking transcription. When lactose becomes available, some gets converted to allolactose (an isomer of lactose). The lac repressor has a binding site for allolactose, and in response to allolactose binding, the repressor's DNA binding site changes shape and no longer binds to DNA, leaving the promoter

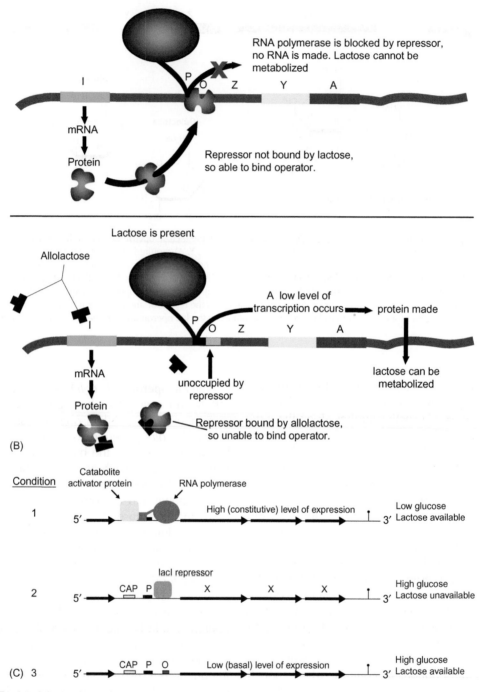

**FIGURE 3.43** Continued

site on the DNA exposed and available for transcription. Thus, allolactose serves as a signal molecule that alerts the cell to the presence of lactose, and it acts as an **inducer** of *lac* operon expression by removing repression.

However, because of the weak lac operon promoter, inducing expression via allolactose only permits a very low rate of expression. If glucose is present as well as lactose (condition 3), this is satisfactory, but if no glucose is present (condition 1), lactose metabolism must be ramped up sharply to meet the cell's energy needs. Therefore, the cell must also sense the level of availability of glucose. This is accomplished via an additional regulatory protein, the catabolite activator protein (CAP, Figure 3.43C). CAP is activated in response to a lack of glucose and binds to a control

---

**Box 3.2   Which Gene Makes Us Human?**

We have discussed many of the basic biochemical processes that account for life, including gene storage and expression, the synthesis of RNAs, and the production of proteins. We have also begun to explore how cells control gene expression and decide which genes are transcribed into RNA, when and why. Knowing the DNA sequence of a gene or a genome is not enough to reveal many secrets about the living organism. Perhaps this is best explained by the controversy surrounding the sequence differences between humans and chimpanzees. At the time of the human genome project completion, the human and chimp genomes were thought to have over 98% the same sequence, inspiring questions such as "Which gene makes us human?" "We've known for a while that the protein coding genes of humans and chimpanzees are about 99 percent the same," said Yale scientist Michael Snyder, "The challenge for biologists is accounting for what causes the substantial difference between the person and the chimp." (*Science* 317: 815-819, 2007).

A more recent study indicates that the suggested 98% similarity in DNA between chimp and humans should be revised to about 95%. Either way, humans and chimps apparently differ by only 2% to 5% of the DNA bases. Conventional wisdom says that if the difference between humans and chimps is not due to the gene sequence, then it must be due to differences in the control of gene expression that influence the production of similar proteins in humans and chimps. Experimental support for this idea was lacking until recent studies reported that closely related species of yeast use very different patterns of gene regulation, even in cases where in the end, the outcome of the gene regulation was the same. Scientists have found that gene regulation is the key to the large differences observed between species. So we know that it is not just the genes that count, it's how (and when) you turn them on that matters.

---

sequence of DNA upstream of the promoter. CAP assists RNA polymerase in binding to the promoter, greatly increasing the rate of transcription and providing plenty of lac operon gene products for deriving energy from the breakdown of lactose. When lactose is no longer plentiful, allolactose levels will drop and the lac repressor will return to the operator DNA, silencing transcription once again.

The lac operon is an illustration of how two sources of information from a cell's environment can be integrated to elicit the appropriate response. Of course, even bacterial cells are subject to more than two signals at a time. In eukaryotes, the integration of signals increases in complexity with the size of the genome and the increasing complexity of multicellular organisms.

## EUKARYOTIC GENE REGULATION

Gene regulation in bacteria is relatively simple compared with the much more complex systems that control gene expression in eukaryotic cells. Part of the issue is the number of genes; a human cell has more than 20,000 genes compared to only about 2500 for a single-celled bacterium. The human body has more than 200 cell types and, consequently, executes vastly different patterns of gene expression in different tissues. For example, in order to function correctly, human liver cells and human brain cells must express very different genes as well as a common set of "housekeeping" genes, those required for any cell

to live. Moreover, a eukaryotic organism undergoes many stages of development (such as embryonic, fetal, adolescent, and adult development) and requires varying regulation of gene expression in cells at different stages of life. The complexity of eukaryotes is reflected in the many opportunities for regulation of the gene expression pathway (Figure 3.44).

Gene expression regulation starts with transcription initiation, and we will take a broad look at how this step is regulated. A detailed study of the many specific processes that regulate this and other steps of eukaryotic gene expression is beyond the scope of this book; interested students can look further by exploring books and web sites listed at the end of the chapter.

In eukaryotes, DNA is wrapped in histone proteins, tightly bundled into nucleosomes and higher-order structures that prevent transcription unless they are disassembled for RNA polymerase initiation complexes to gain access. Chromatin remodeling machines are a necessary part of the RNA polymerase machinery. Because of this default "off" state of genes, transcriptional activators are more common than repressors. Rather than acting singly, eukaryotic transcription factors combine to form complexes that have varying effects depending on their exact makeup (Figure 3.45). These smaller complexes contribute to a much larger picture of **combinatorial control** integrated by the mediator protein that connects the transcription factors scattered over long distances of DNA to the RNA polymerase initiation complex. The initiation of transcription occurs only if the balance of all these inputs falls on the side of activation.

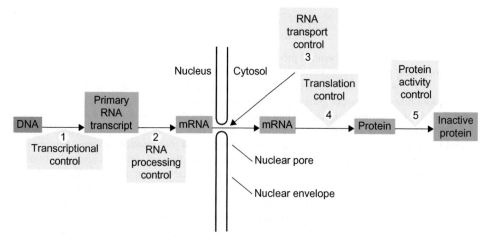

**FIGURE 3.44** Opportunities for regulation of eukaryotic gene expression. The pathway of gene expression in eukaryotes has many stages, thanks to the compartmentalization of the cell and the complexity of mRNA processing. The steps are outlined, and the stages where expression can be regulated are noted. Eukaryotic gene expression can be controlled at all steps; only some of these steps would apply to prokaryotic gene expression because of the lack of a nucleus and the absence of mRNA processing.

Among many additional controls of gene expression is the regulation via small ncRNAs. For example, **micro RNAs (miRNAs)** transcribed by RNA polymerase II contain sequences complementary to parts of mRNAs. The binding of miRNA to mRNA triggers degradation of the mRNA by cellular machinery, providing a way to turn off or dampen gene expression during or after transcription. **Small interfering RNAs (siRNAs)** also exert control over eukaryotic gene expression (see Chapter 11).

Regulation occurs at many points along the gene expression pathway, including the posttranscriptional processing and transport of mRNA molecules. These multiple and complex levels of gene regulation often challenge scientists as they experimentally reproduce and test different aspects of the gene expression mechanisms in the cell. Regardless of complexity, the regulatory mechanisms must be considered, understood, and resolved if DNA technology is to be successfully applied to treating complicated genetic diseases (see Chapter 10) and developing new therapeutic approaches (see Chapter 11), including developing designer drugs (see Chapter 13). The regulatory aspects of gene expression are as important to gene activity as the fundamental processes of transcription and translation.

## SUMMARY

The realization that chromosomes are involved in heredity led scientists to investigate the role of DNA in living systems. Garrod's studies indicated that enzymes and proteins are related to gene activity. Beadle and Tatum showed that a gene regulates the production of an enzyme, a concept that was extended by Ingram's experiments. He concluded that genes regulate protein production and that a mutation that affects the identity of even a single amino acid in a protein can have a devastating effect. RNA was found to act as an intermediary between the information in DNA and the amino acid sequence of a protein chain. Crick's group was instrumental in identifying a three-base sequence in DNA as the basic unit of the genetic code, the codon.

Much of gene expression is the pathway from DNA through the intermediary RNA to the synthesis of proteins; however, DNA also encodes RNA that is a final gene product. Transcription is the act of copying RNA from the genetic messages in DNA templates. Translation is the transfer of the specific genetic message in an mRNA to a corresponding amino acid sequence in the protein. DNA encodes three forms of RNA involved in translation: messenger RNA (mRNA), whose codons transport the genetic message; ribosomal RNA (rRNA), which is used to construct ribosomes; and transfer RNA (tRNA), which transports amino acids to the ribosomes and whose anticodons complement the codons in mRNA. Translation in both eukaryotes and prokaryotes occurs on ribosomes, requiring charged tRNAs, mRNA, and a variety of initiation, elongation, and termination factors that ensure accurate protein synthesis.

Genes in bacteria are frequently organized into operons, which often consist of a single transcription unit including several genes with related functions that are transcribed as one polycistronic mRNA molecule. Because of the absence of a nucleus, a bacterial mRNA can be efficiently transcribed and translated at the same time. In contrast, the typical eukaryotic transcription unit encodes only a single gene, but inherent in the gene itself are options that can change the

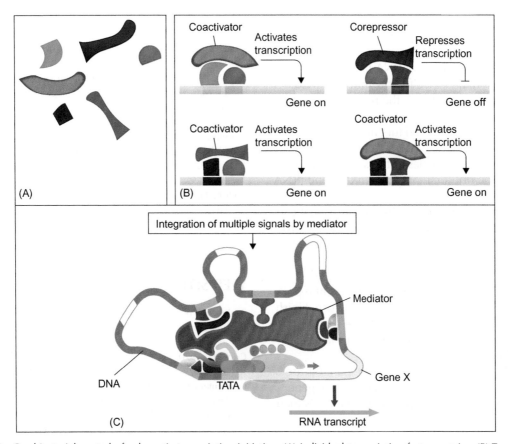

**FIGURE 3.45** Combinatorial control of eukaryotic transcription initiation. (A) Individual transcription factor proteins. (B) Transcription factors in eukaryotes act in combination and can have different effects depending on their partners in a complex. Differing combinations of regulatory sequences form the control regions in DNA that attract various combinations of partners. (C) The combinatorial effect is compounded by the integration of many signals by the mediator protein. The various inputs are themselves combinatorial, vastly increasing the possibilities for fine-tuned, cell-type specific regulation.

nature of the protein product. For example, many genes have alternative RNA splicing patterns that allow the synthesis of more than one protein from a single gene. Additionally, transcription and translation occur in separate compartments of the cell.

Gene expression is controlled by regulatory proteins in response to signals from the cell and from the environment. Differential gene expression is required to preserve cell energy and chemical resources, to respond to environmental changes, and to provide for the growth and development of an organism. In prokaryotic cells, negative control of gene expression occurs when a regulatory protein binds to a regulatory site in the DNA called the operator and inhibits transcription. Positive control occurs when an activator protein stimulates transcription by encouraging the binding of RNA polymerase to the promoter DNA site. Eukaryotic gene expression is complex because of the separation of the nuclear and cytoplasmic compartments and the demands of multicellular organisms. In addition to having promoter regions at the start of genes, eukaryotic genes transcribed by RNAP II have control regions that can lie at a great distance from the genes they regulate. Many signals are integrated to determine whether the transcription of a eukaryotic gene should begin.

## REVIEW

This chapter has focused on the mechanisms by which the genetic message in DNA is expressed in a protein. To assess your understanding of the discussions, answer the following questions:

1. Briefly summarize the insight of Garrod and the work of Beadle and Tatum, and show why both were important to the evolving interest in DNA.
2. What did Ingram's experiments indicate, and why were his results significant?
3. What evidence pointed scientists to the role of RNA in protein synthesis, and how did Crick's experiments verify the three-base codon in the DNA molecule?

4. Name three types of RNA that participate in protein synthesis, and specify the function of each type.

5. Describe the synthesis of RNA taking place in transcription, and specify how mRNA is modified before it is exported from the nucleus of a eukaryotic cell.

6. Explain the structure and function of transfer RNA molecules.

7. Summarize the activities of translation taking place at the ribosome once the mRNA and tRNA molecules have arrived.

8. Compare negative and positive controls that can influence the expression of a gene. Use the processes of repression and activation as guides for your comparison.

9. Explain how the lac operon integrates and responds to signals from the environment reporting the availability of lactose and glucose. Finish by drawing the operon and CAP site under conditions where neither glucose nor lactose is available.

10. Describe in broad terms why gene controls are necessary in a cell and why regulation of gene expression is more complex in human cells than in bacterial cells.

## FOR ADDITIONAL READING

Alberts, B., et al., 2008. Molecular Biology of the Cell, fifth ed.. Garland Science, New York.

Beardsely, T., August 1991. Smart Genes. Scientific American, 72–81.

Britten, R.J., 2002. Divergence between Samples of Chimpanzee and Human DNA Sequences Is 5% Counting Indels. Proceedings National Academy Science 99, 13633–13635.

Clark, D.P., 2005. Molecular Biology: Understanding the Genetic Revolution. Elsevier.

Crick, F., 1970. Central Dogma of Molecular Biology. Nature 227, 561–562.

Fox, M., March/April 1996. Breaking the Genetic Code in a Letter to Max Delbruck. Journal of College Science Teaching 15, 324–325.

Freedman, D.H., July 1991. Life's Off-switch. Discover.

Griffiths, A.J.F., et al., 2008. An Introduction to Genetic Analysis, eight ed.. W. H. Freeman, New York.

Judson, H.F., 1979. The Eighth Day of Creation. Simon & Shuster, New York.

Lodish, H., et al., 2007. Molecular Cell Biology, fifth ed.. W. H. Freeman, New York.

Watson, J., Tooze, J., 1981. The DNA Story. W. H. Freeman and Co, San Francisco.

## WEB SITES

Aminoacyl tRNA Synthetases. In: Essential Biochemistry. <www. wiley.com/college/pratt/0471393878/student/structure/trna_aars/ index.html/ >.

Crick, F., 1970. Central Dogma of Molecular Biology. Nature 227, 561–562 www.nature.com/nature/focus/crick/pdf/crick227.pdf.

DNA Interactive, from the Dolan DNA Learning Center, Cold Spring Harbor Laboratory. <www.dnai.org/ >.

Gene Expression and Regulation, at Scitable: A Collaborative learning space for undergraduates. <www.nature.com/scitable/ topic/Gene-Expression-and-Regulation-15/ >.

MicroRNAs: Definition and Overview. <www.ambion.com/techlib/ resources/miRNA/mirna_intro.html/ >.

RNA Polymerase-Promoter Complex. <www.pingrysmartteam.com/ RPo/RPo.htm/ >.

Steve's Place: Science (see Transcription, Translation sections) <www.steve.gb.com/science/index.html/ >.

# Tools of the DNA Trade

## Designer Enzyme Cuts HIV Out of Infected Cells

Bacterial Enzyme Turns the Tables on Deadly Retrovirus
*Scientific American*, June 28, 2007

By J. R. Minkel

Scientists have constructed a custom enzyme that
reverses the process by which the human immunodeficiency virus (HIV) inserts its genetic material into host
DNA, suggesting that treatment with similar enzymes
could potentially rid infected cells of the virus. In tests on
cultured human tissue, the mutated enzyme, Tre recombinase, snipped HIV DNA out of chromosomes.

Scientists designed an enzyme that can remove the
human immunodeficiency virus (HIV) DNA from the
host genome, as a test to see if it might be possible
to engineer an enzyme to attack the HIV genome in
humans and potentially rid the infected cells of the HIV
virus and the infection. The function of the engineered

Tre recombinase enzyme was first tested in cultured
human tissue cells in the lab. The enzyme cut the cell's
genome DNA at both ends of the integrated HIV virus
DNA, removing it from the chromosome, and rid the
cells of the virus. This exciting advance in the lab was
tempered by the caveat that potential AIDS treatments
can often take a decade to progress to clinical trials. The
media hype about a possible cure for AIDs obscures
the underlying achievement of this novel experiment,
the ability to extend the reach of a traditional biotechnology tool, the restriction enzyme, directly into the
patient.

In this chapter the reader will find out about the
structure and function of restriction enzymes, truly tal-
ented proteins that made research in molecular genetics
and biotechnology possible for nearly 40 years. Each
restriction enzyme acts like a "molecular scissors" that
cuts across the DNA helix. Each restriction enzyme cuts
DNA at only one specific DNA sequence, the restriction enzyme recognition site. Each time the restriction
enzyme finds its recognition sequence in the DNA, it
cuts across both strands of the DNA helix at that site
only, and not at other sequences in the DNA. For decades, restriction enzymes have been used routinely to
produce DNA fragments carrying specific genes and
using recombinant DNA technology moved the genes
into them into vectors for expression in host cells (see
Chapter 5). The large number of biotechnology companies marketing more than 3000 different restriction
enzymes reflects the importance of restriction enzymes
to the progress of biomedical research.

Recently scientists have created designer restriction
enzymes that cut DNA at desired base pair sequences
that are "programmed" into the structure of the cus-
tom-built enzymes. Customized restriction enzymes
are a good example of the many innovative tools and
technologies in use and currently under development
to study DNA and gene products in cells.

## LOOKING AHEAD

This chapter focuses on the scientific discoveries
and developments that create the foundations for the

practical applications of DNA technology that made the DNA revolution possible. When you have completed this chapter, you should be able to do the following:

- Discuss the biochemical tools used in DNA technology, and recognize why microorganisms occupy an important place in this area of DNA science.
- Explain the discoveries and experiments that formed the basis for the key features of modern DNA technology.
- Identify some of the individuals whose work laid the foundations for DNA technology.
- Explain the basic processes used in recombinant DNA technologies including bacterial transformation and cutting and ligation of DNA.
- Discuss the ethical and safety concerns that have been raised about DNA technology over the years, and describe the role of the Asilomar conference.
- Recognize some important terms and concepts used in DNA technology, and broaden your vocabulary to include words common to this area of science.

## INTRODUCTION

During the 1950s and 1960s, scientists made substantial gains in molecular biology as they clarified the role of DNA in the biochemistry of protein synthesis and explained the intricate details of this process. In the 1970s, scientists began to manipulate DNA and devised methods to cut and join DNA fragments to produce recombinant DNA molecules (as we will see in this chapter). These recombinant DNA experiments introduced the era of DNA technology.

The science of DNA technology was a new frontier to explore. Researchers discovered, for example, that they could transplant animal genes into bacterial cells and coax the genes to function in this new environment; they kindled hopes that plants, noted for their ability to produce carbohydrates, could be engineered to produce proteins; and they laid the foundations for treatments for genetic diseases (see Chapter 15). In addition, the future held promise for inexpensive sources of bioenergy, the development of novel vaccines, and the mass production of pharmaceuticals (see Chapter 13). Some saw no limit to what could be accomplished by DNA science.

Today, in the twenty-first century, much of the amazing progress predicted by scientists has come to pass or is within sight. The public benefits from a large selection of proteins produced in genetically engineered organisms including bacteria, yeast, and mammalian cells, which can produce insulin, human hemoglobin, human growth hormone, detergent enzymes, and many other proteins from different species. Many advances

are on the horizon as well. Therapeutic antibodies are now available to combat cancer and immunological diseases, and we have an effective vaccine against the human *papilloma* virus that causes cervical cancer is now available (Guardasil from Merck & Co.). DNA technology and the general field of molecular biology continue to have staggering potential for improving pharmaceutical yields, diagnosing and treating human disease, and creating new genetic testing and diagnosis products for use at home (see Chapters 10, 11, 13).

At the same time, DNA technology has triggered a degree of public concern and confusion. It has created dilemmas for governmental and regulatory agencies, and in general it has raised fears that genetically modified organisms, especially in agriculture, could pose a threat to human welfare. We will discuss many of these issues in the chapters ahead as we consider the foundations for DNA technology and the practical applications of the awesome abilities it gives us. For the present, we will explore the thought processes and discoveries that laid the foundations for modern DNA technology. The fruits of this technology are the subject matter for the remainder of this book.

> As genetic engineering research blossomed in the mid-to-late twentieth century, it raised both hopes and concerns about biotechnology.

## TOOLS OF GENETIC ENGINEERING

Humans have guided the flow of genetic information for thousands of years without realizing it, for example, breeding useful agricultural plants from wild ancestral stock seeds (Figure 4.1). New organisms can be created by breeding organisms that ordinarily do not mate. A nectarine, for example, is the offspring of a cross between a plum and a peach, and a mule is derived from breeding a horse with a donkey. The approaches used in modern DNA technology go well beyond the genetic crosses of the past. Scientists can now intervene directly in determining the genetic fate of organisms, and many of these can be performed outside the cell in test tubes. By taking DNA fragments from different species and connecting the DNA fragments together in new combinations, researchers can now make entirely new chromosomes that profoundly change the character of the organisms. An early example is the human insulin protein expressed in bacteria carrying the human insulin gene DNA on a plasmid. Arguably, this is a new species of bacterium.

During the 1960s and 1970s, the new DNA technology required new approaches to science, new insights, and new materials. Methods to cut DNA and

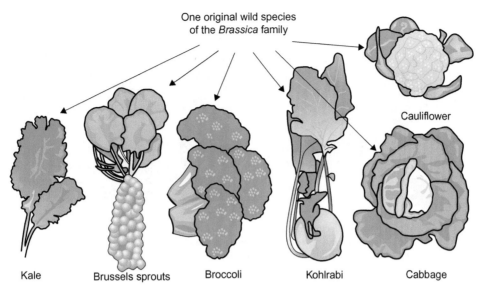

One original wild species
of the *Brassica* family

Cauliflower

Kale    Brussels sprouts    Broccoli    Kohlrabi    Cabbage

**FIGURE 4.1** Guiding the inheritance of genes. Through the centuries, humans used a wild species of a cabbage-like plant to breed all the modern plants shown here. Depending on the climate and the desires of the local population, some varieties were bred to have a hard head as in a modern cabbage, some formed masses of flower buds as in broccoli and cauliflower, and some made clusters of leaf buds as in Brussels sprouts.

connect DNA fragments together had to be found, organisms had to be enlisted to carry the new fragments, and biochemical methods had to be developed to permit expression of the DNA fragments. In the pages ahead, we will explore some products of that imagination as we survey the breakthroughs that formed the basis for the new science of DNA technology.

> Humans have altered the genetic makeup of organisms for centuries by breeding plants and animals to create new organisms with the desired genetic and physical characteristics inherited from the parents.

## Microorganisms Are Used to Streamline DNA Technology

Advances in DNA technology were made possible by seminal work with bacterial cells (Figure 4.2), which are cultivated easily and can be studied conveniently in the research lab. The most popular bacterium used in DNA technology is the "workhorse" ***Escherichia coli (E. coli)***. Having studied this bacterium for decades, scientists now know more about the biochemistry, morphology, physiology, and genetics of *E. coli* than we know about any other organism (including humans). As a result of such intense investigation, *E. coli* has become a model system for research designed to figure out how cells work. *E. coli* is also easy to grow in the lab, and, except for specific dangerous strains such as *E. coli* 0157:H7, *E. coli* is not a serious human pathogen.

The **bacterial chromosome** is particularly well suited for DNA experiments because it is small compared to the large genomes in many eukaryotic cells, and most bacteria have a single double-stranded circular DNA chromosome in each cell. In eukaryotic cells (such as animal cells), each chromosome consists of a very long, linear DNA molecule, and typically occur in pairs. Because there is only one chromosome in a bacterium, its genes are expressed without the influence of genes on a second copy of the chromosome. Bacteria do not have nuclei, but the circular bacterial chromosome is attached to the membrane inside the cell through the activities of special proteins that interact with the lipid membranes and also bind to the chromosome DNA. Eukaryotic chromosomes, by contrast, are contained in the nucleus of the cell at all times except during cell division (see Chapter 9). Because the genetic code is nearly universal, the protein expression machinery in the bacterium can produce any foreign protein.

> The relative simplicity of bacterial cells and chromosomes compared to their eukaryotic counterparts make bacteria a good choice for experimental work.

Viruses have proven to be very useful in many areas of DNA research. Viruses are microscopic particles consisting of RNA or DNA genomes enclosed in a protein coat (called a **capsid**), which in some viruses is also surrounded by a lipid envelope (Figure 4.3). Viruses replicate only inside living prokaryotic and eukaryotic cells where they shed their protein coats

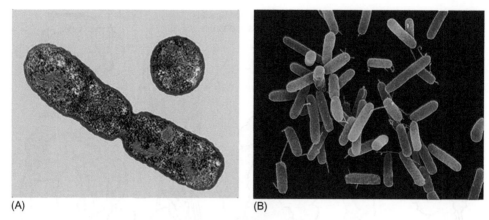

**FIGURE 4.2** Bacteria in service to humanity. Electron micrographs of (A) *E. coli*, probably the most widely used bacterium in molecular biology, and (B) *Pseudomonas*, some strains of which can naturally break down chemical pollutants in the environment. *Pseudomonas* has also been engineered to assist in cleanup after oil spills damage the environment.

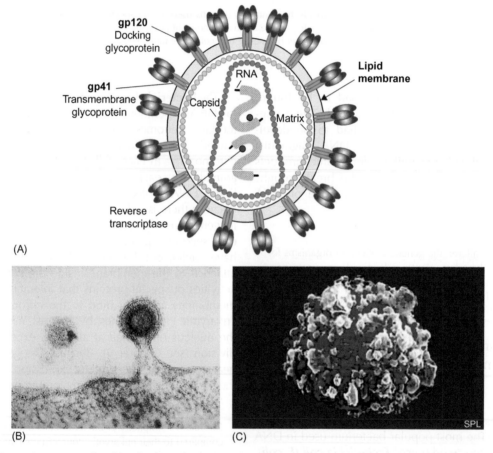

**FIGURE 4.3** Some viruses are enclosed in a lipid membrane. (A) A cross section of an HIV virus particle. (B) HIV virus obtains a lipid envelope (membrane) as it exits a cell. (C) HIV virus (red) attacks an immune system cell.

and use the molecular machinery of the host cell to produce new viruses. The genes of a DNA virus direct the synthesis of new viral proteins. The RNA of an RNA virus often masquerades as a cellular messenger RNA molecule, tricking the cell into producing viral enzymes and other viral proteins.

Some types of viruses do not replicate themselves immediately. Instead, they integrate their DNA into the host's chromosome DNA and become part of the cell's genome. The DNA of a **herpesvirus**, for example, can **integrate** (be inserted) into a nerve cell genome where it can remain within that cell for years, causing

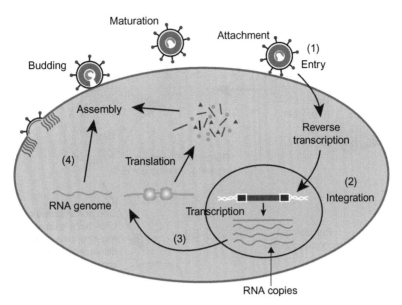

**FIGURE 4.4** Retroviruses integrate into the chromosome. (1) The retrovirus enters the cell. (2) The retroviral genome is made of RNA and is copied into double-stranded DNA (dsDNA) by a viral enzyme called reverse transcriptase. The DNA copy integrates into the host cell genome; the integrated viral DNA is called the provirus. (3) Later the host RNA polymerase transcribes the provirus DNA to make RNA copies of the retrovirus. (4) Once packaged with proteins, the genomes form new retroviruses that exit the cell to infect new cells and continue the cycle.

periodic herpes infections in the body. A different process occurs with the **human immunodeficiency virus (HIV)**, which causes AIDS. Once inside the cell, the RNA from this virus serves as a template for the synthesis of DNA, which is then inserted into the host cell genome (Figure 4.4). Once it is part of the host chromosome, the viral DNA is replicated as part of the chromosome and is expressed as mRNA encoding viral capsid proteins. The ability of viruses to integrate, or insert, their DNA into host cell genomes drew the attention of scientists who began using viruses as vectors to carry foreign genes into host cells. Numerous organisms in addition to viruses and bacteria are used in DNA technology, including yeast cells, various other microbes and viruses, insect and mammalian tissues, and a number of plants. We will mention their names and significance as we encounter them in succeeding chapters.

> Viruses consisting of no more than a protein shell covering a small genome, can be valuable allies in delivering DNA into prokaryotic and eukaryotic cells.

## Microbes Exchange DNA Naturally

Research conducted in the 1950s demonstrated conclusively that genetic recombination occurs among bacteria and could even involve bacteria-attacking viruses, **bacteriophages**. The story began with Griffith's experiments in 1928 (see Chapter 1), which showed

that genetic recombination occurs in bacteria, leading to Avery's identification of DNA as the molecule responsible for genetic recombination. By the 1950s, scientists knew of three ways that bacteria in the wild can change their genetic makeup: **transformation, conjugation**, and **transduction**. In nature, **transformation** occurs when donor bacteria die, break open, and release their contents, including DNA, into the environment. The lucky recipient bacteria in the local neighborhood take up pieces of DNA, which are inserted into the bacterial chromosome (Figure 4.5). In the wild, DNA uptake and transformation take place in less than 1% of a bacterial population, but can bring about profound changes in the population. For example, one mutation might increase the pathogenicity of the organism (as in Griffith's *pneumococci*), whereas a different mutation might affect the efficiency of transfer of bacterial **plasmids** carrying antibiotic resistance genes from one microbe to another microbe in nature. A modified version of this transformation process has become a standard method used to introduce plasmid DNA into bacterial cells (discussed later).

Joshua Lederberg, Francois Jacob, and Elie Wollman established that bacteria can transfer genetic information from one bacterium to another by a process called **conjugation** (Figure 4.6). During conjugation, the donor and recipient bacterial cells are joined to each other by a cytoplasmic bridge structure called a **pilus**. Single-stranded chromosome DNA from the donor bacterium crosses the cytoplasmic bridge into the recipient cell where it either integrates into the

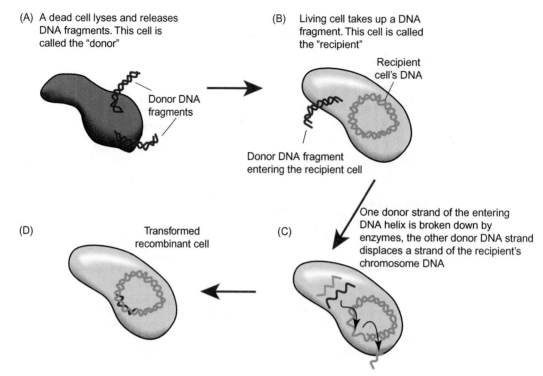

**(A)** A dead cell lyses and releases DNA fragments. This cell is called the "donor"

Donor DNA fragments

**(B)** Living cell takes up a DNA fragment. This cell is called the "recipient"

Recipient cell's DNA

Donor DNA fragment entering the recipient cell

**(D)**

Transformed recombinant cell

**(C)**

One donor strand of the entering DNA helix is broken down by enzymes, the other donor DNA strand displaces a strand of the recipient's chromosome DNA

**FIGURE 4.5** DNA used in transformation is picked up by bacterial cells is integrated into the host chromosome. (A) When a cell dies, it releases its DNA into the surrounding environment (donor cell). (B) A nearby cell takes up the DNA (recipient). (C) One strand of the incoming double-stranded DNA is removed by enzymes, and the resulting single-stranded DNA displaces one strand of recipient's DNA (D) The donor DNA has been integrated into the host recipient chromosome.

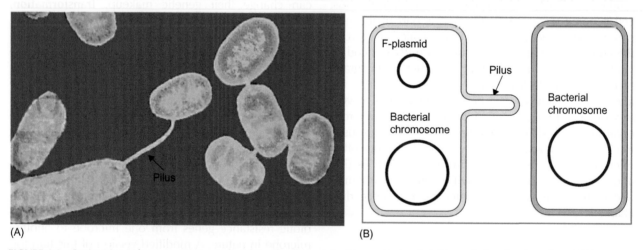

**(A)**

Pilus

**(B)**

F-plasmid

Pilus

Bacterial chromosome

Bacterial chromosome

**FIGURE 4.6** Conjugation in bacteria. (A) Two *Escherichia coli* cells are joined by a cytoplasmic bridge called a pilus that allows DNA to move from a donor cell to a recipient cell. One cell has produced three pili for conjugation with three cells. Conjugation provides an opportunity for the transfer of genetic material from one cell to another, but it is a one-sided transfer—not an exchange. This is one way that bacteria acquire genes for drug resistance. (b) A small circular DNA molecule called the F-plasmid is one of several types of conjugative plasmids that produce the pilus and the single-stranded F-plasmid DNA transfers through the pilus to a recipient bacterium that does not have the conjugative F-plasmid.

recipient's chromosome or it is copied into a double-stranded circular DNA molecule that can replicate like a plasmid. The newly acquired donor genes are then expressed in the recipient cell. It is important to note that conjugation and DNA transfer have been demonstrated between cells from different genera of bacteria such as *Salmonella* and *Shigella* cells.

Bacteria exchange DNA naturally in the processes known as transformation and conjugation. The DNA that is transferred often remains physically separate from the bacterial chromosome and forms a small, independent circle of double-stranded DNA called a plasmid.

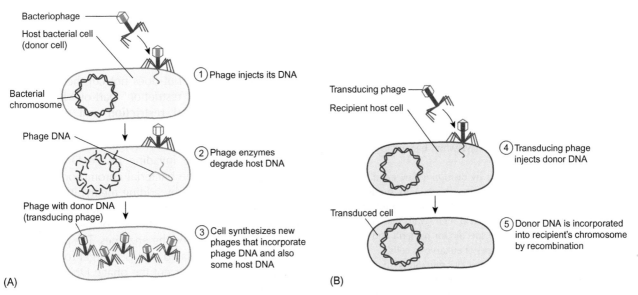

**FIGURE 4.7** Transduction of bacterial DNA by bacteriophages. A phage can incorporate a piece of bacterial host genome into its own genome, spreading bacterial genes along with viral infection. When a phage that has picked up some bacterial DNA infects another cell, the DNA form the previous bacterium is integrated into the subsequent bacterium's genome along with the viral DNA. Thus, genes can be transferred from one species of bacteria to another by bacteriophages.

Joshua Lederberg and Norton Zinder discovered a third form of bacterial DNA transfer in the late 1950s called **transduction** (Figure 4.7). In this process, bacteriophages "accidentally" transfer chromosome DNA among bacterial cells. The bacteriophage (or phage, for short) infects the bacteria by attaching to the cells and entering through the cell wall (see Chapter 5). Once inside the cell, the phage either replicates its DNA, packages it into newly synthesized capsid proteins, and the phage progeny are released, or the phage genome can integrate into and become part of the bacterial chromosome. When the phage DNA is excised from the chromosome, it occasionally brings along a piece of the bacterial chromosome DNA that is connected to the phage DNA. The bacterial DNA is replicated along with the phage DNA and packaged with the phage genome into the new capsids. Later the newly produced phage will be released from the cell and go on to infect other bacteria, continuing the transduction process.

Transduction is an infrequent event in nature, but the potential for transduction is great for viruses that integrate into host cell genomes. In some cases, entire genes are transduced. Pathogens include the *Diphtheria* bacterium, which harbors a bacteriophage that produces a highly destructive toxin protein, and *Salmonella* bacteria, which also produce toxins that can cause serious food-borne infections.

The biological processes of transformation, conjugation, and transduction permit bacteria to naturally acquire new DNA sequences and assume new genetic characteristics. In the 1950s, molecular biologists first began using the word "recombined" to refer to genetically altered bacteria containing foreign DNA. Gradually, the term **recombinant DNA** crept into the scientific lexicon and refers to a DNA molecule containing foreign DNA sequences linked to host DNA sequences. Scientists knew that recombination between the DNA molecules from different species occurs in nature, and they began to wonder whether they could perform similar DNA recombination experiments in the laboratory.

When viruses infect bacteria, they sometimes bring with them DNA from previously infected bacteria along with their own DNA. This means of DNA transfer between bacterial species is known as transduction.

**Box 4.1 Superbugs**

**The Rise of Antibiotic Resistance**

When antibiotics were first introduced into medical practice in the 1930s and 1940s, they were a real miracle. Suddenly it became possible to cure previously incurable widespread diseases caused by pathogenic bacteria. In the early 1900s, tuberculosis (TB), called "consumption" at the time, was responsible for one of every four deaths in England. Because tuberculosis is infectious, people with the disease were isolated in medical facilities to help curb its spread; however, fully one-half the people who entered a sanatorium died from TB.

The horrible toll taken by TB changed dramatically in the Western world with the advent of antibiotics and a vaccine against tuberculosis for children; tuberculosis was no

## Box 4.1   Continued

longer a public menace. But in Africa, TB is still a killer, in part because of scarcer medical resources and higher rates of HIV/AIDS, which renders people more susceptible to TB. Additionally, over the past two decades many pathogenic microbes, including some that cause TB, have become increasingly resistant to antibiotic treatment. We are potentially facing a return to the days when bacterial infections were not easily cured.

Antibiotics kill bacteria by disrupting key cellular processes that are required for the bacteria to survive and reproduce, such as DNA and RNA synthesis, protein synthesis, enzymatic systems, and cell wall maintenance. Importantly, antibiotics are specific for bacteria and do not affect viruses or the processes of eukaryotic cells.

Bacteria can acquire specific mutations that render them resistant to the action of antibiotics. At a rate of about 1 in 1 billion, bacteria experience a chance series of mutations that lead to resistance. This can take several forms, the most common being the modification of an existing bacterial enzyme so that it alters the structure of the antibiotic, rendering it harmless. Other mutations change existing molecular pumps so they eject the antibiotic from the cell, or modify enzymes in the pathways under attack (for example, the enzymes involved in DNA synthesis) that make them resistant to the antibiotic. Once a bacterium is resistant to an antibiotic, it can spread this ability by transferring the appropriate DNA through the processes of transduction, transformation, or conjugation, as well as by cell division.

Recently, large increases in antibiotic resistance exceed the frequency accounted for by the bacterial mutation rate alone. The widespread addition of antibiotics to feed for healthy livestock as well as the tendency for the public to demand antibiotic medications to treat viral illnesses (antibiotics do not treat virus infections) have encouraged the widespread growth of bacteria that are resistant to more than one antibiotic, known informally as "superbugs." Ongoing efforts to solve this problem include restrictions on the routine use of antibiotics, reductions of antibiotics in some livestock feed, and improved public education about the correct use of antibiotics, among others. The race is on, and we must not let the superbugs win.

## Restriction Enzymes Enable Accurate and Controlled Cutting of DNA

In the 1950s, Salvador Luria and his colleagues observed that *E. coli* bacteria could somehow resist infection by bacteriophages by "restricting" (i.e., inhibiting) replication of the invading phage. By 1962, Werner Arber and his research group had found an enzyme system in bacteria that could inhibit phage replication by cleaving the bacteriophage DNA once it entered the cell. The team isolated the DNA-cleaving enzyme from *E. coli* cells and called it an **endonuclease**, because this enzyme cleaves nucleic acids. Endonucleases cut the DNA from within the DNA molecule, while **exonucleases** digest the ends of the DNA molecule. In bacteriophage infections, the *E. coli* endonuclease cleaves the invading phage DNA but does not cut the host bacterial DNA. This is the "restriction" part of what we now know as the **bacterial restriction and modification system**, which protects the bacteria against bacteriophage infection. The endonuclease enzyme can distinguish between the invading phage DNA and the host DNA because the bacterial chromosome is modified during replication by adding methyl ($CH_3$-) groups to specific DNA bases. This methyl "tag" identifies the methylated DNA as the native chromosome and protects it from cleavage by the enzymes. The phage DNA, however, is unmethylated and is a target for the endonuclease to "restrict" or cut the phage DNA to prevent replication of the phage.

> Endonuclease enzymes that occur naturally and function in the restriction and modification system in bacteria and quickly became known as restriction enzymes. These special enzymes play an important role in microbial warfare, but they were also essential for the early development of the infant science of recombinant DNA technology.

In 1970, Hamilton Smith and his colleagues isolated a new restriction enzyme called *Hin*dIII from the bacterium *Haemophilus influenzae*. Daniel Nathans used the *Hin*dIII restriction enzyme to cut viral DNA isolated from simian virus capsids (SV40, Figure 4.8). SV40 is an animal virus with a small circular double-stranded DNA genome, which is quite a bit easier to manipulate and study than the much larger bacterial chromosomes. Smith discovered that the *Hin*dIII restriction enzyme cut each SV40 DNA genome at only one site, even when an excess of enzyme was present. The scientists were amazed to learn that the *Hin*dIII enzyme cut the SV40 DNA in exactly the same location along the DNA in every SV40 DNA molecule that was exposed to the enzyme (Figure 4.9). For their work, Arber, Smith, and Nathans were awarded the Nobel Prize in physiology or medicine in 1978 (Figure 4.10).

As scientists gained more hands-on experience with restriction enzymes in the lab, they learned that restriction enzymes could cut any DNA molecule on demand, regardless of the source of the DNA (except when the DNA was protected by chemical modification). For example, the *Hin*dIII restriction enzyme will readily cleave any DNA molecule that contains its specific *Hin*dIII recognition DNA sequence, regardless of the source of the DNA molecule: virus, plant, animal, or human. It is easy to appreciate why restriction enzymes quickly became essential tools for any type

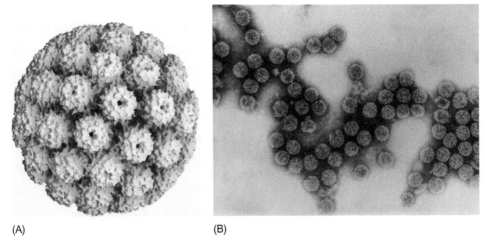

(A) (B)

**FIGURE 4.8** SV40, an animal virus (A) The structure of the SV40 polyoma virus with a small, double-stranded circular DNA genome that infects animal cells. (B) Electron micrograph of SV40 capsids.

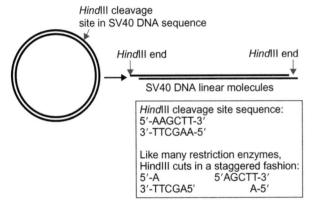

**FIGURE 4.9** Circular SV40 genome is cut by HindIII. The HindIII restriction enzyme cuts the SV40 circular DNA at a specific DNA sequence, which happens to occur only once in SV40 DNA. This converts the circular DNA into a linear double-stranded DNA molecule. The SV40 genome contains about 5000 base pairs (5 kb) (circular or linear).

of gene research and for the emerging field of biotechnology. At first, individual research groups had to isolate their own restriction enzymes from bacteria if they wanted to cut DNA in an upcoming experiment. This was often easier said than done, however, as it is necessary to grow a myriad of different bacterial species to obtain access to the required enzymes. Many microorganisms are notoriously difficult to grow in the lab, often requiring special conditions such as particular additives to the media or extreme temperatures. In response to the increasing demand, it was not long before biotechnology companies began to sell dozens of restriction enzymes for research applications, allowing scientists to plan experiments based on restriction enzymes they could purchase.

Scientists in the late 1990s could choose from thousands of commercially available highly purified

restriction enzymes for use in DNA technology. To avoid confusion about the identity of an enzyme, its cleavage site, or its source organism, restriction enzymes are named according to a system that starts with the name of the bacterium from which they were isolated. The first letter of the enzyme name is the first letter of the bacterium's genus name (in italics), followed by the first two letters of its species (in italics), then a letter representing the bacterial strain, and finally a Roman numeral that signifies the order in which this enzyme was discovered (in organisms that have more than one restriction enzyme). For example, the *Eco*RI restriction enzyme was named as follows:

| | | |
|---|---|---|
| *E* | **E**scherichia | (genus) (in italics) (capitalized) |
| *co* | **co**li | (species) (in italics) (not capitalized) |
| R | **R**Y13 | (strain) |
| I | First one identified | Order the enzyme was identified in bacterium (Roman numerals) |

Each restriction enzyme cuts DNA only at a specific base sequence, called the **recognition sequence** (Table 4.1), which can range from 4 bp to rare enzymes that recognize sites of 16 or 20 base pairs. Any given recognition sequence, especially a longer one, is potentially very rare in a genome of millions or billions of DNA base pairs. So how do enzymes find their recognition sequences? Restriction enzymes adopt specific 3D shapes that progress along nonspecific sequences in a DNA helix, "scanning" until reaching their specific recognition sequence. Most restriction enzymes recognize and cut at a **palindrome** DNA sequence with twofold rotational symmetry. The recognition site has the same base sequence on the top strand (read 5′ to 3′,

(A)                                          (B)                                          (C)

**FIGURE 4.10**   Scientists who won the Nobel prize for discovering restriction enzymes. (A) Werner Arber, (B) Hamilton Smith, and (C) Daniel Nathans, winners of the 1978 Nobel Prize in physiology or medicine for their work with restriction enzymes.

left to right) of the DNA and on the bottom strand (read 5′ to 3′, right to left) of the DNA. In the English language, a palindrome is when the letters of a sentence spell the same words whether they are read forward or backward; for example, "Madam, I'm Adam" or the author's current favorite, "Sit on a potato pan, Otis." Note that in DNA palindromes, both strands of the DNA must be read to reveal the palindrome (see Table 4.1).

The BamHI restriction enzyme attaches to the DNA helix as a dimer (two BamHI proteins bind together) and scans the DNA helix looking for the recognition sequence, 5′-GGATCC-3′ (Figure 4.11A). BamHI cuts the DNA backbone at the same point along the recognition sequence on each strand, but the locations of the cuts are staggered on the DNA helix, leaving overhanging, single-stranded ends after the DNA is cut (Figure 4.11B). *Bam*HI screens the sequences while loosely bound to the DNA helix until the enzyme finds the recognition sequence, then the proteins change conformation so they fit tightly into the DNA helix, which stimulates the enzyme to cut the DNA (Figure 4.12).

Restriction enzymes cut DNA at specific base sequences and as a result produce one of three different types of ends on the DNA: **blunt, 5′ cohesive ends**, or **3′ cohesive ends** (Figure 4.13). The cohesive DNA ends generated by restriction enzymes are also called sticky, staggered, overlapping, and overhanging DNA ends. The single-stranded DNA ends can only form base pairs with the complementary single-stranded ends generated by the same or a different enzyme (Figure 4.14).

> Restriction enzymes cut DNA at specific sequences, regardless of the source of the DNA, and often cut at staggered sites along the double helix, leaving short, complementary "sticky ends" on the DNA.

The use of restriction enzymes as tools to cut and then connect DNA fragments from different organisms is a central approach used in genetic engineering. Restriction enzymes are precise molecular tools that scientists use to cut any kind of DNA regardless of source. When the goal of the experiment is to connect two DNA fragments together, the source of the DNA (bacterium, bird, or buttercup) is not relevant as long as the sticky ends of the DNA fragments are complementary, and can base pair together to connect the DNA fragments together. It is important to note that not all restriction enzymes cut DNA and leave sticky ends. Many restriction enzymes cut the DNA backbone at two sites directly across from each other on the helix and generate blunt-ended DNA fragments that lack single strands.

## Sealing the Deal: DNA Ligases Connect the DNA Backbones

Bringing together the complementary sticky ends on two DNA fragments does not of itself forge a covalent bond between the DNA ends (Figure 4.14). After base pairing occurs, there is a gap or "nick" remaining in the DNA backbones so that the molecules are held together by complementary base pairing. The hydrogen

**TABLE 4.1** Some Restriction Enzymes, Their Sources, and Their Recognition Sites

| Enzyme | Microbial Source | Sequence* |
|--------|------------------|-----------|
| AluI | Arthrobacter luteus | 5'—A—G↓C—T—3'<br>3'—T—C↑G—A—5' |
| BamHI | Bacillus amyloliquefaciens H | 5'—G↓G—A—T—C—C—3'<br>3'—C—C—T—A—G↑G—5' |
| EcoRI | Escherichia coli | 5'—G↓A—A—T—T—C—3'<br>3'—C—T—T—A—A↑G—5' |
| EcoRII | Escherichia coli | 5'↓C—C—T—G—G—3'<br>3'—G—G—A—C—C↑5' |
| HaeIII | Haemophilus aegyptius | 5'—G—G↓C—C—3'<br>3'—C—C↑G—G—5' |
| HindIII | Haemophilus influenzae b | 5'—A↓A—G—C—T—T—3'<br>3'—T—T—C—G—A↑A—5' |
| PstI | Providencia stuartii | 5'—C—T—G—C—A↓G—3'<br>3'—G↑A—C—G—T—C—5' |
| SalI | Streptomyces albus | 5'—G↓T—C—G—A—C—3'<br>3'—C—A—G—C—T↑G—5' |

infected with T4 bacteriophage. The T4 DNA ligase enzyme connects (ligates) the backbones of DNA strands together by forming a covalent **phosphodiester bond** between the phosphate group at the 5′ carbon of one nucleotide and the 3′ carbon of the next nucleotide (Figure 4.15 and Figure 4.16). This covalent bond seals the DNA backbones together and creates a longer DNA molecule. DNA ligase can also link together blunt-ended DNA fragments, but it is much more difficult to join them together because they lack the stabilizing influence of the complementary sticky ends on both DNA fragments.

> DNA molecules with sticky ends base pair with each other temporarily; the DNA ligase enzymes connect the two DNA molecules together end to end by forming covalent backbone bonds between them.

When DNA molecules are cut by restriction enzymes, the DNA fragments produced by the reaction are often analyzed by separating the DNA fragments by **gel electrophoresis**. In this common technique, the DNA sample is loaded into a well (a depression) located on one end of the solid gel. When an electrical current is applied to the gel, the DNA fragments, which have a negative charge, migrate through the gel toward the positive electrode. The shorter DNA fragments migrate more quickly through the gel than the longer DNA fragments. The DNA in the gel is stained with ethidium bromide and visualized under ultraviolet light (Figure 4.11 B).

## Plasmids Are Used as DNA Carriers

In the early 1970s, Paul Berg and his team at Stanford University began to study enzymes that cut *E. coli* chromosome DNA, with the goal of using restriction enzymes to cut and connect DNA from different sources. At the time Berg's group was using a restriction enzyme that leaves blunt ends on the DNA fragment, making the ends difficult to ligate together (at the time, enzymes that created sticky ends were as yet unknown). They overcame this problem by adding artificial sticky ends to the DNA fragments. To one DNA fragment they added a tail of all "A" bases, and they added a tail of all "T" bases on to the end of the other DNA fragment. When the tailed DNA fragments are mixed together, the single-stranded A and T tails formed A-T base pairs, permitting the ends of the DNA fragments to overlap and base pair. Connecting together these two DNA fragments from different organisms produced the first **recombinant DNA molecule**. Paul Berg shared the 1980 Nobel Prize in chemistry for this

bonds between the base pairs together are not strong enough to stabilize the overlapping DNA ends indefinitely at physiological (body) temperatures. This problem is solved using another enzyme isolated from cells, called **DNA ligase**, which catalyzes the formation of the strong covalent chemical bonds in the backbones of the two DNA helices.

In the lab, scientists typically use T4 DNA ligase, which is made in large amounts in bacterial cells

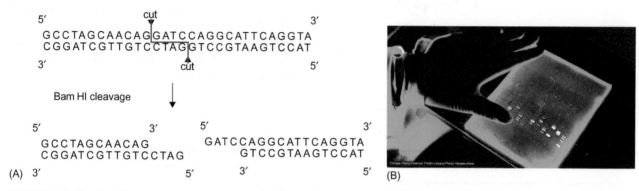

BamHI recognition sequence:
```
5'-GGATCC-3'
3'-CCTAGG-5'
```

BamHI makes a staggered cut across the DNA backbones:

```
     5'                cut                              3'
        GCCTAGCAACAG GATCCAGGCATTCAGGTA
        CGGATCGTTGTCCTAG GTCCGTAAGTCCAT
     3'                cut                              5'
```

Bam HI cleavage

```
     5'              3'    5'                              3'
        GCCTAGCAACAG          GATCCAGGCATTCAGGTA
        CGGATCGTTGTCCTAG          GTCCGTAAGTCCAT
(A)  3'              5'    3'                              5'   (B)
```

**FIGURE 4.11**  The BamHI enzyme recognizes and cuts double-stranded DNA at GGATCC. BamHI, like most restriction enzymes, recognizes and cuts DNA at a site with two-fold rotational symmetry—that is, the base sequence reads the same if read left to right on the top strand of DNA, and right to left on the bottom strand. The BamHI cleavage site is shown in red, and cutting generates sticky ends on the resulting the DNA fragments, also known as cohesive or complementary ends.

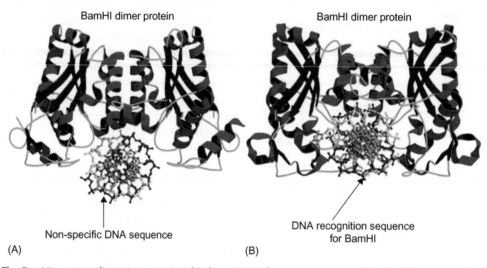

BamHI dimer protein                                    BamHI dimer protein

Non-specific DNA sequence                              DNA recognition sequence
                                                       for BamHI
        (A)                                                    (B)

**FIGURE 4.12**  The BamHI enzyme dimer (two proteins) binds to DNA. The BamHI enzyme dimer binds to nonspecific DNA sequences while scanning the genome for occurrences of the specific BamHI recognition sequence. (A) DNA is shown end-on (looking down the double helix). The BamHI dimer binds to the DNA double helix but cannot cut the DNA without binding to the BamHI specific recognition DNA sequence (B) When BamHI dimer finds the recognition sequence, it binds more closely with the double helix and catalyzes the DNA cleavage event.

amazing feat, signaling the use of plasmids as vectors to carry recombinant DNA molecules.

Meanwhile, for the first time, Herbert Boyer's research group isolated the *Eco*RI restriction enzyme from *E. coli* bacteria and discovered that when *Eco*RI cuts DNA it produces 5' sticky DNA ends. At the same time in Stanley Cohen's laboratory at Stanford University,

scientists had identified small circular double-stranded DNA molecules called **plasmids** (Figure 4.17), which are found in bacteria but not in most eukaryotic organisms (we now know that some eukaryotic cells do carry plasmids). In the cell, plasmid circles adopt a compact, twisted shape (see Figure 4.17). Plasmids contain low, variable numbers of genes (compared to a few

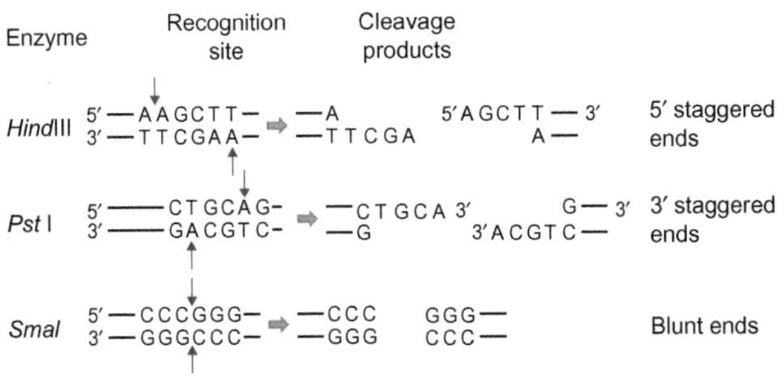

| Enzyme | Recognition site | Cleavage products | | |
|--------|------------------|-------------------|---|---|
| *Hind*III | 5'—AAGCTT—<br>3'—TTCGAA— | —A<br>—TTCGA | 5'AGCTT—3'<br>A— | 5' staggered ends |
| *Pst*I | 5'——CTGCAG-<br>3'——GACGTC- | —CTGCA 3'<br>—G | G—3'<br>3'ACGTC— | 3' staggered ends |
| *Sma*I | 5'—CCCGGG—<br>3'—GGGCCC— | —CCC<br>—GGG | GGG—<br>CCC— | Blunt ends |

**FIGURE 4.13** Cleavage by restriction enzymes create three different types of DNA ends. Different restriction enzymes cut DNA at different recognition DNA sequences. Cleavage generates one of three types of DNA ends: 5' staggered ends, blunt ends, or 3' staggered ends.

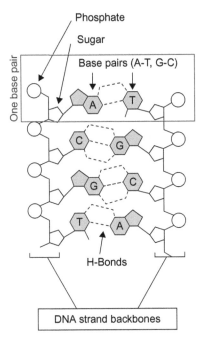

**FIGURE 4.14** Base pairing between DNA bases holds complementary DNA strands together. Hydrogen bonds (H-bonds) form base pairs: A base pairs with T and G base pairs with C as shown. Note that A-T base pairs contain 2 H-bonds and G-C base pairs contain 3 H-bonds.

thousand genes in most bacterial chromosomes) and usually are not essential for bacterial growth; they can be lost without significant harm to the bacterium. Plasmids replicate autonomously (separately from the bacterial chromosome) and do not integrate into the bacterial chromosome (Figure 4.18).

In 1972, Cohen constructed a new DNA plasmid (Figure 4.20) with several useful features, (1) an *Eco*RI recognition site at a single location on the plasmid DNA circle, (2) an **origin of replication**, which allows the plasmid to replicate in bacteria, and (3) a tetracycline antibiotic resistance gene, so that the bacteria carrying this plasmid would be resistant to the effects of the antibiotic tetracycline and survive, whereas bacteria lacking this plasmid would be killed by the antibiotic. In the early days of this emerging technology, scientists named their DNA inventions after themselves, hence this plasmid is called pSC101 ("SC" for Stanley Cohen).

Cohen studied the best way to insert the pSC101 plasmid DNA into the host bacterial cells using an artificial form of the natural process of **transformation**. Cohen grew the bacterial cells to a specific cell density and treated the cells with very cold calcium buffer to make them **competent** to receive foreign DNA. Then the temperature was abruptly raised so that the bacterial cells and the recombinant vector DNA were submitted to a sudden, temporary increase in temperature, or "heat shock," at 42°C for 45 seconds. The exact mechanism behind heat shock is not clear, but of the few bacterial cells that take up a plasmid, almost always the bacteria will pick up only one plasmid DNA. This is a fundamental fact that permits the process of DNA cloning to create large numbers of identical copies of the same plasmid. The conditions for bacterial transformation are surprisingly strict, and little variation is tolerated if transformation is to be achieved (Figure 4.20).

Once inside the cell, the plasmid replicates and forms hundreds of identical copies of the plasmid. Whatever genes are carried on the plasmid DNA, whether of bacterial or foreign origin, are replicated as part of the plasmid. As the bacterium harboring the plasmid rapidly multiplies, each new cell inherits a number of plasmids. Before long, a single bacterium has produced a culture of millions of descendants, each carrying identical copies of the plasmid DNA. Such a population of cells derived from a single parent cell is called a **clone** (Figure 4.20). The cloned population of cells together carries millions of copies of identical plasmids. It did not take too much imagination to realize that plasmids could become the ideal DNA carriers, or **vectors**, for transporting foreign genes. The stage was set for the recombinant DNA experiments that would not only change science forever, but would also alter the course of history.

Plasmid DNA molecules are closed double-stranded circles that can be easily inserted into bacterial cells. Plasmids replicate independently of the bacterial chromosome and make numerous identical copies of the plasmid DNA in the cell.

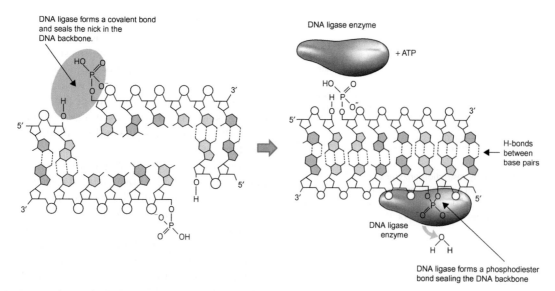

**FIGURE 4.15**   DNA ligase closes the gaps in the DNA backbones. (A) When two DNA fragments are brought together, a covalent bond seals the backbones and prevents the DNA fragments from coming apart easily. (B) The DNA ligase enzyme uses the phosphate on the 5' end of a DNA fragment and the hydroxyl on the 3' end of the other DNA fragment to provide the energy to forge the new phosphodiester bond. The result is a single DNA double-helix molecule.

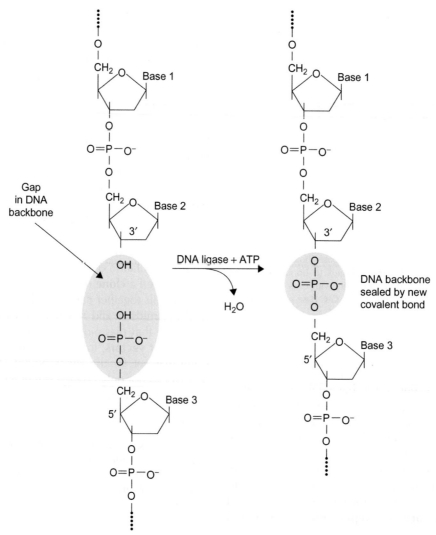

**FIGURE 4.16**   Ligase enzymes seal the phosphodiester backbone. After the single-stranded complementary DNA ends base pair together, ligase enzymes seal the nick that remains in the DNA backbone. Once the nicks in the DNA backbone are sealed, these particular DNA fragments will remain connected until they are cut again with EcoRI. (Other enzymes would cut at different recognition sites.)

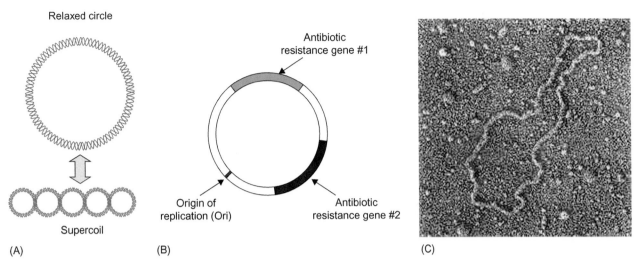

FIGURE 4.17 Plasmids are double-stranded DNA circles. (A) The relaxed circle form of a double-stranded DNA plasmid (upper), which, in a cell, is usually found in a supercoiled form (lower). (B) The map of a simple DNA plasmid containing genes coding for antibiotic resistance and an origin of replication site (Ori) so the plasmid can replicate independently of the host bacterial chromosome. (C) An electron micrograph of a relaxed circular DNA plasmid.

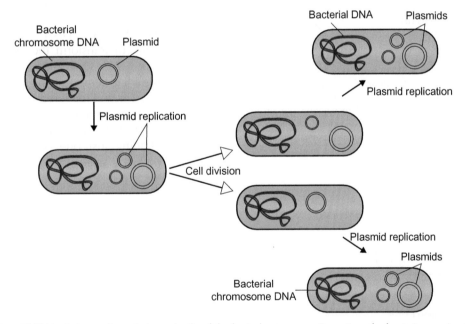

FIGURE 4.18 Plasmid DNA circles replicate independently of the host chromosome. Every time the bacteria carrying the plasmid divide, plasmid copies are transmitted to the offspring cells. Plasmids are often used in recombinant DNA research to transfer genes between cells.

## THE ADVENT OF RECOMBINANT DNA EXPERIMENTS

### Boyer and Cohen Pioneer the Tools to Transfer Genes

One could say that the field of DNA technology came into being over pastrami sandwiches at a deli in Waikiki Beach in 1972. It happened that Herbert Boyer was attending a scientific conference in Hawaii where he gave a talk about the new *Eco*RI restriction enzyme. Fortunately, Stanley Cohen was in the audience. After the meeting session concluded, Cohen invited Boyer to lunch to talk about a possible scientific collaboration. (Both scientists are shown in Figure 4.21.) As it turns out, this lunch meeting would send DNA technology into a new era.

Cohen had been successfully experimenting with plasmids, but he was having technical difficulties cleaving the plasmid DNA, and Boyer's new restriction enzyme, *Eco*RI, seemed like the ideal solution. Boyer agreed, and the bargain was struck. Boyer and Cohen joined forces and used Boyer's *Eco*RI enzyme to cut two different plasmids and then try to ligate the two plasmids together to form a single, much larger plasmid (Figure 4.22). If successful, they would try to produce a recombinant DNA molecule by inserting

foreign DNA from a different species into a bacterial plasmid.

Boyer and Cohen performed these pioneering recombinant DNA experiments in 1973, 20 years after Watson and Crick had published their historic article on the structure of the DNA double helix. In their first experiments, Boyer and Cohen successfully cut and ligated together two different plasmids, pSC101 (containing the tetracycline resistance gene) and plasmid pSC102 (containing a gene for resistance to another antibiotic, kanamycin) and then used the ligated DNA to transform *E. coli* host cells that are sensitive to antibiotics. The role of antibiotic resistance or other methods to select for cells that carry the desired recombinant DNA molecules was a crucial strategy for Boyer and Cohen, and it continues to be an important part of current recombinant DNA experiments. At the start of the experiment, the recipient bacterial cells were sensitive to both antibiotics. If the cells were spread on solid media containing either tetracycline or kanamycin, the bacteria could not survive and grow.

Boyer and Cohen cut both plasmids pSC101 and pSC102 with *Eco*RI, mixed the two cut plasmids together, treated them with DNA ligase, and then used the plasmid mix to transform the antibiotic-sensitive bacteria (Figure 4.23). Once inside the bacteria, the antibiotic resistance genes on the recombinant plasmids made proteins that rendered the bacteria resistant to both tetracycline and kanamycin, therefore these cells could grow in the presence of both antibiotics. Bacteria containing just one parent plasmid, either pSC101 or pSC102, were resistant to one antibiotic,

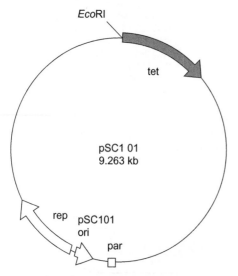

**FIGURE 4.19** Plasmid pSC101. Plasmid pSC101 was the first plasmid constructed with recombinant DNA experiments in mind. Plasmid pSC101 contains a unique *Eco*RI recognition site on the double-stranded plasmid DNA, an origin of replication allows the plasmid to replicate in bacteria, and a gene that confers tetracycline resistance. The modern version of this plasmid has many additional restriction sites and other useful features.

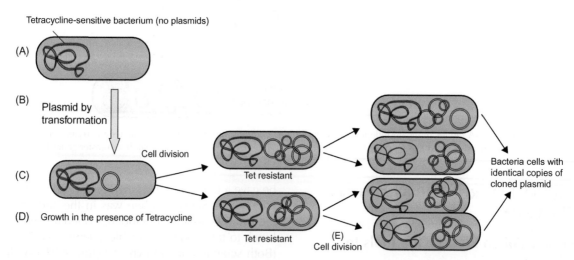

**FIGURE 4.20** Transformation is used to introduce DNA plasmids into bacteria to make a clone. (A) Recipient bacteria without plasmids are sensitive to the antibiotic tetracycline. (B) Plasmid DNA with tetracycline resistance gene are used to transform the bacteria; each recipient cell receives no more than one plasmid, and most bacteria get none. (D) After treatment with the antibiotic, the surviving cells are tetracycline-resistant. (E) The transformed bacteria that are resistant to tetracycline divide and make more bacteria, all of which contain identical copies of the plasmid DNA. The culture is always grown on "selective medium" containing tetracyclin to prevent the growth of any bacteria without a plasmid. The plasmids replicate independently of the bacteria chromosome and make many identical plasmid copies in each cell.

but not to the other. Thus, the cells that were transformed with the recombinant plasmids were selected from the ones with parent plasmids (or no plasmids). This is a basic genetic selection method used in many DNA technology experiments, including those involving eukaryotic host cells. Eukaryotic cells require different vectors and selection methods using the appropriate drugs for use in animal cells.

Boyer and Cohen collaborated to bring together restriction enzymes and plasmids in order to create recombinant DNA molecules in the laboratory. They combined two plasmids carrying different antibiotic resistance genes and screened for bacteria that were resistant to both antibiotics.

(A)                 (B)                 (C)

**FIGURE 4.21** The first gene engineers. (A) Herbert Boyer. (B) Stanley Cohen. (C) Boyer and Cohen.

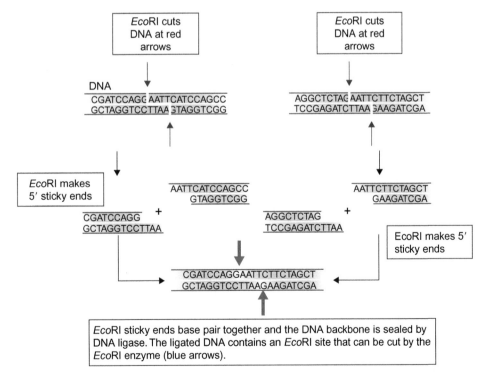

**FIGURE 4.22** Using *Eco*RI enzyme to combine two different DNA plasmids. The restriction enzyme EcoRI is used to cleave the DNA of two different plasmids (blue and yellow) as indicated by red arrows. Each plasmid has a single EcoRI cleavage site so both plasmids become converted into linear molecules by cutting with EcoRI. The Eco RI enzyme cuts DNA at a specific DNA sequence and produces fragments with sticky DNA ends. The cut plasmids are mixed together so that the sticky ends generated by cleavage with EcoRI can join, forming much longer DNA molecules. All the necessary backbone bonds are catalyzed by DNA ligase.

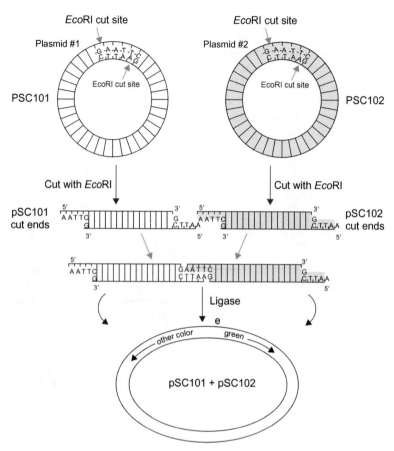

**FIGURE 4.23** Boyer and Cohen cut and recombine two different plasmids. Two plasmids, pSC101 and pSC102, each containing a different antibiotic resistance gene, are cut with EcoRI, creating the same sticky ends on the DNA fragments. The sticky ends base pair together and the DNA ligase enzyme is added to make backbone bonds, sealing the plasmids together.

In a second set of key experiments, Boyer and Cohen decided to clone a gene from the African clawed frog (*Xenopus laevis*) into the pSC101 plasmid DNA vector (Figure 4.24). They cut both the frog DNA and the pSC101 plasmid DNA with EcoRI and mixed the cut DNA fragments together, allowing the EcoRI-cut complementary single-stranded ends to base pair with each other. Once the DNA ligase enzyme had sealed the DNA backbones, the new recombinant plasmid containing the frog DNA was formed and could replicate in the bacterial cells. The researchers called the recombinant plasmid, made up of two different DNAs, a **chimera**, named for the creature of Greek mythology, a combination of lion-goat-serpent.

In the next experiment, the researchers decided to see whether the cells carrying the recombinant plasmid could successfully express the foreign frog gene and direct the synthesis of the frog protein in the bacterial cells. Cohen grew a large number of *E. coli* cells containing the recombinant plasmids and analyzed the proteins made in the bacteria. He found that the *E. coli* cells did produce the frog protein that is normally made

only in frog cells. They had successfully moved the *X. laevis* frog gene into a completely different organism and had successfully expressed a frog protein in bacterial cells. The scientific media and the national press trumpeted that the Boyer and Cohen experiments had breached a theoretical barrier separating biological species and launched the era of recombinant DNA technology. As one observer noted, "Biotechnology used to be BBC (before Boyer-Cohen); now it is ABC (after Boyer-Cohen)." Many years later, in 1996, Boyer and Cohen were honored with the prestigious Lemelson-MIT prize and shared the $500,000 award.

Those in the emerging science of molecular biology were quick to grasp the implications of recombinant DNA technology, and many scientists began to perform their own gene manipulation experiments. Within weeks, scientists were investigating other gene transfer and cloning experiments in an effort to transcend other species barriers. In one experiment, genes from the common skin bacterium, *Staphylococcus aureus*, were transferred into *E. coli* cells. Other scientists attempted to isolate human genes and insert the human DNA into

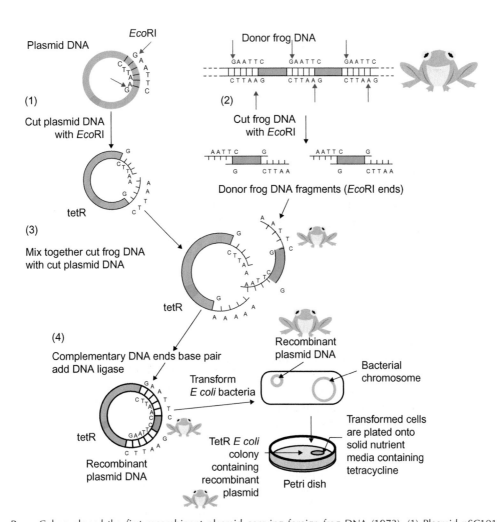

**FIGURE 4.24** Boyer-Cohen cloned the first recombinant plasmid carrying foreign frog DNA (1973). (1) Plasmid pSC101 containing the tetracycline resistance gene (tetR) was cut with EcoRI (red arrows). (2) Donor DNA from the frog genome was also cut with EcoRI. (Note that EcoRI acts at the same recognition sites in the plasmid vector and the donor frog DNA.) (3) Fragments of donor frog DNA are combined with the cut plasmid DNA. (4) Complementary base pairing takes place between the EcoRI-cut DNA ends. DNA ligase is added to seal the DNA backbone to form a recombinant DNA molecule: bacterial plasmid and frog DNA. (5) The recombinant plasmid DNA is introduced into tetracycline-sensitive E. coli bacterial cells by transformation, and subsequently the cells are grown in medium containing tetracycline so that only the transformed cells survive. (6) The bacteria carrying the plasmids divide to make many identical clones. (7) The frog DNA in the plasmid causes the production of frog proteins in the bacterial cells.

plasmids to grow in bacteria. The scientific community worldwide began to speculate on the extremely powerful implications of the first experiments in recombinant DNA technology. People predicted that genes could be inserted into living cells to relieve genetic deficiencies and that rare pharmaceutical proteins could be cheaply produced in mass quantities. Many still dream of inexpensive bioenergy sources and infants free of birth defects. There seems to be no limit to what the new technology might accomplish.

## Initial Reactions: How Safe Is DNA Technology?

Despite the widespread enthusiasm for the new recombinant DNA technology by many scientists, others warned of potentially dangerous consequences and suggested a more cautious approach to recombinant DNA research. Especially alarming to many was the proposal by Paul Berg to insert genes from a cancer virus into bacterial cells. Colleagues emphatically pointed out that if the E. coli cells carrying the cancer virus DNA escaped from the laboratory, the engineered bacteria could easily populate the human intestine (where E. coli normally thrives) and express the cancer genes in a human host. Berg responded to the concerns of his scientific peers and canceled the experiment. But the safety issues extended far beyond the work in Berg's lab, and through the next few months, groups of scientists began to discuss the safety issues surrounding the use of DNA technology and recombinant DNA.

**FIGURE 4.25** Scientists at the Asilomar Conference of 1975. Addressing the concerns of their own community as well as others, scientists drafted principles for containment facilities for DNA technology experiments.

In 1974, in an unprecedented move, Paul Berg and nine other respected scientists wrote a letter that appeared simultaneously in three prestigious science research journals *Science*, *Nature*, and *Proceedings of the National Academy of Sciences*. At the same time, a news conference was held to coincide with the publication of the letter, which clearly pointed out the potential danger:

> Recent advances in techniques for the isolation and rejoining of segments of DNA now permit construction of biologically active recombinant DNA molecules *in vitro*. Although such experiments are likely to facilitate the solution of important theoretical and practical biological problems, they would also result in the creation of novel types of DNA elements whose biological properties cannot be completely predicted.... There is a serious concern that some of these DNA molecules could prove biologically hazardous.

The letter written by Berg and his colleagues asked scientists worldwide to institute a voluntary moratorium on certain types of DNA experiments until after an international conference could consider necessary safeguards. This was the first time that scientists had ever voluntarily restricted research that had not yet proven dangerous. Although the recommendations did not carry the force of law, scientists around the world subscribed to the request. In February 1975, a group of 139 leading researchers from 17 nations assembled for a four-day meeting held at Asilomar, in Pacific Grove, California. The attendees developed guidelines and recommendations for scientists conducting experiments using recombinant DNA technology (Figure 4.25). One goal of this meeting was to reassure an uneasy public that the host bacteria used for recombinant DNA experiments were specifically bred and genetically "disarmed" so they could not survive outside the laboratory and would not grow in the human intestines. Scientists also wrestled with the actual danger

posed by recombinant DNA work, as there was no evidence that either supported or rejected the hazardous nature of DNA work. The scientists opted for caution and set about policing their own research. Different recombinant DNA experiments were classified as low risk, medium risk, or high risk, and very risky experiments (such as those previously proposed by Berg) would not be conducted until better safety methods were developed. They discussed strategies to contain the DNA experiments, including physical and biological containment, and high priority was given to developing strains of bacteria unable to survive outside the laboratory.

The Asilomar recommendations led to the formation of the Recombinant DNA Advisory Committee of the National Institutes of Health (NIH), which soon issued stringent guidelines that paralleled the Asilomar recommendations. However, with additional years of research, it became clear to the scientists and public alike that recombinant DNA technology would not create a celebrated "Andromeda bug," and as a result the regulations were periodically revised and relaxed. The Recombinant DNA Advisory Committee at NIH continues to be the watchdog over DNA technology research.

> Scientists were alert to the potential dangers of gene experimentation and recombinant DNA technology. They organized an international effort to carefully monitor and regulate scientific activity in this field; the NIH Recombinant DNA Advisory Committee grew and has continued to regulate recombinant DNA research for about 35 years.

## Recombinant DNA Techniques Become Widely Applied

The successes in recombinant DNA technology gave rise to the discipline of biotechnology, which focuses on the commercial applications of molecular genetics. Biotechnology is a vast discipline in which the tools and concepts of molecular biology are used to solve a wide range of problems associated with pollution, food production, energy production, and synthesis of new medicines (Figure 4.26). Biochemists soon came to regard bacteria and other microbes as the chemical factories of the future and reasoned that they could be programmed with DNA genes to produce any number of genetically engineered products with industrial, economic, and medical significance.

During the 1970s, some of the promise of recombinant DNA technology was fulfilled, and numerous biotechnology companies began applying the techniques of DNA science to manufacture useful products. By 1980, one company had successfully harvested insulin

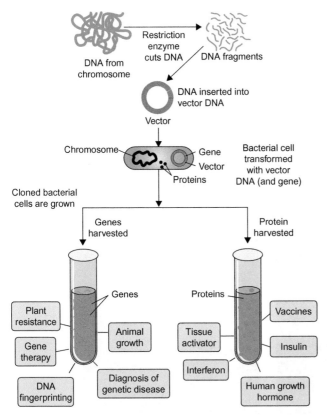

**FIGURE 4.26** The impact of DNA technology and gene cloning. The two major objectives of recombinant DNA technology were to clone genes and accomplish the large-scale production of proteins.

Today, of course, both optimism and concerns remain. Researchers have gone well beyond inserting foreign DNA into bacterial cells and DNA technologies have dramatically impacted many fields (Figure 4.26). For instance, DNA chips allow researchers to monitor the activation or suppression of thousands of genes simultaneously; ordinary people can trace genealogies using DNA; genetic tests can identify risks of some genetic diseases, allowing people to make informed choices about their future; sequencing of entire genomes has led to the birth of a new field, bioinformatics (see Chapter 7), and genome DNA sequencing is almost affordable for everyone. More than anything, the advent of recombinant DNA technologies gave humans the ability to exert control over the very molecules they are made of. The most optimistic scientists envision that humans will come to understand the human body and all living things. Hardly any facet of the human existence will remain untouched by DNA technology, and the social and political questions raised will require solutions on a global scale. What if DNA technology leads to gene tampering and efforts to "improve" the human race? It is clear that ethical and moral dilemmas will abound at every step of the genetic path along which we are all traveling.

In addition to the exciting possibilities offered by recombinant DNA science, new techniques and applications will always introduce some risk. Careful monitoring of DNA research is important to ensure public safety.

Discussions of these and numerous other topics form the basis for the remainder of this book. As we will see in the pages ahead, DNA technology has entered the mainstream of human life and has become one of the most eloquent applications of scientific research. We conclude the book with a chapter that discusses our current understanding of human genes and the human race.

from bacteria that carry human insulin genes. Other research groups constructed recombinant plasmids in bacteria that could produce human interferon (a virus inhibitor) and human growth hormone, as well as many other products such as vaccines and enzymes.

The 1980 and 1990s saw the development of many other products of DNA technology. Recombinant bacteria were constructed that could dissolve oil spills, dispose of toxic waste, and dissolve clogs in drains. Recombinant bacteria were constructed to produce medically relevant proteins such as the human clot-dissolving enzyme, urokinase, and a kidney hormone, erythropoietin. A new form of forensic science called DNA fingerprinting won acceptance in the court system, gene therapy trials in humans began, and the first cloned mammal, Dolly, made her debut. Hundreds of companies worldwide were working on the industrial applications of DNA technology, and many scientists spoke optimistically of a future in which fertilizers would be obsolete, plants could use microbial toxins to drive off pests, and crops would be cultivated without the danger of frost damage. At the same time, new concerns surfaced about the safety of biotechnology research and development, especially with regard to the use of genetically modified organisms in food.

## SUMMARY

Genetic alterations using DNA technology permit scientists to intervene directly in the fate of living organisms. DNA technology has been made possible in part by knowledge about microorganisms that result from decades of studying bacteria. Interest in DNA technology was heightened considerably with the discovery of restriction enzymes because these enzymes cut a DNA molecule at a target sequence regardless of the source of the DNA. Moreover, some restriction enzymes such as *Eco*RI leave cohesive DNA ends that scientists use to unite the DNA fragments from different sources to construct recombinant DNA molecules. The DNA

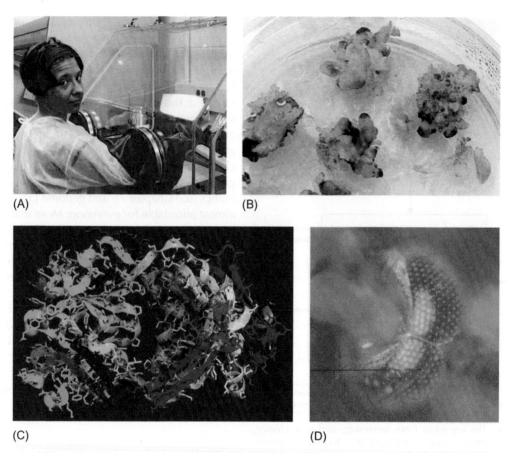

(A)                                             (B)

(C)                                             (D)

**FIGURE 4.27** Facets of DNA technology. (A) Viral vector production. While producing viral vectors for gene therapy, scientists are protected by the biocontainment insulator. (B) Agricultural biotechnology. Cloned sundew plants grew in a Petri dish on solid media that were started from the single cells of a parent plant (clones). (C) Biomedical biotechnology. Herceptin is a drug made by recombinant DNA technology that is designed to treat breast cancer. (D) Biomedical disease research. Transgenic mosquitoes can be used to fight malaria. These insect larvae carry a new gene that is expressed in their cells as visualized by the fluorescent green color.

ligase enzyme forges permanent bonds in the DNA backbone at the junctions between the cohesive ends of the DNA molecules.

Another key development was the use of plasmids as vectors to carry foreign DNA fragments. Plasmids are circular double-stranded DNA molecules that grow naturally in bacteria. In 1973, Stanley Cohen and Herbert Boyer inserted the DNA from a frog into a bacterial plasmid and then grew the recombinant plasmids in host bacteria. The bacteria replicated the recombinant plasmids and expressed the foreign frog protein in the cells, as well as the usual bacterial proteins. The era of recombinant DNA technology was launched.

Safety concerns about recombinant DNA science brought scientists together at the Asilomar conference to discuss and establish guidelines for conducting future experiments in recombinant DNA technology. The modern Recombinant DNA Advisory Committee, the governing body for DNA technology, is a result of the Asilomar conference. As suggested in the 1970s, recombinant DNA technology has had profound consequences on basic research as well as on the practical applications of DNA technology. Pharmaceutical products, agricultural advances, genetic testing, and forensic science are just a few fields that reflect the impact of DNA cloning technology as described in later chapters of this book.

## REVIEW

This chapter concentrated on the scientific discoveries of the 1960s and 1970s that formed the basis on which DNA technology was developed. To test your knowledge of the chapter's contents, consider the following review questions:

1. Describe some of the attributes of a bacterium and its DNA that make it a useful organism for studying DNA function.
2. What are viruses, and how are they useful to DNA technology experiments?
3. Explain the process of genetic recombination in bacteria and describe the relationship to recombinant DNA technology.

4. Discuss the sources and nomenclature of restriction enzymes, and explain how they are used in the laboratory to study DNA.

5. What are ligases? What functions do they perform? How are they used in DNA technology?

6. Describe some of the characteristics of plasmids, including their source, composition, method for insertion into bacteria, and value to DNA technology.

7. Summarize the experiments performed by Boyer and Cohen that set the foundations for the development of DNA technology.

8. Describe an example of how an experiment in DNA technology can be dangerous, and explain how scientists addressed safety in the 1970s.

9. What are the functions of the Recombinant DNA Advisory Committee of the National Institutes of Health?

10. List some of the possible uses of DNA technology for resolving human problems of a practical nature, and indicate some of the theoretical implications of DNA technology.

## ADDITIONAL READING

Alberts, B., et al., 2003. Essential Cell Biology, Second ed.. Garland, New York.

Brownlee, C. Restriction Enzymes and DNA Mapping: Altering the Fabric of Life, PNAS at 100: Classics of the Scientific Literature. <www.pnas.org/misc/classics3.shtml./>

Elmert-Dewitt, P., January 17, 1994. The Genetic Revolution. Time 143, 46–53.

Karp, G., 2007. Cell and Molecular Biology: Concepts and Experiments. John Wiley & Sons, New York.

Malacinsk, G.M., 2003. Essentials of Molecular Biology, fourth ed.. Jones & Bartlett, Sudbury, MA.

Miller, R.V., January 1998. Bacterial Gene Swapping in Nature. Scientific American 278, 66–71.

Pommerville, J.C., 2006. Alcamo's Fundamentals of Microbiology, eight ed.. Jones & Bartlett.

Raven, P.H., Johnson, G.B., Singer, S., Losos, J., 2004. Biology, seventh ed.. W. C. Brown/McGraw Hill, Dubuque, IA.

Yim, G. Attack Of The Superbugs: Antibiotic Resistance. In: The Science Creative Quarterly. www.scq.ubc.ca/attack-of-the-superbugs-antibiotic-resistance.

## WEB SITES

Dolan Learning Center, Cold Spring Harbor Laboratory, CSH, NY. http://www.dnalc.org, http://www.dnalc.org.

This is an excellent resource to learn more about basic DNA science and recombinant DNA technology

# Working with DNA

## Scientists Swap Genes in Bacteria

Associated Press, Updated: June 30, 2007

By Lauran Neergaard

Talk about identity theft: Scientists changed one species of bacteria into another by performing a complete gene swap.

It's a step in the quest to one day create artificial organisms, part of a bigger project to custom-design microbes that could produce cleaner fuels. But the way it was performed, dubbed a "genome transplant," has genetics specialists buzzing.

"This is equivalent to changing a Macintosh computer to a PC by inserting a new piece of software," declared genome-mapping pioneer J. Craig Venter, senior author of the new research published Thursday by the journal *Science*.

For years, scientists have moved single genes and even large chunks of DNA from one species to another. But Venter's team transplanted an entire genome, all of an organism's genes, from one bacterium into another in one fell swoop.

Scientists have been introducing foreign genes and fragments of DNA into bacteria to clone, study, and exploit for the good of humankind since the 1960s. This chapter outlines the strategies and methods that scientists use to achieve a desired end in the laboratory. The experiment described in the preceding article represents a step well beyond routine lab methods. But just as transferring a fragment of DNA between organisms was once very unusual and exotic, methods for transferring whole genomes will likely become routine in the near future.

To capture a genome and transfer it wholesale into a different organism, scientists at the J. Craig Venter Institute carefully extracted the entire DNA genome of an organism from closely related species of bacteria. They coaxed a tiny number of bacteria to pick up the new genome DNA using a chemical cocktail that encourages bacteria to merge their membranes, and a few rare bacteria lost their own genome in favor of the new genome. As yet, the genome transfer mechanism involved is not fully understood.

This experiment represents a novel and important step along the path to making synthetic life. The long-term goal of synthetic life is to create living cells from scratch by synthesizing an entire genome that will dictate everything about the host organism, including where it can live, what it eats, what it produces, how it reproduces; all its genetic characteristics will come from its "designer genome." This approach might work to design microbes that can gobble up excess carbon dioxide ($CO_2$) in our atmosphere, or produce bio-fuels cheaply and efficiently, or churn out medicines by the ton (see Chapters 14 and 15).

The next challenge will involve actually synthesizing an entirely artificial bacterial genome in the laboratory; even the simplest live organism will probably require a chromosome of more than 500,000 base pairs of DNA. This goal is a technological challenge, as to date the longest DNA molecule made in the lab is only about 35,000 base pairs long. Venter's team expects to synthesize an entire genome, and then it will be a matter of getting the long synthetic DNA genome into the recipient bacterium. At that point, Venter claims, he will have created the first synthetic life, taking the relatively new field of synthetic biology a fundamental step forward.

## LOOKING AHEAD

This chapter describes the methods used by DNA technologists to clone and introduce genes and other DNA fragments into cells and stimulate those cells to produce the encoded proteins. On completing the chapter, you should be able to do the following:

- List some criteria for selecting vectors to carry genes or DNA fragments and for selecting host cells to produce proteins.
- Describe two laboratory methods used to introduce foreign DNA into cells.
- Conceptualize the steps involved in gene expression, understand the types of problems that can arise along this pathway, and explore how DNA technologists resolve those problems.
- Explain the concept of a DNA library, and explain why in some cases it is necessary to use complementary cDNA versus genome DNA libraries.
- Discuss how DNA probes are used to screen for genes in a DNA or cDNA library.
- Describe the process of polymerase chain reaction (PCR), its advantages, and potential limitations.

## INTRODUCTION

In the real world of the research and development laboratory, DNA technology is a diverse blend of molecular biology, genetics, chemistry, physics, mathematics, and high-tech biotechnology. Producing proteins and manufacturing molecules are among the most elegant endeavors of the cell and of biological science in the lab; both are widely used in applied and experimental contexts now, which will continue in the future.

The science of **recombinant DNA cloning** allows scientists to use the natural biochemical processes of cells to do the work of copying DNA and producing specific proteins. The biochemistry behind DNA technology allows scientists to introduce new genes into cells. The cells then exhibit new biochemical activities as directed by the new gene that they carry. As the cells reproduce by cell division, they replicate the new gene along with their "original" chromosome DNA, producing millions of progeny cells containing millions of copies of the gene. In this way a gene is cloned and many identical copies are created.

The extremely powerful method of **polymerase chain reaction (PCR)** is routinely used to clone genes and other specific regions of DNA sequence *in vitro* (outside the cell), without copying the entire DNA genome. PCR harnesses the routine activity of heat-stable DNA polymerase enzymes that normally replicate DNA in special species that live in high-temperature environments. Some of the enzymes made by organisms growing in hot environments are thermally stable and retain enzyme activity even in the high temperatures required for PCR. The enzyme technologies used to manipulate and study DNA and RNA are often based on the ability to understand the processes normally performed by the enzymes in the cells, then repeat the processes outside the cells using purified enzyme proteins and appropriate substrates.

## THE BIOCHEMISTRY OF RECOMBINANT GENE EXPRESSION

Experiments involving cloning genes and expressing proteins require the use of **host cells** to receive the foreign **cloned gene**. Some experiments use microorganisms or "microbes," single-celled prokaryotes such as *E. coli* and *Bacillus subtilis*, and eukaryotes such as the budding yeast *Saccharomyces cerevisiae* (Figure 5.1). Microbes are often chosen as hosts for DNA cloning because they are relatively easy to grow in a laboratory, have been grown and studied extensively for decades, and have well-understood genetics that can be manipulated to make them appropriate hosts. Many types of cells can be converted into biochemical factories to produce various kinds of biomolecules. *E. coli* and *B. subtilis* are both commonly used as host cells for DNA cloning. Fortunately, humans have become very experienced at cultivating microbes cheaply and efficiently on large and small production scales. Over the centuries brewers and bakers have learned to employ yeast cells to manufacture beer, bread, and related food products (Figure 5.2). In terms of impact on human health, probably the most important products made by bacteria for human consumption are antibiotics. Since the 1940s, antibiotics have been mass-produced from microorganisms grown by the ton.

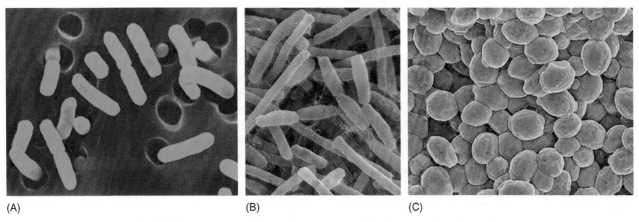

**FIGURE 5.1** Some of the microbes important in DNA technology. (A) An electron micrograph of *Escherichia coli*, the common bacterium in the human and animal intestine. (B) *Bacillus subtilis*, a bacterium often isolated from the soil environment that is not a pathogen. (C) *Saccharomyces cerevisiae*, the nonpathogenic budding yeast commonly used in genetic research, alcohol fermentation, and baking.

The Brewer, designed and engraved, in the sixteenth century, by J. Amman.

(A)

- New and improved
- Built-in pressure gauge
- Bar-style tap

(B)

**FIGURE 5.2** (a) Uses of *S. cerevisiae* through the ages. Humans have used brewer's yeast (*S. cerevisiae*) for millennia to make beer and ale, and still do. (A) Brewers of the 1500s. Engraving is from the sixteenth century, by J. Amman. (B) A modern, personal-size brewing machine.

## What Is DNA Cloning?

Recombinant DNA technologies use enzymes that cleave or copy DNA in living cells. The purified enzymes can perform DNA manipulations *in vitro* in recombinant DNA experiments.

A clone is an identical copy. At its most basic level, cloning DNA involves "cutting and pasting" DNA from its original genome into a convenient carrier DNA molecule called a **vector**. The vector molecule is then introduced into a suitable host cell. Reproduction of the host cells includes replication of DNA, thereby creating identical copies of the inserted DNA along with the organism's DNA. The details for accomplishing the goals of any DNA cloning procedure depend on the characteristics of the DNA of interest and the conditions required for transferring the cloned DNA of interest into the host cell or organism.

Several important factors contribute to the success of a DNA cloning experiment. The target DNA coding for the gene of interest is isolated from the cells. Scientists often choose the vector and host cells together since they work together based on the specific requirements of the experiment. For example, if the plasmid vector encodes a tetracycline (antibiotic) resistance gene, it is necessary to use a recipient host strain of *E. coli* that is tetracycline sensitive so that the bacterial cells cannot grow in the presence of tetracycline (see Chapter 4). Once a foreign gene is introduced into a host cell on a vector, the expression of the foreign gene is turned on and transcribed into mRNA. The mRNAs are translated into protein by ribosomes in the cell.

> Cloning genes is a process where fragments of DNA are transferred from one organism to another, usually carried on a DNA vector. Microbes like bacteria are convenient carriers and hosts for cloning DNA.

## Vectors Are DNA Carrier Molecules

Vectors are DNA molecules into which fragments of DNA inserted, cloned, and, in many cases, expressed as RNA and protein. The vector carries the foreign DNA attached to the vector DNA. The host cells must support replication of the vector DNA. Replication is initiated and controlled by the **replication origin** DNA present on the vector. Vectors with appropriate DNA elements can replicate and produce large quantities of the gene (and the vector), and can be easily isolated for DNA sequencing and other gene studies. A **promoter**, or transcriptional control DNA region, carried on the vector can be designed to promote high levels of RNA expression from the cloned foreign gene, resulting in large amounts of the cloned protein. Because the DNA sequences required to initiate DNA replication (and transcribe and translate gene products) are different in prokaryotic and eukaryotic cells, it is important that the necessary DNA elements carried on the vector match the functions required by the host cells.

In the early days of DNA research, the choice of vectors was limited to naturally occurring DNA molecules that carry a few genes. For instance, plasmid vectors

permit the host cells to survive in the presence of antibiotics because the plasmids carry antibiotic resistance genes. By now there are many hundreds of modern vectors available in a variety of sizes, containing many different DNA elements used to control replication and transcription functions in both prokaryotic and eukaryotic host cells. Modern vectors offer every conceivable combination of selection genes, replication origins, cloning sites, and coding regions, all tailored to function in a large range of host cells. Most types of vectors are either commercially available or can be obtained on request from the scientist that constructed the vector.

Cloning vectors contain special replication origins that permit the vector DNA to replicate and produce many copies of the vector inside growing bacterial cells. The replication origin that enables the vector to replicate also influences the vector's host range. If the replication control elements in the vector do not match the type of DNA replication machinery used in the host cells, the vector cannot replicate and will be diluted out of the population as the cells grow.

> DNA vectors can carry a fragment of DNA from one organism to another and are specifically designed to perform the critical functions in the host cells such as replication of the vector and expression of the gene carried by the vector.

## Expression Vectors Can Produce Foreign Proteins

Expression vectors contain different types of promoter elements that regulate the amount and level of RNA transcription of a gene carried on the vector. Promoters are often controlled by the cell to regulate RNA expression, which in turn controls the amount of the protein produced. Expression vectors are designed to carry a cloned gene into cells and then produce the encoded protein. Promoter DNA elements vary between organisms and have different DNA sequences in prokaryotes and eukaryotes. Promoter information is important to consider when choosing a vector. Some vectors also include elements that function to enhance the stability of the mRNA and increase protein expression levels.

Vectors are often supplied from a commercial vendor complete with all of the appropriate gene expression functions needed. The researcher can further modify the vector using standard DNA cloning methods to insert the desired DNA fragment, followed by **transformation** into bacteria. The efficiency of bacterial transformation (the number of transformed cells per amount of incoming DNA) depends on the ability of the treated bacterial cells to take up DNA from the environment, which is affected by many experimental factors. Scientists often obtain vectors from colleagues doing related research, which is especially helpful in cases where vectors are designed for specific research purposes. Scientists can also obtain vectors and host organisms from sources such as the American Type Culture Collection (ATCC), a unique private, nonprofit resource dedicated to the collection, preservation, and distribution of scientifically relevant microorganisms, DNA libraries, and mammalian tissue culture cell lines.

> The choice of vector is determined by the goals of the experiment and by the characteristics of the host cells that will carry the vector and express the foreign DNA.

## Plasmid Vectors

**Plasmids** are circular, double-stranded DNA molecules that replicate independently from the host chromosome (Figure 5.3; see Chapter 4). In nature, plasmids play a central role in transferring antibiotic resistance from one bacterium to another by exchanging plasmids between the two bacteria. Plasmids commonly serve as the starting DNA vector used to construct more complex types of DNA vectors. Plasmids are ideal to work with in the lab because they are small, circular double-stranded DNA molecules that are not easily broken by physical manipulation or by cycles of freeze-thawing during storage in the lab. In contrast, vectors made from long, linear DNA molecules (such as the lambda phage genome) are easily broken by physical manipulation during experiments. Small plasmids are taken up more efficiently by bacterial host cells than are larger plasmids, and small plasmids are less likely to be damaged during purification and handling.

The maximum length of foreign DNA that can be carried by a vector without interfering with vector replication is important to consider. Plasmid vectors can often carry DNA fragments ranging from hundreds of base pairs to much longer fragments of many thousands of base pairs. A unit of 1000 DNA base pairs is called a **kilobase pair (kb)**. In the 1970s, Stanley Cohen and his research coworkers discovered that plasmid DNA circles could be cut at specific DNA sites by restriction enzymes and that certain DNA fragments could be connected together to form **recombinant DNA plasmids**, or **chimeras** (Chapter 4). In the early days of DNA cloning few plasmids were available for experiments in molecular biology. At the time, pBR322, a circular, double-stranded DNA plasmid was commonly used in DNA cloning experiments. The entire DNA sequence of pBR322 is known, which means that all of the restriction enzyme recognition

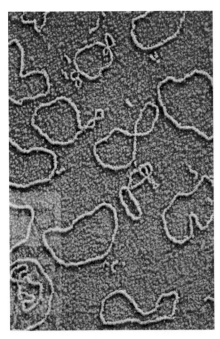

**FIGURE 5.3** Circular DNA plasmids derived from *E. coli*. Plasmids are circular double-stranded DNA molecules that replicate separately from the main chromosome. This electron micrograph shows many copies of a plasmid commonly used for recombinant DNA cloning work, pBR322, which is approximately 4300 base pairs (4.3 kb) in length. Many vectors in use today are derivatives of pBR322.

sites are known. A simple restriction enzyme map of the pBR322 plasmid shows that inserting a DNA fragment into pBR322 can change the function of one of the antibiotic resistance genes encoded by the plasmid (Figure 5.4).

The strategy to insert a DNA fragment into the BamHI site in the pBR322 plasmid begins by cutting the pBR322 plasmid DNA and the foreign DNA to be cloned with the same restriction enzyme, in this case, *Bam*HI. This restriction enzyme cuts double-stranded DNA and leaves complementary overhanging ends (sticky ends) on the DNA molecules. The BamHI DNA sticky ends base pair together, temporarily connecting the pBR322 vector DNA to the foreign DNA fragment through weak hydrogen bonds between the complementary bases (A with T and C with G). To make the union permanent, the **DNA ligase enzyme** joins the sugar-phosphate backbones of the plasmid DNA to the backbone of the inserted DNA fragment, closing the circular recombinant DNA plasmid.

The "parental" pBR322 plasmid DNA encodes antibiotic resistance genes for ampicillin (ApR) and tetracycline (TcR) (see Figure 5.4). Cells carrying the parent pBR322 plasmids can grow in medium containing both antibiotics. The BamHI enzyme recognizes only one recognition site in the pBR322 plasmid DNA, which is located within the gene encoding resistance to tetracycline. The location of the BamHI site is a key feature

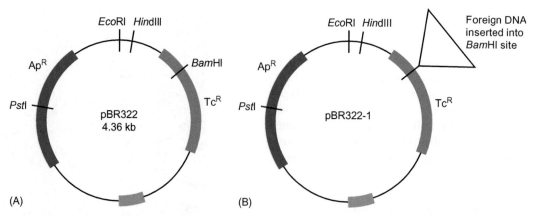

**FIGURE 5.4**  pBR322 plasmid DNA is used as a vector to clone a foreign DNA fragment. (A) pBR322 (ApR, TcR) is shown with a very simple restriction enzyme map indicating the sites where different restriction enzymes can cut the pBR322 DNA. The ampicillin and tetracycline (antibiotic) resistance genes are indicated in red (ApR) and blue (TcR), respectively. The positions of the DNA replication origin and BamHI cloning site inside the TcR gene are noted on the plasmid map. (B) This pBR322-1 map shows the site where the foreign DNA fragment (indicated as a triangle) was inserted into the BamHI site to create the new pBR322-1 vector. The DNA insertion into the BamHI site of pBR322 inactivates the tetracycline resistance gene (TcS). Bacteria cells carrying the new pBR322-1 vector are sensitive to tetracycline because the inserted DNA disrupted the tetracycline resistance gene, but remains resistant to ampicillin.

in this cloning approach, because inserting a DNA fragment into the BamHI site will disrupt the tetracycline resistance gene on the plasmid. Cloning into the BamHI site offers scientists a way to select for and detect the rare cells that have picked up a plasmid DNA molecule. The cells that pick up the pBR322-1 DNA (the vector containing the inserted DNA) are genetically different; they can grow in the presence of ampicillin, but they cannot grow in tetracycline.

## Transferring DNA to Bacterial Cells in the Lab

The genetic alteration of a cell resulting from the uptake of foreign DNA is called **transformation.** In a typical transformation experiment in the lab, the actively growing bacteria are collected by centrifugation, and then suspended in ice-cold calcium buffer. Bacterial cells treated in this way are said to be "competent" for transformation with DNA. The recombinant vector carrying the foreign DNA fragment (for example, pBR322-1) is added to the competent cells, and the tube is subjected to a sudden increase in temperature, called a heat shock (see Chapter 4). Under these conditions, a very small number of bacterial cells take up the plasmid vector DNA. The low efficiency of plasmid uptake is partly offset by the rapid division rates of the transformed bacteria carrying the plasmid vectors. These cells survive an appropriate antibiotic selection screen because they carry the recombinant plasmid DNA containing an intact antibiotic resistance gene.

To be able to select for vectors in different types of host cells, vectors carry at least one **selectable marker gene**. Selectable genes allow the scientists to detect which cells have taken up a plasmid and which cells have not. When using antibiotics for selection, cells that survive the DNA transformation procedure but do not receive a plasmid cannot grow because they do not have the antibiotic resistance gene, and will succumb to the actions of the antibiotic. In this way, the scientists can select for the growth of antibiotic resistant cells, those that harbor the plasmid DNA vector.

> Selectable marker genes allow detection of the rare bacterial cells that pick up the target vector DNA and become transformed. When antibiotics are used for selection, only the transformed cells carrying the vector are antibiotic resistant and can grow in the presence of the antibiotic.

## Viral Vectors

Plasmids are great vectors used in bacterial cells for DNA cloning purposes and for some types of protein expression studies. But plasmids have some limitations in comparison to other vector options, including viral genomes. Many different kinds of viral genomes have been engineered for use as vectors to carry foreign DNA; several of these are described in later chapters covering gene therapy and genetic diseases (see Chapters 10 and 11). Viral vectors can carry long fragments of DNA (more than 20,000 base pairs, or 20 kb) but the fragment length is limited by the genome packaging capacity of the virus protein coat or **capsid**. Viruses use efficient mechanisms to infect bacterial cells. When artificial viruses transfer a recombinant DNA vector genome into

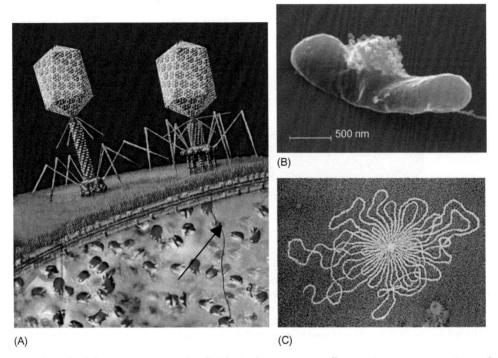

(A)

(B)

(C)

**FIGURE 5.5** Bacteriophage lambda (λ) preys on *E. coli* cells. The λ phage uses naturally occurring receptor proteins on the surface of the bacteria as landing sites to invade the cell. (A) An artist's drawing of two λ phages attached to the surface of a bacterium. The virus on the right has just injected its viral DNA genome into the cell (arrow). (B) Newly made phages are bursting out of a ruptured *E. coli* cell. (C) An electron micrograph showing one double-stranded DNA genome from a single λ bacteriophage.

cells (in place of the virus genome), the transformation process is called **transfection**.

Different types of viruses enter cells and launch infections using different mechanisms. Many viruses attach to protein receptors on the cell surface, where they are taken into the cell by endocytosis. Other viruses actively inject the viral genomes directly through the membrane and into the cell. Once inside the cell, the viral genome usually follows one of two major pathways inside the cell: (1) The **lytic** pathway, in which the virus commandeers the host cell transcription and translation machinery to make viral proteins and copy the viral genome. The host cell processes are suspended by the virus and each genome is packaged individually into a viral capsid and released as a virus from the dying cell, which further spreads the virus infection. (2) The **lysogenic** pathway, where the viral genome DNA integrates directly into the host chromosome DNA where the virus genome can remain indefinitely as a **provirus**.

The lambda phage genome was used to generate many types of bacteria vectors. **Lambda (λ)** is a well-studied bacteriophage that infects *E. coli* cells (Figure 5.5). Phage λ latches onto receptor proteins that serve as docking stations on the outside of the bacterial cell. The viral proteins build a channel through the cell membrane and the phage λ DNA genome is injected into the bacterial cell through the membrane channel. Once

inside the cell, the viral genome takes over the protein-making machinery and uses host cell enzymes to make the viral gene proteins and eventually more viruses.

To maximize the efficiency of gene transfer into bacterial cells, scientists in the 1960s mimicked the normal mechanisms used to package the λ genome into a capsid *in vivo*, and used the same approach to package and deliver lambda vectors. The scientists isolated the viral proteins from infected bacteria and created a test tube mixture of cellular components that works as an *in vitro* **virus packaging extract**. The *in vitro* packaging extract rapidly packages the lambda vector DNA, along with whatever foreign DNA has been inserted, into virus particles that will efficiently transfer the DNA from phage to bacteria.

The vector DNA is applied to a plate of *E. coli* bacteria in a dilute solution so that the recombinant phages infect only nearby cells. Each infected cell releases new phages that infect nearby cells. Gradually a single lambda phage makes a **plaque**, a clear circle in a lawn of bacterial cells where the infected cells have died and the virus has infected the surrounding cells (Figure 5.6).

Methods offering high efficiency transfection are particularly important for studies where the odds for successful cloning are low, for example, when manipulating large mammalian genomes and genome fragments. Success or failure in these cases can depend

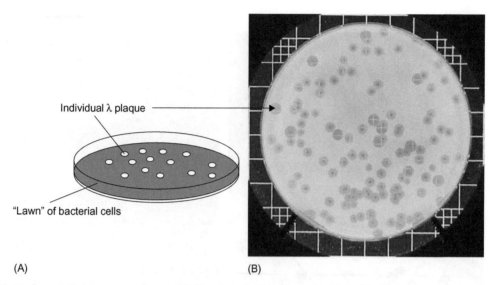

Individual λ plaque

"Lawn" of bacterial cells

(A)                                                                    (B)

**FIGURE 5.6**   Phages form viral plaques on a plate containing a layer (lawn) of bacteria. (A) Healthy bacteria are grown in a continuous layer or lawn across the solid agar in a Petri dish to use as a substrate for phage production. The phage stock is diluted in a solution, and applied to the plate so that a single phage infects a single bacterial cell. The new phage are released from the dying cell and infect nearby bacteria in the lawn. The death of a small neighborhood of infected bacterial cells causes a clear circle called a **plaque** that appear in the cloudy background lawn of uninfected bacterial cells. (B) This plate contains viral plaques that appear as cloudy circles on the lawn of bacteria across the plate.

entirely on the efficiency of DNA transfer attained in the experiment.

> Viral vectors and packaging extracts offer scientists the option of packaging recombinant DNA into virus particles for efficient delivery into the bacterial host cells, a process called transfection.

## Cosmid Vectors

The phage λ genome has the form of a double-stranded linear DNA molecule during most of the phage life cycle, but at some point the linear λ genome temporarily converts to a circular form. Surprisingly, the phage λ genome circularizes using the complementary sticky ends (**cohesive end sites, or cos**) on the viral genome DNA (Figure 5.7A). The single-stranded DNA bases on each end of the λ genome resemble the sticky ends left by restriction enzyme cleavage, but the cos ends are much longer. Because the cos cohesive ends are long, when the cos ends base pair together they make a stable (though noncovalent) connection and form a stable double-stranded circle until the ligation reactions join the DNA backbones. In this way the λ cos sites convert the double-stranded, linear genome into a stable circular genome. Scientists used the λ cos DNA sequences to create **cosmid vectors**.

A cosmid is a small double-stranded plasmid that has been engineered to contain the *cohesive end sites* (the "cos" of cosmid) that naturally occur on the lambda phage genome. The recombinant cos sequences in the

cosmid vector have the same function as they do in the phage; the cos sites convert the double-stranded linear DNA into a very large circular DNA molecule.

During transformation, bacteria do not take up large DNA plasmids very efficiently; in addition, large plasmids are unstable and are broken easily by manipulation. Cosmid vectors combine the advantages of plasmids and viruses. Cosmids can be inserted into bacterial cells using *in vitro* packaging with high efficiency, but also enables the vector to replicate as a plasmid, independent of the genome (Figure 5.7B). Cosmid vectors can be used to carry DNA fragments that are much longer (around 40 kb) than those typically carried by virus or phage vectors.

> Cosmid vectors enable us to take advantage of the helpful features of both plasmids and viral vectors. Cosmid vectors can carry very long DNA fragments, enter the cells with high efficiency, and replicate as independent plasmids in the cells.

## Artificial Chromosome Vectors

A different approach to vector design began with the use of artificial chromosomes in the 1990s. Artificial chromosomes mimic native chromosomes and must include all of the DNA control elements necessary for the vector to behave just like a natural chromosome. This includes the appropriate **DNA replication origin** for host cells, where DNA replication is initiated

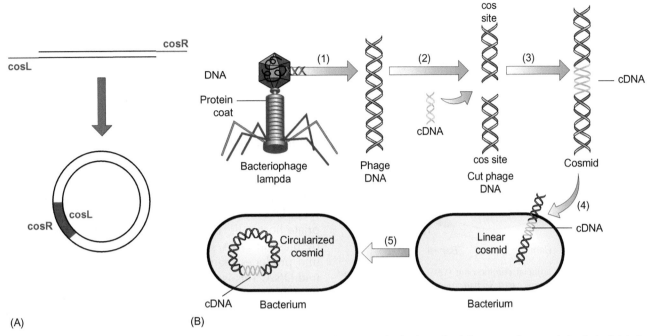

**FIGURE 5.7** The lambda genome and the cosmid vectors both use "sticky" cos ends. The linear lambda genome forms a circle via its "sticky" cos ends. (A) Inside the infected cell the lambda phage DNA genome exists either as a linear double-stranded molecule *(top)* or as a circular molecule *(bottom)*. The linear double-stranded DNA genome contains single-stranded sticky ends, cosL and cosR, on the left and right ends of the genome, respectively. The cos single-stranded DNA ends are complementary; the single-stranded ends can base-pair with each other. When it is the appropriate time in the lambda life cycle, the cos DNA ends base pair to each other and the viral genome forms a double-stranded circle *(bottom)*. (B) Cosmid vectors have the advantages that they can be packaged like lambda and they can form DNA circles resembling plasmids when grown in bacteria in the lab. Cloning into cos vectors involves these steps: (1) The lambda DNA is cut once with a restriction enzyme that leaves short cohesive ends at the DNA cut site. (2) Each phage genome is cut into two DNA fragments; each fragment contains short cohesive ends left by the restriction enzyme on one end and the long single-stranded cos ends from the lambda genome on the other end. (3) A specific DNA fragment or a cDNA is ligated to the short cohesive ends of both DNA fragments to create a recombinant molecule called a cosmid. (4) The cosmid DNA vector and its cargo remain linear until it is inside the bacterium. (5) Once inside the cos ends of the linear DNA base pair to each other and convert the linear cosmid DNA into a double-stranded DNA circle.

and controlled, as well as functional **centromere** and **telomere** DNA sequences. Centromere DNA serves to attach the chromosome to the mitotic spindles during cell division, and its function is essential for correct chromosome segregation during mitosis. Without functional centromere DNA the offspring cells can inherit an unbalanced number of chromosomes, a condition that is usually lethal. The telomere ends of eukaryotic chromosomes play a key role in the specialized DNA replication mechanism that is necessary when cells replicate the ends of the chromosomes; telomeres protect the chromosome ends from becoming shortened during each cell cycle replication event (see Chapter 9).

> Artificial chromosomes carry long DNA fragments and contain all the DNA elements needed to ensure that they will replicate only during the DNA synthesis phase of the cell division cycle, along with the native chromosomes.

The first artificial chromosome was created from an unusually large circular bacterial plasmid called F′ (F prime) whose natural function is to allow a bacterium to transfer its chromosome to another bacterium. The F′ plasmid was adapted for use as a **bacterial artificial chromosome** (BAC) vector, which can carry up to 350 kb (350,000 base pairs) of additional DNA.

The **yeast artificial chromosome** (YAC) was the first linear eukaryotic artificial chromosome ever made and was designed to carry DNA in budding yeast cells. Artificial chromosome vectors have now been developed that function in other model organisms as well, such as zebra fish and mouse. The uncut YAC vector is a circle of double-stranded DNA that becomes a linear chromosome in the yeast cells (Figure 5.8). YAC vectors have both bacterial and yeast replication origin sequences, enabling them to perform as **shuttle vectors** that move DNA between a prokaryote and a eukaryote.

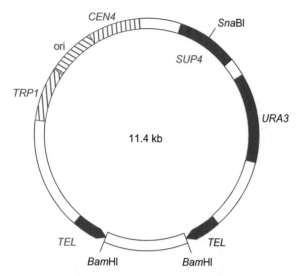

**FIGURE 5.8** Yeast artificial chromosome (YAC) vector. In this case yeast refers to the budding yeast *Sacharromyces cerevisiae*. YAC vectors are typically built using DNA elements derived from *S. cerevisiae* chromosomes that are necessary for a chromosome to behave properly inside the budding yeast nucleus. The circular YAC molecule converts into a linear chromosome in yeast cells, which replicates and segregates when the yeast cells divide, just like a native linear chromosome. To function like a native chromosome, YAC vectors usually include: TEL: Telomere sequences are authentic yeast telomere DNA engineered into the circular YAC vector so that the vector can function as a linear chromosome inside the yeast cells. CEN: Centromere DNA (CEN4) is derived from authentic yeast centromeres and is required to ensure proper chromosome segregation (distribution) during mitosis (cell division) just as if the YAC were a native chromosome. ORI: Origins (ori) are required for the YAC DNA to replicate normally when the yeast cells divide. Selectable genes and/or markers: URA3, SUP4, and TRP1 are yeast genes that are used to select for the YAC in yeast cells. Genes are also included to select for YACs in bacterial cells.

Preparing a YAC requires the handling, separation, and purification of extremely long DNA molecules, including the yeast chromosomes, which range from 225 kb to nearly 2000 kb (2 million base pairs) in length. This can be accomplished using a special type of gel electrophoresis that permits very long DNA molecules to be resolved. Other species-specific artificial chromosomes have been developed, including **human artificial chromosomes (HACs)**, which hold particular promise for applications in gene therapy (see Chapter 11). Usually when DNA is added to human cells it tends to integrate randomly into the cell genome, potentially disrupting important genes, or alternatively the foreign gene might be inserted into a **transcriptionally silent** region of the genome that is not expressed. These disadvantages are avoided when the gene is present on an artificial chromosome, because it does not integrate into the existing DNA and it replicates independent of the main genome. Additional advantages are that HACs offer a way to study aspects of chromosome function and to explore the 98% of human DNA that does not encode gene products and whose functions are not yet well understood.

Artificial chromosomes can carry large DNA fragments and replicate independently along with the native chromosomes, by virtue of DNA sequences that are necessary for centromere, telomere, and replication functions.

## Choosing Host Cells for Cloning

Choosing a suitable host cell and vector combination depends on the specifics of the planned cloning experiment. If the goal is to clone a gene to make a lot of DNA for further studies, then bacteria are appropriate host cells. Bacterial cells are easily transformed by plasmid DNA, and once inside the cell, the plasmid produces many copies of the recombinant plasmid DNA. This is one way to make a large quantity of plasmid vector DNA that is easily isolated from the bacterial cells.

Because vectors carry DNA control elements that must function in the host cells, the vector and host cells are often considered together. Vectors carry DNA elements that function as replication origins, which are sites on the genome where DNA replication begins. Although the fundamental mechanism of DNA replication is the same in prokaryotes and eukaryotes, the origin sequences involved in initiating DNA replication are very different. So a plasmid outfitted for replication in *E. coli*, for example, would not be able to replicate in yeast or mammalian cells.

Host cells are easily cultivated in the laboratory, and the cells carry the genes and control elements as appropriate to accommodate the other requirements of the experiment. In addition, scientists must have a genetic selection or screen that will permit detection of the transformed cells containing the recombinant vectors.

*E. coli* was used in the earliest DNA technology experiments, and by the early 1950s the genetics of *E. coli* cells were well established and played a key role in the genetic conjugation experiments in the 1960s. In addition, *E. coli* bacteria were used to solve the puzzles surrounding the process of protein synthesis in experiments performed during the 1960s and 1970s. *E. coli* bacteria continue to be the major type of host cell used for DNA cloning and protein expression experiments and are still widely used in part because bacterial cells divide rapidly. Under ideal growth conditions in the wild, a bacterial cell can reproduce and make two cells every 20 minutes or so, compared to 90 minutes to two hours for wildtype budding yeast cells and 24 hours or more for mammalian cells. An alternative prokaryotic organism host is the bacterium *Bacillus subtilis* (see Figure 5.1). This rod-shaped, nonpathogenic bacterium was first used in 1958 and since then, its genetics have

## Box 5.1 The *E. coli* Family Tree

### The Good, the Bad, and the Ugly

The bacterial species *E. coli* is actually a family of related groups of bacteria called **bacterial strains**. Many strains of *E. coli* are beneficial to humans. One of the most common strains of *E. coli* lives peaceably and beneficially in the human intestines where it helps with food digestion and represses harmful bacteria. The strain of *E. coli* so widely used in the laboratory is harmless to humans and is an excellent host for recombinant DNA vectors.

Other strains of *E. coli* have attracted attention because of their devastating and sometimes gruesome effects on human health. One is an extremely rare strain that rapidly destroys soft tissue, known as "flesh-eating" bacteria. But this rather shocking term is not correct—the bacteria do not eat flesh, but they do secrete toxins that kill skin cells and other tissues, rapidly spreading the infection throughout the body. The flesh-killing *E. coli* is a rare strain; there are other bacterial species that dominate this grisly category.

Another infamous *E. coli* strain, O157:H7 (the numbers and letters of strains refer to antibody reactions used to identify them), lives in the intestines of about half of all cattle used for meat products. *E. coli* O157:H7 is harmless to cows, but not to people, although the meat is harmless if processed and cooked correctly. Infections caused by the *E. coli* O157:H7 are often deadly in children, the elderly, and immune-compromised individuals. Cases of human illness have been traced to contamination of the public water supply with *E. coli* O157:H7, usually as a result of careless methods of animal slaughter and poor waste management. Outbreaks of *E. coli* O157:H7 have been linked to drinking water sources contaminated by runoff water from a cattle farm.

What makes the *E. coli* O157 strain deadly for people? The *E. coli* O157:H7 cells secrete the Shiga toxin, which kills cells while spreading throughout the body, ultimately causing organs to shut down. As yet there is no effective antidote or cure for *E. coli* O157:H7, but with appropriate medical care, most healthy people recover from an infection in 5 to 10 days.

How did some ancestor of the well-trusted *E. coli* bacteria living in our intestines acquire the toxin gene necessary to make such a devastating pathogen? During a bacteriophage infection the phage genome DNA, including the Shiga toxin gene, are integrated into the bacterial chromosome, establishing a lifetime provirus copy of the phage DNA and the toxin gene in *E. coli* O157:H7.

Ironically, the introduction of low doses of antibiotics into cattle feed in the 1950s might have activated the dormant provirus copies lurking in the *E. coli* O157:H7 chromosomes and triggered toxin production. The 1950s also saw the first cases of an infection in children, which we now know was caused by a pathogen related to the *E. coli* O157:H7 bacteria.

Many restaurant menus now issue a warning that your steak or burger is only guaranteed to be safe when cooked medium-well. This is because one way to kill O157:H7 is to cook the beef to 160°F throughout. Research is under way to find a vaccine that would protect cattle from the *E. coli* O157:H7 bacteria.

---

been thoroughly studied. The naturally occurring plasmids in *B. subtilis* have been scrutinized in detail, and the bacteriophage viruses that attack it have been well studied, making *B. subtilis* an attractive cloning host. Strains of *B. subtilis* actively export proteins out of the cell by secretion, and scientists have used vector DNA sequences to direct the cloned foreign proteins for secretion, a useful alternative to *E. coli* for many protein expression studies. As a result, *B. subtilis* has been used to produce antibiotics, insecticides, and industrial enzymes, among other products.

> Vectors and host cells are often chosen together, since the vector must replicate inside the host cells. Prokaryotic cells like *E. coli* and *B. subtilis* divide quickly and are easily grown in the lab, making them commonly used host cells for DNA cloning experiments.

Although *E. coli* has been the prokaryotic workhorse of molecular genetics for many years, bacteria have some limitations as expression systems for eukaryotic proteins. Indeed, certain experiments in DNA technology require eukaryotic cells, or, in some cases, mammalian

cells to be successful. All proteins must fold properly into specific three-dimensional shapes in order to function in the cell. Eukaryotic proteins often fold incorrectly when expressed in prokaryotic cells, resulting in biologically inactive proteins. Furthermore, bacteria lack the enzymes needed for the **posttranslational modifications** of proteins that normally take place in eukaryotic cells. These modifications are chemical alterations such as the permanent attachment of phosphates, sugars, carbohydrates, lipids, and other types of small molecules to the newly made proteins. These protein modifications do not occur in prokaryotic cells.

> Many eukaryotic proteins fold incorrectly or fail to be modified correctly when expressed in bacteria, and therefore expression of cloned eukaryotic proteins often requires a eukaryotic host.

Scientists study the expression of eukaryotic genes in various host cells or organisms such as mouse, yeast (*S. cerevisiae*), fruit fly (*Drosophila melanogaster*), and roundworm (*C. elegans*) (Figure 5.9). Gene expression

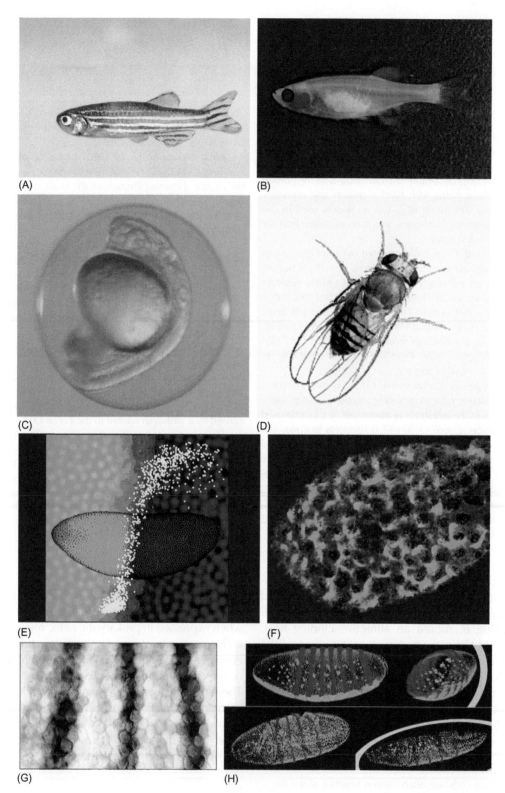

**FIGURE 5.9** Model organisms are often used to study gene expression and development. (A) Zebra fish—adult. (B) Transparent zebra fish. This adult zebra fish was engineered to be transparent to permit scientists to have a better view inside the functioning fish. (C) Scientists can observe embryo development directly in this transparent zebra fish embryo. (D) *Drosophila* (fruit fly) is a very important model system for studying genes and development. (E) Image of *Drosophila* embryos just after fertilization shows the dynamics of morphogen molecules that communicate positional information to individual nuclei in the embryo. The nuclei were fluorescently labeled to show the Bicoid protein (blue), the Hunchback protein (green), and DNA (red). (F) The red dots in this *Drosophila* embryo image indicate the positions of specific RNA:protein complexes (U7snRNPs) (stained red), which are involved in mRNA metabolism during embryo development. (G) Early *Drosophila* embryo shows that the segmentation genes coding for the Ftz protein (brown) and the Odd RNA (blue) protein are expressed in a series of alternating stripes that appear fuzzy because this embryo has a mutation that causes the expression patterns to overlap. (H) Confocal microscopic images of *Drosophila* embryos with DNA (white), with different genes expressed shown in red and green. The tilted embryo (top right of panel) reveals the interior of the embryo.

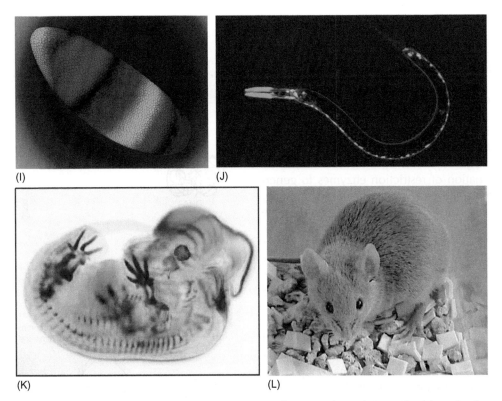

(I)                                              (J)

(K)                                              (L)

**FIGURE 5.9** (Continued) (I) This *Drosophila* embryo shows the expression of gene regulators during early embryo development. Each color represents the expression of one protein: Knirps (green), Kruppel (blue), Giant (red). The darker areas of the embryo contain cells that do not express these genes and the yellowish areas are regions of cells that express both Knirps and Giant. (J) This *C. elegans* roundworm carries a GFP gene that expresses the Green Fluorescent Protein in certain cells near one end of the roundworm. (K) This mouse embryo contains a reporter gene that makes beta-galactosidase and a blue color in the cells that normally express the mouse noggin gene as revealed by the blue color in the brain and in the skeleton. (L) An adult lab mouse at the National Institutes of Health (NIH).

is also studied in human tissues and cell lines. For each type of organism or cell, specific experimental protocols are available to introduce vector DNA into the host, ranging from calcium and heat shock in bacteria to injecting plasmid DNA into some large mammalian cells and amphibian egg cells, to creating transgenic animals and plants (see Chapter 14 and Chapter 15). The characteristics of the vectors used for protein expression are considered later in this chapter.

## DNA LIBRARIES STORE CLONED DNA SEQUENCES

Scientists who work with genes would face an almost impossible task if genes or other DNA sequences being studied had to be reisolated from the original cells or tissues every time anyone wanted to perform an experiment on them. For this and other gene cloning experiments, DNA libraries were developed to allow scientists a convenient way to store and catalog DNA fragments (see Chapter 7). Libraries make it easier to identify DNA fragments and to retrieve a

cloned copy of the gene from the collection. A DNA library is a collection of host cells containing vectors with cloned DNA fragments that represent all of the DNA sequences in question. There are two main types of DNA libraries: the **genomic DNA library** and the **complementary DNA (cDNA) library**.

## A Genomic Library Contains All the Sequences in a Genome

A genomic DNA library contains all of the DNA fragments from the entire genome of an organism. Thus, in addition to the protein-coding genes, a genomic library also includes the 98% of the human genome that contains non-protein-coding DNA sequences such as promoters, introns, and highly repeated DNA sequences. In contrast, cDNA libraries contain only DNA that has been transcribed into RNA from the genome of interest, in the cell type used to generate the library. The bacteria that carry this type of DNA library contain cloned cDNA copies of mRNA sequences that were expressed and translated into the proteins in the cells used to make the library. A cDNA library is always much smaller than a genomic library from the same organism, because only the genes that were being actively expressed in the cells are included in the cDNA library. Because eukaryotic genomes contain much more noncoding DNA than prokaryotic genomes, a cDNA library is essential for isolating genes from eukaryotes.

To begin the process of making a genomic library, the genome DNA is isolated from the organism of interest and is cut into fragments by a restriction enzyme or by a combination of restriction enzymes to generate the desired range of DNA fragments for cloning. The fragment size range desired depends on the size of the genome and the intended vectors for the library. For example, if the target genome is to be cloned in a bacterial vector and host (Figure 5.10), the DNA is cut into fragments of about 10 to 15 kb, and then the vector and the DNA fragments are ligated together and used to transform *E. coli* cells. The bacteria are grown in medium containing an appropriate antibiotic for selection of transformed cells. Because the plasmid vector carries an antibiotic resistance gene, cells that pick up plasmids during the transformation process will survive in the presence of the antibiotic, whereas cells without plasmids will not survive. Each bacterium in the library contains a vector carrying one of the specific DNA fragments cut from the genome, and each bacterium will make many identical copies, or clones, of the vector containing the fragment. Genomic libraries are redundant; each library contains the equivalent of at least three copies of every DNA sequence in the genome, which guarantees that the DNA inserts in the library will overlap in the genome.

Large DNA fragments can be cloned using a YAC vector. First the YAC DNA is cut with both BamHI and SnaBI (Figure 5.11A). Cleavage by BamHI occurs at two sites, releases the two telomere (TEL) DNA ends, and produces a BamHI fragment that is lost. The left arm of the YAC contains the trp1 gene and the right arm carries the ura3 gene. The YAC DNA is mixed with the fragmented genome DNA, and the ligation reaction joins the two arms via the DNA fragment inserted between them. Yeast cells are transformed with the DNA ligation mix and then distributed onto agar plates in medium that will select for cells that have been transformed by the ligated DNA. The genetic phenotypes of the transformed yeast cells are tested and compared to the host yeast cells, which have mutations in the ura3, trp1, and sup4 genes. The yeast cells transformed with the linear YAC vector but *lack* a DNA insert are positive for ura3, trp1, and sup4. These cells can be distinguished from the transformed cells carrying YAC vectors containing DNA inserts, which test positive for ura3 and trp1 but are defective in sup4

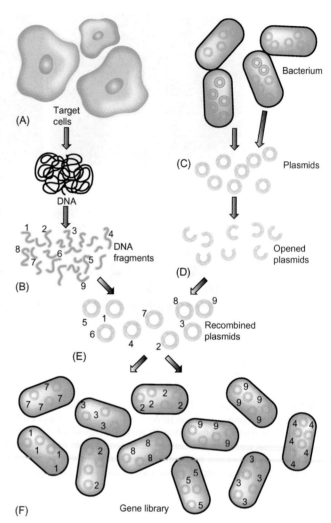

**FIGURE 5.10** Constructing a genomic library. (A) Genome DNA is isolated from target cells. (B) The genome DNA is cut into DNA fragments using a restriction enzyme. (C) Plasmid DNA vectors are grown in and isolated from bacterial cells. (D) Plasmid vector DNAs are cut with the same restriction enzyme as in part B. (E) DNA fragments are mixed with the cut plasmid DNA, the single-stranded sticky ends base pair, and the DNA ends are ligated together to yield recombinant plasmids. (F) The recombinant plasmid DNA is used to transform the bacterial cells (from antibiotic sensitive to antibiotic resistance); the plasmids replicate independently of the bacterial chromosome during cell division so that many additional plasmid copies are made in each cell. During transformation, each bacterial cell picks up one molecule of recombinant DNA, becomes resistant to the antibiotic used to select for transformed cells, and grows into a colony made up of cells carrying identical recombinant plasmids. The entire population of bacterial cells containing all of the recombinant plasmids represents an entire genome library.

function because the DNA insertion into the SnaBI site disrupts sup4 gene function. A special type of gel electrophoresis can be used to separate large DNA molecules cloned using YACs, including the native yeast chromosomes (Figure 5.11B).

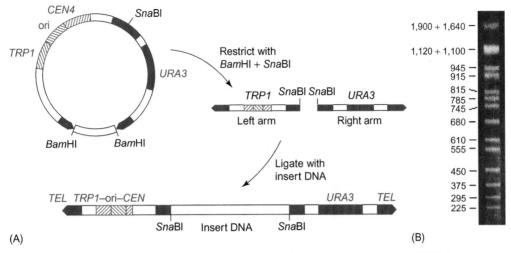

**FIGURE 5.11** DNA cloning using a YAC vector. (A) The cloning procedure shows how a double-stranded DNA fragment is inserted into the *Sna*BI site in the YAC DNA, which physically connects the left and right arms of the YAC DNA and generates a linear YAC molecule. The linear YAC chromosome replicates along with the native chromosomes in the cell. The vector contains sequences for centromere, telomere, and replication origins, as discussed earlier. The YAC vector DNA is cut with *Sna*B1 and *Bam*HI. Insertion into the *Sna*BI site disables function of the SUP4 gene, which allows scientists to select for the transformed cells. (B) Pulsed-field gel electrophoresis is used to separate native, uncut yeast chromosomes. The white bands on the PFGE image represent full-length double-stranded DNA molecules—the yeast chromosomes—listed by length, ranging from 1900 kb (1,900,000 bp) (1.9 Mb) to 225 kb (225,000 bp) in length.

To construct a genomic library, the genome DNA of the organism to be cloned is cut into overlapping DNA fragments that are inserted into the vector. The vectors and the inserted fragments are introduced into host cells by transformation. Taken together, these cells, vectors, and genomic DNA inserts constitute a recombinant DNA library.

it comes to searching for a target gene encoding a protein, but they cannot be used to isolate nontranscribed DNA regions such as regulatory sequences and introns. cDNA libraries are far more compact and carry discrete copies of each expressed gene, rather than restriction fragments of genomic libraries.

A cDNA library contains copies of expressed genes only (genes transcribed into mRNA). The DNA copies cloned in cDNA libraries are typically shorter than the DNA fragments cloned in a genomic library because the noncoding DNA sequences such as introns are not included in the cDNA libraries.

## A Complementary DNA (cDNA) Library Stores Expressed Genes

A cDNA library contains only transcribed sequences, eliminating the nontranscribed sequences and introns from the eukaryotic DNA cloned into the cDNA library. Most eukaryotic genes in vertebrate genomes, including the human genome, contain introns and exons. When expressed in the cell, an interrupted gene is copied into a precursor RNA that is much longer than the fully processed mRNA. Cellular RNA splicing complexes remove the intron sequences in the precursor RNA, producing a final, mature mRNA (Figure 5.12). The mature mRNA contains contiguous exon sequences, which are translated to make the encoded protein. The cDNA copies of the mRNA sequences are inserted into the vector and transformed into a prokaryotic host such as *E. coli*.

The bacteria that make up the library contain vectors carrying complementary DNA copies of the mRNAs present in the cells at the time the library was constructed. cDNA libraries have major advantages when

The process of copying RNA into DNA was virtually unknown until the 1970s when Howard Temin and David Baltimore discovered **reverse transcriptase** (RT), an enzyme that uses RNA as a template and synthesizes a complementary DNA strand from the mRNA template. The name "reverse transcriptase" comes from the fact that RT copies RNA into DNA, instead of the usual process of transcription in gene expression in which DNA is used as a template to make mRNA. This pioneering research on reverse transcriptase and retroviruses led to the Nobel Prize for physiology or medicine awarded to Temin and Baltimore in 1975.

Scientists routinely use commercially available reverse transcriptase to copy mRNA molecules into a single-stranded DNA copies that can then be made

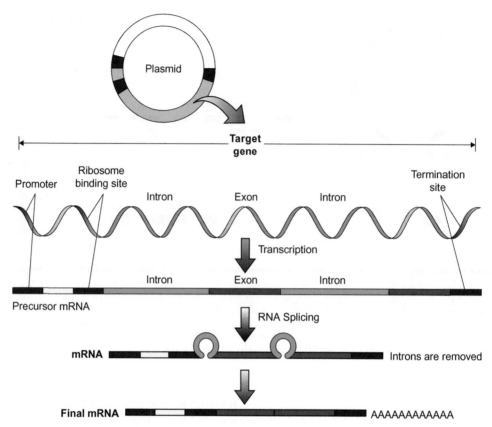

**FIGURE 5.12** RNA splicing. Splicing is the process that removes introns from precursor mRNAs to produce mature mRNAs. Interrupted genes contain introns and exons, which are expressed as long precursor RNAs that contain both introns and exons. The RNA splicing process removes the introns from the precursor RNAs and produces the fully processed mRNA molecules.

into double-stranded DNA and cloned into a cDNA library. Scientists often use cDNA libraries to clone copies of expressed eukaryotic genes. Genomic DNA libraries are frequently used to identify and study the genomic sequences flanking certain eukaryotic genes and the associated transcriptional control elements.

## Making a cDNA Library

For a cDNA library to be scientifically useful, it is essential that the RNA molecules are isolated from the specific cell type that expresses the protein of interest. For example, to clone the human gene for the insulin protein, it is necessary to start the experiment with human pancreatic islet cells that are actively expressing the insulin gene.

In the first step of the procedure, the target cells (such as the pancreatic islet cells, for cloning the insulin gene) are harvested, broken open (lysed), and the RNA is separated from the other cellular components. This **total RNA fraction** contains mRNA, transfer RNA (tRNA), ribosomal RNA (rRNA), and small nuclear (and nucleolar) RNAs; the mRNA actually represents a very

small percentage of the total RNAs. Ribosomal RNAs are usually the most abundant RNAs in an active cell because the cell needs to make large amounts of ribosomes to satisfy the need for protein synthesis.

## Catching mRNA by Its Tail

RNA is chemically unstable compared to DNA and in the cell RNA is rapidly degraded by **ribonucleases** (RNAse). Scientists developed a sterile, one-step purification process to enrich for the relatively minor number of mRNAs present in the total RNA population without risking degradation of the RNA. This widely used procedure takes advantage of the poly-A tail attached to almost all mRNA molecules, which clearly distinguishes mRNA from any other type of RNA in the cell. Because only the mRNAs have poly-A tails, scientists devised a way to use the poly-A tails as a tool to separate the mRNAs from the other RNAs in the total RNA fraction.

The mRNA molecules are purified using a form of **affinity chromatography**, which "grabs" the mRNAs by their poly-A tails. The scientist prepares a small **column**

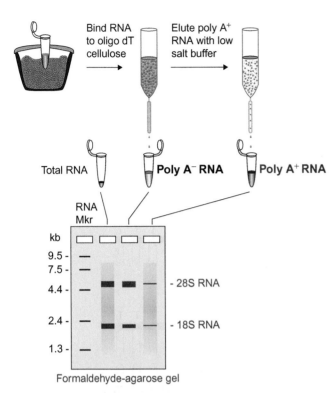

Total RNA    **Poly A⁻ RNA**    **Poly A⁺ RNA**

**FIGURE 5.13** Purification of mRNA by affinity chromatography. Scientists often want to analyze the RNAs expressed from genes in a particular cell type, where the mRNAs are usually a very minor part of the total RNA in the cells, making it advantageous to start by purifying the mRNAs away from the other RNAs. To do this, the total RNA prepared from the cells stored in ice is applied to an oligo-dT cellulose column. The mRNA molecules contain poly-A tails that base pair with the oligo-dT and remain on the column, while other RNAs move through the column and are collected (called the poly-A-minus RNA fraction). Following a wash to remove any remaining non-mRNA from the column, the mRNAs with poly-A tails are eluted from the oligo-dT column with low salt buffer and collected (the poly-A+ RNA fraction). Samples of the RNA fractions are visualized by electrophoresis on a denaturing agarose gel. The major species in the poly-A-minus fraction are the 28S and 18S rRNAs, whereas the poly-A+ fraction is enriched for the mRNAs (compared to the rRNAs). A specific poly-A+ mRNA can be detected in the gel by Northern blot hybridization. www.biochem.arizona.edu/classes/bioc471/pages/Lecture9/Lecture9.html.

filled with a matrix of cellulose beads, each with attached single stranded DNA consisting entirely of T nucleotides, called oligo-dT (Figure 5.13). The poly-A tails on the mRNA form base pairs with the T strands on the oligo-dT cellulose, capturing the mRNAs on the column, while the RNAs lacking poly-A tails flow through the matrix of beads in the column. Following a wash to remove any remaining non-mRNAs in the column, the mRNAs are released using a low-salt buffer that disrupts the hydrogen bonds between the A-T base pairs of the poly-A tails with oligo-dT. The mRNA molecules are collected in the solution as it exits the column.

The next step in constructing a cDNA library is to convert the mRNA of the cell into complementary DNA using the reverse transcriptase enzyme. The single-stranded DNA made by reverse transcriptase is then converted into double-stranded DNA in preparation for ligation into vectors. Like DNA polymerase, reverse transcriptase (RT) also requires a primer with a free 3′ OH group to initiate DNA synthesis. The poly-A tail provides a solution because the synthetic oligo-dT can base pair to the poly-A tail and serve as the primer for RT (Figure 5.14). The RT enzyme copies the entire mRNA molecule into a complementary DNA strand, resulting in a DNA:RNA **heteroduplex** molecule; an RNA strand base paired to its complementary DNA strand. Next the mRNA strand of the DNA:RNA heteroduplex is removed either by the RNaseH enzyme or by alkaline treatment, both of which degrade the RNA strand only, leaving behind single-stranded DNA. The primer to initiate synthesis of the second DNA strand is provided by a short, complementary 'hairpin' that forms at the end of the DNA strand, or by short random DNA primers added to the reaction.

## cDNA Is Different from Genome DNA

The double-stranded cDNA molecule is different from genomic DNA in one very important, fundamental way: the cDNA sequence is a copy of a fully processed and spliced mRNA, which means that the cDNA sequence contains only exons and lacks the introns that are removed during mRNA splicing. In comparison, the equivalent genomic DNA contains introns and exons, and regions flanking the genes, including promoters and other control elements. The chromosome DNA sequences that are not expressed as RNA in the cell will not be included in a cDNA library.

A complementary DNA (cDNA) contains just the sequences that represent a copy of a mature mRNA. If provided with a promoter element, the cDNA can be used to express the protein encoded in the exon sequences, avoiding the need to properly process

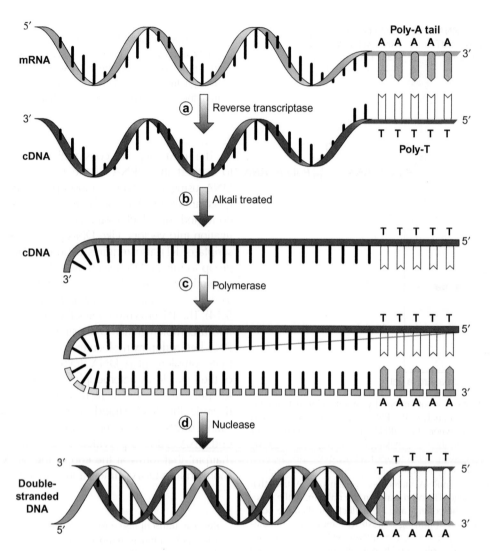

**FIGURE 5.14** Reverse transcription of mRNA to cDNA. (a) Mature fully processed mRNAs with poly-A tails are isolated from cells and copied into an mRNA:cDNA heteroduplex by reverse transcriptase. (b) The RNA strand is removed from RNA:DNA heteroduplex with alkali or RNase H activity. (c) The cDNA loop (or added random premers) start DNA synthesis by DNA polymerase to make a double-stranded DNA product. (d) The hairpin loop is cut with a nuclease that leaves a double-stranded cDNA molecule that is suitable for insertion to a cloning vector. The cDNA is a copy of the mRNA and contains only exons not introns.

or splice the RNA product in the cells. For this reason, mammalian expression vectors are specifically designed to express proteins from cloned cDNA genes. This approach is particularly important in cases where expressing the genomic copy of a gene is not practical or when the host cells do not support the necessary splicing events needed to produce fully processed mRNA (e.g., bacteria). Usually it is not practical to try to manipulate the large genome copy of a eukaryotic gene containing many introns and exons, which can encompass many tens of thousands of base pairs (kilo base pairs; kb) of DNA.

## Screening a DNA Library

Whether constructing a cDNA or genomic DNA library, the appropriate DNA fragments are mixed with the cut vector DNA and ligated together to form recombinant DNA molecules. The ligated DNA molecules are used to transform *E. coli* cells; the transformed cells containing the recombinant vector grow in the presence of an antibiotic. To use a DNA library productively, it is then necessary to use a DNA probe to find the gene or sequence of interest among the cells that contain the cloned DNA fragments.

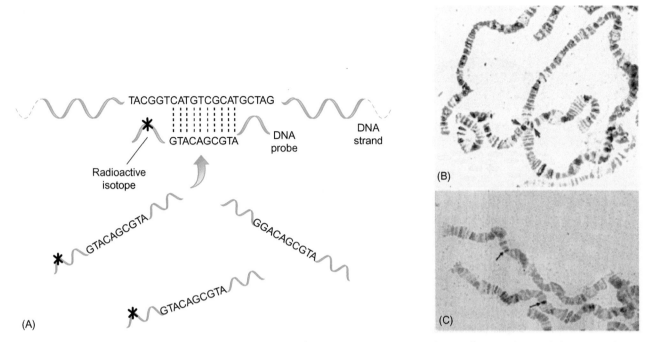

**FIGURE 5.15** DNA probes identify target DNA sequences in chromosomes. (A) A DNA probe is a short, single-stranded segment of DNA that base pairs preferentially to a complementary DNA sequence. DNA probes are used to find specific DNA sequences in DNA libraries, on chromosomes in Southern and Northern blots, and in many other applications where a target sequence is sought. (B) An example of an experiment using DNA probes: in this micrograph, a DNA probe has base paired (hybridized) to specific genes located on the fruit fly salivary chromosomes (arrows) (C) Arrows point to DNA probes base paired to specific DNA regions on the chromosomes (arrows).

## DNA Probes

Scientists use **DNA hybridization probes** as tools to identify specific target DNA sequences hidden among hundreds of thousands of other DNA sequences in a genome or library (Figure 5.15). DNA probes are used in many different contexts in addition to screening DNA libraries. For example, DNA probes can be used in thin sections of tissue, on chromosome spreads, or in microarrays (see Chapter 13) to study different gene expression patterns (see Figure 5.9).

DNA probes work by **hybridization**, which is the ability of one DNA strand (in this case, the DNA probe) to base pair with a complementary DNA strand (the DNA target). To design a DNA probe, it is essential to know at least part of the target DNA sequence. This sequence information can be obtained from various sources including DNA databases. Sometimes the DNA sequences used as probes are deduced from the known amino acid sequence of a protein of interest; other probes are based on the sequence of a gene that has a similar function to the target gene, requiring the use of a probe with a sequence that is **homologous** (similar) to the target gene.

A DNA probe is a short (about 20 bases), single-stranded DNA molecule with a chemical marker, fluorescent "tag", or radioactive label that can be detected once the DNA probe has hybridized to the target DNA in the library or other biological preparation. In the early days of recombinant DNA, the only way to label DNA probes was with radioactivity, but other forms of nonradioactive detection have been developed since then, including fluorescence and chemiluminescence.

> A short, single-stranded DNA molecule called a DNA probe is used to locate a specific gene or sequence of interest. DNA probes are synthesized with a chemical "tag" or label that is needed to detect the probe after the probe has base paired to its complementary target DNA.

During a DNA library screening process (described later), the probe DNA is exposed to the single-stranded DNA of the library under conditions that vary the level of hybridization **stringency**. Under high stringency conditions the probe will base pair only to sequences that are fully complementary to the probe sequence; whereas low stringency allows the probe to hybridize to sequences that are similar to the probe but are not exact complements. Higher temperatures and lower salt concentrations increase stringency, allowing the probe to base pair only to sequences that are exact complements.

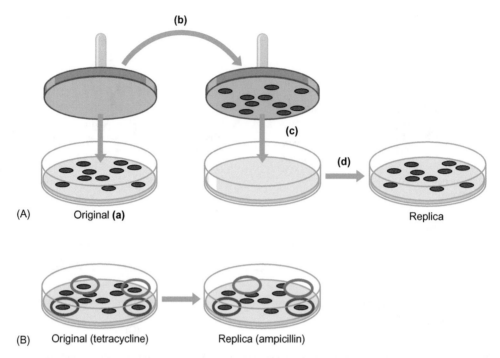

**FIGURE 5.16** The replica plating technique for screening bacterial colonies. (A) Bacteria grow in colonies on solid agar in Petri plates (dishes). (a) A circle of sterilized velvet fabric is placed directly over the colonies. (b) Some bacterial cells from each colony are transferred to the fabric. (c) The fabric circle is laid onto a fresh Petri dish of agar, transferring cells on the velvet to the growth medium. (d) The plates are incubated and the colonies grow, making a copy, or "replica plate," of the original plate of colonies. (B) The replica plating technique can be used to screen for antibiotic resistant bacteria that contain antibiotic-resistant genes. A replica of bacterial colonies is made on a plate without antibiotics (left) and on a plate containing the antibiotic ampicillin (right). All of the colonies can grow on the master plate copy lacking antibiotics. Only the colonies that are resistant to ampicillin can grow on the replica plate containing ampicillin (circled in red), whereas colonies that are not resistant to the antibiotic cannot grow on the plate containing ampicillin (circled in blue).

## Replica Plating

The purpose of a cDNA library screen is to find the cells that contain the target DNA so that the cells can be grown and the target DNA isolated or expressed. Because exposure to the DNA probe involves killing the cells, it is necessary to keep a record to allow the scientist to relocate the appropriate target cells. This is often accomplished using a replica plating method that creates a master plate of library colonies (Figure 5.16). The transformed *E. coli* cells are diluted and spread on solid nutrient agar in Petri dishes so that each bacterial cell has room to grow into a pure colony of millions of identical, cloned cells generated by reproduction of a single transformed bacterium. A replica of each Petri dish is made for the purpose of identifying the colony containing the target cloned DNA, which are available on the original master plate.

The **replica plating** technique used to make copies of Petri dishes containing growing microbes or phage plaques was originally developed in 1952 by Joshua and Esther Lederberg. A circle of sterile velvet cloth is pressed firmly onto a Petri dish of tiny bacterial colonies (Figure 5.16A). The velvet is carefully peeled off and then gently pressed onto the agar of a fresh Petri dish. The velvet cloth picks up some cells from each colony on the first dish and transfers the cells to the second fresh plate, the replica plate, which now becomes a copy of the original plate. The colonies on the copy plate can be analyzed without damaging the original library master plates. A replica copy of the bacterial colonies can be made onto plates containing an antibiotic such as ampicillin. All of the colonies grow on the copy of the master plate without antibiotics, but only colonies that are resistant to ampicillin can grow on the replica plate containing ampicillin (Figure 5.16B).

The replica plating technique is still important in modern molecular genetics labs for screening DNA libraries (Figure 5.17A–D). Special materials that bind DNA tightly, such as nitrocellulose filters or nylon membranes, are used in place of velvet to make replica copies on the cells on a Petri dish. The replica filters are then soaked in a sodium hydroxide solution at a high temperature (65°C) to break open the cells in the colonies. Importantly, the DNA remains fixed to the filter or membrane at the same spot as previously

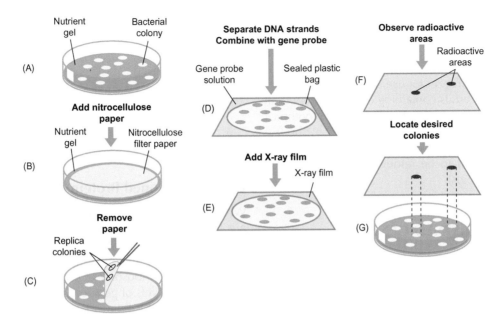

**FIGURE 5.17** Screening a DNA library. (A) Bacterial cells containing the cloned gene are plated onto nutrient agar, where they form colonies. (The same approach can be used to screen the DNA or RNA in plaques made by viruses or phage.) (B) A circle of nitrocellulose filter paper is gently applied to the surface to obtain a replica copy of the colonies on the plate. (C) The nitrocellulose paper is peeled off the gel surface complete with bacteria transferred to the filter from the colonies. (D) The cells on the filters are disrupted in place and the DNA on the filter is separated into single strands that remain on the filters. The treated filters are placed in a sealed plastic bag in a hybridization buffer solution containing a single-stranded DNA probe that will specifically base pair with a short segment of the target DNA in the very few colonies in the library where the DNA fragment is present. (E) The filters are exposed to x-ray film (or are processed appropriately for alternative systems) to detect the DNA probe and consequently the target DNA. (F) Dark areas appear on the x-ray film where positive clones have emitted radioactivity. (G) The x-ray film is compared with the original master plate to determine which colonies contained the DNA that emitted the radioactivity. These are colonies of bacteria carrying the target DNA sequence.

occupied by the cells. The solution also denatures the DNA (separates the two strands) so that the single-stranded DNA remains fixed to the filter in the exact same pattern as the cells on the original plate. These "filter replicas" of all the plates in a DNA library are then screened with DNA probes to identify the location(s) of potential positive clones.

### Finding the Target DNA

The replica filters are placed in a sealed plastic bag in a solution of buffer containing millions of copies of a specific DNA probe (Figure 5.17D). The DNA probe has access to the single-stranded DNAs located on the filter circles in the positions previously occupied by the colonies. The probe will bind specifically only to a complementary DNA sequences on the filters. Under conditions of high stringency, the probe will base pair to its exact DNA complement; under low stringency conditions, the probe will base pair to similar sequences as well.

The filters are removed from the bag and any nonspecific probe is washed away. Then, if the probe is radioactive or uses a light detection system, the dried filters are placed against to a piece of x-ray film in the dark so that any light signals or radioactive emissions

from the filters will create dark spots on the film (Figure 5.17E–G). Each dark spot indicates the positions on the filter where the DNA probe base paired with the target DNA and thereby indicates the position of that specific positive colony on the corresponding original plate. The live cells on the original plate can be harvested and grown to continue the process of obtaining the target DNA.

Once positive colonies have been identified on the master plates, the cells from each "positive" colony are carefully transferred from the master plate into separate liquid cultures for further study and for storage. The methods for cultivation and maintenance of most host organisms are well established, and instructions are provided by commercial sources of DNA libraries. Scientists often purchase a commercially prepared cDNA library instead of making the library in the lab. An expensive library can be rendered useless if the cells are grown incorrectly, because this can allow a subpopulation of vectors without inserts to overgrow the total library cell population. For this reason it is can be worth the investment to either purchase a well-characterized library or contract with a company to create specialized libraries from RNA preparations provided by the client.

Cells containing DNA libraries have been kept frozen for years or for decades when lyophilized (freeze-dried), which avoids the danger of contamination when the frozen culture of original library cells is removed from the freezer for use. To grow more library cells, a few ice crystals containing cells are scraped off the top of the frozen culture with a sterile toothpick and spread out to thaw on a sterile agar dish. The frozen cells begin to grow once again, providing cloned genetic information to be retrieved when needed by the DNA technologist.

## EXPRESSING CLONED GENES

Once a gene has been identified, isolated, and cloned from the appropriate genome or identified in a DNA library, the next step could involve expressing the protein encoded by the gene. To accomplish this feat, a copy of the gene is inserted into an expression vector that is designed to promote transcription and translation, producing the encoded protein in the cell. A key question in protein expression experiments involves which type of host cell will best express the protein product, and there are many choices. Many mammalian proteins are expressed in bacteria, but bacterial cells do not provide posttranslational modifications, which are common in eukaryotic proteins, although posttranslational modification is available to different degrees in eukaryotic tissue culture cells, insect cells, and yeast cells.

Regardless of the source of the gene under study, the choice of expression vector will depend on the characteristics of the recipient cells. For example, expression vectors contain different types of promoter sequences to ensure that the inserted gene (or cDNA) will be transcribed properly in the host cells. (Promoters are naturally occurring DNA sequences that are usually located before the start of a gene and help to govern the frequency and level of transcription of the gene; see Chapter 3.) There are different types of promoter sequences, and the promoters used in prokaryotic cells are very different from the eukaryotic promoters.

Basic types of promoters used to drive transcription include **constitutive** promoters, which are "on" all the time and lead to continual expression of the gene. Constitutive promoters usually express high levels of a desired protein, but they do not shut off, which may have damaging consequences to the host cells. **Inducible** promoters regulate gene expression by responding to signals; *in vivo*, the signals come from the metabolism or environment of the cell, and *in vitro*, the scientist supplies the signal in the form of a chemical or other stimulus whenever expression is desired. Such signals might include the presence or absence of a small molecule in the growth medium or the presence or absence of a phosphorylated protein.

> Expression vectors have promoters to drive gene expression: either constitutive promoters, for a constant level of expression, or inducible promoters that can be turned on or off in response to various cellular signals.

## Collecting Cloned Protein Products

When expressed in eukaryotic host cells, cloned genes are transcribed along with the endogenous genes in the nucleus. The mRNAs are transported out of the nucleus and into the cytoplasm for translation into proteins by the ribosomes. Expression vectors often promote high levels of gene expression, so the mRNAs and proteins build up in the host cells. Because the recombinant gene product is foreign to the cell producing it, the host cells sometimes destroy the foreign protein. Even simple bacteria make protease enzymes that degrade foreign proteins inside the cells. To avoid having the desired cloned protein degraded by host cell enzymes, scientists constructed host bacterial strains that are genetically deficient in the major host protease enzymes.

When expressed in bacterial cells, the cloned foreign proteins sometimes accumulate in insoluble masses called inclusion bodies. Although inclusion bodies are an artifact of protein accumulation, they can be useful in protecting the newly synthesized proteins from protease attack, and they can be easily harvested from the bacteria. Insoluble protein can be used to make antibodies against the cloned protein, which in turn can be used to track protein expression in various ways. To be functional, however, cytosolic proteins must be rendered soluble and refold into the three-dimensional shape required for activity, and after being in an insoluble state, many proteins cannot refold correctly and are nonfunctional. Other cloned proteins are soluble when expressed in the bacterial cells and so are targeted for purification strategies involving protein affinity chromatography.

Many different types of protein expression vectors offer the efficient production of cloned protein products. In some cases the cloned proteins are conveniently secreted out of the host cell and can be purified from the medium. Genes encoding proteins that are normally secreted from the cell contain a DNA **signal sequence** and encode a corresponding amino acid signal peptide located at the beginning of the synthesized protein. The signal peptide ensures that the protein will be sent along the correct pathway to be secreted from the cell. A signal sequence is included on many expression vectors so

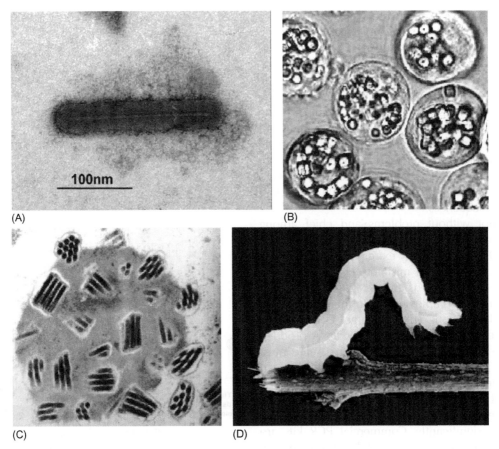

**FIGURE 5.18**  Baculovirus is a very efficient protein expression system. (A) The rod-shaped baculovirus. (B) Insect cells infected with baculovirus growing in tissue culture. (C) Rod-shaped baculoviruses are enveloped in the cell nucleus and covered by a protein matrix. (D) Larvae (caterpillars) infected with baculovirus vectors produce the largest amounts of a target protein.

that the resulting protein, whether originally destined for secretion or not, can be efficiently targeted for secretion from the cell, potentially increasing the yield and ease of protein recovery. The signal peptide can subsequently be removed during purification of the protein.

> Cloned proteins are usually foreign to the host cell and may be degraded or, in bacteria, may form insoluble inclusion bodies of inactive protein. Targeting a protein for secretion from the host cell is one way to avoid these problems.

**Baculovirus** (Figure 5.18) normally infects insect cells but recently has been used to transfect mammalian cells as well. It is useful for expressing protein in eukaryotic cells because of the large amount of protein that can be obtained and because the host cells perform the most important eukaryotic posttranslational processing steps that are lacking in *E. coli* and other types of prokaryotic host cells. In the wild, the rod-shaped baculovirus attacks more than 500 different species of insects. Baculovirus vectors are particularly good at producing large amounts of a cloned protein in the insect larvae, exceeding the productivity of foreign proteins expressed from vectors in mammalian cells.

## THE POLYMERASE CHAIN REACTION (PCR)

Kary Mullis developed PCR in 1984 while working for the Cetus Corporation, and he was awarded the Nobel Prize in chemistry in 1993 for this achievement. PCR allows a scientist to amplify a single DNA molecule into a billion identical copies of DNA in a few hours, using just DNA primers and a special DNA polymerase enzyme. Since its discovery, PCR has become the central method used on literally thousands of diagnostic tests in genetics and medicine, and in police fingerprinting and forensic analyses, with additional applications continually being developed (see Chapters 8, 10).

PCR can start with a tiny amount of DNA, even as little as the chromosomal DNA in a single hair follicle. PCR is catalyzed by remarkable thermostable (heat-stable) enzymes that survive the extremely high temperatures used to separate the DNA strands between PCR cycles. PCR is a rapid way of "cloning" identical DNA fragments directly from a genome DNA preparation without using a vector or host cells and can be used to prepare the DNA probes used in DNA screening to identify a clone in a library. As long as the scientist has a sample of the DNA of interest and knows some part of the target DNA sequence or neighboring sequences, PCR can shorten the steps in recombinant DNA cloning procedures. As we will see, PCR permits the scientist to locate the molecular equivalent of a DNA needle in a veritable haystack of different DNA sequences.

PCR is not without problems and chief among these is the risk of contaminating the PCR sample with extraneous DNA. As with any extremely powerful tool, PCR can be misused and misinterpreted; if the sample DNA is contaminated, any DNA sequences that are coincidentally complementary to the PCR primers will be amplified along with the target DNA, ending up as a large fraction of the final PCR DNA product. To eliminate contamination artifacts, great care must be taken in preparing samples for PCR testing. Such preparation tends to be labor intensive and costly. The development of uniform rules for handling DNA and specialized equipment have reduced the contamination problem significantly. Commercial PCR kits are readily available and are very useful in preparing the necessary reagents and enzymes to perform PCR in research and many other applications.

> PCR is used to amplify a target DNA sequence that is present in tiny amounts—as little as a single molecule—while surrounded by large amounts of nontarget DNA. It is critical to avoid contamination with extraneous DNA in order to achieve reliable results with PCR.

## PCR Cycles and DNA Amplification

To amplify a region of DNA using PCR, the following biological reagents are required (in addition to a PCR thermocycling machine):

- A DNA molecule (usually double stranded) containing the target sequence to be amplified.
- Many copies of two short DNA "primers", which are complementary to the DNA sequences flanking the target DNA. The primers are used by the enzyme to initiate DNA replication (Figure 5.19).
- The most well-known thermostable DNA polymerase enzyme used in PCR is *Taq* polymerase, which is derived from the thermophilic (heat-loving) bacterium *Thermus aquaticus*, first isolated by Thomas Brock in the 1980s from the hot springs in Yellowstone National Park.

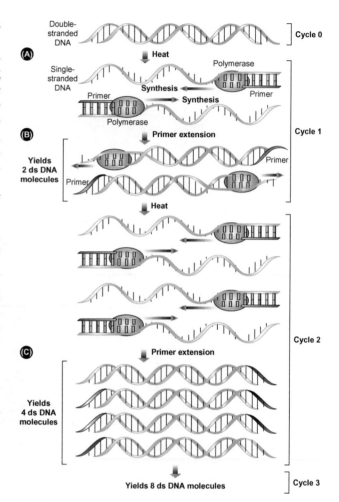

**FIGURE 5.19** The polymerase chain reaction (PCR). (A) Heat is used to separate the double-stranded (ds) DNA into single strands of DNA. Cycle 1: Nucleotides, a heat stable polymerase, and a DNA primer are mixed together. (B) The polymerase extends the primer and produces two ds DNA molecules. (C) The process is repeated; at the end of cycle 2, four dsDNA molecules are present. Repeating the process in cycle 3 yields a total of eight short ds DNA molecules. The short strands are amplified, increasing the number of copies geometrically in future cycles, amplifying the target DNA.

- Nucleotides to build new DNA strands during DNA replication.
- A buffer solution containing the required reagents and chemical environment for the synthesis of DNA.

PCR DNA amplification involves three steps that are performed as a continuous cycle (Figure 5.19) in a PCR machine:

First, a tiny amount of solution containing the target double-stranded DNA sample is heated, which breaks the hydrogen bonds between the base pairs, denaturing double-stranded DNA molecules into two separate single strands.

Second, the single-stranded DNA primers are added and the temperature is lowered so the primers, which are complementary to specific base sequences on both sides of the target DNA sequence, can base pair to the single-stranded target DNA. The primers essentially function as "start" and "stop" signals during each round (cycle) of DNA synthesis, acting as molecular brackets to identify the region of the DNA molecule to be copied in the amplification step (see Figure 5.19). The primers greatly outnumber the target DNA strands, so the target strands do not base pair with each other at this stage.

Third, the thermostable *Taq* DNA polymerase initiates DNA synthesis at the primers that are base paired to the template DNA, synthesizing the complementary DNA strand and thus making a double-stranded DNA molecule (Figure 5.19). Using a heat-stable polymerase enzyme is very important because the heat used to denature the DNA strands in each cycle would destroy an ordinary enzyme, requiring that the enzyme be replaced after each heating step, making PCR expensive and impractical.

At the end of step 3, two double-stranded DNA molecules exist where there had been just one starting DNA molecule. As the PCR process continues, the cycle (steps 1, 2, and 3) is repeated 30 to 60 times, each cycle beginning with a reheating of the DNA mixture to denature the DNA into single strands. A full PCR cycle takes one to two minutes, and in each cycle, each newly synthesized DNA molecule serves as an additional template for DNA synthesis. Thus, the number of DNA copies increases exponentially and millions of identical copies of the same DNA strand are made, all starting from as little as one DNA helix molecule.

> PCR consists of steps that specifically replicate a target DNA sequence exponentially, resulting in millions of copies of the sequence in a few hours, effectively cloning the target DNA sequence without vectors or host cells.

The usefulness of PCR extends beyond making millions of identical DNA copies in a lab. PCR technology is fundamental to the development of many important methods such as making specific DNA mutations, diagnosing diseases, detecting infections, identifying individuals, and classifying pathogenic organisms, to name just a few of the many applications of PCR technology in modern science. All these applications rely on the specificity of DNA base pairing in PCR—the ability to bracket a sequence of interest accurately and replicate it faithfully and quickly.

## SUMMARY

DNA is cloned routinely in laboratories where scientists are doing research on the fundamentals of genes and cells, as well as in labs that apply science to practical goals such as discovering new medicines. Coaxing a cell to manufacture a desired gene product often begins with the isolation of the gene. The gene is cut from the genome by restriction enzymes and is then ligated into a molecule called a vector, which carries the foreign DNA into the host cell. The vector must have DNA sequences that enable it to replicate, in the host cells, appropriate restriction sites for cloning, and an appropriate antibiotic resistance gene to survive antibiotic selection methods.

So far, the *E. coli* bacterium is the most widely used host cell for routine DNA cloning experiments and for many protein expression strategies as well. The bacterium *B. subtilis* and the single-celled eukaryotic yeast *S. cerevisiae* are also commonly used for protein expression. More complex eukaryotic cells can be used as hosts, although these cells are more difficult to grow in the lab and to store for long periods of time unchanged. cDNA has become an indispensable tool for studying eukaryotic genes and expressing eukaryotic proteins. In some cases it is possible to engineer the bacteria to export the gene product out of the cell; many expression vectors therefore contain protein secretion signal sequences.

DNA libraries are collections of host cells that act as storage systems for whole genomes (genomic DNA libraries) or for sets of genes that are expressed from a particular cell type (cDNA libraries). Genomic libraries contain virtually all the DNA of an organism, coding and noncoding, whereas cDNA libraries contain DNA copies of the mRNAs expressed at the time the target cells were harvested, and no other DNA sequences. Searching a DNA library for a sequence of interest requires a DNA probe that will bind specifically to the desired target DNA.

Under some circumstances, where even part of a target DNA sequence is available, PCR is the technique of choice for amplification and cloning of DNA sequences. PCR is more than just another tool in the DNA technologist's toolbox. The impact of this technology has become so widespread that PCR is routine in labs big and small around the world.

## REVIEW

To test your knowledge of the chapter's contents, consider the following review questions:

1. List three naturally occurring cellular processes that scientists have adapted for use in recombinant DNA technology, and explain how each one is used in DNA cloning.
2. Name three possible DNA vector types, and describe the advantages and limitations of each as vectors.

3. Describe a method used to introduce recombinant DNA into *E. coli* cells.

4. *E. coli* has been called the "workhorse" of molecular genetics. What features made this bacterium such a useful laboratory organism?

5. Explain the differences between a genomic DNA library and a cDNA library.

6. How is a cDNA molecule different from the original gene in the DNA genome?

7. Explain the important characteristics of a DNA probe.

8. Explain in what circumstances a researcher would decide to use a cDNA library instead of a genomic library to find a gene.

9. In screening a DNA library, a scientist forgets to denature the DNA on the replica filters but continues on and finishes the experiment. What will the scientist see on the x-ray film?

10. Imagine that your PCR machine has completed three PCR cycles, which amounts to nine steps of PCR (three PCR steps, repeated three times). Assuming that you started with a single target DNA molecule, how many copies of the single target DNA molecule will you have? Draw and label the target DNA, primers, and newly synthesized strands as they appear after the first, fourth, and seventh steps.

## ADDITIONAL READING

Bloom, M.V., Freyer, G.A., Micklos, D.A., 1997. Laboratory DNA Science. Benjamin Cummings, Pearson Education, Upper Saddle River, NJ.

Clark, D.P., 2005. Molecular Biology: Understanding the Genetic Revolution. Elsevier Academic Press, NY.

DNA Interactive: Manipulation. Dolan DNA Learning Center, 2009. www.dnai.org/b/index.html. Modules: Revolution, Techniques, Production. For an animation of PCR, go to Techniques, Amplifying, PCR animation.

Kimball's Biology Pages, 2008. http://users.rcn.com/jkimball.ma.ultranet/BiologyPages/R/RecombinantDNA.html

Miklos, D.A., Freyer, G.A., Crotty, D.A., 2003. DNA Science: A First Course, second ed.. Cold Spring Harbor Laboratory Press, Cold Spring Harbor, NY.

Wade, N., July 8, 2007. Genetic Engineers Who Don't Just Tinker. New York Times www.nytimes.com/2007/07/08/weekinreview/08wade.html?ex=1341547200en=7ab8d39ab463112cei=5088partner=rssnytemc=rss

# Human Genomics

## Monogamy Gene Links Men's DNA to Happily Ever After in Marriage

Bloomberg.com, www.bloomberg.com/apps/news?pid=20601124&sid=a5kGdZ7L7vMl&refer=home

By Michelle Fay Cortez, 2008

Traits like faithfulness and commitment in married men may depend on the same gene that keeps prairie rodents committed to their mates, according to studies on the "monogamy gene." Early studies on small prairie rodents called voles indicated that the levels of vasopressin hormone (made in the hypothalamus) influenced the monogamous and polygamous behavior. In 2002, Larry Young, a behavioral neuroscientist at Emory University (Atlanta), reported on the highly developed social behaviors of the Prairie voles and the Mountain voles (Figure 6.1). The two voles look very similar, but underneath the two animals exhibit very different "family values." Prairie voles are very sociable, mate for life, and the parents share care for the young. Mountain voles on the other hand are typically solitary, promiscuous, and only the mother mountain vole cares for the young.

**FIGURE 6.1** The vole story of the monogamy gene. The prairie voles and mountain voles look very similar, but they have very different "family values": prairie voles are sociable, they mate for life, and parents share care of the young. Mountain voles are solitary, promiscuous, and the mother alone cares for the young.

Continued

In 2008, Swedish scientists (Karolinska Institute) extended this work to humans and they found that vasopressin in men contributes to making a happy marriage. Researchers performed genetic tests on the men and conducted surveys to gather personal information about the participants. This study showed that men carrying a DNA variation in the vasopressin control gene, or who had extra copies of the variant gene, scored much lower on a scale measuring partner bonding. Even in humans it seems that complex social behaviors such as personal relationships depend on biology and genes. Variations in the vasopressin control gene are also implicated in the social deficits of autism, suggesting a possible connection between autism and a gene involved in forming strong personal relationships.

A major author of the study, Paul Lichtenstein, strongly cautioned people not to over-interpret these results, insisting that this gene alone can't predict successful relationships, "It gives you a predisposition, but it doesn't determine how successful you will be in marriage." In the overall population, people with this gene variant will probably tend to have more trouble in their marriages, but we know that the genotype alone cannot be used to predict any trait including marriage potential, because many other factors contribute to complex human behaviors.

We do not yet understand how biological processes in the human brain are affected by variations in vasopressin and influence human bonding and relationships. In rodents, the interactions between voles activated gene expression in the reward and reinforcement centers in the brain, regions that also control addiction. In the human brain vasopressin regulates water retention by the body and blood pressure and is also linked to aggression.

## LOOKING AHEAD

This chapter deals with the monumental effort made by scientists to learn the DNA base sequences of all 46 chromosomes and to pinpoint the locations of all the genes in the human genome. On completing this chapter, you should be able to do the following:

- Explain the most surprising features of the human genome as revealed by the DNA sequence of the human genome.
- Describe the technical and ethical challenges that faced the Human Genome Project scientists.
- Describe how geneticists and DNA technologists go about developing gene linkage maps and physical maps of genomes such as the human genome.
- Summarize the processes by which the sequence of bases is determined in a particular gene.
- Identify the organisms whose genomes have already been sequenced and relate what surprising facts were revealed by each new DNA genome sequence.

- Describe how the sequence of the dog genome helped us to understand more about certain human diseases.

## INTRODUCTION

On a January afternoon in 1989, a group of biologists, ethicists, industry scientists, engineers, and computer experts gathered in a conference room at the National Institutes of Health (NIH) and listened to Norton Zinder, a molecular biologist from Rockefeller University (New York), announce the the official beginning of the most ambitious scientific endeavor ever undertaken. "Today we begin," declared Zinder, "Today we are initiating an unending study of human biology. Whatever else [happens] … it will be an adventure, a priceless endeavor."

Those words launched the Human Genome Project (HGP), a monumental scientific effort that rivaled in scope and importance the Apollo program, which put humans on the moon. The goal of the HGP was to determine the specific sequence, or the "order," of every DNA letter (base) in the human genome. This effort was of paramount importance because it gave scientists access to the human master plan, all the genetic information needed to make a human being. The success of this ambitious goal has tremendous implications for the future, and represents an awesome gift for our children and our children's children.

The typical adult human body is made up of about 220 different types of cells. Each cell in the body contains chromosomes that carry human genome DNA. All the cells in the same human body carry the same human genome DNA. Each human chromosome contains one very long, linear double-stranded DNA helix molecule extending from one end of the chromosome to the other end. As discussed earlier (see Chapters 2 and 3), the genetic instructions in the human genome are written in the DNA molecular letters A, G, C, and T, and are arranged in DNA sentences that convey the genetic information needed to make specific proteins to create, build, and maintain each unique, individual human being (Figure 6.2).

In this chapter we will find out about the amazing Human Genome Project, the people involved, and the secrets revealed. The goal of the HGP was to determine the DNA sequence of the entire "human genome" which is created when a sperm fertilizes an egg. The resulting fertilized egg, the zygote, inherits two copies of the human genome, one from Mom (in the form of 23 chromosomes) and one from Dad (23 chromosomes). The two parental human genomes are brought together to form the offspring's new human genome (46 chromosomes) (Figure 6.3).

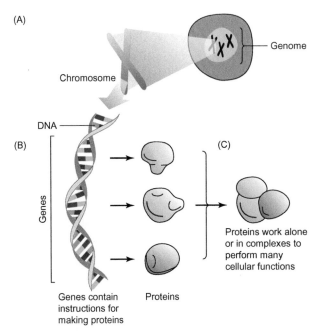

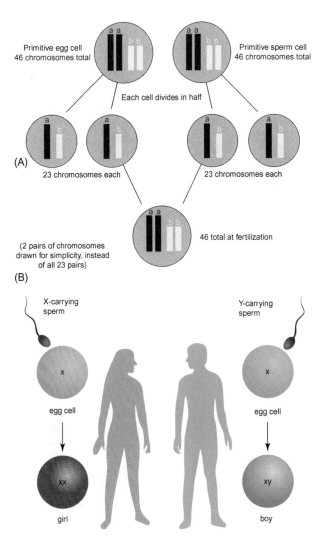

**FIGURE 6.2** Human genome instructions, written in the DNA language and expressed as proteins. (A) The human genome instructions are written in the DNA language and carried in chromosomes. (B) Genes are written in DNA letters (A, G, C, and T). (C) Genes encode information to direct the synthesis of proteins. The proteins interact with each other and with other components in the cell to build the diverse molecular machines needed to create a unique human being.

The goal of the Human Genome Project was to determine the DNA sequence of the human chromosomes, giving the world access to the human DNA master plan, the genetic information needed to build a living human being.

## MODEL ORGANISMS ARE FUNDAMENTAL TO GENOMICS

Major advances in automated DNA sequencing technologies and computer software innovations have rapidly increased the number of genome sequences available for study. Now, the genomes from many different organisms are accessible online with instant links to information on genes, DNA mutations, and related diseases; just click on your favorite organism's genome sequence and reveal the secrets written in the DNA.

Scientists studying genomes other than the human genome made many important contributions to the eventual success of the HGP. Research on smaller genomes helped scientists to develop the technological advances required to determine the DNA sequence of the much larger human genome years later. For decades before the HGP began, scientists were studying genomes from prokaryotic organisms such as the *E. coli* bacteria and more recently the genome of the white lab mouse was sequenced (Figures 6.4 and 6.5).

**FIGURE 6.3** Humans inherit 23 chromosomes from Mom and 23 chromosomes from Dad. (A) Each egg and sperm cell contains 23 chromosomes (different chromosomes are colored blue and yellow). (B) When a human sperm fertilizes an egg, a total of 46 chromosomes (two copies of each of 23 chromosomes) are inherited by the zygote. (C) Human gender is determined by the X and/or Y chromosomes: XX (female), XY (male).

The budding yeast *Saccharomyces cerevisiae* is a model single-celled eukaryote that is essential for modern research in genetics and cell biology (Figure 6.6). People in the ancient world used natural breeding methods to produce strains of yeast cells that make bread and wine. Research on the nematode roundworm *Caenorhabditis elegans* has had a huge impact on our understanding of nerve cell development in all organisms including humans. The origin, development, and fate of all the cells in the adult *C. elegans* worm have been traced from the beginning of the embryo to the adult, providing scientists with insights into the development of the worm's body systems, including nerve development (Figure 6.7). The

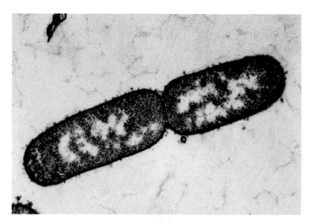

**FIGURE 6.4** *E. coli* bacteria is a model procaryotic organism, it lacks a nucleus. This *E. coli* cell has almost finished dividing into two cells. The dark staining material around the inside perimeter of each cell is the circular *E. coli* DNA chromosome that is associated with the inner cell membrane.

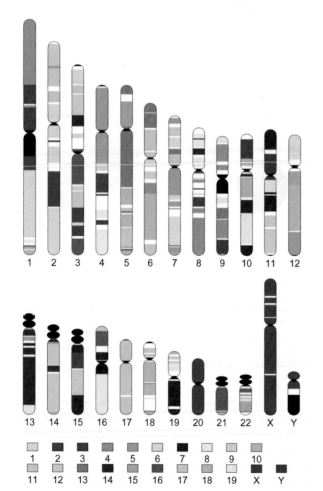

**FIGURE 6.5** Comparison of human and mouse chromosome DNA. The DNA sequences of human and mouse chromosomes are compared. The colored blocks represent segments of the human genome containing at least two genes that are in the same order in the mouse genome and in the human genome. Each color corresponds to a specific mouse chromosome as indicated by the color key at the bottom of the figure. Shown in black are the centromeres, pericentric regions, and the repeated DNA on some short chromosome arms.

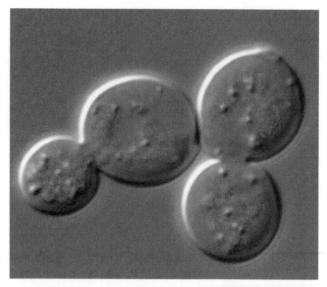

**FIGURE 6.6** The budding yeast, *Saccharomyces cerevisiae,* is a model organism used in many areas of basic research. The two budding yeast cells shown contain vacuoles *(red),* which are special compartments that transport proteins and lipids in the cell.

fruit fly, *Drosophila melanogaster* (Figure 6.8), and the zebra fish, *Danio rerio* (Figure 6.9), are both key model systems that are central to research on cell differentiation and specialized cell development. The major plant systems used in research are mustard weed (*Arabidopsis thaliana)* and corn (*Zea mays*) (Figure 6.10).

In 1996, genome research changed our understanding of the organization of life on earth (see Chapter 7, Figure 7.3). The existence of a third major branch of life, the **archaea** (halophiles and thermophiles), in addition to the bacteria (cyanobacteria and heterotrophic bacteria) and eukaryota (plants, animals, fungi) branches, was confirmed by DNA sequence analysis of the *Methanococcus jannaschii* genome, a thermophilic organism found growing in submarine thermal vents. By the early 2000s, many genomes were sequenced including genomes from fish, cats, and catfish.

Even before the start of the HGP, scientists studied and sequenced genomes from many model organisms including bacteria, single-celled eucaryotes, plants, and even some simple animal genomes.

## Dog Genes Hold the Secrets to Many Human Diseases

The DNA sequence of the dog genome was completed by the National Human Genome Research Institute (NHGRI) and released in 2005. The DNA sequence of the dog genome was particularly important to

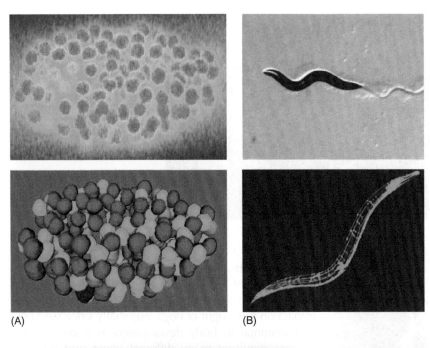

**FIGURE 6.7** Development of the *C. elegans* embryo. (A) Original 3D image of cells in a *C. elegans* embryo (top). Surface rendering of the segmented *C. elegans* embryo with cells randomly colored (bottom). (B) Adult nematode worm swimming (top). Adult *C. elegans* worm containing cells in the nervous system that are expressing green fluorescent proteins (GFP) (bottom).

(A)

(A)

(B)

**FIGURE 6.9** Zebra fish are model organisms for research on cell and tissue development. (A) Baby zebra fish. (B) Adult zebra fish.

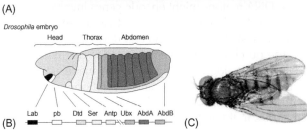

(B)      (C)

**FIGURE 6.8** Gene expression patterns in fruit fly embryo. (A) This *Drosophila* embryo shows the gene expression patterns early in embryo development. Each color represents the production of these proteins: Knirps *(green),* Kruppel *(blue),* and Giant *(red).* The yellowish areas indicate the cells that express both Knirps and Giant. The darker areas of the embryo contain cells that are not expressing any of these three genes. (B) Diagram of a *Drosophila* embryo with the sections labeled. (C) Adult fruit fly.

scientists who are searching for genes involved in cancer and other human diseases. At least half of the 300 genetic diseases inherited by dogs also affect humans, suggesting that these diseases probably involve similar genes in both dogs and humans. Genetic diseases are more common in certain purebred dogs, such as kidney cancer in German shepherds and eye problems in border collies. As we learn more about the genomes of purebred dogs, scientists have been able to use this information to identify genes involved in these disorders in dogs and humans.

Dogs are an amazing species; they exhibit more diversity in body size and characteristics than any other mammal. Just note the dramatic size difference between an Afghan hound and a Chihuahua (Figure 6.11).

(A)

(B)

(A)

(B)

**FIGURE 6.11** Dogs exhibit the greatest diversity in size and shape of all mammals. (A) These Afghan and Chihuahua dogs illustrate the dramatic size differences between breeds. (B) The DNA genome from the boxer, Tasha, was sequenced.

This is not the case for people; adult humans tend to grow to about the same height, between 5 and 6 feet tall. Different breeds of dog vary in size and exhibit diverse body characteristics, but adult dogs of the same breed usually grow to about the same height. For example, all adult Golden Retriever dogs stand between 20 and 24 inches at the shoulders.

To understand more about the genes that control height in dogs, scientists have focused on the insulin-like growth factor (IGF) gene, which determines overall body size in mice. Scientists decided to study Portuguese water dogs (the same breed as President Obama's dog, Bo) because this breed normally produces both small- and large-sized adult dogs. They analyzed the DNA genomes of 463 Portuguese water dogs and found two different forms (alleles) of the dog IGF gene, the "small" dog IGF allele and the "large" dog IGF allele, which differ by only two DNA base pairs. They discovered a correlation between genes and height: almost all of the small Portuguese water dogs carried the small dog IGF gene allele, whereas the large Portuguese water dogs carried the large dog IGF gene allele. This result was

supported by a study of 3000 genomes from 143 different dog breeds, which showed that the IGF gene also controls height in many other breeds. Perhaps it is surprising to find that a single gene can have such a large impact on the height of dogs, especially since we know that mammalian body development is a complicated process requiring many different genes and proteins. Whereas the IGF gene determines the adult height of specific breeds of dogs, other canine genes are involved in attenuating the impact of the IGF gene within a breed, for example, making one Great Dane grow a little bit taller or shorter than another Great Dane, but all Great Danes are very tall dogs.

In 2005, a team led by geneticist Kerstin Lindblad-Toh (Broad Institute of Harvard and MIT) determined the DNA sequence of the genome of a female boxer named Tasha (Figure 6.11). About 5% of the 2.4 billion base pairs of Tasha's genome DNA are identical to sequences in the human and mouse genomes, suggesting that these DNA regions might encode genes involved in fundamental biochemical processes that are shared by all mammals, such as body plan and development.

The purebred dogs we know today originated with the selective dog breeding programs made popular in Europe in the 1800s. Normally evolution works along with the environment to select for the survival of the "fit" dogs in the wild so that only certain dogs successfully transmit their genes to offspring. However, when people began to selectively breed dogs to enrich for traits such as short legs or longer snouts, the dogs were pampered and protected from the challenges of living in the wild, which effectively made them immune to the effects of the environment. However, in addition to desired traits, this breeding process selected for mutant genes that could negatively affect the health of the purebred dogs and would probably have been lethal in wild animals. These detrimental mutations persist in the gene pool of modern purebred dogs in the form of genetically inherited diseases and disorders such as hip dysplasia in large dogs and blindness in other breeds.

The DNA changes that alter the characteristics of a species or create an entirely new species take place

## Box 6.1 The Platypus Genome Is Really an Odd Duck

The strange looking duck-billed platypus resembles a creature made by combining characteristics from birds, mammals, and reptiles (Figure 6.12). Not only does the platypus have a duck-like bill, webbed feet, and lay eggs, but it also sports a fur coat, has a tail like a ping-pong paddle, and makes milk to feed its young. In the nineteenth century, the platypus was classified as a mammal, but unlike typical mammals that have lost their reptilian features during evolution the male platypus has retained the ability to produce venom in the spurs on its hind legs.

The platypus evolved from a lineage that branched off of the mammalian lineage about 166 million years ago, which gave rise to animals with features common to both mammals and reptiles, such as the ability to lay eggs. Two of these egg-laying mammals, called monotremes, still exist today, although most would agree that the "family resemblance" between the platypus and the echidna (spiny anteater) is difficult to appreciate (Figure 6.12). The unusual traits of the platypus raised important questions about the organization and evolution of the platypus genome.

An international group of scientists funded by the NHGRI analyzed the genome DNA from a female platypus from Australia named Glennie. Analysis of the platypus genome sequence proved to be very difficult because about 50% of the platypus genome is made up of repeated DNA sequences, which make it difficult to identify the original positions of the repeated DNA sequences in the native genome. The scientists found that the platypus genome contains about 18,500 genes that are distributed over 52 chromosomes, including some large chromosomes, many small chromosomes, and 10 sex chromosomes. Interestingly the DNA sequences in the platypus X chromosome are similar to the DNA sequences in a sex chromosome found in birds, in support of an evolutionary relationship between the platypus and birds.

Scientists found additional evidence that that the peculiar mix of platypus traits is reflected in the DNA of the platypus genome, including reptile-like genes that direct venom production, but evolved independently in modern venomous reptiles. The platypus genome encodes genes required for egg laying and milk production (lactation). The female platypus makes nutritious milk containing sugars, fats, calcium, and some milk proteins that closely resemble milk proteins made by mammals. These features suggest that the sophisticated lactation processes in humans and platypuses evolved in mammals before the Jurassic period.

The platypus and the echidna both descended from the most basic form of very early mammal and the platypus genome shares about 82% of its gene sequences with the echidna, human, mouse, and dog genomes. This is not surprising as all eukaryotes are made of cells with similar genomes that encode the proteins needed to make a functional cell.

(A)

(B)

**FIGURE 6.12** The egg-laying mammals, the platypus and echidna. (A) The platypus is a duck-billed, web-footed, egg-laying, venomous mammal. (B) The echidna (spiny anteater) is also an egg-laying mammal.

over millions of years and occur in short, rapid evolutionary bursts that accompany an increase in the accumulation of DNA mutations in the genome. Studies in dogs show that changes in the repeated DNA sequences located in key genes are correlated with certain characteristic traits. Repeated DNA sequences are short sections of DNA sequence that are repeated many times in tandem in the genome. Scientists studied the tandem DNA repeats in the genomes of 92 dog breeds and found a strong connection between the number of tandem DNA repeats in a particular gene and the function of the protein product expressed from that gene. The diverse characteristics common to different breeds of dogs is linked to the rapid accumulation of tandem DNA repeats at specific locations in the canine genome.

Tandem repeat mutations often result from mistakes made when the DNA genome duplicates (replication); sometimes the enzyme copying the genome DNA accidentally inserts a few extra DNA base pairs into the genome, which amplify into many tandem repeats with successive genome replication. Tandem DNA repeats are common in the genomes of most species; in humans DNA repeats play a key role in Huntington's disease (see Chapter 10).

The selective breeding of dogs for many generations influenced the evolution of the dog genome in ways that differ from standard evolutionary processes. Further research will investigate the mechanisms responsible for the rapid evolution of our canine friends.

## EARLY HUMAN GENOME MAPS

### Human Gene Linkage Maps

Before automated DNA sequence analysis was routinely available in the 1990s, maps depicting the human chromosomes were drawn by geneticists using genetic information to determine the relative locations of genes and other landmarks along a DNA molecule (Figure 6.13). These gene linkage maps show the genes on the chromosome DNA based on the frequency of recombination events that occur between genes when the cells go through meiotic cell division. Meiosis is the process whereby cells reduce the number of chromosomes by half, to produce germ cells like sperm and eggs. Recombination, also called crossing over, takes place when DNA sequences are physically exchanged between two chromosomes. This process can involve the exchange of only a few base pairs or many thousands of DNA base pairs (Figure 6.14). Recombination is a normal genetic process that purposefully "shuffles genes" during cell division, which contributes to genetic variation and helps to ensure that each human being inherits a unique human genome (except for identical twins).

The distance between two genes on the linear DNA in a chromosome is related to the frequency of meiotic recombination events that occur between the two genes. If the two genes are physically located close together on the chromosome, then the length of the DNA between the genes is too short to undergo frequent DNA exchanges, resulting in a low recombination frequency. In this case these adjacent genes are said to be "linked" and are almost always inherited together by individual offspring. In contrast, genes that are located far apart on the chromosome DNA undergo frequent recombination exchange events and as a result are inherited as independent, "unlinked" genes. The genetic distance determined from the frequency of recombination between two genes is

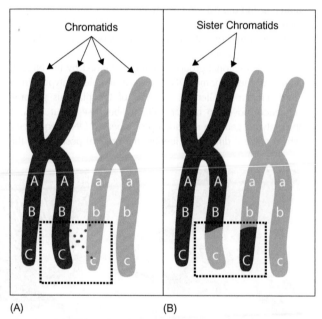

(A)                                    (B)

**FIGURE 6.14** Genetic linkage maps are based on recombination events (genetic recombination). (A) Two homologous chromosomes are shown, one dark blue, one light blue; each chromosome contains two chromatids. The light blue and dark blue chromosomes pair side-by-side during meiosis. The dark blue chromatids carry the normal (wildtype) gene alleles: A, B, and C. The light blue chromatid DNA carries the mutant versions (alleles) of the same genes: a, b, and c (shown in lowercase). When DNA exchange events (*red dotted line*) occur between chromosomes with very similar DNA sequences, the dark and light blue chromatids can exchange DNA helices (*box with dotted lines*). (B) After the dark blue and light blue chromatids exchange DNA (*box with dotted lines*), the order of the A, B, and C genes on the chromosome is not changed but the DNA strands encoding the normal wildtype gene (C) and the mutant (c) gene have been swapped by exchanging the DNA between chromosomes. The normal A, B, and C alleles start out on the dark blue chromatids, with the mutant a, b, and c alleles on the light blue chromatids. After the crossover event, the wildtype C allele (dark blue chromatid) has swapped positions with the mutant c allele (light blue chromatid), linking the light blue a and b alleles to the dark blue wildtype C gene. The other recombined chromatid contains the a and b alleles linked to the dark blue C allele.

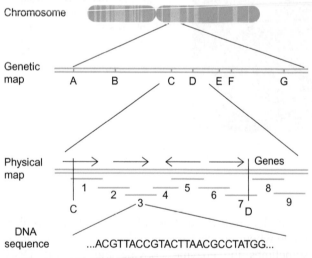

**FIGURE 6.13** Navigate the human genome using chromosome maps. The banded condensed chromosome at the top shows the constricted centromere region and displays the banding patterns along the chromosome arms, visible after staining for cytology. Chromosome maps show the relative positions of chromosome landmarks (centromere, telomere, etc.) and genes on the DNA molecule, which extends from one end of the chromosome to the other end. Genetic maps or gene linkage maps are derived from the results of meiotic recombination studies, which reveal the relative positions of the genes on each chromosome DNA helix. Physical maps are based on the most updated sequence analysis of the chromosome DNA (often from sequence analysis of overlapping genomic DNA fragments). The physical map of a chromosome is at a very high resolution and is based on the DNA sequence of the chromosome.

measured in centimorgans (cM), named for the famous early geneticist Thomas Hunt Morgan. A meiotic recombination frequency of 1% means that the two genes in question are tightly linked on the chromosome and are inherited together in 99% of all meiotic cell divisions. By definition, these genes are separated on the chromosome by a genetic distance of 1 cM, which is equivalent to a DNA distance of about 1 million base pairs.

## Down to Details: Physical Maps of the Human Genome

Physical chromosome maps are often based on known DNA sequences and accurately depict the positions of genes along the DNA at very high resolution (Figure 6.14). In 1992, a team of 35 coauthors published the first physical map of the smallest human chromosome 21, predicting the locations for genes involved in amyotrophic lateral sclerosis (Lou Gehrig's disease), epilepsy, and Alzheimer's disease.

A real milestone was reached in 1999 when scientists completed the sequence of the first human chromosome (chromosome 22); later that year researchers celebrated sequencing the billion(th) base pair of human genome DNA. David Page and colleagues (Whitehead Institute for Biomedical Research) published the first high-resolution physical map and sequence of the smallest human chromosome, the Y sex chromosome, in *Science* magazine (Figure 6.15).

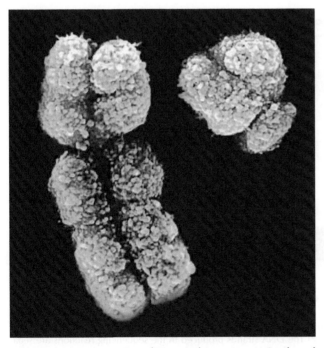

**FIGURE 6.15** Human X and Y sex chromosomes. Condensed mitotic human sex chromosomes are shown; X *(left)* and Y *(right)*.

Early maps showed the positions of genes on the chromosomes but contained little information about the sequence or molecular characteristics of the human genome. This changed with the development of rapid, automated DNA technologies that were essential to the huge success of the Human Genome Project.

## DETERMINING THE DNA SEQUENCE OF THE ENTIRE HUMAN GENOME

### History of the Human Genome Project (HGP)

Officially the Human Genome Project (HGP) was a 13-year (1990–2003) research effort staffed by U.S. and international scientists with the goal of determining the DNA sequence (the exact order of the DNA bases) of the entire human genome. This amazing feat required the combined skills and dedicated cooperation of hundreds of research scientists working in private and public laboratories around the world. In the end, the public- and private-sector corporate scientists published the human genome sequence at the same time in two top scientific journals.

The first documented proposal to sequence the human genome DNA was made by Robert Sinsheimer in 1985. Many noted scientists were strong advocates of the HGP, including Nobel Laureates Walter Gilbert and Paul Berg. The Human Genome Organization (HUGO) was established in 1988 to begin to organize genome sequencing efforts, followed by the first annual conference on human genome mapping and sequencing held at the Cold Spring Harbor Laboratory, Cold Spring Harbor, New York. Because of extensive government red tape and funding issues, the HGP was not fully underway until 1990.

### Public Support and Free Access to the HGP DNA Sequence

During the 1990s, many established and new corporations moved into the biotechnology arena and began to participate directly in human genome research. The fact that most of the U.S. funding for Human Genome Project came from taxpayer funds earned the HGP the nickname "the public human genome project," which easily distinguished the taxpayer-funded public HGP from the corporate-sponsored "for-profit private-sector HGP." The international race to sequence the human genome promoted stiff competition between these two groups, the consortium of publicly funded HGP scientists and the privately funded research sponsored by the biotechnology companies.

James D. Watson, co-discoverer of the DNA double helix, was appointed to be the first director of the public HGP (see Chapter 2). Watson reassured the public, Congress, and the scientific community that credible science, and not politics, would be the primary focus of the taxpayer-funded HGP research. Watson also committed HGP funds to support studies on the ethical, legal, and social issues (ELSI) that arise as a result of human genome research. In 1994, the HGP ELSI Working Group supported the federal Genetic Privacy Act, legislation designed to regulate the collection, analysis, storage, and use of human DNA samples and personal genetic information. ELSI established free online access to DNA information resulting from HGP research, but the human DNA information generated by the private research corporations was not accessible to the public without charge. ELSI continues to support programs to educate the U.S. public about the human genome and the impact of the human genome sequence information on society.

The next HGP director, Francis Collins, realized that the private biotechnology companies were fast becoming much more competitive in the race to sequence the human genome (Figure 6.16). Collins successfully refocused the goals of the public HGP and made important changes that dramatically increased the productivity of the public research efforts. The technologies used in public HGP labs were modified to take advantage of the improved DNA sequencing and cloning methods used by the private labs.

Techniques to sequence DNA were first developed many years before scientists conceived of sequencing the entire human genome.

In the 1970s and 1980s, DNA sequencing experiments were performed by hand in the lab, which required the use of methods that were expensive, labor

**FIGURE 6.16** Francis Collins, director of the HGP, refocused the public HGP goals. As director, Collins made changes that increased the productivity of the public research efforts to sequence the human genome.

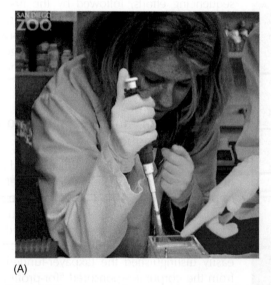

(A)

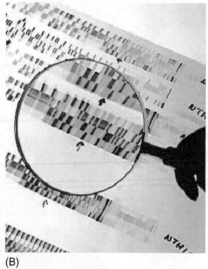

(B)

**FIGURE 6.17** Before DNA sequencing became automated, DNA sequence analysis was performed "by hand" and read "by eye." (A) Scientists use a pipetting device to add a small DNA sample into the well of a gel. During electrophoresis, the DNA molecules separate in the gel according to their length. The longer DNA strands migrate slowly and are near the top of the gel, and the shorter DNA strands migrate quickly in the gel and run near the bottom of the gel (see Chapter 3). (B) The DNA strands produced in the sequencing reactions are separated on a gel, and the bands of DNA are detected by exposing the gel to an X-ray film. The order of the DNA bands in the four sequencing lanes indicates the DNA sequence (see Chapter 3).

intensive, and required using dangerous chemicals and radioactive reagents (Figure 6.17). However, by the mid-1980s the Sanger chain termination method of DNA sequence analysis, named after Fred Sanger, who developed the technique and won the Nobel Prize for science and medicine twice, was fast becoming the method of choice everywhere. The Sanger chain termination method of DNA sequence analysis is an ingenious experimental approach that utilizes the same enzyme that is used by cells to make copies of DNA. When a DNA strand is copied or replicated, the enzyme adds nucleotide building blocks to the new growing strand of DNA (see Chapter 2). Sanger used four substitute "dideoxy-" nucleotides to the reaction mix, which were different from the normal building blocks because the dideoxy- nucleotide building blocks lack an essential chemical group required for DNA synthesis, the 3' hydroxyl group (—OH) (Figure 6.18). The dideoxy-DNA building block causes the DNA strand to stop growing (terminate) at a specific spot on the DNA template strand. Analysis of the collection of terminated DNA strands will reveal the sequence or order of the bases in the DNA strand. The quality of DNA sequence analysis depends on having a good DNA template strand to be copied by the enzyme in the sequencing reactions.

Early studies on human genes required that the individual genes be painstakingly identified by years of research using genetic, biochemical, and molecular methods to find and clone the specific DNA of interest. At that time researchers studying human genes needed enough information to be able to determine the location of a single gene in the human genome, even though they did not know the DNA sequence of the gene or even the chromosome location of the gene. For example, the search for the specific human gene that causes Huntington's disease involved genetic studies on families with this inherited disease (see Chapter 10).

Scientists developed a method called shotgun DNA cloning to permit easy preparation of template DNA for sequencing analysis. Shotgun cloning and Sanger chain termination (dideoxy-) DNA sequencing became standard approaches used together to determine the sequence of very large DNA molecules including genomes (Figure 6.19). The long genome double-stranded DNA molecules are cut into shorter DNA fragments, which are then inserted, or cloned, into plasmid vectors (see Chapters 4 and 5). From then on the plasmids will carry the inserted DNA fragments and will replicate as part of the plasmid DNA molecules. The scientists then use the Sanger chain termination method to sequence the foreign DNA inserted into the vectors. The DNA sequence information is sent to the computer programs for analysis (Figure 6.20). To actually reassemble an entire genome sequence, the

scientists must piece the sequenced DNA fragments back together, bit by bit, until they have reconstructed the original genome sequence. Computers play an essential role in this entire process.

The Sanger chain termination method is very well suited for use in automated DNA sequencing technology. But even the best DNA sequence method yields only several hundred bases of readable DNA sequence per sequencing reaction.

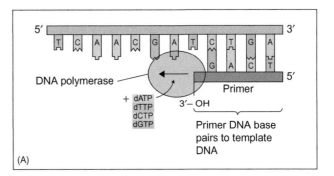

(A)

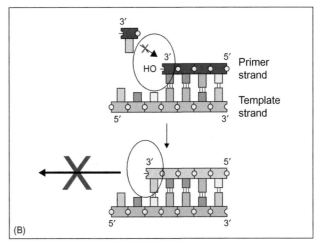

(B)

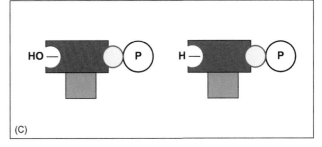

(C)

**FIGURE 6.18** Dideoxy DNA-building blocks stop DNA synthesis. (A) For DNA synthesis to begin, a short DNA primer must be base paired to the template DNA strand. The DNA primer contains a "free" 3' OH (hydroxyl) group that is required to add the next building block during synthesis of the new DNA strand. (B) When a dideoxynucleotide chain terminator *(circled)* is added to the 3' end of the primer, the lack of the 3' OH means that DNA synthesis must stop (chain termination) because the polymerase enzyme cannot add to the primer. (C) The differences between the deoxy *(left)* and dideoxy *(right)* nucleotides are fundamental to chain termination DNA sequencing.

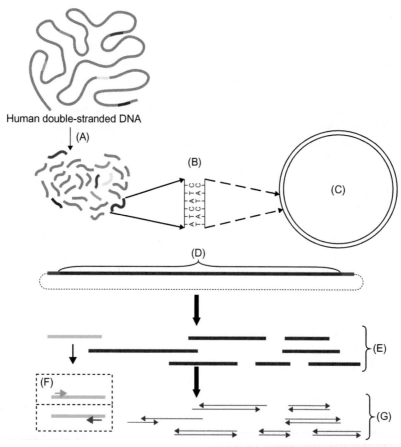

**FIGURE 6.19** Shotgun cloning used to sequence DNA genomes. (A) Shotgun DNA cloning and sequencing is an efficient way to determine the sequence of large genomes like the human genome. (B) The random fragments are cloned randomly into plasmid vectors, without identifying or characterizing the DNA fragments until after DNA sequence analysis is complete. (C) The recombinant plasmids are introduced into bacterial cells where they replicate to provide enough cloned human genome DNA for sequencing. More recently, polymerase chain reaction (PCR) permits scientists to skip the cloning steps entirely and directly sequence the genome DNA itself (see Chapter 5). (D) The human DNA fragment *(blue line)* is cloned into a plasmid vector *(dotted line)*. (E) The double-stranded human DNA is cut into shorter double-stranded DNA fragments. (F) Each individual DNA sequencing reaction *(dotted box)* contains a DNA template to be sequenced *(light blue line)* and a DNA primer *(green arrow)* that is complementary to *one* region of the DNA fragment. In a separate reaction, a different primer *(red arrow)* is used to determine the sequence of the opposite strand of that DNA fragment. (G) Many different short primers *(purple arrows)* are used to determine the sequences of all of the DNA strands *(dark blue lines)*. The results from sequencing are sent to computers that analyze and align the DNA sequences and reconstruct the sequence of the original cloned DNA fragment, eventually revealing the sequence of an entire human chromosome.

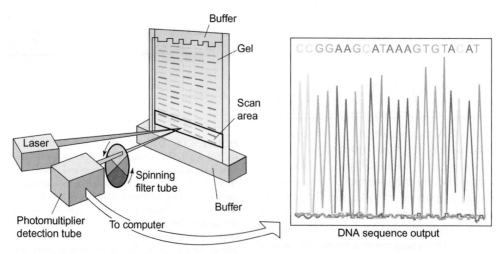

**FIGURE 6.20** Automated DNA sequence analysis requires a DNA sequencing machine. This DNA sequencing machine automatically performs the chemical reactions, resolves the DNA products of the reactions on a gel or using a capillary system, and sends the DNA sequence information directly to the computer.

(A)           (B)           (C)

**FIGURE 6.21** Decoding the human book of life. (A) Craig Venter and Francis Collins were interviewed in the popular press, including *Time* magazine (2000); (B) Craig Venter, President Bill Clinton, and Francis Collins; (C) the public and private Human Genome Projects DNA sequences were published in two scientific journals at the same time. Celera and Craig Venter published in *Science,* whereas the International Consortium and Francis Collins published in *Nature.*

In 1988, J. Craig Venter and Mike Hunkapillar founded Celera Genomics, which focused on developing the sophisticated techniques needed to manipulate and sequence large genomes. They built the first DNA sequencing instruments and pioneered the new technologies needed for automated DNA sequencing. Many scientists were skeptical of Venter's focus on automated DNA sequencing analysis, including James Watson, co-discoverer of the structure of the DNA helix (see Chapter 2), who was then director of the public Human Genome Project. Watson challenged Venter's priorities in public and Venter responded by stating that Celera Genomics would not only finish a draft DNA sequence of the human genome by 2001, but they would spend only $200 million to complete the project. With this exchange the race between the public and the private human genome scientists to sequence the entire human genome began in earnest.

At Celera Genomics Venter and Hunkapillar built a huge DNA computer facility the size of a football field to analyze DNA sequence information, which was at the time the largest civilian supercomputer in existence. They outfitted the Celera labs with hundreds of newly developed automated DNA sequencing machines that transferred the DNA sequence information directly to the computer (Figure 6.20). In time the revolutionary technological advances made by the Celera scientists were adopted by most Human Genome Project research groups, both private and public, which had an enormous positive impact on the progress of the work to complete the human genome sequence.

As DNA sequence analysis became automated and much less expensive, many biotechnology companies began to offer DNA sequencing services in addition to customized recombinant DNA cloning and gene library services. Today modern research labs usually send DNA out to companies to be sequenced or use DNA sequencing kits, which provide the reagents, controls, buffers, and the necessary enzymes to perform DNA sequencing reactions in research labs and in many teaching labs as well. Some biotechnology companies have focused on developing therapeutic agents derived from the results of human genome research, while others have worked on the commercial computer software used to construct and maintain DNA and protein sequence databases (see Chapter 7).

In 2001 nearly fifty years after the discovery of the DNA double helix, Francis Collins and Craig Venter, the leaders of the public and private HGP groups, announced the release of the first version of the human genome DNA sequence, ahead of schedule and under budget. The two competing research groups agreed to publish their research simultaneously in two equally prestigious scientific journals, the British *Nature* and the American *Science* (Figure 6.21).

## WHAT WE LEARNED FROM THE HUMAN GENOME SEQUENCE

### Only 2% of Human Genome DNA Codes for Proteins

A new human DNA genome is made when the zygote inherits a complete genome complement of 46 chromosomes, 23 chromosomes from Mom and 23 chromosomes from Dad (Figure 6.22). How similar are the DNA sequences of the two different versions of the same human chromosomes, for example, chromosome 21 from Mom and chromosome 21 from Dad? Since

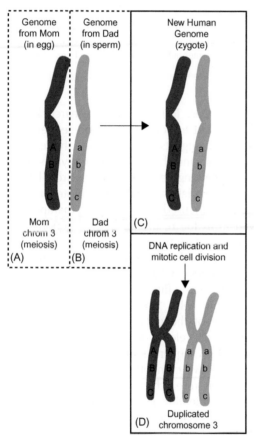

**FIGURE 6.22** A new human DNA genome is created when the human zygote inherits a complete set of chromosomes from Mom and Dad. (A) Mom has one copy of each chromosome in each egg cell, including chromosome 3; meiosis cell division reduces the number of chromosomes when producing egg and sperm cells. (B) Dad has one copy of each chromosome in each sperm cell, including chromosome 3, meiosis cell division reduces the number of chromosomes when producing egg and sperm cells. (C) The DNA in chromosome 3 from Mom and the DNA in chromosome 3 from Dad have similar sequences, but the genomes are not identical; Mom and Dad's chromosome 3s encode different alleles of the A, B, and C genes. The dark blue chromosome 3 DNA carries the normal (wildtype) A, B, and C genes. The light blue chromosome 3 DNA carries the mutant alleles of the a,b, and c genes. (D) When the zygote begins to divide by mitosis to make more cells, all of the cellular components, including the chromosome DNA, are duplicated to prepare for the new cells. Mitosis maintains the number of chromosomes in the cells, whereas meiosis reduces the number of chromosomes by half.

the first draft of the human genome sequence was released in 2001, the genome sequence was updated and in 2008 the human genome contained 3165 million (3.2 billion) base pairs of DNA carried on 23 chromosomes. Most human cells (except sperm and egg) contain 23 chromosome pairs (46 chromosomes), carrying 6.2 billion bp of DNA in each nucleus.

The human genome contains noncoding DNA, regions of the genome DNA that are not copied into RNA and do not code for proteins. Still it was a big surprise to learn that about 98% of the human genome does not code for proteins expressed in the cells.

All of the genes needed to make a human being are encoded in only 2% of the human genome DNA. This means that only a relatively small number of DNA base pairs in the genome carry the instructions for a human master plan.

## How Many Human Genes Are in the Human Genome DNA?

For years scientists have estimated the total numbers of genes in the genomes of various organisms using different methods, but accurate numbers are still often difficult to find, even though genome sequencing projects are now routine. Part of the problem is that the definition of a "gene" has changed as scientists learned more about the structure and function of different types of genes. It was especially important to include the fact that the majority of human (and most other eukaryotic) genes contain intron and exon sequences (Figure 6.23). This complicates the issue because in many cases the exon of one gene can serve as the intron for a different gene. In addition, one DNA coding region in the genome can serve as the template for RNA transcripts, which are processed in such a way that they can actually code for a number of different proteins. DNA sequence analysis shows that the budding yeast genome encodes 5,770 genes, the roundworm contains 19,427 genes, and the fruit fly genome has 13,379 genes. It seems entirely reasonable that humans should need

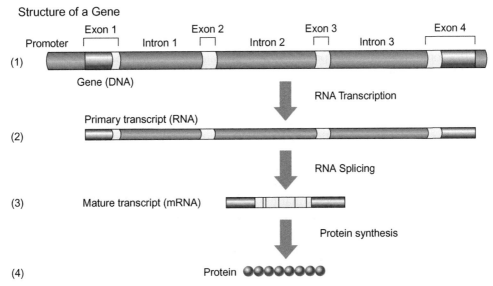

**FIGURE 6.23** Most human genes contain introns and exons encoded in the genome DNA. (1) When the gene is expressed in the cell, the DNA is first copied into a long precursor RNA transcript. (2) This precursor RNA contains the introns and exons in the coding region, which dictates the order of amino acids in the new protein chain. (3) Both exons and introns are included in the long precursor RNA transcript, but RNA splicing enzymes remove the introns from the precursor RNA to make a messenger RNA. (4) The final messenger RNA (mRNA) sequence contains only the exons precisely linked together to resurrect the complete protein coding region and is ready for transport to the cytoplasm and translation.

a significantly larger number of genes than single-celled yeast or fruit flies (Figures 6.8–6.11).

Before the start of the HGP, most people estimated the total number of human genes at between 80,000 and 100,000. So imagine the surprise in 2001 when the HGP scientists announced that the human DNA genome encodes only between 30,000 and 35,000 genes. The recent updates of the human genome DNA sequence provided a much more accurate picture of the size and complexity of the human genome. In 2008 the total number of different human genes was estimated at between 20,000 and 25,000.

Many types of eukaryotic cells, including human cells, use the process of RNA splicing to permit cells to express more than one protein product from a single gene (see Chapter 3). The genome DNA containing the intron and exon sequences in a gene are copied into long precursor RNAs containing both introns and exons (Figure 6.23). RNA splicing enzymes remove the introns and produce final mRNAs containing only the exons needed to be translated into an amino acid protein chain (see Chapter 3).

Differently spliced RNAs copied from the same gene can carry different RNA information that can be translated into different proteins. This gene expression strategy increases the number of different proteins available to the cell without substantially increasing the total number of genes encoded by the genome.

An average human gene occupies about 3000 bp (3 kb) of genome DNA, but because the intron sequences vary considerably in length, almost all human protein-coding genes are significantly longer than the actual protein-coding region of the genome. At this time the largest known human gene encodes the dystrophin protein. Mutant dystrophin proteins cause muscular dystrophy (see Chapter 10). The dystrophin gene covers 2.4 million bp of genome DNA, including many exon and intron sequences. The long precursor RNA copied from the dystrophin gene is spliced to remove the introns, which converts each dystrophin RNA precursor into a much shorter messenger RNA (mRNA) containing only the exons needed to encode the dystrophin protein.

> Most protein-coding genes in the human genome are significantly longer than needed to account for the protein product of the gene. This is explained because the genes in the genome DNA contain intron as well as exon sequences.

Scientists have long studied genes that are copied into RNA but are never translated into proteins; instead these RNA molecules perform important functions in the cell as RNA strands, not proteins. For example, transfer RNA genes code for the transfer RNA (tRNA) molecules that carry specific amino acids to the ribosomes during protein synthesis (see Chapter 3). Another example is ribosomal RNA (rRNA) genes that encode small rRNA products. These rRNAs form extensive base-paired regions and single-stranded loop regions that function in ribosome assembly and protein synthesis (Figure 6.24). Discovered more recently are the small nuclear RNA (snRNA), small nucleolar RNA (snoRNA), and other small RNA genes, which produce a variety of short RNAs that function in a long list of essential cellular processes in addition to RNA splicing.

The human genome DNA sequence gave scientists the first detailed view of how genes are organized along the chromosomes. The sequence shows that the protein-coding genes are not evenly distributed along the human chromosome DNA. Instead the genes are arranged in random gene clusters separated on the chromosome from other gene clusters by vast regions of noncoding DNA. Regions with clusters of human genes are often flanked by sections of noncoding DNA containing CG base pairs repeated over 30,000 times [-CGCGCGCGCGCGCG-], which are called CpG islands. It is possible that the CpG islands in the genome form transcription barriers to prevent the RNA polymerase enzymes from copying past the end of a gene and into the flanking noncoding DNA.

## NINETY-EIGHT PERCENT OF THE HUMAN GENOME IS NONCODING DNA

### Jumping DNA Elements Move around the Genome

The human genome DNA sequences that code for human proteins were the focus of intense study for years before the HGP began. With the completion of the human genome sequence scientists have increased research on the more mysterious noncoding DNA sequences that make up the remaining 98% of the human genome DNA.

Believe it or not, about half of the noncoding DNA in the human genome contains transposable DNA elements that can physically move (jump or translocate) from one DNA location to another site in the human chromosome DNA (Figure 6.25). This amazing behavior is not restricted to the human genome. Many different mobile DNA elements have been discovered in prokaryotes and other eukaryotes. These transposable DNA elements move around the genome using different molecular mechanisms. Sometimes transposable elements are called "selfish DNA" because these DNA elements act like parasites that are reproduced along with the organism's genome. Some transposable DNA elements have their own genes that also move when the DNA element moves. Transposable elements sometimes insert into active genes that are transcribed,

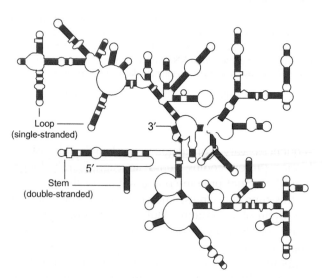

**FIGURE 6.24** Base-paired secondary structure of ribosomal RNA (rRNA). The secondary structure of this rRNA forms a complex pattern of short double-stranded stems and unpaired single-stranded loops and bubbles. The structure shown is a continuous single strand of RNA, starting with the 5' end and finishing at the 3' end. The rRNA folds into a structure with other rRNAs and many special proteins to make a ribosome.

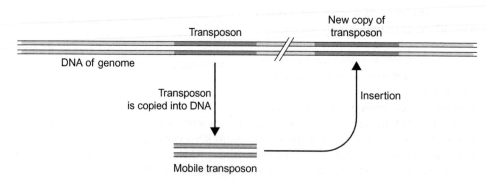

**FIGURE 6.25** Transposable DNA elements can jump around the genome. Some transposons can move from one location in the genome to another site in the genome; some use a "copy, cut, and paste" mechanism.

and they can disrupt gene expression. Fortunately, transposable elements rarely insert into active genes in the human genome.

## Repeated DNAs Are Very Common Genome Sequences

About half of the noncoding DNA in the human genome is made up of repeated DNA sequences; the human genome has more repeated DNA than the fruit fly (3%), worm (7%), and mustard weed (11%) genomes. Examples include very short satellite DNA sequences that are repeated thousands of times in tandem in the human genome. Highly repeated short DNA sequences are commonly located near the telomeres and the centromeres of most eukaryotic chromosomes (Figure 6.26). DNA control elements in the genome, which regulate gene expression and control the cell cycle, occupy a small fraction of the overall human genome DNA. These control elements include transcriptional promoter and enhancer DNA elements that regulate gene expression and "DNA replication origin" elements involved in the initiation of DNA replication and chromosome duplication. DNA control elements are located at many critical positions in the genome where they influence cell functions by binding to special regulatory proteins made in the cell.

## INDIVIDUAL GENOMES AND GENETIC VARIATION

## Genetic Variation in the Human Genome

The release of the first version of the human genome sequence promoted researchers to find out more about the sequences of individual human genomes.

The development of new DNA technologies and rapid automated DNA sequencing machines made it possible to determine the DNA sequences of hundreds of human genomes since 2001. DNA sequencing analysis proves that each person carries a unique genome DNA sequence (see Chapter 8). Except in the case of identical genetic twins, an individual's genome DNA sequence is detectably different from the sequences of all the other human genomes on earth.

Initial sequence comparison studies suggest that the DNA sequences of different human genomes are almost identical, 99.9% the same. So no matter how different two people might look on the outside, their genomes have almost identical DNA sequences. Later genome studies show that human genomes might vary by as much as 4.5 percent.

The rare differences between the DNA sequences of individual human genomes, called DNA polymorphisms, have become extremely useful genetic markers for human genome studies. There are correlations between the susceptibility to certain diseases and the inheritance of genomes containing specific DNA variations. Detecting the small DNA differences between human genomes is the basis for the powerful DNA fingerprinting technology used in DNA forensics to identify suspects (see Chapter 8).

Restriction fragment length polymorphisms (RFLP, pronounced "rif-lip") were the earliest type of detectable DNA differences used in gene analysis (see Chapter 10). RFLP "chromosome markers" occur when the DNA of one genome is different at a location that is normally recognized in the unchanged genome and is cut by a highly specific restriction enzyme. The DNA base pair sequence that is altered by the polymorphism prevents the restriction enzyme from cutting at that specific DNA site in that genome. A DNA genome lacking this specific polymorphism can be cut

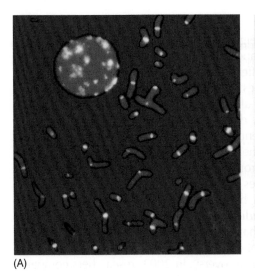

(A)

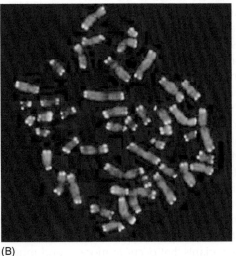

(B)

**FIGURE 6.26** Chromosomes with centromeres *(yellow)* and telomeres *(light blue)*. (A) A human interphase nucleus *(top left)* is surrounded by many metaphase chromosomes with the repeated DNA surrounding the centromere regions stained yellow. (B) The repeated DNAs at the ends or telomeres of these metaphase human chromosomes are stained light blue.

by the enzyme at that DNA sequence in the genome. The RFLP marker at that site in the genome reflects the different action of the restriction enzyme, whether or not it cuts the different genomes at a specific RFLP site in the genome. Cleaving the genome DNA will generate DNA fragments of different but predictable lengths that can be routinely detected by gel electrophoresis and Southern blot hybridization (see Chapter 5).

## Single Base and Copy Number Variations in the Human Genome

Single nucleotide polymorphisms (SNPs) are the single base pair differences between the genomes of different individuals. The HapMap Project and the Public SNP Consortium are cataloging the millions of single base differences between individual human genomes.

Researchers at the Wellcome Trust Sanger Institute (Cambridge, United Kingdom) compared the DNA sequences of genomes from 270 people from different ethnic groups: Yoruba (Nigeria), European descendants (United States), Han Chinese (Beijing), and Japanese (Tokyo). The team measured the number of copies of specific genes (**gene copy number**) in each genome using powerful microarray screening techniques to locate copy number variations (CNVs) in the genome DNA (see Chapter 13). They identified 1447 cases of CNVs, which involved about 12% of the human genome DNA, which is a surprisingly large amount of gene copy number variation. The 1000 Genomes Project began in January 2008, when an international consortium of research scientists announced their goal of using SNPs to provide a very high-resolution map showing the positions of biomedically relevant genetic variations in the human genome. This detailed map of genetic variation in the human genome will provide an important new tool for medical research that will help scientists to identify genes that are linked to the different SNPs and other genetic changes and to make the data available to the international scientific community through free public databases.

The scientists working on the 1000 Genomes Project focused on studying the small fraction of the genome DNA that differs between individual human genomes, looking for valuable clues to find out why individuals differ in their susceptibility to diseases, drug responses, and sensitivity to environmental factors. The project started with the comprehensive catalog of human genetic variations compiled by the International HapMap Project. The information in the original HapMap catalog was fundamental to the identification of more than 130 DNA variants that are genetically linked to common human diseases. However, the HapMap data are limited to detecting only the genetic variants that occur in more than 5% of the human genomes.

The new technologies of the 1000 Genomes Project can detect rare genetic variations in the human genome sequences. Scientists continue to search for possible new correlations between these rare genome DNA variations and observed differences in behaviors, talents, or personalities among individual people.

## HUMAN AND CHIMPANZEE DNA: WHAT MAKES US HUMAN?

Each species has its own unique genome DNA sequence. These special genome sequences distinguish between different species. Scientists are studying the similarities and differences among the genomes of different species to try to determine what sets humans apart from every other species. Comparisons among eukaryotic genomes show that the human genome contains DNA sequences that are also found in the genomes of other eukaryotic organisms. Some DNA sequence similarities were found when the human, monkey, and mouse genomes were compared, suggesting that these similar genes might be needed for biological processes common to all eukaryotes. For example, all mammals have genes that control the development and location of tissues and organs in the body. In terms of body structure, mouse and man are actually similar, one head, body trunk, and four appendages. All eukaryotic organisms are made up of cells with nuclei and DNA genes that must divide to make more tissues for the organism to grow.

All eukaryotic cells need different enzymes to replicate the genome DNA, control the cell cycle, process nutrients, and make proteins. These biological processes are directed by very similar genes in humans, mice, and bananas.

Six to eight million years ago, the lineages two groups of apes separated and evolved independently into the different species, humans and chimpanzees. The genomes of modern-day chimps and humans carry a record of all of the DNA changes that make humans biologically different from chimpanzees, written in the DNA sequence.

Researchers are analyzing the DNA sequences of genomes from many different organisms, including chimpanzees as well as organisms that are very different from humans, such as puffer-fish. Initial comparison studies of the human and chimp genomes indicated that the chimp and human DNA are about 98.5% identical, an observation that raised questions and much controversy. In 2003, researchers compared the DNA sequences of human chromosome 21 with the corresponding chromosome from the nonhuman primates, chimpanzee, orangutan, rhesus macaque, and the woolly monkey. In all cases, the DNA was rearranged much more frequently during the evolution of the primate genome than was previously thought, and large regions of the human and chimp genomes were found to have very different DNA sequences. Considering the genetic similarities between chimps and humans, understanding the differences between the DNA of these two species promises to reveal more important information about what makes humans human, how we can talk, walk upright, and read.

How can scientists find the specific DNA differences between the human and chimp genomes that might identify a genetic variation with an impact on evolution? Important clues have come from studying genes that are involved in developing characteristics that are distinctly human compared to chimps. For example, humans typically have larger brains than chimps, suggesting that scientists might find functionally important DNA differences (polymorphism) between the human and chimp genomes if they analyzed the sequences of the genes involved in promoting growth of the human brain (Figure 6.27). One such potentially important DNA difference is located in the MYH16 gene, which is mutated in the human genome, but is unchanged in the chimp genome, suggesting that this specific DNA change in the MYH16 gene could have played an important role in the evolution of the human brain.

The normal MYH16 gene codes for myosin, a muscle protein that is made in both humans and chimps. A mutation affecting MYH16 in humans produces defective myosin proteins that actually weaken the jaw muscles. When muscles change in physical strength, the bones attached to the muscles often change as well. Over evolutionary time the weak jaw muscles could have reshaped the human skull sufficiently to accommodate a larger human brain. Possibly the mutation in the human MYH16 gene allowed the subsequent evolutionary changes to occur. The MYH16 gene mutation appeared in the human population about 2.4 million years ago, which is about 400,000 years before human ancestors developed smaller jaw muscles and the human brain started to grow larger.

We know that humans can talk and chimps cannot use language, but scientists do not know when humans actually acquired speech. People who inherit a specific DNA mutation in the FOXP2 gene have speech difficulties, indicating that the FOXP2 protein is important for human speech. Studies show that two DNA changes occurred in the FOXP2 gene between 100,000 and 200,000 years ago, which was after humans diverged from the chimpanzee. Researchers think that these changes in the FOXP2 gene have contributed to the continual improvement in human speech over the past 200,000 years.

**FIGURE 6.27** The human brain is larger than the chimp brain. Chimp brain and skull *(top)* compared to the larger human brain and skull *(bottom)*.

Studies on genetic variations in the genome sequences of human populations around the world might reveal important information about human evolution and assist in identifying genes involved in human development and human diseases.

## WHAT WE STILL NEED TO LEARN ABOUT THE HUMAN GENOME

Despite the great advances made in genome research and DNA biotechnology, there is still quite a lot about the human genome that remains a mystery. It is important to know much more about how our own genes work so we will be better able to develop effective treatments for diseases that occur when our genes go wrong. The Human Genome Project gave the world the first full "printout" of the human DNA master plan, but it still left many unanswered questions.

Scientists continue to work on finding the genome locations and biological functions of all the human genes. In addition, research is needed to learn more about the impact of gene copy number and changes in gene regulation when cells are exposed to different environments and stress conditions. Researchers will also continue to learn more about chromosome organization and the relationships between DNA sequences and landmark chromosome structures such as centromeres and telomeres. The vast amount of noncoding DNA sequences in the human genome continues to raise many questions about the distribution, information content, and functional importance of noncoding DNA sequences. Of particular interest are DNA elements that move or jump around the human genome as well as further work on the many types of repeated DNA sequences in our DNA.

## Benefits of Future Genome Research

Future human genome research has much to offer biomedical science with the successful correlation of SNPs (single-base DNA variations among individual genomes) with specific inherited diseases and traits such as disease-susceptibility, and diverse individual reactions to drugs and other treatments. Scientists are using many approaches including molecular genetics to take advantage of the wealth of information generated by the HGP to develop new drugs, some of which are custom-designed drugs based on the genetic profile of specific individual patients (personalized medicine; see Chapter 13). In the future, scientists will be able to routinely evaluate the health risks associated with environmental exposure to radiation, chemicals, and toxins.

Large-scale genome sequencing and comparison studies on thousands of individual human genomes continue to reveal more about the inheritance of genes that control complex traits and the specific proteins involved in multigene diseases. Genome-wide gene expression profiles of human cells contribute to our understanding of the genes that control cell and tissue differentiation and cancer cell development. Studies show that specific epigenetic changes made to the human genome at fertilization are responsible for the development of embryonic stem cells into various adult cell types. These changes also play a key role in the development of new stems cells that resemble embryonic stem cells, but which were developed from adult cells, not embryonic stem cells (see Chapter 12).

Microbial genomics research is increasingly important for our understanding of deadly pathogens and ways to rapidly detect and treat diseases caused by these microbes. The impending threat of global warming and the increases in oil and gas prices have focused public support on research to develop new sources of "green" sustainable energy, as well as better ways to detect and monitor environmental pollutants. Microbe-based methods are under development that will safely and efficiently clean up toxic waste spills, which would protect the environment and save billions of dollars in future toxic cleanup costs.

Genome DNA studies (genomics) will continue to reveal new insights into evolution, whereas the comparison of genome and protein sequences will tell us more about evolutionary conservation of protein structure and function.

## SUMMARY

In this chapter we learned about the Human Genome Project, including some of the amazing people involved and the secrets revealed about human DNA and genes. The adult human body contains more than 220 different types of cells, and each cell contains chromosomes that carry human genome DNA. All of the cells in the same body contain the same genome DNA sequence. The genetic instructions in the human genome are written in a DNA language that conveys the genetic information needed to make specific proteins that build and maintain each individual human being.

The goal of the international Human Genome Project was to determine the DNA sequence of the entire "human genome," the collection of 46 chromosomes that is created when an egg is fertilized by a sperm. Research on genomes other than the human genome predated the HGP and contributed in many ways to the eventual success of the HGP. Studies on small genomes helped scientists to develop the technological advances required to determine the DNA sequence of very large DNA genomes.

The HGP research revealed surprising information about the organization of DNA sequences in the human genome. Possibly most intriguing is the finding that only 2% of the human DNA actually codes for protein genes. The rest of the human genome contains large amounts of repeated DNA sequences and transposable DNA elements that have the ability to move around to different sites in the chromosome DNA. The human genome sequence information has had a huge impact on the development of genetic approaches to the diagnosis and treatment of human diseases and disorders. The success of the Human Genome Project gave the entire world access to the human master plan, the genetic information needed to create and build a living human being.

## REVIEW

1. Explain which advances in DNA technology and instrumentation made the biggest impact on the success of the Human Genome Project.
2. Describe the most surprising discovery revealed by the human genome DNA sequence.
3. Describe what the DNA sequence of the dog genome revealed about how dogs evolved.
4. Explain the basis for the legal, social, and ethical issues that were raised by the Human Genome Project research.
5. Describe how geneticists create gene linkage maps and physical maps of chromosomes in the human genome.
6. Summarize the process by which the sequence of DNA bases is determined using the Sanger chain termination method.
7. Identify some genomes that have already been sequenced and explain what surprising facts were revealed by the new DNA genome sequence.
8. Describe the properties of non-coding DNA and explain the impact of this finding on the structure and function of the human genome.
9. Explain why genetic variation plays an important role in understanding genes involved in human diseases.
10. Some noncoding DNA in the human genome contains elements that can move in the genome. Explain how this is accomplished.

## ADDITIONAL READING

Candille, S. How to turn an ape into a human. Stanford University. <www.thetech.org/genetics/news.php?id=8>.

Brownlee, C., 2004. Kibble for thought: dog diversity prompts new evolution theory. Science News, December 18.

Monogamy gene links men's DNA to happily ever after in marriage. <www.bloomberg.com/apps/news?pid=20601124&sid=a5kGdZ7L7vMI&refer=home>.

Platypus genome explains animal's peculiar features; holds clues to evolution of mammals. Science Daily (May 7, 2008).

Woese, C.R., Kandler, O., Wheelis, M.L., 1990. Towards a natural system of organisms: proposal for the domains Archaea, Bacteria, and Eucarya. Proc. Natl. Acad. Sci. USA 87, 4576–4579.

National institutes of health. <www.nih.gov>.

NHGRI division of intramural research. <www.genome.gov>.

NIH's National Center for biotechnology information: genome sequences including platypus, mouse, dog, cow, etc. <www.ncbi.nih.gov/Genbank>.

## WEB SITE

SNP Fact sheet. http://www.ornl.gov

# Bioinformatics

---

**Warfarin Dosing Based on Genetics**
**Genetic Testing Cited for Blood Thinner**

Associated Press, August 16, 2007
  By Andrew Bridges
  Federal health officials are stopping short of recommending genetic tests for patients on the blood-thinner warfarin, even though they have said such screenings could prevent thousands of complications each year.

Warfarin, sold under the brand name Coumadin and in generic forms, on Thursday became the first widely used drug to include genetic testing information on its label. The information can help doctors determine how best to prescribe the drug.

"This means personalized medicine is no longer an abstract concept but has moved into the mainstream," the Food and Drug Administration's clinical pharmacology chief, Larry Lesko, said in announcing the label change.

The updated label for warfarin suggests that lower doses may be best for patients with variations in two specific genes. One produces an enzyme that helps the body metabolize warfarin and other medicines; the second produces the blood-clotting protein that warfarin blocks.

The FDA has not changed its dosing recommendations for the drug, and tailoring the proper dosage remains largely a matter of trial and error.

A patient's age, weight, diet, and other prescription drug use all play a role in determining a proper dose. Patients taking too much warfarin can bleed to death. If people take too little of the drug, it can fail to protect them from deadly blood clots and stroke.

Genetic testing can reveal which patients may require less of the drug and lead doctors to recommend doses closer to the lower end of the scale, FDA officials said.

Rebecca Burkholder, vice president of health policy for the National Consumers League, said the FDA's action was a good first step. But she said that once patients are on the drug, they still must have regular blood tests to see if it is working properly.

It is clear that the age of personalized medicine is here. Based on a patient's genetic profile, personalized doses of warfarin can be prescribed (see Chapter 13). Similarly, several types of cancer treatments have also been shown to be more or less effective based on a patient's genetic profile. The emerging field of pharmacogenomics, which deals with the influence of genetic variation in an individual's response to drug therapy among other things, is vastly dependent on the emerging field of bioinformatics. The use of bioinformatics to find genes associated with diseases and treatments is strongly promoted by international partnerships such as the HapMap Project. This group of scientists is working to locate, identify, and disseminate information about the 10 million or so sites of genetic variation in human chromosomes (see Chapter 6). The information available through the science of genetic variation combined with the power of bioinformatics will have an impact on moving personalized medicine more rapidly into the doctor's office and clinic.

## LOOKING AHEAD

Bioinformatics is a new area of science that functions at the interface of basic laboratory research and the computer. It has become an essential research tool to investigate the enormous amount of biological data, including sequence information, resulting from genetic research. On completing the chapter, you should be able to do the following:

- Describe the need for and importance of archiving biological data in computerized databases.
- Be familiar with different sequence databases, including the basic features offered and the types of sequence information available.
- Understand the tools available for analysis of the vast amount of available biological data.
- Successfully carry out a basic BLAST search.
- Understand connection between gene studies and sequence analysis.

## INTRODUCTION

Over the past few decades, significant developments in molecular biology and genomics have generated staggering amounts of biological information, especially in the form of nucleic acid (DNA and RNA) and protein sequences. This huge amount of human genome information has helped to propel bioinformatics into a critically important interdisciplinary field that develops and uses computational methods to store, organize, analyze, and manipulate vast amounts of biological data. Those involved in the field of bioinformatics make use of biology, computer science, mathematics, genetics, statistics, and several other areas of expertise to analyze DNA and protein sequence data, to study genomes, to predict nucleic acid and protein structure and function, and to apply these data to understand the workings of biological organisms. Bioinformatics has made it possible for laypeople to have access to information from researchers around the world, and it has brought together a global scientific community connected by the need to examine, manipulate, and understand the message carried in the universal biological language of DNA.

Although bioinformatics has really expanded since the late 1990s, the basics of modern bioinformatics were put into practice more than 50 years before then by Margaret Oakley Dayhoff (1925–1983), who is

**FIGURE 7.1** Margaret Oakley Dayhoff, founder of the field of bioinformatics.

widely considered to be the founder of the field of bioinformatics (Figure 7.1). Starting with her Ph.D. work in 1948, she integrated biological science and chemistry with the new field of computer science. Dayhoff created the first computerized banks of protein and DNA sequences, and she developed many of the tools we use today for designing and using computerized databases. In 1965, Dayhoff published the first Atlas of Protein Sequence and Structure, which contained all 65 protein sequences known at the time. The original Atlas and its subsequent volumes were extremely valuable both as reference works and as a foundation of molecular sequences used to study biological questions.

Many biological questions are asked and answered using the tools available in bioinformatics. For example:

- What is the amino acid sequence of the protein encoded by the DNA sequence of this gene?
- Is this protein sequence similar to other known amino acid sequences? Does this sequence show an evolutionary relationship to any others?
- What is the biological significance of the sequence under study? What is the structure and function of the protein?
- When, where, and under what conditions is the gene expressed (copied into RNA)?
- Does the gene sequence contain a mutation that might cause a disease in humans?

## AN EXPLOSION OF DATA FUELED THE RISE OF BIOINFORMATICS

Throughout the 1970s and 1980s, the rapidly increasing number of protein and nucleic acid sequences stored in computer databases led to the development of software programs designed to search for sequence characteristics such as identity or sequence similarities.

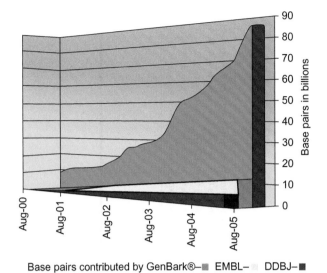

Base pairs contributed by GenBark®–■ EMBL–  DDBJ–■

**FIGURE 7.2** Growth of the international nucleotide sequence database collaboration.

research and has allowed scientists to solve problems and understand many aspects of basic biology that we would have thought impossible even as recently as the 1980s. Imagine a researcher studying the *Escherichia coli* bacterial chromosome, which is tiny in size compared to the genomes of humans and other multicellular eukaryotes. The roughly 4 million base pair genome in an *E. coli* bacterium is only about 0.1% as big as the human genome, yet even the most diligent and hardworking research scientist would not make much progress understanding the *E. coli* DNA sequence information without the use of bioinformatics.

Due to user-friendly online programs that provide free access to the computational tools of bioinformatics, both students and researchers routinely use these resources to understand how biological data are organized in databases and use programs to explore the online information.

The increase in genomic research has dramatically increased the number of genomes sequenced since the mid-1990s, and caused a large increase in the amount of nucleic acid sequence and protein sequence and structure data available. By August 2005, these databases had collected and dispersed 100 billion bases of sequence data, representing both individual genes and partial and complete genomes of more than 165,000 organisms (Figure 7.2). Of course many genomes have been sequenced since 2005, including studies that require comparisons of thousands of genome sequences from thousands of individuals (see Chapter 6).

For this tremendous collection of biological data to be useful, scientists need to have easy access to the information and to the methods by which the sequence data can be manipulated, compared, and analyzed. This is where bioinformatics comes into play. Most biologists, and also many researchers who are experts in their own scientific area, are not trained as computational scientists in terms of developing and analyzing the computer algorithms or understanding the step-by-step procedures needed to search and compare data. However, the ability to manipulate nucleic acid and protein sequences by computer is now an absolutely essential skill for all students of the biological sciences. Most of us will never become bioinformatics experts, but thanks to the development of user-friendly sequence analysis computer programs, it is reasonable to expect that, by now, both students and researchers should understand how biological data are organized, how to access the necessary databases and programs, and how to analyze the relevant data to explore various biological questions of interest.

The availability of collective contributions of data from many scientists has changed the way we carry out

## SEQUENCE SIMILARITIES SUGGEST PROTEIN FUNCTION AND EVOLUTIONARY RELATIONSHIPS

Sometimes scientists know the amino acid sequence of a protein, but not the function of the protein in the cells. However, comparison of the sequences of many known proteins in a database can often reveal sequence similarities between the unknown and known proteins. The sequences and function information about other, well-understood proteins provide clues as to the possible function of the unknown protein. A good example is the complex that carries oxygen in the blood, called hemoglobin. The globin proteins that make up hemoglobin have similar amino acid sequences, structures, and functions in many species besides humans. A newly discovered protein with an amino acid sequence similar to that of a globin protein is likely to have a similar function to hemoglobin. In addition, proteins can have different functions but still share some subset of sequences needed for a specific aspect of their overall job. For example, a specific subset of amino acids is found in proteins that bind to guanosine triphosphate (GTP). This subset of amino acids, which confers on the entire protein the ability to bind GTP, is found in many proteins that function by binding to GTP. Obviously a scientist finding this sequence in a newly discovered protein would have a potential clue to the function of the unknown protein. It most likely binds to GTP to do its job.

The principle of sequence similarity applies to DNA and RNA sequences as well as proteins. The sequences of newly found genes are compared to various databases that help to determine the identity and

**TABLE 7.1** Sequence similarities among genomes of model organisms

| Organism | Number of genes (estimated) | Percentage of genes similar to human genes | Description of the organism |
|---|---|---|---|
| Human (*Homo sapiens*) | ~20,500 | — | |
| Mouse (*Mus musculus*) | ~25,000 | 90% | Mice are used to study many human diseases (see Chapters 6, 10). A mutation in a single gene found in both mice and humans makes the mouse on the right have 5X as much fat as her sister on the left (Figure 7.4). |
| Zebra fish (*Danio rerio*) | 16,456 | 85% | Zebra fish are used to study cell and tissue development and human diseases (see Chapter 10). |
| Fruit fly (*Drosophila melanogaster*) | 13,379 | 36% | For 100 years fruit flies have been a model organism for the study of genetics. The fruit fly is important to study animal development, and retinal degeneration in flies and humans. The normal fly eye has 800 regular units (left). Putting a human gene into the fly causes the retina of the fly eye to rapidly degenerate (right). The closely related fly and human genes are used to study retinal diseases in humans (Figure 7.5). |
| Thale cress (*Arabidopsis thaliana*) | ~28,000 | 26% | *Arabidopsis* is a model flowering plant, but has a basic cell biology that is similar to that of humans and animals (see Chapter 15). |
| Yeast (*Saccharomyces cerevisiae*) | 5,770 | 23% | Single-celled yeast is a eukaryotic model organism for genetic research that played a key role in understanding cell cycle regulation (see Chapter 9). |
| Roundworm (*Caenorhabditis elegans*) | 19,427 | 21% | The roundworm provides an excellent model system to study tissue development and to analyze genes involved in aging and neurological diseases. A single gene determines whether a worm becomes a loner or prefers to eat in company. This research provides clues to the basis of social behavior in humans. The green color is from the green fluorescent proteins (GFP) expressed by the worms (Figure 7.6). |
| Bacteria (*Escherichia coli K12*) | 4,377 | 7% | *E. coli* is the king of the model organisms, which has revealed a lot about the basic processes of DNA replication, transcription, and translation in prokaryotes. |

*Gene estimates and numbers updated 2009*

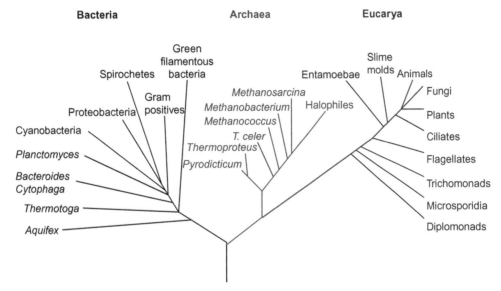

FIGURE 7.3 Phylogenetic tree of life. Phylogenetic tree shows the separate domains for bacteria, archaea, and eukaryotes. This tree was based on RNA sequence comparison studies and was proposed by scientist Carl Woese. The exact relationships among the three domains are still controversial, but the existence of three domains is supported by a large amount of data available in biological databases.

function of the new gene. Even though people look very different from mice or roundworms, humans share a remarkable number of genes with these and other **model** organisms (Table 7.1). The term "model" refers to the fact that model organisms such as the mouse, fruit fly, bacterium or human, are the focus of very active research and have been very well characterized. Although the genes in mouse and man are not identical, many of the proteins encoded by these organisms have similar functions, suggesting that the genes and their products are comparable. The similar DNA sequences described in this example are considered to be **homologous,** because the two (or more) genes have almost identical DNA sequences (same linear order of the DNA bases along the strand).

Sequence similarity is often an indication that the genes in question could have originated from a distant common ancestor. As genes evolve (through random mutations in the DNA), natural selection leads to the retention of sequences that are critical to the survival of the individual. Genes and the proteins they encode retain functionally important sequences through evolution with few changes. Many genes with similar sequences are found in seemingly distant species, such as the *E. coli* bacterium and a human being. Genes and other DNA and RNA sequences are routinely used to deduce previously unknown evolutionary relationships among biological organisms. In fact, extensive studies comparing gene and protein sequences from many organisms ultimately led to the reclassification of all biological organisms from the previous two divisions of prokaryotes and eukaryotes into three new domains: *Archaea, Eubacteria,*

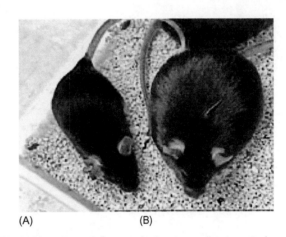

FIGURE 7.4 Genes influence obesity. A mutation in a single gene found in both mice and humans makes the mouse on the right have about five times as much fat as her sister on the left.

and *Eukarya* (Figure 7.3). The analysis of ribosomal DNA sequences, the genes that encode the ribosomal RNAs (rRNAs), has enhanced our understanding of the relationship between land animals and aquatic ancestors. Mitochondrial DNA sequences such as the cytochrome b and the rRNA genes were used to reveal the familial relationships among the giant tortoises on Darwin's Galapagos Islands.

The research on genome sequences from many organisms has led to important medical and agricultural advances. For example, studies conducted on nematodes (*Caenorhabditis elegans*) and fruit flies (*Drosophila melanogaster*) provided the scientific community with fundamental information about the structure and function of

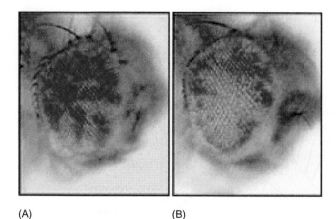

(A)  (B)

**FIGURE 7.5** (A) A normal fly eye has equal amounts of normal (red) and mutated (white) tissues. (B) Flies with mutations in growth restriction genes have a larger proportion of white, mutated tissue than normal red tissue in their eyes.

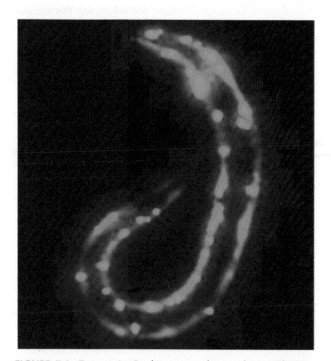

**FIGURE 7.6** Transgenic *C. elegans* roundworm glows with green fluorescence. A single gene determines whether a worm becomes a loner or prefers to eat in company. This research is designed to provide clues to the basis of social behavior in humans. The green color indicates the fluorescent proteins expressed by foreign green fluorescent GFP genes in the worms.

cellular receptors and the process of intracellular signal transduction. These cellular receptors often have counterparts in humans; a specific receptor found in both humans and fruit flies is implicated in an inherited form of colorectal cancer. Research on the plant *Arabidopsis thaliana* revealed new information about the synthesis of vitamin E that will allow researchers to increase the vitamin E production in soybean crops.

Studies on cell receptor proteins in many organisms helps scientists to better understand how cell receptor proteins function in humans. A mutation in a specific receptor protein found in humans and fruit flies is implicated in colorectal cancer, providing a new model system to study this cancer.

## BIOLOGICAL DATA ARE ORGANIZED IN COMPUTER DATABASES

A **biological database** is simply a collection of biological data that is organized in a specific and useful way. Bioinformatics databases are very large, accessible by computer on the Internet, and must be continuously updated with new information, revisions, and corrections in order to be maximally useful. The computerized interfaces to the databases are designed to be user friendly and they allow researchers to ask for and receive information from the database online. In the terminology of bioinformatics, a request to the database is known as a **query** and the information obtained from a query to the database is a **result**.

The advent of bioinformatics databases has led to a new research approach called **database mining**, which is similar to mining for gold as it involves sifting through a tremendous amount of starting material to find comparatively tiny amounts of valuable "nuggets." In bioinformatics, the starting material is the vast amount of information in the database(s), and the nuggets are the few pieces of data that are of interest to a particular researcher. In bioinformatics research, each bit of data is a potential nugget; different researchers sifting through the same starting material are looking for very different nuggets. Most important, however, to make it possible to sift through information to find valuable nuggets, the data must be present in an orderly database.

Biological databases are maintained by a number of private and government organizations, both in the United States and elsewhere in the world. Biological databases typically contain collections of nucleic acid sequences (DNA or RNA), protein sequences, genome sequences, and literature resources, but many databases also contain information relevant to answering specific biological questions. These databases have become invaluable tools for researchers worldwide.

### Nucleic Acid Sequence Databases

Nucleic acid sequences (DNA and RNA) are stored in three comprehensive databases on the Internet, which are free and easily accessible to the public, allowing scientists from all over the world to share and compare data.

- GenBank (www.ncbi.nlm.nih.gov/Genbank/index.html): Maintained at the National Center for Biotechnology Information (NCBI) in Maryland, United States
- EMBL (European Molecular Biology Laboratory; www.ebi.ac.uk/embl): Maintained at the European Bioinformatics Institute in Cambridge, United Kingdom
- DDBJ (DNA Database of Japan: www.ddbj.nig.ac.jp): Maintained at the National Institute of Genetics in Mishima, Japan

The GenBank, EMBL, and DDBJ databases contain the International Nucleotide Sequence Database collaboration, which stays current by sharing updated information daily. Within each database, sequence files, called "records," are available as online web pages. Researchers can submit queries about specific sequences in a database and have quick and easy access to the results. In addition to the actual DNA or RNA sequences, the databases also contain information about gene function, encoded proteins, mutations, regulatory DNA sites, references, and links to other web sites where even more information can be found. The query results contain any information available in the database and linked to the sequence being queried.

GenBank was one of the first sequence databases available to researchers, and is a good example of how sequence databases work. The result of a query about the cystic fibrosis transmembrane conductance regulator (CFTR) gene from mouse shows part of the GenBank record page (Figure 7.7). The mouse CFTR protein is very similar in sequence to the human version. The CFTR functions in the transport of chloride ions across cell membranes. When the human CTFR does not function correctly, the consequence is the serious genetic disease, cystic fibrosis. The CFTR record demonstrates some of the key features of all GenBank entries:

- Gene name (cystic fibrosis transmembrane conductance regulator).
- Accession number or unique identifier for the sequence record (NM_021050 and XM_622568). This number can be used to access the sequence record quickly.
- Source organism (*Mus musculus*) and taxonomy.
- Bibliographical references along with a link to a journal article resource for each published sequence in the database (this record originally had 10 references). Summary of information known about the sequence and its encoded protein.
- Coding region sequence (CDS), which indicates the specific nucleotides corresponding to the amino acid sequences in the encoded protein (bases 138–4568 are coding region) and the amino acid sequence of the protein translated from the CDS.

- Origin: the original nucleotide sequence submitted to the database.
- To view the current CTFR record online right now, go to www.ncbi.nlm.nih.gov. In the drop-down menu following "Search," choose CoreNucleotide, and type the accession number NM_021050 into the search blank.

The major databases share and integrate the data from different sources, and each database record provides links to other databases and other online sources with additional information (Figure 7.2). This feature is an advantage for researchers because it makes easy connections between database entries. For example, some records contain information about regions of biological significance (e.g., known mutations or functional domains of encoded proteins) and include additional information about the sequence or the protein product (i.e., the DNA or protein is likely modified). Much of the information at the NCBI site has direct links to other relevant NCBI pages, web sites, and resources for further information.

> GenBank information is usually accessed through Entrez: The Life Sciences Search Engine web page (www.ncbi.nlm.nih.gov/gquery/gquery.fcgi.) Entrez is a user-friendly interface that provides easy access to the GenBank database as well as to a variety of other useful biological databases.

The Entrez search web page shown in Figure 7.8 provides links to numerous cross-referenced databases that are available without cost to the public. Each of the databases listed on the web page also links to a small popup window with a short description of the web page offering. Scientists from many different areas of the life sciences use these databases for a variety of reasons. The Entrez page also provides access to other DNA, RNA, and protein sequence databases.

The wealth of online information is a real gold mine to a scientist who has just found a new protein that plays an important role in a disease but who has no idea what the protein might do in the cell. To try to uncover the possible function of a protein, it is useful to find homologs, proteins with known functions that are similar but not identical in amino acid sequence to the new protein. Using the Entrez search page, a scientist looking for homologs of a new protein can use the HomoloGene link to search among the genes of several completely sequenced eukaryotic genomes or can access the Conserved Protein Domain Database (CDD) link to look for clusters of conserved amino acids among known proteins. Sequence similarities between proteins (the entire sequence or short regions) often provide clues about the function of a new protein or about the functions of structural domains in the new protein.

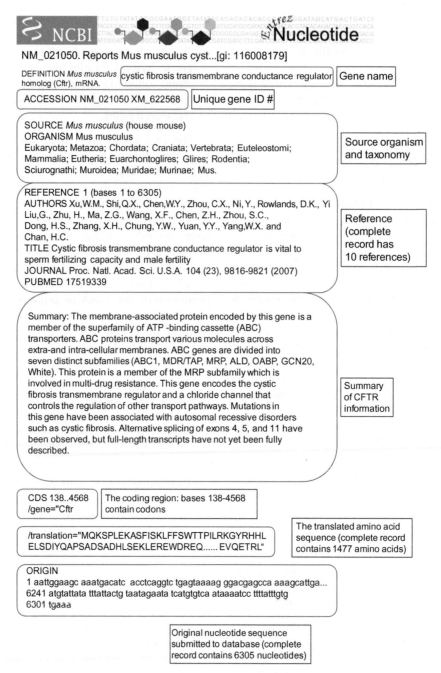

**FIGURE 7.7** Partial GenBank entry for a mouse cystic fibrosis transmembrane conductance regulator gene. This example shows the key features of GenBank entries, including gene names/identifiers, nucleotide sequence, translated amino acid sequence (using single letter abbreviation), basic information about the encoded protein, information about the source organism, and links to literature resources.

In 2007, Entrez added a link to three-dimensional macromolecular structures. The Molecular Modeling Database (MMDB) contains more than 40,000 three-dimensional macromolecular structures, including DNA, RNA, and proteins. The MMDB is linked to the other NCBI databases, including sequences, bibliographical citations, taxonomic classifications, and sequence and structure neighbors. This new feature at Entrez permits a query search to provide everything from simple primary sequence to final three-dimensional molecular structures, linked to other relevant information all along the way.

The data manipulations and databases described here are fundamental to understanding the secrets underlying linear sequences of DNA letters or amino acids. Although fairly basic, the examples described here demonstrate the power and potential of the tools available to scientists through bioinformatics.

**FIGURE 7.8** The Entrez Search Engine home page. The Entrez Search Engine home page illustrates the vast number and variety of databases that can be used in biological research.

In addition to access to sequence data retrieval and data analysis tools, Entrez also contains a link to the scientific literature database at the National Library of Medicine (NLM) called PubMed. This link provides access to the NLM biomedical literature citations and abstracts, and to PubMed Central, which allows free access to many selected full-text journal articles. Entrez and PubMed provide the scientific community and the world with literally millions of published scientific articles and references and a vast array of resources that are available without ever leaving the computer.

## Protein Sequence Databases

Information about DNA sequences is an important starting point for research but it is the proteins encoded by the genes that determine the structure, function, and behavior of an organism. Mutations in genes can lead to the production of defective proteins, which can lead to disease and disorders. For example, the cystic fibrosis transmembrane conductance regulator protein (CFTR, Figure 7.7) is normally found in a variety of epithelial tissues in the body, including the lungs.

**Box 7.1    Literature Databases**

Through NCBI or through Entrez a researcher can link to several literature databases, type in key words and phrases to begin a search, and provide hints for narrowing subsequent pubsearches. **PubMed** accesses the more than 15 million citations found on Medline, an indexing service for medical research (provided by the National Library of Medicine). Each abstract has a "related articles" link that allows a researcher to easily expand the search.

- **PubMed Central** is an archive of journal literature encompassing the life sciences. This database provides free access to the full text of more than 150 life science journals.
- **Bookshelf** contains online biomedical textbooks.
- **Online Mendelian Inheritance in Man (OMIM)** is a comprehensive and continuously updated listing of human genes and genetic disorders.
- **Online Mendelian Inheritance in Animals (OMIA)** is a dataset of traits, genes, and inherited disorders in animals species other than mouse and human.
- **Coffee Break** provides brief reports usually based on recently published peer-reviewed literature. The work is first put in a broad context and then narrows to potential applications. Links are provided to demonstrate how bioinformatics tools were used in the research process.

to query and analyze sequence information and provide links to access additional information about the protein of interest. These results include descriptions of functions, domain structures, posttranslational modifications, mutations, and much more. UniProt also provides links to other pages offering abundant external resources for the scientific researcher.

The UniProt record for the *Drosophila melanogaster* (fruit fly) opsin protein indicates that the opsin proteins are part of the pigment components in the eye that absorb light and are necessary for vision.

The UniProt opsin record demonstrates some of the key features available using the UniProt data files:

- Protein name (Opsin Rh3)
- Accession numbers (P04950; Q9Tx53)
- Source organism (*Drosophila melanogaster*) and taxonomy
- Amino acid sequence of the protein with associated information
- Bibliographical references along with links to articles
- Comments about protein function, sub-cellular location (if known), and other relevant information about the protein
- Database cross-references and links to additional resources and information
- Special features specific to the opsin protein such as identifying the amino acids that span the membrane, and noting which end of the protein (amino- or carboxyl terminus) is extracellular and which end resides in the cytoplasm.

Powered by ATP, the CFTR protein transports chloride ions ($Cl^-$) across epithelial cell membranes. In the respiratory tract, the passage of $Cl^-$ is followed by the passage of water, which makes the mucus lining the airway thin and fluid. When the CFTR gene is mutated, the mutant CFTR protein cannot fold into its normal three-dimensional shape. Scientists now know that protein function almost always depends on the ability of the protein to fold into its correct three-dimensional structure. In cystic fibrosis patients the mutation in the CFTR protein prevents the $Cl^-$ ions from being transported and the airways fill with thick, sticky mucus. This leads to severe respiratory problems, one of the hallmarks of this genetic disease (see Chapter 10).

Dayhoff's *Atlas* and more recent research on protein evolution led to the development of the Protein Sequence Database (PSD), which since 1984 has been maintained by the Protein Identification Resource (PIR) at Georgetown University (www.pir.georgetown.edu), which provides free information and sequence analysis tools.

In 2002, PIR and its international partners, the European Bioinformatics Institute (EBI; www.ebi.ac.uk) and the Swiss Institute of Bioinformatics (SIB; www.isb-sib.ch), created UniProt (www.pir.uniprot.org), a single, worldwide, comprehensive database dedicated to protein sequence and function. UniProt allows researchers

## Genome Databases

The scientific discipline called **genomics** greatly benefitted from the technical advances that allowed entire genomes to be rapidly sequenced, which contributed to the avalanche of data available for biological research. Genomics refers primarily to studies on whole sets of genes or entire genomes. Genome databases have been developed for many model organisms including the bacterium *Escherichia coli*, budding yeast (*Saccharomyces cerevisiae*), fission yeast (*Schizosaccharomyces pombe*), nematodes (a type of roundworm, *Caenorhabditis elegans*), fruit flies (*Drosophila melanogaster*), the mustard plant (*Arabidopsis thaliana*), mice (*Mus musculus*), zebra fish (*Danio rerio*), and many other prokaryotic and eukaryotic organisms (including humans) and viruses (Table 7.1). Computer databases are required for scientists to manage, organize, and use vast amounts of genomic information. By mid-2007, the NCBI database contained whole and partial genome sequence data for more than 1000 organisms. Database tools allow scientists to identify genes and gene families within specific genomes, to localize (or "map") genes to specific

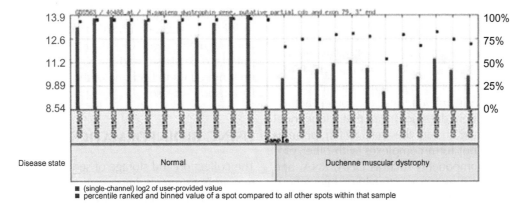

**FIGURE 7.9** Expression profile of the dystrophin gene. This data set examines the results of skeletal muscle biopsies analyzed to compare the expression of the dystrophin gene in biopsies from Duchenne muscular dystrophy (DMD) patients compared with unaffected individuals.

locations in a genome, and to analyze evolutionary relationships between organisms by comparing entire genomes (46 chromosomes in humans).

Researchers formed consortiums dedicated to "maintenance and repair" of the genome sequences in the databases and to maintain the molecular genetics of specific model organisms. The web sites maintained by these consortiums, in addition to containing a multitude of links to other sequence and literature databases, also contain information about members of the scientific community and instructions on obtaining lab strains or stocks of a particular model organism. The high degree of integration among the genomic databases and their information has made sharing scientific knowledge and materials easy, commonplace, and expected. Many journals follow the rule that once published, the reagents (solutions and other substances) including biological reagents like strains used in the experiment are in the public domain and should be made available free of cost (except for shipping).

> Researchers send specific DNA, RNA, or protein amino acid sequences to online sites that rapidly compare, search, or otherwise manipulate the query sequences. Online databases provide extensive information about gene function and mutations, protein structures, DNA control regions, literature references, links to other web sites, and more.

## Gene Expression Databases

The phenotype of an organism—its physical characteristics, the way it interacts with the environment, and its diseases—depends on which genes are expressed throughout the lifetime of the organism. Gene expression varies in different cell types and changes in any given cell over time, based on its developmental stage and the environment. High-throughput microarray methods have been developed to permit patterns of genes expressed in many cells under different conditions (see Chapter 13). **Expressed sequence tags** (ESTs) are DNA copies made from the ends of mRNAs. These short DNA strands represent a collection of genes expressed in a set of cells (see Chapter 6). Full-length DNA copies of messenger RNA transcripts are called complementary DNAs (cDNAs). ESTs and cDNAs are used as molecular tools to identify genes, coding sequences, and patterns of gene expression in specific cells and tissues. **Microarray analysis** permits the simultaneous detection of all of the mRNAs transcribed from thousands of genes in a genome at any one point in time (see Chapter 13). This allows scientists to study how the expression levels of different genes are regulated over time, in different cells and tissues, and in response to hormone changes and environmental signals. In addition to databases containing gene and protein sequences, online databases also store, organize, access, and analyze RNA and protein gene expression data generated from EST and microarray tests.

The Gene Expression Omnibus (GEO) repository at NCBI stores and freely distributes microarray, EST, and other forms of high-throughput data that are submitted by scientists around the world. In addition to its archival function, GEO also provides data mining tools that allow researchers to query, retrieve, and download data relevant to their personal research interests. An example of data mined through GEO shows the expression of the human dystrophin protein in Figure 7.9. The dystrophin protein is important for muscle structure and defects in the dystrophin gene (and protein) cause muscular dystrophy disease (see Chapter 10).

GEO is the largest fully public online gene expression resource with more than 120,000 samples from more than 200 organisms (representing 3.2 billion individual measurements).

## Proteome Databases

Understanding gene expression is a key step toward understanding how a cell or an organism works, but gene expression does not tell the whole story. The mere observation that a gene is transcribed into RNA (as indicated by ESTs, microarrays, or another method to detect transcription) does not mean that a functional protein is produced inside the cell. Often proteins do not act alone but form complexes with other proteins or bind to components such as DNA, RNA, and membranes. Scientists are studying interactions between proteins and with nucleic acids, studying all of the interactions required for a cell to function. The new, exciting field of **proteomics** is the study of the **proteome**. The term "proteome" is analogous to the term "genome" but (so far) is less well defined. Scientists working on the Human Genome Project referred to the proteome as the "proteins expressed by a cell or organ at a particular time and under specific conditions." However, the term has also been used to refer to the complete set of proteins expressed during the complete developmental stages of an organism. It is important to read carefully when the word proteome is used to understand how the author is using the term.

The major goals of proteomics are to identify, catalog, and understand the structure and function of a set of proteins. The fact that there are far fewer human genes than human proteins indicates that the protein complement of an organism cannot be fully characterized by gene expression analyses alone, making proteomics a necessary tool to understand the complexities of living cells (see Chapter 6).

The current protein expression data in computerized databases is still very small compared to the amount of gene expression data. There are probably two reasons that protein data has not yet caught up with gene expression data. First, working with proteins is difficult in general, and the methods used to analyze protein expression and protein-protein interactions (such as two-dimensional electrophoresis and two-hybrid analysis) are technically challenging and time consuming. In addition, other methods, such as mass spectrometry, are relatively new techniques applied to protein expression analysis, with limited data yet online. The proteomics databases are maintained by the NCBI and the European Bioinformatics Institute, as well as specialized databases established by model organism consortiums and various research groups. The international collaboration coordinated by the Human Proteome Organization (HUPO) is cataloging all human proteins, their functions, and protein-protein interactions, so coming years are likely to see a large increase in the amount of protein expression data available in online databases. Without the benefit of modern online bioinformatics tools, it took nine years to identify the cystic fibrosis transmembrane conductance regulator gene (CFTR). Later, with access to computerized databases and microarray technology, it took only nine days to find one of the human genes causing Parkinson's disease.

## USING BIOINFORMATICS DATABASES

The collection and storage of sequence and other information is extremely important, but it means nothing if people cannot find and analyze the data to help answer biological questions. Successful use of the various databases takes practice and the initial experience can be a bit overwhelming. But with experience, people can develop precise and useful queries and access information that leads to the answers for biological questions.

The next part of this chapter will not make the reader an expert in searching databases and analyzing data, but will help everyone to become familiar with the basics needed to navigate the computer databases. The databases often contain tutorials that help the reader to become skilled at finding information. NCBI has a comprehensive list of tools for data mining available on its site (www.ncbi.nlm.nih.gov/Tools).

## How to Translate a DNA Sequence into an Amino Acid Sequence

Given that most genes encode proteins, scientists usually first determine the DNA sequence and then deduce the amino acid sequence of the protein encoded by the gene. In the cell the gene coding for a protein product is first transcribed (copied) from DNA into mRNA, and then the 3-base RNA codons are translated by the ribosome into a specific order of the amino acids in the protein. As a result of knowing the genetic code, it is possible to scan through all of the possible reading frames in a given DNA coding region and determine the most likely amino acid sequence(s) predicted by the gene sequence. Although for short genes this is possible to do by hand, it is time consuming and highly error prone. Fortunately, bioinformatics tools in the form of computer programs can quickly predict the most likely gene product (i.e., the amino acid sequence) using the genetic code and the three reading frames possible on each DNA strand (Figure 7.10). Two of the three reading frames can often be eliminated from consideration by the presence of stop codons early in the sequence.

### Searching a Database for Similar Sequences

The amino acid sequence of a protein does not necessarily reveal the identity of the protein or the

Computer DNA Sequence File

```
5′ ATGTCCAC  GCGGTCCTG  GAAAACCCAG  GCTTGGGCAG
GAAACTCTCT  GACTTTGGAC  AGGAAACAAG  CTATATTGAA
GACAACTGCA  ATCAAAATGG  TGCCATATCA  CTGATCTTCT
CACTCAAAGA  AGAAGTTGGT  GCATTGGCCA  AAGTATTGCG
CTTATTTGAG  GAGAATGATG  TAAACCTGAC  CCACATTGAA
TCTAGACCTT  CTCGTTTAAA  GAAAGATGAG  TATGAATTTT
TCACCCATTT  GGATAAACGT  AGCCTGCCTG  CTCTGACAAA
CATCATCAAG  ATCTTGAGGC  ATGACATTGG  TGCCACTGTC
CATGAGCTTT  CACGAGATAA  GAAGAAAGAC  ACAGTGCCCT
GGTTCCCAAG  3′
```

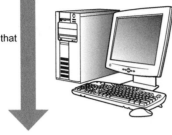

Run a computer program that translates this sequence into an amino acid seguence in all three reading frames.

**(A)**            Possible amino acid sequences

5′ → 3′ Frame 1
**Met** Ser Thr Arg Ser Trp Lys Thr Gln Ala Trp Ala Gly Asn Ser Leu Thr Leu Asp Arg Lys Gln Ala Ile Leu Lys Thr Thr Ala Ile Lys **Met** Val Pro Tyr His STOP Ser Ser His Ser Lys Lys Lys Leu Val His Trp Pro Lys Tyr Cys Ala Tyr Leu Arg Arg **Met Met** STOP Thr STOP Pro Thr Leu Asn Leu Asp Leu Leu Val STOP Arg Lys **Met** Ser **Met** Asn Phe Ser Pro Ile Trp Ile Asn Val Ala Cys Leu Leu STOP Gln Thr Ser Ser Arg Ser STOP Gly **Met** Thr Leu Val Pro Leu Ser **Met** Ser Phe His Glu Ile Arg Arg Lys Thr Gln Cys Pro Gly Ser Gln

5′ → 3′ Frame 2
Cys Pro Arg Gly Pro Gly Lys Pro Arg Leu Gly Gln Glu Thr Leu STOP Leu Trp Thr Gly Asn Lys Leu Tyr STOP Arg Gln Leu Gln Ser Lys Ser Trp Cys His Ile Thr Asp Leu Leu Thr Gln Arg Arg Ser Trp Cys Ile Gly Gln Ser Ile Ala Leu Ile STOP Gly Glu STOP Cys Lys Pro Asp Pro His STOP Ile STOP Thr Phe Ser Phe Lys Glu Arg STOP Val STOP Ile Phe His Pro Phe Gly STOP Thr STOP Pro Ala Cys Ser Asp Lys His His Gln Asp Leu Glu Ala STOP His Trp Cys His Cys Pro STOP Ala Phe Thr Arg STOP Glu Glu Arg His Ser Ala Leu Val Pro Lys

5′ → 3′ Frame 3
Val His Ala Val Leu Glu Asn Pro Gly Leu Gly Arg Lys Leu Ser Asp Phe Gly Gln Glu Thr Ser Tyr Ile Glu Asp Asn Cys Asn Gln Asn Gly Ala Ile Ser Leu Ile Phe Ser Leu Lys Glu Glu Val Gly Ala Leu Ala Lys Val Leu Arg Leu Phe Glu Glu Asn Asp Val Asn Leu Thr His Ile Glu Ser Arg Pro Ser Arg Leu Lys Lys Asp Glu Tyr Glu Phe Phe Thr His Leu Asp Lys Arg Ser Leu Pro Ala Leu Thr Asn Ile Ile Lys Ile Leu Arg His Asp Ile Gly Ala Thr Val His Glu Leu Ser Arg Asp Lys Lys Lys Asp Thr Val Pro Trp Phe Pro

**(B)**

**FIGURE 7.10** Translating a gene DNA sequence into a protein amino acid sequence. Computer programs are designed to use the genetic code to translate a gene sequence to an amino acid sequence. Although this can be done by hand, it is extremely time consuming. (A) Coding strand of a DNA sequence. (B) Three possible translation reading frames. The top sequence is the correct reading frame because it lacks translation stop codons. Met is translation start; STOP indicates translation termination.

**(A)**
Sequence 1: N-G-C-A-N-N-C-T-T-A-G-C-N-T-A-A-G-C-G-C
Sequence 2: N-G-C-A-N-N-G-A-T-A-A-C-N-T-A-A-G-C-G-C

**(B)**
Sequence 1: N G C A N N C T T A G C N T A A G C G C
Sequence 3: N G C - - - G A T A G C N T A A G C G C

**(C)**
Sequence 1: N G C A N N C T T A G C N T A A G C G C
Sequence 2: N G C A N N G A T A A C N T A A G C G C
Sequence 3: N G C - - - G A T A G C N T A A G C G C

**FIGURE 7.11** Sequence alignments. (A) Generalized pair-wise alignment of two closely related DNA sequences. (B) Sequence 1 in a pair-wise "gapped" alignment with another sequence. The "gaps" occur as a result of varying mutations in the different sequences throughout evolution. (C) Multiple sequence alignment: N = any nucleotide; shaded regions show differences between sequences. Adapted from Brooker *et al.*, *Biology,* McGraw Hill, New York, 2007.

perform the same or similar functions. The regions of similar amino acids might indicate a possible function for the new protein or might signal an evolutionary relationship between the genes.

The most straightforward way to determine the degree or percent of similarity between two or more nucleic acid or protein sequences is to perform a computer comparison by **sequence alignment**. Comparing one sequence directly with one other sequence is called a **pair-wise sequence alignment**; comparing one sequence in parallel with several other sequences is a **multiple sequence alignment** (Figure 7.11).

As with many analyses now carried out by computers, sequence alignments were originally performed by hand, by visually searching for stretches of two amino acid or DNA sequences that were similar to each other. Doing the alignments is challenging because changes such as base pair substitutions or deletions occur in gene sequences throughout evolutionary time. A base pair substitution results in a difference between sequences, and a deletion of base pairs results in a gap in one sequence relative to another in the alignment. In other words, while the genes may still be significantly related to each other, the alignment of the sequences becomes difficult without the help of a computer. Modern database tools accommodate sequence inconsistencies and gaps, and some can be used to readily perform multiple sequence alignments.

The Basic Local Alignment Search Tool (BLAST) is one of the principal online tools used for sequence alignments. BLAST can be accessed through NCBI and GenBank home pages. It contains several protein, nucleotide, and genome sequence databases and is used to find regions of similarity when comparing sequences in nucleic acid and protein databases. Using BLAST is relatively simple: a researcher submits

actual function of the protein in the cell. Scientists use several clues to figure out the role of a protein in a cell. For example, sequence comparison will reveal if a gene sequence (or the translated amino acid sequence) is similar to a known gene or protein sequence. If the sequences are similar, subsequent experiments are based on the idea that the two proteins might each

a sequence query (the sequence of interest, which can be copied and pasted into the web page) and BLAST compares that query sequence against the sequences in selected databases, and finds matching sequences among the millions of sequences available in the database. Imagine trying to do this without a computer, and you can see why many scientists consider BLAST to be the most important tool in the bioinformatics workshop. The BLAST search home page allows a scientist to select from several different programs, depending on the sequence being entered and the needs of the researcher. The researcher can choose to run nucleotide comparisons, protein comparisons, or can have the program translate the nucleotide sequence into its predicted protein product and compare the predicted protein to the sequence databases of known proteins. After the program searches and finds similar sequences in different databases and organisms, it calculates the statistical significance of each match. This information assists the researcher in deciding the relative biological importance of the different sequence similarities found in the searches.

These results might help the scientist to decide the direction that the research experiments take in the future. For example, if the two proteins contain a region of similar amino acids, and this region of the known protein binds to actin, it is logical to infer that the unknown protein also binds to actin. This clue to the possible role of the protein inside the cell can now be tested experimentally. Researchers could follow up using actin-binding assays to determine if the unknown protein can in fact bind to actin. If it does, then the researchers could produce a mutant protein containing a mutation that alters the region of amino acid similarity in the unknown protein.

## Using BLAST to Compare Nucleotide Sequences

The BLAST programs are useful for basic research and will be discussed in detail. NCBI offers helpful tutorials for review of the basics and for moving beyond (www.ncbi.nlm.nih.gov/Education/BLASTinfo/tut1.html).

In this example, a researcher has the following partial gene sequence she or he wishes to begin to characterize. The sequence is entered using a computer file (or by hand) as a continuous string of lower case letters, which are separated into blocks of 10 bases shown here but are not separated when submitted as a query.

attgtgattg gcgctataat agtcgtctcg gttttacaac cctacatctt cctag-
caacg gtgccagtgc tagtgacctt tattttactg agggcctact tccttcacac
attacagcag ctcaaacaac tggaatctga aggcaggagt ccaatttttca cccac-
cttgt tacaagctta

The sequence is submitted online as a query to BLAST, using the initial nucleotide blast program to compare the nucleotide query against nucleotide databases (with parameters set to search through all nucleotide databases and to find all "somewhat similar" sequences (Figure 7.12). As a researcher becomes experienced with database searches and data mining, search parameters can be refined, perhaps limiting the search to sequences within specific organisms or choosing to search only certain databases.

The results of the BLAST search are presented in three ways:

- A graphical overview
- A listing
- Pair wise sequence alignments of the top (best) sequence matches

Each of these representations of the data is associated with links to additional online information about the sequences that match the query. The results also provide a **Score (S)** that is assigned to each match (the higher the score, the better the match) and an **Expect value (E)** that measures the number of hits (matching sequences) that can be expected by random chance with a particular query (the lower the E-value, the more likely that the sequences are related to each other). For example, an E-value of 10 indicates that the search could have yielded 10 matches just by chance. Thus, the lower the E-value, the better the chance that the query sequence is related to the sequence found by the search. The E-values shown in Figure 7.13 are very low numbers indeed; the best match has an E-value of 4e-71 (or $4 \times 10^{-71}$). This very small value means it is likely that the query sequence, the unknown gene the researcher wants to characterize, encodes a cystic fibrosis transmembrane conductance regulator (CFTR) protein (see Chapter 10). The high level of sequence similarity indicated by the E-value can also be visualized by looking at the pair-wise sequence alignment produced by comparing the query sequence with the *Mus musculus* CFTR sequence (Figure 7.13).

## Using BLAST to Compare Nucleotide Sequences with Protein Sequences

It is not unusual to use a nucleic acid sequence as a query and obtain results containing only unacceptably high E-values or a message stating that no significant similarity was found. This does not mean that the available electronic databases do not contain the desired information. Because the genetic code is redundant, nucleic acid sequences can differ over time, a process called **divergence**; but because the code is redundant, the protein sequences encoded by two divergent genes can remain relatively unchanged. Also, if the partial

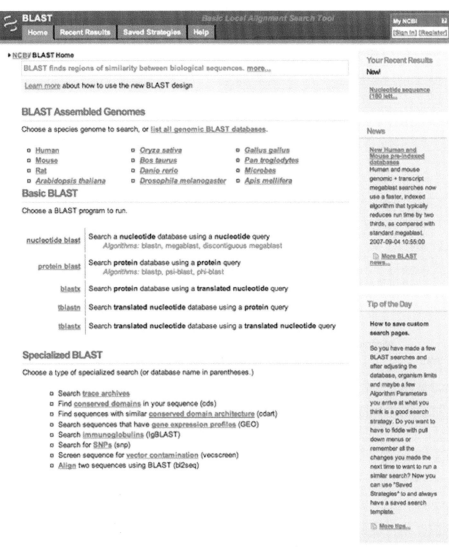

(A)

(B)

**FIGURE 7.12** BLAST home page online. (A) From this site a researcher can carry out BLAST searches. (B) The researcher enters a sequence of interest into the query box and the computer will compare it to selected nucleotide and protein databases. A variety of other databases are available for the advanced user in addition to BLAST searches.

Distribution of top 10 hits using sequence query:

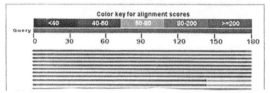

(A)

Identities of the top 10 hits:

| Accession Numbers: | Description: | E value: | | Links to other resources: |
|---|---|---|---|---|
| NM_021050.2 | Mus musculus cystic fibrosis transmembrane conductance regulator homolog (Cftr), mRNA | 4e-71 | 93% | U E G |
| M69298.1 | Mouse cystic fibrosis transmembrane conductance regulator (CFTR) mRNA, complete cds | 4e-71 | 93% | U E G |
| M60493.1 | Mouse cystic fibrosis transmembrane conductance regulator (CFTR) mRNA, complete cds | 4e-71 | 93% | U E G |
| XM_001059206.1 | PREDICTED: Rattus norvegicus cystic fibrosis transmembrane conductance regulator homolog (Cftr), mRNA | 3e-60 | 89% | G |
| XM_001062374.1 | PREDICTED: Rattus norvegicus cystic fibrosis transmembrane conductance regulator homolog (Cftr), mRNA | 3e-60 | 89% | U G |
| NM_031506.1 | Rattus norvegicus cystic fibrosis transmembrane conductance regulator homolog (Cftr), mRNA | 3e-60 | 89% | U E G |

(B)

Alignments of top 2 hits:

```
>ref|NM_021050.2| U E G Mus musculus cystic fibrosis transmembrane conductance regulator
homolog (Cftr), mRNA
Length=6305

  Score =  275 bits (304),  Expect = 4e-71
  Identities = 169/180 (93%), Gaps = 0/180 (0%)
  Strand=Plus/Plus

Query  1     ATTGTGATTGGCGCTATAATAGTCGTCTCGGTTTTACAACCCTACATCTTCCTAGCAACG  60
             ||||||||||  |||||||||||||||||||  ||||||||||||||||||||||||||||
Sbjct  3120  ATTGTGATTGGAGCTATAATAGTCGTCTCGGCATTACAACCCTACATCTTCCTAGCAACG  3179

Query  61    GTGCCAGTGCTAGTGACCTTTATTTTACTGAGGGCCTACTTCCTTCACACATTACAGCAG  120
             |||||||| |||||||  |||||||||||||||||||||||||||  ||| ||||||||||
Sbjct  3180  GTGCCAGGGCTAGTAGTCTTTATTTTACTGAGGGCCTACTTCCTTCATACAGCACAGCAG  3239

Query  121   CTCAAACAACTGGAATCTGAAGGCAGGAGTCCAATTTTCACCCACCTTGTTACAAGCTTA  180
             ||||||||||||||||||||||||||||||||||||||||||||||||||||||||||||
Sbjct  3240  CTCAAACAACTGGAATCTGAAGGCAGGAGTCCAATTTTCACCCACCTTGTTGACAAGCTTA  3299


>gb|M69298.1|MUSCFTR U E G Mouse cystic fibrosis transmembrane conductance regulator (CFTR)
mRNA, complete cds
Length=6299

  Score =  275 bits (304),  Expect = 4e-71
  Identities = 169/180 (93%), Gaps = 0/180 (0%)
  Strand=Plus/Plus

Query  1     ATTGTGATTGGCGCTATAATAGTCGTCTCGGTTTTACAACCCTACATCTTCCTAGCAACG  60
             ||||||||||  |||||||||||||||||||  ||||||||||||||||||||||||||||
Sbjct  3119  ATTGTGATTGGAGCTATAATAGTCGTCTCGGCATTACAACCCTACATCTTCCTAGCAACG  3178

Query  61    GTGCCAGTGCTAGTGACCTTTATTTTACTGAGGGCCTACTTCCTTCACACATTACAGCAG  120
             |||||||| |||||||  |||||||||||||||||||||||||||  ||| ||||||||||
Sbjct  3179  GTGCCAGGGCTAGTAGTCTTTATTTTACTGAGGGCCTACTTCCTTCATACAGCACAGCAG  3238

Query  121   CTCAAACAACTGGAATCTGAAGGCAGGAGTCCAATTTTCACCCACCTTGTTACAAGCTTA  180
             ||||||||||||||||||||||||||||||||||||||||||||||||||||||||||||
Sbjct  3239  CTCAAACAACTGGAATCTGAAGGCAGGAGTCCAATTTTCACCCACCTTGTTGACAAGCTTA  3298
```

(C)

**FIGURE 7.13** BLAST results from comparing nucleotide sequences. (A) A graphical overview of the top 10 matches from the database. The matches are aligned to the query sequence and the score of the alignment is indicated by color. A researcher can "mouse over" a bar, revealing the definition of the gene and the exact score. (B) A list of the top 10 matches generated from the query. (C) A pair-wise sequence alignment with the top matches.

gene sequence used as a query is located in an intron or a noncoding region of a gene, the sequence has probably diverged to a much greater extent than sequences in the exons or coding regions of the gene (see Chapter 6). For these and other reasons, a researcher who is not successful using a simple nucleotide-nucleotide search may get more favorable results by querying a nucleotide sequence against a protein database. Specific BLAST searches (blastx) accomplish this by translating the nucleotide query sequences into amino acid sequences and comparing them with protein sequence databases. The following sequence query was generated as part of

an undergraduate classroom project, in which cDNAs were copied from the mRNAs expressed in red clover (*Trifolium pratense*) and were partially sequenced.

aagtgtataa aggtcaaatt attggcatcc atcaacgccc tggggacttg gccttgaatg tttgcaagaa aaaagctgca acaaacattc gttccaacaa ggaacaatca gtgattcttg atacaccatt ggattacagt ctggatgact gcatt-gagta catccaagaa gatgaactag tagagatcac cccccaaagt

When a search was carried out using "nucleotide blast" (RNA or DNA query used to search the nucleotide database), no significant matches were found (all E-values were greater than 1). The same DNA sequence, when

gi|68057657|gb|AAX87910.1| GTP-binding protein TypA/ BipA [Haemophilus influenzae 86-028NP]
gi|68249458|ref|YP_248570.1| GTP-binding protein TypA/ BipA [Haemophilus influenzae 86-028NP]
Length=616

Score = 75.5 bits (184), Expect = 6e-13
Identities= 35/67 (52%), Positives = 50/67 (74%), Gaps = 1/67 (1%)
Frame = +3

Query 3    V Y K G Q I I G I H Q R P G D L A L N V C K K K A A T N I R - S N K E Q S V I L D T P L D Y S L D D C I E Y I Q E D E L    179
           V Y + G Q I I G I H - R - - D L - + N - - + K - - - T N + R - S - K + - + + + L - T P + - + S L + - - I E + I - + D E L
Subjct 524 V Y E G Q I I G I H S R S N D L T V N C L Q K K L - T N M R A S G K D D A I V L T T P V K F S L E Q A I E F I D D D E L    583

Query 180  V E I T P Q S  200
           V E + T P + S
Subjct 584 V E V T P E S  590

**FIGURE 7.14**  DNA Sequence Alignment (blastx search). This sequence alignment was generated by blastx, which uses a translated nucleotide query to search protein databases. In this example the query is a nucleotide sequence from an unknown red clover cDNA sequence. The cDNA is made by copying mRNA into DNA (see Chapter 5). In the alignment shown, the amino acid sequence is located between the query sequence and the subject line (Sbjct). The returned subject indicates the position of the exact amino acid matches between the two sequences (single-letter amino acid designations are shown). Conservative amino acid changes, in which the R groups of amino acids are chemically similar, are represented by a (+). When the R groups of amino acids are very dissimilar, they are represented by a dash. Similar to a nucleotide BLAST search, graphical alignments and a listing of significant matches are returned but are not shown in this figure.

submitted to blastx (the DNA query was translated into an amino acid sequence and used to search the protein database), resulted in matches that showed significant similarities to the sequence of a putative bacterial GTP-binding protein TypA. Since the sequencing and BLAST search was initially performed in 2003, many more GTP-binding proteins from a variety of organisms have been added to the international sequence databases. Most sequences do not show significant nucleotide sequence similarity to the sequence derived from the clover cDNAs shown above, but using blastx, more than 150 of the GTP-binding protein entries are returned with E-values of at least 2e-10. Blastx searches, similar to nucleotide blast searches, return a graphical overview of sequence alignments, a list of matches, and pair-wise sequence alignments of matches. A pair-wise alignment of the translated red clover sequence with a bacterial GTP-binding protein is shown in Figure 7.14.

> Researchers often use the computer to compare two or more protein sequences or parts of protein sequences. BLAST programs can also predict the amino acid sequence of a protein from the DNA sequence of the gene.

## Using BLAST to Compare Protein Sequences with Other Protein Sequences

Here we have focused on using BLAST to demonstrate sequence similarity searches, but BLAST is not the only online program available for this purpose. Another example is FASTA (**FAST-A**ll), which compares protein or nucleotide sequences. BLAST and FASTA use different computer algorithms to carry out sequence alignments

and calculate the E-values in different ways. BLAST is generally faster than FASTA, but FASTA may be a better choice if the researcher is trying to compare very different sequences.

Pair-wise sequence alignments are very useful, but multiple sequence alignment (MSA) is also an important tool in studying proteins. MSA can provide information about amino acid sequences that are conserved among many different proteins, sometimes called sequence **motifs**. The amino acid sequences that encode protein regions important for the structure, function, or regulation of the protein are sometimes **conserved** relatively unchanged during evolution, and as a result these motifs are often present in all proteins with similar functions. Information regarding conserved DNA and protein sequences is very useful in experimental design, because it allows researchers to target specific regions of a protein for further investigation, as described earlier for the actin-binding site motif.

Several online programs are available for making multiple sequence alignments, including the popular ClustalW (www.ebi.ac.uk/clustalw/#, also accessed through PIR). As with BLAST, ClustalW is easy to use and tutorial help is available on the site.

## APPLIED BIOINFORMATICS

## Phylogenetics: Discovering Evolutionary Relationships

The ability to search for identity and similarity among sequences is an important skill for scientists engaged in many fields including **phylogenetics,** the study of evolutionary relationships among and between organisms.

## Box 7.2   Activity: Perform BLAST Searches

The following sequences belong to a gene that causes a human disease when mutated. Your goal in this activity is to perform a simple nucleotide BLAST search to identify the gene, explore various features of the gene, and determine which genetic disease is caused by a mutation in the gene.

1. First carry out your BLAST search:
   • Visit the NCBI site (http://www.ncbi.nlm.nih.gov/).
   • Click on the box labeled "BLAST."
   • On the BLAST web page, choose "nucleotide blast."
   • In the box under "Enter Query Sequence," type in one of the partial sequences shown below (or copy and paste sequence in as a query).
   • Click the BLAST button at the bottom of the page.

   A few tips:

   • The default settings permit a search of the human genome/transcript databases for highly similar matches. Leave the default settings in place for this search.
   • Searches can take from only a few minutes to much longer, depending on the number of users on the site. If the search is taking more than a few minutes, it is advisable to save the Request ID (RID) associated with your search. This number is near the top of the page and can be used to access the results of the search at a later time.

   Partial query sequences:

   1. GTGTGTGATGAGCGGACGTCCCTAATGTCGGCCGA
      GAGCCCCACGCCGCGCTCCTG
   2. CTTCTAATGGTGATTATGGGAGAACTGGAGCCTTCA
      GAGGGTAAAATTAAGCACAGTGGAAGAATTTCATTC
      TGTTCTCAGTTTTCCTGG
   3. AAGTTACTGGTGGAAGAGTTGCCCCTGCGCCAGGG
      AATTCTCAAACAATTAAATGAAACTGGAGGACCCG
   4. CAGGGAAACGGCATACACTGGAGAAGAATGTGTTG
      GTTGTCTCTGTAGTCACACCTGGATGTAACCAGCT
   5. CGCCCTGGGATTTACCGTGCTTTTAGCGTCCTACAC
      GAGCCATGGGGCGGACGCCAATTTGGAGGC
   6. CCATGGATTCTGAATGTGCTTAATTTAAAAGCCTTTG
      ATTTTTACAAAGTGATCGAAAGT

2. When the BLAST search is complete, the results will appear on the computer screen. You will see a table that lists "Sequences producing significant alignments." Click on a link under "transcripts" to access your report (sequence similarity data) and answer the following questions:
   • What is the identity of your gene?
   • What is the accession number?
   • What is the E-value associated with your match? What does this E-value tell you?
   • Look in the summary section. What information is given about this gene? If stated, what normal function of the protein encoded by the gene? What disease is caused when this gene is mutated?
   • On which chromosome is the gene located? (Click on "Genome View" in the results page to get positions of the BLAST hits in the human genome.)

3. Repeat this activity with two more of the sequences listed above.

4. Choose one of the three sequences used for the basic search and do another BLAST search, this time changing the search parameters. Instead of only searching the human databases, choose "Other," which will allow a search of all nucleotide databases.
   • Is this gene found in any other organism? Which one(s)?
   • What is (are) the E-values in the other organism(s)? What does this tell you?
   • If this gene is found in organisms besides humans, click on one or more accession number(s) of the nonhuman gene. Is there any indication that this gene is implicated in disease in other organisms?

5. Repeat the BLAST search, and in addition to choosing "Other" databases, also choose "somewhat similar sequences" under Programs.
   • How does the number of matches compare to your first search?
   • In what other organisms is the gene found?
   • What are the E-values in the new organisms?
   • What does the E-value tell you about the evolutionary relatedness of the organisms to humans?

Biologists use comparative sequence data (information revealed from sequence alignments) to construct **phylogenetic trees** (see Figure 7.3). These are diagrams showing the evolutionary relationships among various species believed to have a single common evolutionary ancestor. The common ancestor species of rodents, for instance, eventually gave rise to species of mice and rats. After the ancestral species diverged into two separate species, each branch gives rise to an independent lineage of organisms that accumulate, by chance, different mutations, usually small base pair changes in the genome DNA. A comparison of a short sequence from the beta-globin gene from modern-day mice and rats

shows that the sequences are similar but not identical (Figure 7.15). The beta-globin genes in mice and rats are homologous, because they were derived from the same ancestral gene and have similar functions in both mice and rats (see Chapter 10).

The redundancy of the genetic code means that a change in a nucleotide might not lead to a change in an amino acid and a mutation that leads to a change in an amino acid does not always change the function of a protein (see Chapter 10). Evolutionary biologists often look at slowly evolving proteins as a "molecular fossil record" that can be used to study the relationships between organisms. These biologists want

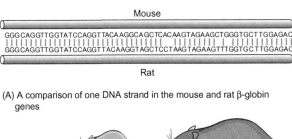

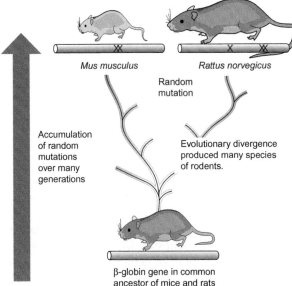

(A) A comparison of one DNA strand in the mouse and rat β-globin genes

(B) The formation of homologous β-globin genes during evolution of mice and rats

**FIGURE 7.15** Comparison of short segments of the β-globin gene in mice and rats. (A) Identical bases are connected by a vertical line; differences in bases are shown in red. The sequences are similar but not identical because mice and rats have accumulated different mutations after their lineage diverged from a common rodent ancestor. (B) Simplified representation of the formation of divergent forms of β-globin. Adapted from Brooker *et al.*, *Biology*, McGraw Hill, New York, 2007.

to know the amount of sequence divergence between two proteins, because the more similar the two protein sequences are to each other, the more closely related to each other the two organisms are likely to be. More distantly related organisms have the most differences in protein sequences. The similarities between gene or protein sequences often provide important clues about the function of an unknown protein. Proteins with significant sequence similarity are conserved and are predicted to be members of the same protein (or gene) family. The combined powers of molecular biology, genetics, and bioinformatics have allowed researchers to establish functional relationships among proteins and to propose how the proteins and their genes evolved.

## Bioinformatics in Modeling Protein Structure

The function of a protein depends on the three-dimensional structure adopted by folding the linear chain of amino acids after translation in the environment of the cell (see Chapter 10). Traditionally the shapes of protein structures have been determined by x-ray crystallography or nuclear magnetic resonance spectroscopy (NMR). These techniques yield the actual physical positions in three-dimensional space of each atom of a protein molecule, but these methods are time consuming and require expensive specialized equipment and personnel. However, researchers using these techniques have substantially increased the numbers of three-dimensional protein structures available in the world's public databases. Bioinformatics provides tools that simplify the process of visualizing the potential three-dimensional protein structures by using databases of x-ray crystallography and NMR information to perform online **protein modeling** that permits scientists to predict the shapes of proteins.

Using the database information, protein modeling software can sometimes predict the structures of uncharacterized or unidentified proteins. This is possible because the structure of a protein depends at least in part on the sequence of the amino acids in the protein, and the sequence of amino acids depends on the gene sequence. When the amino acid sequence is known, it is possible to use protein modeling programs that compare the unknown sequence with the identified protein structures in available databases. Proteins with similar amino acid sequences are likely to fold into similar three-dimensional structures and probably have similar functions.

Even if two proteins do not have identical sequences overall, it is still possible for the proteins to share short stretches of similar (or even identical) amino acids. These similar amino acid sequence motifs might fold in a similar manner in two different proteins, suggesting that the similar protein motifs have a common function, even though they have different amino acid sequences. For example, the CFTR protein has two regions that serve as (ATP)-binding domains (Figure 7.16). Other proteins containing similar amino acid motifs might also bind to ATP as part of their function. For example, the sequence of a motif in the multidrug resistance associated protein (which confers resistance to anticancer drugs) is similar to known ATP-binding domains (Figure 7.16). Thus, it is possible that the multidrug resistance associated protein also binds to ATP, a fact that could possibly be exploited in an anti-cancer treatment strategy; blocking ATP binding might inhibit growth of the cancer cells. Many pattern recognition and protein modeling programs are available through UniProt and NCBI, which allow researchers to perform extensive online analyses of protein domains, protein families, and potential three-dimensional structures. As the genome projects across the world produce more and more DNA sequences from different organisms, as more protein families are being

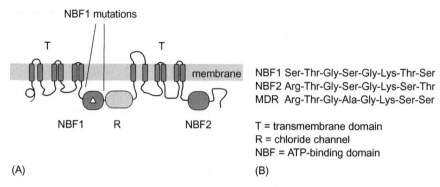

**FIGURE 7.16** CFTR gene mutations cause cystic fibrosis disease. (A) CFTR protein showing the ATP-binding domains (NFB1 and NFB2). (B) Comparison of the amino acid sequences of the NFB1 and NFB2 domains show they have similar sequences. The similarity in sequence with part of the human multidrug resistance (MDR) protein suggests that MDR might also contain an ATP-binding site.

identified by the presence of common sequence motifs. As novel protein domains and protein folding patterns are discovered, the task of online protein modeling will become an increasingly important skill for biologists studying protein function.

## Analysis of Gene Expression

As discussed earlier, genome sequencing projects have generated gigabases of sequence data available in a variety of databases. However, scientists agree that knowing the genome DNA sequence of an organism is really just the beginning of the study of the structure and function of the organism. Take the case of the little mustard plant, *Arabidopsis thaliana*, considered to be a weed by many. The *A. thaliana* DNA genome is relatively small (~114.5 megabases) and contains about 25,000 genes. Somehow scientists need to identify the locations of genes in a genome without necessarily knowing the entire genome sequence. Many years of research on genes in bacteria, humans, and now plants have revealed some almost universal DNA and RNA sequences that indicate landmarks in the genome DNA such as the start and end points of a gene and the locations of introns and exons (see Chapter 6). The field of bioinformatics has translated these molecular signals in the sequence into computer tools that help scientists to answer important questions about sequence data. As a result of basic research on many genes and through analysis of many gene sequences from many organisms, computer programs have been developed that recognize the specific DNA sequences of potential promoters, transcription start sites, transcription termination signals, coding regions, and so on, to help researchers identify the locations of putative genes in the genome. Of course, the presence of a possible gene sequence does not automatically mean that the

sequence is an actual gene, and sequence alone is not evidence that a gene is actually expressed as mRNAs in the cells being studied. As discussed earlier in the chapter, EST and microarray data can be used to identify genes that are expressed in certain cells, with an ever-increasing amount of gene expression data available in computerized databases.

The model organism, *C. elegans*, is widely used to study programmed cell death, or apoptosis (see Chapter 9). Apoptosis is an important process in humans (and other organisms) that is required for normal embryo and tissue development and physiology. For example, apoptosis is used to degrade cells in the tail on developing human embryos, to remove the tissue growing between developing human fingers, and to destroy selected nerves to regulate the number of neurons in a developing nervous system. Several genes important in human cell apoptosis were elucidated from studying *C. elegans* mutants. One of the genes, *ced-1*, encodes a protein that is involved in engulfing the debris released by cells that die due to apoptosis. Because *ced-1* is important in cell death in *C. elegans*, it is quite possible that a gene similar to *ced-1* performs a similar function in other organisms. A BLAST search reveals that gene sequences similar to *ced-1* are present in other genomes and that the *ced-1-like* gene is expressed. Using "*ced-1*" as a query to the EST databases through NCBI yields almost 90 hits, which indicate that genes similar to *ced-1* are expressed in specific tissues of several organisms, including human embryonic brain, rat (*Rattus norvegicus*) adult brain, kidney, placenta, aorta, and heart. This information is central to a better understanding of the *ced-1* gene and the process of apoptosis in other organisms and will be extremely valuable in designing experiments to answer further questions about the function of the *ced-1* protein.

## Box 7.3 Huntington's Disease: The Power of Animal Models and Gene Expression Data

Huntington's disease (HD) is an incurable and fatal hereditary neurodegenerative disorder caused by a mutation in the gene that encodes the huntingtin protein. The altered huntingtin protein destroys specific brain motor neurons, leading to its lethal decline in motor and cognitive abilities. Recent studies have indicated that the mutant huntingtin protein alters the transcriptional activity of other genes in the affected neurons. As gene expression changes in the nerves of HD patients, monitoring the different patterns in gene expression over time provides a new avenue for research to understand the mechanisms of the disease and to devise possible new treatments.

In one study, scientists measured the gene expression patterns in seven different transgenic mice used as models of HD. The transgenic mice strains were engineered to exhibit the symptoms associated with the different stages of HD disease (Figure 7.17). Analyses of the transgenic mice indicated that different forms and different amounts of the huntingtin protein had varying effects on the gene expression profiles. After completing the study in mice, the researchers then developed computational methods to compare the gene expression patterns in mice with gene expression in the human patients. The idea is to use this information to test possible drugs that affect gene transcription in mice with eventual use in human HD patients. This type of study underscores the importance of bioinformatics to store, analyze, and retrieve gene expression data as well as the fact that mouse models are a valuable tool to study human disease and its potential treatment.

**FIGURE 7.17** Transgenic mice show signs of Huntington's disease (HD). (A) Footprints of a normal mouse at age one year. Gray represents hind paws; black represents front paws. (B) Footprints of a mutant mouse with staggering gait at age one year. As the gait worsens, the mutant mice develop clumps of protein in the brain. The mice have a long stretch of repetitive DNA in the gene that encodes the huntingtin protein, the same mutation that occurs in humans with HD. Because the mice share so many similarities with humans, the HD mice can help scientists to understand the molecular basis for the neurodegenerative mechanism in HD.

## Analysis of Gene Mutations

Although it is important to know the sequence of a gene and the structure of the encoded protein, it is often also important to understand what happens when a particular gene (and possibly the encoded protein) is altered by a mutation. Although some mutations cause no change in a protein product, other mutations have a detrimental effect and can lead to disease. As many as 10,000 human genes can cause a genetic disease when altered by mutations (see Chapter 10). For example, mutations in CFTR can lead to cystic fibrosis, mutations in the dystrophin protein lead to muscular dystrophy, and mutations in huntingtin lead to Huntington's disease. Using a combination of scientific approaches including bioinformatics, it is routine to analyze and sequence mutant genes to determine if the mutant gene is the underlying cause of the disease. If the gene mutations do lead to a disease, then the sequence information is useful to develop treatments, cures, and preventions based on this knowledge.

Researchers studying a specific human disease often rely on the use of genomic information in the human **genomic map,** which is a diagram showing the relative positions of every gene indicated along the linear DNA

molecule contain in each chromosome. Various computer tools available through NCBI and other databases allow researchers to visualize entire chromosomes and genomes. NCBI offers a site to "Browse Your Genome" (Figure 7.18). To find the fibrosis transmembrane conductance regulator gene on chromosome 7, click on chromosome 7 and view the map of chromosome 7, which indicates the location of known genes including CFTR. If you click on CFTR, you will have access to a large amount of information on this gene.

The use of bioinformatics to find disease genes and to develop preventions, treatments, and cures is enhanced by the HapMap Project (see Chapter 10). The HapMap Project is a partnership of scientists from many countries and funding agencies that work to compare the genomic sequences from many different individual humans to identify shared regions of genetic variation (RFLPs, SNPs) among human chromosomes. The HapMap project goal is to identify, and make available online, the locations and characteristics of approximately 10 million single nucleotide polymorphisms present in the human genome (see Chapter 6).

The HapMap collection of cataloged SNPs is very valuable for research to understand human disease. For example, researchers are trying to understand the

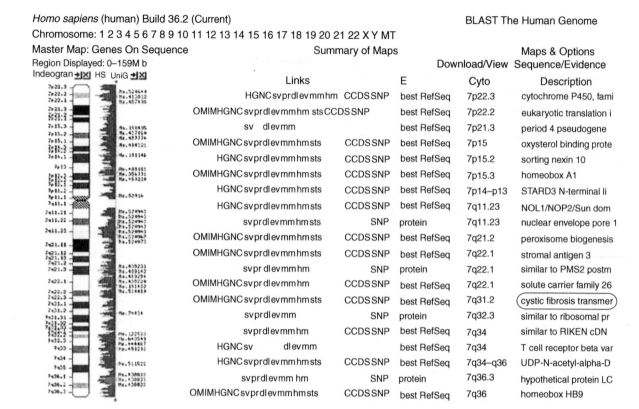

*Homo sapiens* (human) Build 36.2 (Current)                                                 BLAST The Human Genome

Chromosome: 1 2 3 4 5 6 7 8 9 10 11 12 13 14 15 16 17 18 19 20 21 22 X Y MT

Master Map: Genes On Sequence                        Summary of Maps                          Maps & Options

Region Displayed: 0–159M b                                                Download/View Sequence/Evidence

| Links | | E | Cyto | Description |
|---|---|---|---|---|
| HGNC svprdl evmmhm | CCDS SNP | best RefSeq | 7p22.3 | cytochrome P450, fami |
| OMIM HGNC svprdl evmmhm sts CCDS SNP | | best RefSeq | 7p22.2 | eukaryotic translation i |
| sv dl evmm | | best RefSeq | 7p21.3 | period 4 pseudogene |
| OMIM HGNC svprdl evmmhm sts | CCDS SNP | best RefSeq | 7p15 | oxysterol binding prote |
| HGNC svprdl evmmhm sts | CCDS SNP | best RefSeq | 7p15.2 | sorting nexin 10 |
| OMIM HGNC svprdl evmmhm sts | CCDS SNP | best RefSeq | 7p15.3 | homeobox A1 |
| HGNC svprdl evmmhm sts | CCDS SNP | best RefSeq | 7p14–p13 | STARD3 N-terminal li |
| HGNC svprdl evmmhm sts | CCDS SNP | best RefSeq | 7q11.23 | NOL1/NOP2/Sun dom |
| sv prdl evmmhm sts | SNP | protein | 7q11.23 | nuclear envelope pore 1 |
| OMIM HGNC svprdl evmmhm sts | CCDS SNP | best RefSeq | 7q21.2 | peroxisome biogenesis |
| OMIM HGNC svprdl evmmhm sts | CCDS SNP | best RefSeq | 7q22.1 | stromal antigen 3 |
| sv pr dl evmmhm | SNP | protein | 7q22.1 | similar to PMS2 postm |
| OMIM HGNC svprdl evmmhm | CCDS SNP | best RefSeq | 7q22.1 | solute carrier family 26 |
| OMIM HGNC svprdl evmmhm sts | CCDS SNP | best RefSeq | 7q31.2 | (cystic fibrosis transmer) |
| sv prdl evmm | SNP | protein | 7q32.3 | similar to ribosomal pr |
| sv prdl evmmhm | CCDS SNP | best RefSeq | 7q34 | similar to RIKEN cDN |
| HGNC sv dl evmm | | best RefSeq | 7q34 | T cell receptor beta var |
| HGNC svprdl evmmhm sts | CCDS SNP | best RefSeq | 7q34–q36 | UDP-N-acetyl-alpha-D |
| sv prdl evmm hm | SNP | protein | 7q36.3 | hypothetical protein LC |
| OMIM HGNC svprdl evmmhm sts | CCDS SNP | best RefSeq | 7q36 | homeobox HB9 |

**FIGURE 7.18** Browse your genome. You can browse the human genome at www.ncbi.nlm.nih.gov/genome/guide/human. From this graphical view, a researcher can select any chromosome to get additional detailed information about that chromosome. An example of the vast amount of data available for chromosome 7, CFTR is circled, and clicking on CFTR allows a researcher access to specific information about the molecule.

genetic variations in the human genome that increase susceptibility to cardiovascular disease in some people. The scientists can search the HapMap databases for SNPs that are shared in the genomes of people at risk for (or currently diagnosed with) heart disease. Specific SNPs shared in common among genomes from at-risk people probably indicates that an important gene is located near the SNP. In addition, if a researcher had candidate genes that might be important in heart disease, it would make sense to look for the presence of SNPs in or around those genes.

## SUMMARY

Bioinformatics, the marriage of biological data and computer science, allows scientists to access and analyze the huge amount of biological information available worldwide. Biological databases are maintained by a number of private and government organizations and store information about nucleic acid and protein sequences, protein structures, genome organization, and gene expression. This chapter introduced some basic types of data analysis using online resources and the tools of bioinformatics. The search for sequence similarity usually starts with a simple BLAST search and can then begin to address more sophisticated questions about protein structure and function and evolutionary relationships among proteins. Although the field of bioinformatics is still relatively new, it has already enhanced our understanding of basic biological processes, it has changed the way we design and carry out experiments, and it has the potential to enhance the development of medical treatments and advances in all areas of biotechnology.

## REVIEW

To test your knowledge of the chapter's contents, consider the following review questions:

1. Define bioinformatics.
2. Why is bioinformatics an important tool in the study of modern biology?
3. What types of information can be stored in and accessed from biological databases?
4. What are four main types of information you can find in a GenBank record?
5. Compare and contrast this with the information that can be found in UniProt records.

6. If you carried out a BLAST search using the query ATCGA, your results would indicate no significant matches. Explain why this would happen.

7. You carry out a nucleotide-nucleotide BLAST search with a 100-nucleotide sequence and receive results indicating no significant matches. You use the same sequence to carry out a search for a translated protein sequence and find several matches. Explain why matches were found by the search of the protein database but not from the search of the nucleotide database.

8. The genomes of many model organisms have been sequenced, and the information is stored in genomic databases. What is a model organism, and why is information from model organisms useful to understanding human biology?

9. Define the terms genome, proteome, and database.

10. What does it mean to say that two DNA sequences are homologous?

## ADDITIONAL READING

Markram, H., 11 January 2007. Bioinformatics: industrializing neuroscience. Nature 445 (7124), 160–161.

Moody, G., 2004. Digital Code of Life: How Bioinformatics is Revolutionizing Science, Medicine, and Business. Wiley, New Jersey.

Westhead, D.R., Parish, J.H., Twyman, R., Twyman, R.M., Howard Parish, J., 2002. Instant Notes in Bioinformatics. Taylor & Francis, Inc. Oxford, UK.

Freeland, S.J., Hurst, L.D., April 2004. Evolution encoded. Sci. Am. 290 (4), 84–91.

Collins, F.S., Barker, A.D., March 2007. Mapping the cancer genome. Sci. Am. 296 (3), 50–57.

Ezzell, C., April 2002. Proteins rule. Sci. Am. 286 (4), 40–47.

Constantine, D., 23 January 2007. Close-Ups of the Genome, Species by Species by Species. New York Times.

Brown, K., April 2003. Working weeds. Sci. Am. 288 (4), 34. Chapter 7 Bioinformatics figures were adapted from Brooker et al., Biology, McGraw Hill, New York, 2007.

Cohen, M.S., Zhang, C., Shokat, K.M., Taunton, J., 27 May 2005. Structural bioinformatics-based design of selective, irreversible kinase inhibitors. Science 308, 1318–1321.

Hermann, J.C., Marti-Arbona, R., Fedorov, A.A., Fedorov, E., Almo, S.C., Shoichet, B.K., Raushel, F.M., 16 August 2007. Structure-based activity prediction for an enzyme of unknown function. Nature 448 (155), 775–779.

Woese, C.R., Kandler, O., Wheelis, M.L., 1990. Towards a natural system of organisms: proposal for the domains Archaea, Bacteria, and Eucarya. Proc. Natl. Acad. Sci. USA 87, 4576–4579.

## WEB SITES

Bioinformatics for Dummies, www.ebi.ac.uk/2can/home.html. A guide for novices and intermediates to navigate and use many databases, online tools, and other resources.

Bioinformatics Portal, www.hsls.pitt.edu/guides/genetics/obrc. For biologists who say "so much data, so little time," this site provides brief descriptions of more than 1500 free bioinformatics databases.

Bioinformatics Successfully Predicts Immune Response to One of the Most Complex Viruses Known, www.sciencedaily.com/releases/2006/06/060616091145.htm. Bioinformatics has been used to predict the human immune response to the vaccinia virus, which is used in the smallpox vaccine.

Flu Gene Database Speeds ID and Tracking of Emerging Influenza Strains, www.sciencedaily.com/releases/2007/04/070426122415.htm. A free, searchable genetic database will improve testing and genetic tracking of influenza viruses as well as facilitate rapid response to emerging viruses.

Health Tips from Your Own DNA, http://abcnews.go.com/WN/story?id=3506203&page=1.

Largest Study of Human "Interactome" Reveals a Novel Way, www.sciencedaily.com/releases/2006/02/060228180911.htm. Study of the interactome, or the interactions between proteins in our cells, could help us identify genes implicated in human genetic disease.

T. Rex Protein Sequenced in Mass Spec Tour de Force, http://focus.med.harvard.edu/2007/042007/proteomics.shtml. Protein samples from a Tyrannosaurus rex are closely matched to proteins of present-day chicken, shedding light on the evolutionary relationship between birds and dinosaurs.

Uncovering Secrets of the Deep, www.sciencedaily.com/releases/2007/07/070731160156.htm. Metagenomics, the large-scale genomic analysis of microbes recovered from environmental samples, has its own database.

# DNA Forensics

---

### The Stranger Within

*New Scientist,* November 15, 2003

    By Claire Ainsworth

    Explain this. You are a doctor and one of your patients, a 52-year-old woman, comes to see you, very upset. Tests have revealed something unbelievable about two of her three grownup sons. Although she conceived them naturally with her husband, who is definitely their father, the tests say she isn't their biological mother. Somehow she has given birth to somebody else's children. This isn't a trick question—it's a genuine case that Margot Kruskall, a doctor at the Beth Israel Deaconess Medical Center in Boston, MA, was faced with five years ago. The patient, who we will call Jane, needed a kidney transplant, and so her family underwent blood tests to see if any of them would make a suitable donor. When the test results came back, Jane was hoping for good news. Instead she received a huge shock. The letter told her outright that two of her three sons could

not be her biological children. What was going on? It took Kruskall and her research team two years to crack the riddle. In the end they discovered that Jane is a chimera, a mixture of two individuals—nonidentical twin sisters—who fused in the womb and grew into a single human body. Some parts of Jane's body were derived from one twin, whereas other tissues in Jane were derived from the other twin. It seems bizarre that this can happen at all, but Jane's is not an isolated case. About 30 similar cases of chimerism have been reported, and there are probably many more people who are chimeras but who will never discover this fact.

## LOOKING AHEAD

DNA testing is an extremely powerful tool that has been used successfully to investigate many areas of scientific research, and to solve problems in biology. If the tests are performed correctly, the results of DNA testing almost always point scientists in the right direction, even when the DNA test results contradict a mother's biological connection to her child, as in the case of the **chimera** woman called Jane in Boston.

    Different types of DNA tests are appropriate for specific applications in forensics to solve crimes, in medicine to identify the bacteria and viruses responsible for human diseases, and in research to help reconstruct the evolution of the human race. First this chapter will investigate how forensic scientists use DNA testing to establish paternity, solve violent crimes, probe bioterror attacks, trace family relationships, track the spread of disease-causing pathogens, and prosecute poachers hunting illegally in the wild. The second half of this chapter reveals how **anthropologists** use DNA testing to identify the origins of the human race and follow the early migrations of human populations. Upon completing this chapter, you should be able to do the following:

- Explain how variability in the human genome DNA sequences of different individuals provide the molecular basis for forensic DNA testing.

- Describe the means by which investigators use forensic DNA testing to establish paternity.
- Describe how investigators use DNA testing to identify criminal suspects, and explain how they assess the probability of a coincidental match between an innocent suspect and the evidence.
- Explain why DNA testing is considered to be a versatile tool, and cite examples of situations other than violent crimes in which DNA analysis has helped solve mysteries.
- Identify the two types of DNA that are present in human cells, and explain the fundamental differences between the two genomes.
- Explain how anthropologists use DNA sequences to estimate the time to the most recent common ancestor for two populations.
- Describe which of the three hypotheses (uniregional, parallel evolution, or multiregional) is currently the most universally accepted by authorities in the field, and explain why.
- Describe the timeline and the migratory movement of the people who first populated the Americas.

## INTRODUCTION

You only need to glance around a group of people to witness the variability of human biology. On any city sidewalk you can see the complete spectrum of heights, hair colors, body types, and other physical characteristics of the human race. Like your physical characteristics, your DNA sequence is unique; even your closest relative has a different DNA sequence than you do. In the case of identical twins, DNA testing can find differences in the twins' DNA genome sequences, even though identical twins were thought to have identical DNA genomes. The genomes of identical twins show some regions of variability in DNA sequences that are far less variable than the DNA genomes of different unrelated individuals. The DNA sequence is especially variable at certain locations in the human genome, and it is these variable regions of the DNA genomes that are used to distinguish the DNA from one person and another person, and to determine whether or not they are identical twins.

The results of DNA testing often have consequences that profoundly affect people's lives. Some results reveal the identity of biological parents, while other forms of DNA testing are used to convict people of crimes that carry serious consequences including the death sentence. DNA testing has been subjected to a great deal of scrutiny by the scientific and legal communities and is studied considerably more than any of the other commonly used forensic investigation techniques.

DNA testing and the laboratories that conduct DNA testing have improved greatly in the past decade, and the system has emerged stronger because of the scrutiny. Modern DNA testing methods are highly reliable, and the databases of DNA sequences that are used for criminal and anthropological investigations are already impressive in size and still growing. Large DNA databases are important in order to avoid population biases that affect DNA testing results. DNA testing can be performed on any life form, including plants, animals, bacteria, and viruses, making it a versatile tool capable of answering a diverse array of perplexing questions.

## FORENSIC DNA TESTING: A POWERFUL AND VERSATILE TOOL

A woman went out on a date with a man she had recently met. During the course of the evening, they went to his apartment, where he raped her. She tried to fight him off and lost one of her contact lenses during the struggle. After her assailant let her go, the woman went to the police. When the police arrived at the man's apartment to investigate the allegation, the man told the police that he had indeed had sex with the woman that evening, but that the sex was consensual, and there had been no struggle.

The police noticed that the accused rapist had just cleaned his apartment within the past few hours. This appeared to be a suspicious coincidence, so the police examined the contents of the man's vacuum cleaner bag. They found several broken shards from the woman's contact lens. Forensic analysts were able to get enough of the woman's DNA from the contact lens to obtain her DNA profile. This finding corroborated her statement that she had been forcibly raped and exposed the man's lie. This shredded contact lens confirmed the violence of their interaction that night and clearly connected the contact lens to that specific woman. The man was convicted of rape.

There are many such stories that demonstrate the amazing power of forensic DNA testing to solve crimes. Forensic analysts can produce a DNA profile from a fallen hair, a licked envelope, a cigarette butt, and even a few broken shards of a contact lens. DNA testing has enabled investigators to solve hundreds of cases, including decades-old "cold cases" for which they had exhausted all leads. DNA testing serves the cause of justice as the perfect impartial tool; not only does it have the power to implicate the guilty, it also has the power to exonerate the innocent. DNA databanks, in which the DNA profiles of convicted offenders are stored, have helped law enforcement agents identify serial criminals, even when their crimes have been committed in different states or occurred many years ago.

On the other hand, DNA evidence has helped secure the release of hundreds of wrongly convicted individuals, some of whom had been sentenced to death.

## Capitalizing on the Variability of the Human Genome DNA Sequence

The DNA genome contains human genes that encode proteins, but it also contains a lot more than just the genes. In fact, only about 2% of the human DNA sequence encodes protein-coding genes; the other 98% of the sequences do not code for proteins (see Chapter 6). The **protein-coding sequence** of a gene contains a specific DNA sequence that codes for a specific protein. Changes in the DNA sequence coding for the protein will change the amino acid sequence of the protein, making a mutant product. When a mutant gene produces a mutant protein, that does not work properly, then the individual might inherit a genetic disorder or genetic disease (see Chapter 10).

The large amount of **non-protein-coding DNA (noncoding DNA)** in the human genomes does not produce proteins, but some noncoding DNA regions do have functions in the cell. However, changes in, the noncoding DNA is usually silent in terms of traits; even drastic changes in the sequence of the noncoding DNA might not affect the individual. These stretches of DNA do not encode proteins, so the differences in the noncoding DNA sequence might not cause differences in the individual's development or health. However, it would be incorrect to think that the noncoding DNA in the human genome is entirely without function or does not carry genetic information. Scientists know much less about the characteristics of the noncoding human DNA, and relatively little research has been focused on noncoding DNA sequences compared to the research efforts on the regions that code for proteins (see Chapter 6).

The sequence differences between human genomes are located at sites in the genome in both the coding and noncoding DNA, where the genomes of humans sometimes differ in DNA sequence.

The locations in the DNA sequence where the genomes from individuals are different are referred to as **polymorphisms** (from the Latin *poly* = many and *morph* = form). There are a great many polymorphisms in the human DNA genome, and now that the entire sequence of the human DNA molecule has been determined, these polymorphic changes have been identified and the positions located on the chromosomes (see Chapter 6). Because the positions of these polymorphisms on their respective chromosomes are known, the genetic researchers can use these polymorphisms to map the positions of the genes in the genome. This approach

has been instrumental in allowing researchers to draw maps of genes positioned on the different chromosomes, showing the relative locations of different genes on each chromosome. The locations of the polymorphisms on the human chromosome maps are known, so these polymorphisms serve as (and are referred to as) genome **markers**, just like mileposts on a roadmap do. Many of these polymorphisms are actually *single nucleotide polymorphisms* (SNPs) (single base pair differences) that have become an essential tool for many aspects of human genetic analysis (see Chapter 10).

Human chromosomes are arranged in 23 pairs inside most human cells, so that each individual person has two copies of each gene and two copies of each polymorphic marker. The exception to this, of course, is that males have one copy of the X chromosome and one copy of the Y chromosome and females have two X chromosomes but no Y chromosome. Each of the two copies of a gene can exist in more than one form depending on the DNA sequence of the gene. Each form of the gene is called an **allele**. The different forms of each polymorphic marker are also alleles of that marker in that individual. For each gene (or marker), one allele is located on one chromosome, and the second allele lies in the same location on the other member of that chromosome pair. Together, the two alleles for any gene or marker in an individual's genome are referred to as the individual's **genotype** for that marker. An individual's **DNA profile** consists of all the genotypes for all the chromosome markers for which data were obtained.

The standard forensic testing kits usually include reagents to detect 11 to 13 chromosome markers in the human genome. They often also include reagents to detect a polymorphic marker from the amelogenin gene, which are located on both the X and Y chromosomes (Figure 8.1). A region of DNA sequence encoding the amelogenin gene on the Y chromosome copy contains six DNA base pairs that are not present in the amelogenin gene located on the X chromosome. This small difference in DNA sequence between the X and Y chromosomes allows investigators to determine the sex of the individual who is the source of the evidence sample being tested. Forensic scientists test regions of DNA sequence referred to as markers that differ between individual human genomes. Each individual human carries two copies, or alleles, of each gene or marker, which constitute the individual's genotype for that marker; all of the markers tested comprise the individual's DNA profile.

The non-protein-coding human DNA contains a lot of **repeated DNA sequences**, including a type of repeated sequence called a **short tandem repeat (STR)**. As the name implies, in an STR, a short DNA sequence is repeated in tandem, without sequences in between

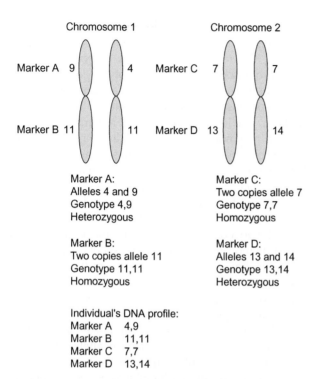

Marker A:
Alleles 4 and 9
Genotype 4,9
Heterozygous

Marker C:
Two copies allele 7
Genotype 7,7
Homozygous

Marker B:
Two copies allele 11
Genotype 11,11
Homozygous

Marker D:
Alleles 13 and 14
Genotype 13,14
Heterozygous

Individual's DNA profile:
Marker A    4,9
Marker B    11,11
Marker C    7,7
Marker D    13,14

**FIGURE 8.1** Relationships among chromosomes, markers, alleles, genotypes, and DNA profiles. Chromosome 1 has two alleles for marker A: 9 and 4, which are named for the number of repeats of a specific repeated sequence that exists at that position in the chromosome 1 DNA. The genome is heterozygous for the repeat at marker A because the alleles have different numbers of specific repeats. marker C on chromosome 2 is homozygous at marker C because each allele has seven copies of the specific repeated sequence.

Tetranucleotide repeat: Allele number 3, with three repeats

...ATGTGGTT**GCATGCATGCAT**CGCTGAAGGAT...
...TACACCAA**CGTACGTACGTA**GCGACTTCCTA...

Tetranucleotide repeat: Allele number 6, with six repeats

...ATGTGGTT**GCATGCATGCATGCATGCATGCAT**CGCTGAAGGAT...
...TACACCAA**CGTACGTACGTACGTACGTACGTA**GCGACTTCCTA...

**FIGURE 8.2** A hypothetical tetranucleotide repeat sequence. The hypothetical tetranucleotide repeat sequence shown here exhibits the features of a tetranucleotide repeat in the genome, including sequences preceding and following the repeat, which are seemingly random nucleotide sequences. The variability in the number of tetranucleotide repeats allows forensic investigators to identify individual DNAs from more than one individual; two different individuals will have different numbers of repeats of the tetranucleotide sequence.

the repeats. The type of STR that is used for most forensic DNA testing is the **tetranucleotide repeat**, in which a four base pair (4 bp) sequence repeat occurs between 5 and 50 times in the human genome (Figure 8.2). The DNA sequences preceding and following the tetranucleotide repeat are seemingly random sequences.

The variability in the nucleotide sequences of a tetranucleotide repeat allows forensic investigators to distinguish one individual's DNA from another individual's DNA. Quite simply, the genomes of different individual genomes will have different numbers of tetranucleotide repeats at some locations. In some people, the tetranucleotide repeats are so variable that a single individual will have a different number of tetranucleotide sequences repeated in each of his or her two alleles. When an individual has the same gene or marker allele on both copies of the chromosome, the individual is said to be **homozygous** for that marker. If the two alleles have a different number of repeats, then the individual is **heterozygous** at that position on the chromosome. The different tetranucleotide alleles are named for the number of repeats located at a given position on the chromosome (**locus**). For example, if the individual had 9 repetitions of the repeated sequence at one locus for marker A, and 4 repetitions of the repeated sequence at the other locus for marker A, that individual's genotype for marker A would be 9,4 (see Figure 8.1).

Tetranucleotide repeat markers are commonly used in forensics DNA testing for most criminal investigations. This marker is a stretch of DNA in which a 4 bp sequence is repeated between 5 and 50 times; the different alleles are identified by the number of 4 base pair repeats at that site.

Forensic analysts use the polymerase chain reaction (PCR) (see Chapter 5) to obtain a DNA profile using field evidence or a suspect's reference sample. Even tiny amounts of DNA can be amplified by the PCR method to produce millions of copies of a predetermined DNA sequence specified by the analyst. In the case of an STR test, the analyst chooses specific DNA primers that allow the amplification of the tetranucleotide repeat DNA plus some of the sequence flanking it on either side. Because the sequence flanking the repeat does not vary from one person to another, the size of the PCR DNA product will directly reflect the number of repetitions of the repeated sequence that exist at that particular locus. For example, if a specific allele containing 10 repetitions of a tetranucleotide repeat yields a PCR product that is 200 base pairs in size, a different version of that allele containing 15 repetitions of the 4 bp sequence will yield a PCR product that is 220 bp in size because it contains an extra 20 nucleotides (5 more repetitions of a 4 bp sequence). By determining the lengths of the PCR DNA products obtained from a sample, the forensic analyst can figure out how many repeated sequences exist in each of the individual's alleles for each of the markers tested. This is part of the information that the forensic analyst uses to create an individual's DNA profile.

**TABLE 8.1** Four-marker DNA profile test determines paternity

|          | Child | Mother | Father 1 | Father 2 | Father 3 |
|----------|-------|--------|----------|----------|----------|
| Marker 1 | 21,30 | 30,36  | 31,34    | 21,29    | 21,27    |
| Marker 2 | 4,10  | 4,9    | 6,15     | 10,11    | 8,13     |
| Marker 3 | 16,24 | 11,24  | 8,10     | 16,20    | 6,22     |
| Marker 4 | 6,12  | 12,16  | 6,10     | 6,14     | 7,10     |

DNA analysts use PCR to amplify regions of the genome DNA that contain tetranucleotide repeats. The length of the PCR products for each genome marker indicates the number of 4 bp repeats in each of the individual's two alleles.

## Using DNA Testing to Establish Paternity

Individuals inherit one chromosome from each chromosome pair from the mother and one chromosome pair from the father (see Chapter 6). Therefore, for any of the polymorphic markers in the genome, each individual inherits one allele from the mother and one allele from the father. A paternity DNA test can easily establish the identity of a child's biological father by determining and comparing the DNA profiles of the mother, the father, and the child to examine which alleles the child inherited from the biological father.

Consider the following hypothetical example using the genetic information provided in Table 8.1. A four-marker DNA profile is shown for the child, the mother, and the three men who might be the child's biological father. The child has inherited the number 30 allele of marker 1, the number 4 allele of marker 2, the number 24 allele of marker 3, and the number 12 allele of marker 4 from his or her mother. The father, therefore, has contributed the number 21 allele of marker 1, the number 10 allele of marker 2, the number 16 allele of marker 3, and the number 6 allele of marker 4 to the child. As Table 8.1 shows, although potential fathers 1 and 3 each possess one or two of the alleles the child inherited from his or her father, only father 2 possesses all four alleles that the child inherited from his or her father. Father 2 is therefore identified as the biological father of the child.

Paternity tests often involve testing the variable-number tandem repeats (VNTRs, pronounced "vinters") that can affect coding as well as noncoding DNA sequences in the human genome. A VNTR occurs when the number of DNA base pairs located between two fixed positions on a DNA molecule varies between genomes due to the presence or absence of a different number of DNA repeats. The different VNTRs are distributed in the genome in unique patterns that are specific to each individual human genome. DNA fingerprinting technology

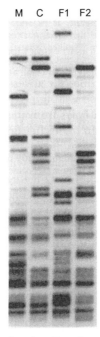

M   C   F1   F2

**FIGURE 8.3** The results of a DNA fingerprinting analysis using the 33.1 VNTR DNA probe. The lanes contain (M) mother DNA, (C) child DNA, (F1) DNA from possible father #1, and (F2) DNA from possible father #2. Comparison of the DNA bands in the different lanes indicates that the child's biological father is father #2.

is based on the detection of these types of DNA differences that make human genomes unique (Figure 8.3) (see Chapter 10).

To establish paternity, the DNA analyst determines the DNA profile of the child and mother, then "subtracts" the alleles the child inherited from the mother to reveal the alleles that the child inherited from the biological father.

## Using DNA Testing to Identify Criminal Suspects

Any time two people make physical contact with each other, some biological material is transferred from one person to another. The material may range from a stray hair falling on the coat of the person in the next seat at

**TABLE 8.2** DNA test results from a rape case

|  | Vaginal swab | Fingernails | Victim | Suspect 1 | Suspect 2 | Suspect 3 |
|---|---|---|---|---|---|---|
| Marker 1 | 19,22,24,26 | 22,24 | 19,26 | 19,29 | 20,26 | 22,24 |
| Marker 2 | 4,6,9 | 6,9 | 4,6 | 5,8 | 7,11 | 6,9 |
| Marker 3 | 13,22,24,27 | 13,24 | 22,27 | 15,17 | 14,19 | 13,24 |
| Marker 4 | 8,10,11,13 | 8,10 | 11,13 | 9,11 | 9,12 | 8,10 |

lunch to semen that a rapist has deposited in the vagina of his victim. The people who perpetrate crimes often leave biological materials at the crime scene, challenging forensic investigators to extract the perpetrator's DNA from the evidence and identify the DNA profile.

In this example, forensic investigators are investigating an alleged rape with three reasonable suspects. The vaginal swab samples from the victim, as well as scrapings of skin from under her fingernails, are available for testing. The results of these DNA tests are presented in Table 8.2, along with reference blood samples from the victim and the three suspects.

The vaginal swab sample contains a mixture of DNA from the cells lining the victim's vagina and the rapist's sperm cells, and it therefore contains both of their DNA profiles. You can see that, for each marker, the alleles that were found in the vaginal swab sample represent the alleles possessed by the victim plus the alleles possessed by suspect 3. These findings clearly implicate suspect 3 as the rapist. Suspects 1 and 2 are exonerated, because they each possess alleles that are not present in the vaginal swab sample. In addition, the DNA profile of the skin that was scraped from under the victim's fingernails clearly matches the DNA profile of suspect 3 at all four markers tested.

> To implicate a suspect as the source of crime scene evidence, the DNA analyst compares the DNA profiles of the suspect and the evidence sample. Some evidence samples contain material from the perpetrator only, whereas others contain a mixture of DNA from the perpetrator and the victim.

The preceding example postulated a crime for which there were three probable suspects, but in real life, there are usually no identifiable suspects, just the perpetrator's DNA in evidence left at the crime scene. The nation's network of **forensic DNA databanks** is one of the most powerful tools used by law enforcement agents to help solve crimes. The databanks contain the DNA profiles of people who have been convicted of crimes, as well as DNA profiles from evidence samples collected at unsolved crimes (Figure 8.4). The Federal Bureau of Investigation's **Combined DNA Index System (CODIS)** enables local,

state, and federal law enforcement officers to search the forensic DNA databanks of participating law enforcement agencies across the country. Investigators can search for DNA profiles from previously convicted offenders that match DNA evidence from crimes currently under investigation or for evidence that the perpetrator of a solved crime has also committed past unsolved crimes. By law, previously convicted individuals do not enjoy the same level of privacy protection as people without criminal records. As a result, once an individual's DNA profile has been entered in a forensic database, law enforcement agents can include the individual's DNA profile to investigate the possibility that the individual was involved in another crime, without needing additional evidence that gives them probable cause to believe that the individual might have been involved in the second crime.

These DNA databanks have been used to successfully solve hundreds of crimes, including many "cold hits" or "cold cases" for which investigators had exhausted their leads. The success of these databanks has prompted law enforcement agents to support the idea of collecting DNA samples from people convicted of nonviolent crimes, or possibly even from people who were arrested but not convicted of certain crimes. Data from states such as Virginia, which collects DNA from people convicted of certain nonviolent crimes as well as from people convicted of violent crimes, reveal that about half of the violent crimes were solved using DNA databanks with DNA from nonviolent offenders. Some public officials, including the former mayor of New York City and 2008 Republican presidential candidate Rudy Giuliani, took a more extreme position and advocated collecting DNA profiles from all U.S. citizens to include in law enforcement databanks. This could be accomplished if DNA profile testing were included among the blood tests already performed on newborn babies to test for a number of metabolic disorders and genetic diseases (see Chapter 10). With time, these databanks would obviously provide law enforcement officers with a powerful tool that would enable them to identify many violent offenders as soon as they committed a crime. However, this proposal and the long-term storage of personal DNA information obviously raise

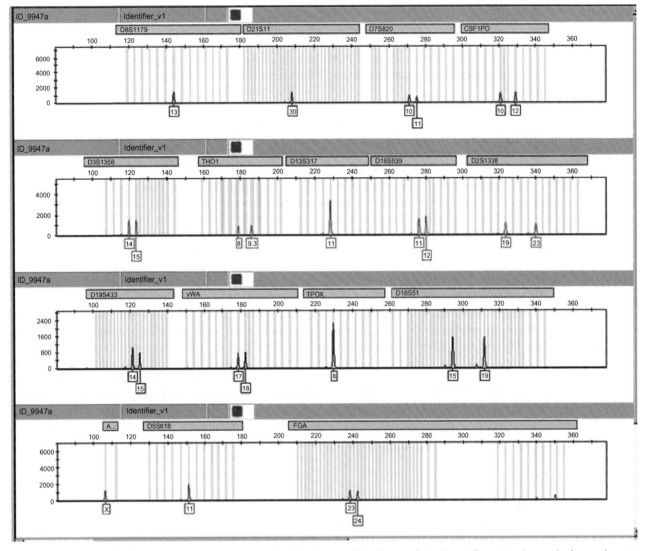

**FIGURE 8.4** Results of a forensic DNA test: a DNA profile. The DNA profile of a sample is shown illustrating the results from a forensic DNA test. The sample's DNA profile contains the following genotypes (start at top panel and read left to right):

| | | |
|---|---|---|
| D8S1179—13,13 | D21S11—30,30 | D7S820—10,11 |
| CSF1PO—10,12 | D3S1358—14,15 | TH01—8,9.3 |
| D13S317—13,13 | D16S539—11,12 | D2S1338—19,23 |
| D19S433—14,15 | vWA—17,18 | TPOX—8,8 |
| D18S51—15,19 | D5S818—11,11 | FGA—23,24 |

serious ethical issues, and some people suggest that storing the DNA profiles of people who are not convicted offenders constitutes an illegal invasion of privacy. Others argue that collecting DNA from the public for the purpose of storing DNA profiles in the databank is fair and unbiased as long as no individual or group is singled out for inclusion in the database.

## A DNA Match Does Not Prove That a Suspect Is Guilty

A match between the DNA profile derived from an evidence sample and that of a suspect does not automatically prove that the suspect is the perpetrator of the crime. There are several reasons why an innocent person's DNA profile might match the DNA profile of evidence found at a crime scene. A reported match may not be a true match and should be checked for human errors such as mislabeled samples, degraded DNA in a sample may produce an inaccurate DNA profile, or unscrupulous investigators may have tampered with the evidence. It is also possible that the suspect's sample could have been left at the crime scene innocently before or after the crime was committed.

Sometimes a DNA match is coincidental because the crime may have been committed by a close relative whose initial DNA profile, using a limited number of markers, is the same as the defendant's. Forensic DNA testing typically analyzes only a few genetic markers at first, which makes it possible to obtain what appear to be identical DNA profiles for two close relatives. However, DNA testing can easily distinguish the DNA from close relatives when the DNA testing involves enough genetic markers.

For the prosecutors in a legal case to prove beyond a reasonable doubt that they have identified the perpetrator of the crime, the investigators usually rely not only on the DNA evidence but also on witness testimony, the circumstantial evidence surrounding the crime, alibis, and other evidence that should either confirm or deny the suspect's involvement in the crime. The lawyers and investigators work to find and fit together the pieces of the puzzle before they can conclude with confidence that the perpetrator of the crime has been correctly identified. Even if the suspect's DNA profile matches that of the evidence perfectly, if the suspect has an alibi that clearly establishes that he or she was elsewhere when the crime was committed, it is important for the legal system to seek an explanation for the DNA match other than the suspect having committed the crime.

> A DNA match is strong evidence but alone does not prove a suspect is guilty beyond a reasonable doubt. There are many possible explanations for a coincidental match between an innocent suspect and the evidence, or there may be an innocent explanation for the suspect's sample being at the crime scene.

One of the most difficult issues surrounding the use of forensic DNA testing evidence in criminal trials involves the question of a **coincidental match**. Although it is highly likely that a given suspect DNA profile is unique in the human race, this is essentially a statistical argument unless in a specific case the forensic analysts have a database that includes the DNA profile of every person in the suspect pool. Without such a database, the analyst cannot definitively state that nobody else in the population has the same DNA profile as the suspect. In most cases the analyst must determine the probability that the same genotype will be selected at random from the reference population in the database.

The American legal system assumes the suspect is innocent until proven guilty. When the DNA profile of the suspect matches that of the evidence, one is obligated to consider the possibility that the suspect is innocent but coincidentally has the same DNA profile as the true perpetrator. If there is no witness or other evidence that can narrow down the list of possible suspects, the possible suspect pool is usually large. Investigators may consider the population of the city, the county, or even the country in which the crime was committed. The reasonable suspect pool helps to limit the size of the database to screen for matches to the suspect's DNA profile. The larger the reasonable suspect pool, the better the chance that the investigator might find an innocent person whose DNA profile matches that of the actual perpetrator.

The probability of a coincidental match is usually referred to as the **random match probability (RMP)**, which represents the probability that the DNA profile in question would be found in a person who was randomly selected from the same racial/ethnic group to which the defendant belongs. This is a difficult concept for jurors, lawyers, judges, and even expert witnesses to comprehend correctly. Many people, including expert witnesses, have committed logical fallacies or made misstatements in court regarding the meaning of the RMP.

To calculate the RMP for the DNA profiles for the suspect and the evidence, analysts must determine how common the suspect's marker genotypes are in the general population. Law enforcement agencies have collected DNA profiles from many people who were not involved in crimes but who volunteered to give their DNA to law enforcement agencies to determine the frequencies of common marker alleles and genotypes in the general population. These **reference databases** enable the analyst to determine how frequently the different genotypes for the different markers appear in the general population, or in other words what the probability is that one might find this same genotype in an individual who was selected at random from the reference population in the database.

Because some genotypes are more commonly found in one racial or ethnic group than in others, most forensic analysts maintain separate reference databases for DNA profiles of Caucasians, African Americans, Southeastern Hispanics, Southwestern Hispanics, Asians, and different Native American tribes. The reference database that most closely matches the suspect's ethnic heritage will be analyzed first to determine the frequencies of the genotypes of the DNA profile in the general population. Then the analyst applies the **product rule** of probability theory to calculate the RMP. The product rule states that the probability of a series of independent events happening is equal to the product of the probabilities of the individual events. For example, if the probability of an event is 20% (0.2), then the probability that a series of three independent events will happen is equal to the product of the probabilities of the individual events (0.2 × 0.2 × 0.2 = 0.8). In a forensic investigation, the analyst determines the probability of finding each of the marker genotypes for each DNA profile in the reference population, then the analyst multiplies the probabilities of the individual genotypes together to determine the overall probability of finding that DNA profile in an individual who was randomly selected from that reference population.

The more genetic markers that are available for the forensic analyst to evaluate, the smaller the RMP will be. The random match probability is the likelihood that a person who was randomly selected from the defendant's racial/ethnic group would coincidentally have the same DNA profile as the DNA profile of the defendant and the DNA profile derived from the DNA in the evidence. In a case involving a degraded DNA sample, reliable results are obtained from only a few markers, which increase the possibility of a coincidental match. In most cases, however, if data from nine or more genetic markers are available, the RMP is infinitesimal—possibly as low as 1 in 1 quintillion (1 in 1,000,000,000,000,000,000 or 0.000000000000000001). To appreciate how large a number 1 quintillion is, consider the fact that if a person counted to 1 quintillion at a rate of one number per second, it would take approximately 31,700,000,000 (31 billion, 700 million) years.

> When a DNA profile match is presented in court, the RMP for that DNA profile is included. The RMP represents the probability that the DNA profile would be found in a random person from the racial/ethnic group to which the defendant belongs. The smaller the RMP number, the more likely it is that the defendant is the source of the DNA evidence.

## Additional Applications of Forensic DNA Testing

Forensic DNA testing has emerged as a versatile tool that can be applied to many situations other than the investigation of violent crimes. Consider the need to identify the human remains of military casualties or victims of mass disasters such as the explosion of Airline Flight 800 over Long Island in 1996 or the attack on the World Trade Center on September 11, 2001. In cases where investigators do not have access to teeth to conduct a dental identification, such as in mass disasters, investigators can use DNA testing to identify the victims. In these cases, the concept of the RMP does not apply because the pool of people to whom the recovered remains might belong is limited to the servicemen and servicewomen who were involved in the battle, the people who were on the airplane, or the people who were in the World Trade Center at the time of the attack. In the future, DNA testing will ensure that all military casualties will be positively identified. The remains that lay unidentified for years in the Tomb of the Unknown Soldier in Arlington, Virginia, were finally positively identified as belonging to an Air Force pilot who was shot down over Vietnam in 1972.

DNA testing was also used to establish family relationships in a number of high-profile cases, as well

### Box 8.1  DNA Testing and the Anthrax Letters

DNA testing was very important to the investigation to find the terrorist who sent letters containing *Bacillus anthracis* anthrax spores to senators Tom Daschle (D-South Dakota) and Patrick Leahy (D-Vermont) as well as to the *New York Post* and NBC News. Anthrax bacteria can cause a serious disease in humans and livestock; the letter bioattack in 2001 killed five people. There are a few research laboratories in the United States that study anthrax, so it was important for investigators to determine the source of the anthrax. Like all DNA genomes, the anthrax genome DNA mutates and changes over time, so when someone moved an anthrax sample from one lab to another, it might be possible to trace the movement of the anthrax and eventually identify the lab source of the anthrax spores.

The Federal Bureau of Investigation (FBI) was convinced that Steven Hatfill, a biodefense researcher in the government labs at Fort Detrick, Maryland, was the anthrax bioterrorist, but Hatfill maintained his innocence and eventually filed a lawsuit, claiming invasion of privacy and harassment. By the summer of 2008, as the Justice Department finally settled Hatfill's suit for nearly $6 million, the FBI had already shifted to focus on U.S. Army researcher Bruce Ivins as the main anthrax suspect. After Hatfill was exonerated, Ivins committed suicide, raising new questions about Ivins's connection to the deadly anthrax letters. In August 2008, the FBI reported that they had used DNA testing and electron microscopy to study

the anthrax spores from the letters and they had found a link to anthrax cultures used by Ivins.

The Center for Disease Control discovered that the original lab "stock" of anthrax bacteria that was used to generate the anthrax spores contained a small number of anthrax mutants that grew into colonies with varied textures, colors, and sizes compared to the wildtype anthrax bacteria. Investigators wanted to follow the anthrax spores from the lab source to the contaminated letters by tracing the changes in the DNA sequences of the anthrax genomes. To identify anthrax cells carrying the rare genome mutations, researchers spread the spores on to many dishes containing solid media in the lab so that each bacterial spore could germinate and grow into a colony of bacterial cells. The rare mutant spores in the population formed colonies that appeared to be very different from the majority of the wildtype anthrax colonies, making it easy to distinguish the mutant anthrax colonies from the wildtype colonies. Scientists sequenced the genome DNA from the rare anthrax mutants and compared the mutant DNA sequences with DNA genome sequences from the initial lab strain of *Bacillus anthracis*. This study identified four rare DNA mutations in the anthrax genome that are passed on from one generation of anthrax cells to the next. The genomes from the lab stock of anthrax spores accessed by Ivins contained the same four rare mutations, a potentially important link between the anthrax spores in the attack letters and the anthrax stock in Ivins's lab.

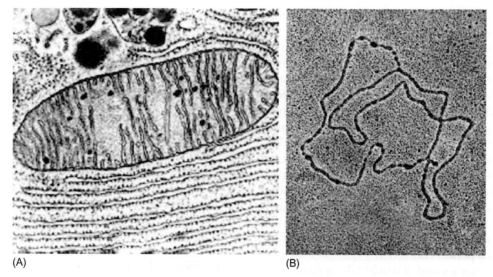

(A)                                                              (B)

**FIGURE 8.5**   A mitochondrion and the mitochondrial DNA genome (mtDNA). (A) This electron micrograph shows the highly folded internal membranes inside the mitochondrion. (B) This electron micrograph shows the circular mtDNA undergoing DNA replication.

as thousands of other less publicized cases. A well-known example involves the claims by several women that they were the Princess Anastasia, daughter of Tsar Nicholas II and Tsarina Alexandra of the Romanov family. The Romanovs were executed during the Bolshevik revolution, and their remains were dumped into a mass grave near Yekatarinaburg. Forensic anthropologists and DNA analysts have identified the remains of the tsar, tsarina, and several of their children, but none of the remains could be positively identified as belonging to their daughter Anastasia. Through the years, several women have come forward claiming to be Anastasia and therefore entitled to the Romanov family fortune. Their claims can be tested, however, by comparing their mitochondrial DNA (mtDNA) sequences to that of Prince Philip, husband of the present Queen Elizabeth II of England and grand-nephew of Tsarina Alexandra.

There are two major types of DNA in human cells, **nuclear DNA** (nDNA) and **mitochondrial DNA** (mtDNA) (Figure 8.5). In human cells, the nuclear DNA encompasses all 23 or 46 human chromosomes, and each human chromosome contains a linear double-stranded DNA molecule that extends from one end of each chromosome to the other end of the chromosome. In contrast, the **mitochondrial DNA (mtDNA)** is a double-stranded circular DNA molecule that contains only 37 genes; it is located in the mitochondrion, a specialized organelle in the cytoplasm that generates energy in the cell. The enzymes in the mitochondrion catalyze the biochemical reactions that allow the cells in the human body to harvest energy from the carbohydrates and fats in the diet.

During fertilization, the sperm contributes little more than its DNA. The rest of the material used

to create the child, including the mitochondria and the mtDNA, is stored in the egg. The mtDNA is therefore inherited solely from the mother to the child (son or daughter). People who are related through a line of female relatives will all carry the same mtDNA chromosome sequence. This was indeed the case for Princess Anastasia and Prince Philip, who were connected through a line of female relatives. The authentic Anastasia will have mtDNA sequences that will be identical to the mtDNA carried by Prince Philip (Figure 8.6).

Another well-publicized case involving DNA testing to determine family relationships tested the idea that Thomas Jefferson had biological children with his black slave Sally Hemings. This was certainly plausible since it is well known that slave owners in the South had children with some of their slaves. The descendants of a man named Eston Hemings had claimed for generations that Eston Hemings was the biological son of Thomas Jefferson and Sally Hemings. In fact, the descendants of a man named Thomas Woodson also claimed that he was the son of Thomas Jefferson and a slave. It was appropriate to use Y chromosome DNA testing in this case, because the individual in question was thought to be related to Thomas Jefferson by a line of male relatives. Jefferson's only son by his wife died in childhood. It was possible to trace the Y chromosome using a DNA sample obtained from a man who had descended from Jefferson's uncle Field Jefferson (Jefferson's father's brother) by an uninterrupted line of males.

DNA testing revealed that the Y chromosome sequences in the descendant of Eston Hemings were identical to those that were found in the descendant of Field Jefferson. This suggested that Eston Hemings was

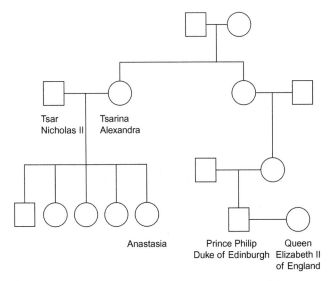

**FIGURE 8.6** Pedigree shows the relationship between the Princess Anastasia and Prince Philip. Princess Anastasia and Prince Philip are connected by a line of female relatives (circles), which means that their mtDNA sequences should match.

in fact the biological child of Thomas Jefferson and Sally Hemings. Because the Y chromosome sequence is shared by all men who can be connected by male relatives, however, this does not definitely prove that Thomas Jefferson was Eston Hemings's father. Jefferson's brother Randolph Jefferson was a frequent visitor at Monticello and also had interactions with the slaves there. There is a large body of historical and anecdotal evidence that indicates that Thomas Jefferson had an extraordinary fondness for Sally Hemings, however, and he is most likely to have been the father of Eston Hemings. Conversely, the DNA testing also disproved the claims of the descendants of Thomas Woodson.

In addition to criminal investigations, DNA testing is also used for biomedical investigations and is a common tool in basic DNA research. Like all DNA genomes, viral DNA genomes mutate, so that the specific interactions between DNA probes and the altered sequences can be used to identify differences in the sequences of the human immunodeficiency virus (HIV) genomes that infect different people. Investigators use the differences in genome sequences to determine the source of the virus in a specific infected patient. In 1992, the analysis of HIV DNA sequences produced results that implicated a Florida dentist as the source of HIV infections in six of his patients. The dentist was believed to have transmitted the virus inadvertently, however, and was not convicted of a crime. On the other hand, in 1997, a woman accused a physician of trying to kill her by injecting her with an HIV-infected patient's blood after she ended their romantic relationship. DNA testing results showed a match between

the strain of HIV that infected the woman and the HIV strain carried by one of the physician's patients. The physician was convicted.

Medical investigators also use DNA testing to identify specific pathogens that cause epidemics in a population. In 1994, a viral epidemic broke out in the "four corners" region of the United States where the borders of Arizona, Colorado, New Mexico, and Utah converge, and it spread quickly among the Native Americans living in the region, killing 35 people. Subsequent DNA testing determined that the little-known hantavirus (originally discovered in the Hantaan river region in Korea) was responsible for this epidemic.

DNA testing can also reveal important information about the mechanism that is used to spread viral infections. The famous 1918 influenza epidemic was the most virulent epidemic of all time, killing approximately 30 million people worldwide. Scientists used DNA investigations to identify the specific strain of influenza virus that was responsible for the 1918 infections. They also demonstrated that the virus could infect pigs and inside the pig host the virus acquired characteristics that enabled the virus to infect humans. A similar investigation revealed that the avian flu virus that affected a number of people in Hong Kong in 1997 had been transmitted through bird hosts. These investigations and many more have revealed important insights regarding the mechanism used by these viruses to make people sick, and revealed information about the mechanisms used to spread the viruses. The H1N1 swine flu virus genome contains human, pig, and avian sequences and caused a pandemic in 2009.

DNA testing has also answered questions regarding the spread of infectious diseases by early North American explorers. For example, many people think that the explorers who came to the New World in the 1500s brought the tuberculosis bacterium with them from Europe. In 1994, however, a research team (University of Minnesota) recovered the DNA from *Mycobacterium tuberculosis* (Figure 8.7), the organism that causes tuberculosis, which was found in the preserved lung tissue of a Peruvian mummy from the year A.D. 1000. This discovery proved conclusively that the tuberculosis disease already existed in the Americas many years before the European explorers arrived in the sixteenth century.

In addition to using forensic DNA testing to investigate homicides, forest rangers and game wardens have begun to depend on DNA testing to identify and prosecute poachers who hunt illegally. Investigators determine the specific DNA profile of an illegally killed animal from the carcass, then they match that DNA profile to the meat stored in someone's freezer or to the trophy animal head hung on a wall. DNA analysis is an excellent way to ensure that the meat being sold for human consumption actually originated from the expected animals and not from unauthorized sources. Investigators now use DNA testing to identify merchants that sell meat from endangered humpback whales or who market disks of skate meat as scallops.

DNA testing is a powerful and versatile tool that can be used to solve many different types of crimes, track the spread of disease-causing pathogens, determine paternity and identify the original source of animal food items.

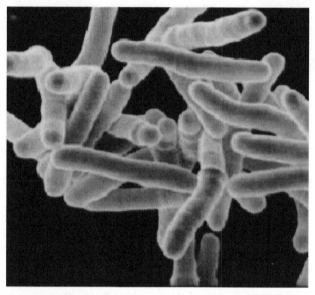

**FIGURE 8.7** *Mycobacterium tuberculosis*. This organism causes tuberculosis (TB).

## USING DNA ANALYSIS TO RECONSTRUCT THE ORIGINS OF THE HUMAN RACE

### Mitochondrial DNA Enables Researchers to Analyze Ancient Specimens

The circular mitochondrial DNA genome contains very little noncoding DNA in comparison to the large amount of noncoding sequences in the nuclear genome (see Figure 8.5). From the standpoint of DNA testing, this means that there are few locations in the mtDNA where changes in the DNA sequence can be tolerated without seriously altering the function of the mitochondrion. The nuclear genome is the DNA of choice for forensic testing, but mtDNA is often most appropriate for tracing maternal inheritance (from the mother) and for use in anthropological research. The mtDNA offers many fewer polymorphic sites that the forensic analyst can use to identify the source of crime scene evidence. Mitochondrial DNA undergoes mutations at a higher

---

**Box 8.2  History's Most Virulent Epidemic Explained. 1918: Could It Happen Again?**

There have been numerous epidemics during the course of human history, but none have involved a disease that was more virulent than the 1918 influenza epidemic. Victims initially presented with the typical flulike symptoms—fever, aches, chills, and a cough—but unlike most influenza infections, which run their course in a week or two, this one often killed its victim within a week after the victim first presented symptoms of the infection. Approximately 700,000 people died in the United States, and between 20 and 30 million people died worldwide.

Each time researchers analyze the DNA or RNA genomes of different viruses and study the proteins encoded by these infectious agents, they gain valuable insights about the mechanism by which an infectious agent causes a particular disease, as well as more about how infectious diseases spread from individual to individual. These insights then guide the development of research strategies to prevent and treat the disease, as well as to prevent the spread of the disease in the event of another outbreak.

Identifying the virus that causes an outbreak can be difficult because the immune system often eradicates the virus quickly from the body, but the patient is vulnerable to other pathogens. This was the case for the influenza epidemic in 1918, when the influenza victims often died of a secondary infection of

bacterial pneumonia. If the investigators could have performed DNA tests in 1918, they might have had a hard time obtaining the DNA evidence of the virus that caused the disease, because there would be little of the virus left in the body but a considerable amount of bacteria in the lungs. Any attempt to obtain the DNA or RNA from the virus that caused the epidemic is hampered by the fact that there is far more bacterial DNA present in the infected tissues than viral genome.

Lung specimens from 70 victims of the 1918 influenza epidemic had been stored at the Armed Forces Institute of Pathology (AFIP) in Washington, D.C. Seven of these victims were reported to have died quickly, thereby raising the possibility that viral DNA could be recovered from these lung samples. Although researchers were unable to obtain viral DNA from six of the victims, a team led by Dr. Jeffrey Taubenberger was able to isolate fragments of viral RNA from one of the seven specimens. These findings suggested that they had found a novel virus, previously undiscovered, that was related to the better-known swine flu virus.

Researchers are now trying to piece together the entire sequence of the novel virus RNA genome, hoping to discover which of the virus genes endow it with tremendous virulence. Researchers will use this information to do discover to develop an effective vaccine to protect people against this virus.

rate than nuclear DNA, which means that mtDNA provides a more powerful tool for use in DNA studies that track changes in DNA sequences over time. Whereas forensic analysts often compare two DNA samples to look for a match with a specific individual, anthropologists often need to track the changes in the DNA sequence of a population over long periods of time.

Two advantages offered by mtDNA include durability and abundance. The biggest obstacle encountered when analyzing DNA from ancient specimens is that DNA molecules degrade over time. The linear DNA molecules residing in chromosomes in the nucleus are much less stable than the mtDNA molecules because the circular nature of the mtDNA protects the mtDNA from some enzymes that degrade DNA. In addition, there are only two copies of each chromosome in the nucleus, but a single mitochondrion can contain more than 10 copies of the circular mtDNA molecule. There are up to several hundred mitochondrial organelles in a single cell, so there may be thousands of mtDNA copies in a single eukaryotic cell. The superior abundance and durability of mtDNA, coupled with the higher mutation rate, makes mtDNA an optimal choice for investigations involving ancient specimens.

> mtDNA contains fewer polymorphisms than the nuclear genome DNA, and is very useful for maternal inheritance studies. The mtDNA is much more abundant and durable than nuclear DNA and has a higher mutation rate, all advantages that make mtDNA an ideal choice for studying ancient specimens.

## Tracking Human Migration by Tracking DNA Sequence Mutations

Genomic research has provided estimates of mutation rates for mitochondrial DNA genomes and for nuclear chromosome sequences. Any DNA sequence with a known mutation rate can serve as a "molecular clock" that anthropologists use to determine how far back in time an investigator must go to find a common ancestor for people from different geographic regions. Scientists can determine the number of differences in the DNA sequences between the genomes of two populations and use the mutation rate of that sequence to calculate the time to **the most recent common ancestor (TMRCA)** for the two populations. Anthropological geneticists compare the DNA sequences recovered from fossilized human remains or from people who currently live in different parts of the world to the DNA genome of our closest evolutionary primate ancestor, the chimpanzee. The DNA sequences derived from the fossilized specimens of the evolutionarily older **hominids**

(hominids = human-like two-footed primates) are more similar to the chimpanzee DNA genome sequence than to the evolutionarily newer present-day human genome sequences. In addition, present-day humans contain some of these evolutionarily older sequences in their DNA, along with sequences that have arisen in the genome DNA more recently.

Calculating the TMRCA for different population groups helps anthropologists to reconstruct the migrations of early humans (Figure 8.8). In the figure, the circles marked with letters A to D represent different groups of people, living in different geographic regions. In this scenario, the people from whom the present-day groups A and D are descended were part of the same population. Sometime in the distant past, two groups of people split off from that population, one settling in region A and one settling in region D. Sometime later, two groups of people split off from group A. One migrated to region B, whereas the other migrated to region C.

When people first populated regions A and D, they had similar DNA sequences because they came from the same original population. However, each time a sperm or egg cell is made, the DNA is replicated, and mistakes occur during the process of DNA replication, altering the new DNA copy at one or more sequences. As a single family expands through several generations, the DNA sequences of the younger generations will differ from the DNA sequences of the older generations. The farther away an individual is from the original generation, the more differences in DNA sequences exist between the genomes of the current generation and

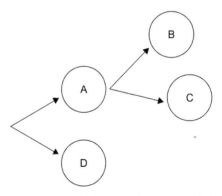

**FIGURE 8.8** TMRCA for different population groups helps to reconstruct early human migrations. This diagram illustrates a hypothetical scenario in which two population groups (A and D) diverged from a common population. Later, groups B and C diverged from group A. The letters A–D represent different groups of people, living in different geographic regions. The people from whom the present-day groups A and D are descended were at one time part of the same population. Sometime in the distant past, two groups of people split off from that population, one settling in region A and one settling in region D. At a later time, two groups of people split off from group A, one migrated to region B, and the other migrated to region C.

those of the family members who originally settled the area. Over evolutionary time, the DNA sequences of the people who live in regions A and D became slightly different.

When groups B and C first split from group A and populated their respective geographic regions, they had similar DNA sequences. With time, however, the DNA sequences of people from regions B and C became different, with measurable differences between the DNA sequences of people from regions A, B, and C. However, the sequences of people from regions A, B, and C will be noticeably more similar to each other than they will be to the DNA sequences of people who live in other geographic regions and have descended from different population subgroups, such as the people in region D. A study of the DNA sequences from people in geographic regions A, B, C, and D would reveal a TMRCA for groups B and C that is considerably more recent than the TMRCA for groups A and D, B and D, or C and D (Figure 8.8).

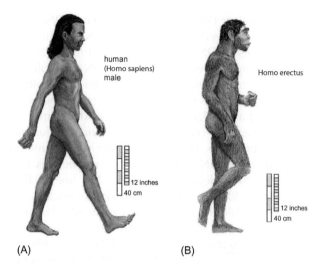

(A)                                    (B)

**FIGURE 8.9** *Homo sapiens* and *Homo erectus*. (A) Human being (*Homo sapiens*), male. (B) Artist's rendering of *Homo erectus*, which lived from approximately 1,700,000 to 200,000 years and migrated out of Africa and into Europe and Asia approximately 2 to 1 million years ago (2.0–1.0 mya).

---

Any DNA sequence for which the mutation rate is known can serve as a molecular clock. Anthropologists determine the number of differences that exist between the DNA sequences of two groups and use the known mutation rate of the sequence to calculate the time to the most recent common ancestor (TMRCA) for those two groups.

---

## The Debate over the Origins of the Human Race

The origin of the anatomically modern human (AMH), *homo sapiens*, is a subject of significant controversy, but there are also several points of agreement. Most experts agree that *homo sapiens* evolved from *homo erectus*, a human-like primate who walked upright on two feet. Most also agree that a wave of migration took *homo erectus* out of Africa and into Europe and Asia approximately 2 million to 1 million years ago (2.0–1.0 mya). Finally, most authorities also agree that another wave of migration out of Africa took place about 100,000 to 200,000 years ago. These recent African migrants encountered different populations of *homo erectus* as they migrated into Europe and Asia, and they coexisted in several locations.

The debate concerning the evolution of *homo erectus* into the anatomically modern human (AMH) *homo sapiens* (Figure 8.9) involves three main competing theories: the **uniregional model** (recent African origin or African replacement), the **multiregional model,** and the **assimilation model**. The primary difference among the three theories involves whether the recent African migrants replaced the other hominids they encountered without interbreeding with them or whether

there was significant interbreeding between the recent African migrants and other hominids. Scientists are using DNA analysis to determine whether the DNA of the human (AMH) *homo sapiens* is derived solely from the DNA of the most recent wave of African migrants or whether it includes evidence of breeding with the other hominids that the African migrants encountered as they migrated into Europe and Asia.

Experts that support the **uniregional** theory think that the recent African migrants did not interbreed with the other hominids they encountered in Europe and Asia. Instead, these other hominid groups all became extinct, and the more recent African migrants replaced them (perhaps by being better able to survive in the environment, perhaps by killing them) and became the sole forerunners of the modern human race. According to these theorists, the recent African migrants provided all of the DNA sequences from which the DNA genomes of the AMH *homo sapiens* were descended.

The **multiregional model** proposes that the original *homo erectus* population migrated into Europe and Asia and then *homo erectus* evolved into *homo sapiens* simultaneously in several different geographic regions. This theory postulates that the multiregional evolution of *homo sapiens* involved significant interbreeding between the more recent African migrants and the different hominid groups that they encountered, as well as interbreeding between groups that had populated different geographic regions. This hypothesis accounts for the idea that hominids in all geographic regions developed certain characteristics in order to survive and evolve into *homo sapiens*, including some physical traits that were not essential for survival, such as facial features, which evolved differently in different geographic regions. Extensive interbreeding between these different populations would preserve the gene mutations that allowed offspring to better adapt to the environment and thrive in all of the different geographic regions. In contrast, traits that are not essential to the survival of the individual but do affect the selection of a mate and survival of the species, such as facial features, probably evolved differently in the different geographic regions. This might explain why some of the cranial and facial features observed in fossilized specimens are also present in the people who live in the same geographic region where the ancient fossil was discovered. For example, the robust cheekbones are seen in both the modern Australian aborigines and in the fossilized *homo erectus* specimens found in Southeast Asia.

The **assimilation model** proposes that the recent African migrants did interbreed with some other hominid groups, but that the degree of interbreeding varied greatly from one geographic region to another and from one time period to another. This model can explain some of the contradictions that appear in the research literature. For example, it was firmly established that the Neanderthals of Europe did not contribute mtDNA to the pool from which the mtDNA of the human (AMH) *homo sapiens* had descended. However, several Australasia studies suggest that the ancient hominids who settled this area did contribute some DNA sequences to the pool from which the DNA of the AMH *homo sapiens* descended.

Three major models have been proposed to explain the evolution of the AMH *homo sapiens*. The earliest DNA studies supported the African replacement model, and most experts currently agree that this model is the most credible explanation of how the AMH *homo sapiens* evolved. When anthropologists study the DNA sequences of people living in Africa, Asia, and Europe, the vast majority of the oldest sequences are found in Africa. In addition, many studies report that modern human DNA contains only DNA sequences that have descended from African migrants, but no DNA sequences that descended from the other hominid groups.

One important example involves the study of mtDNA from the Neanderthal specimens that were found in Europe. It was once believed that the AMH evolved from the Neanderthal, and many textbook models depicted linear evolution in which the Neanderthal was the direct ancestor of the AMH. However, DNA studies have shown that the sequence of the Neanderthal mtDNA is very different from that of the AMH mtDNA. These results indicate that the Neanderthals became extinct and were replaced by more recent African migrants, prompting theorists to suggest that, rather than

a linear model, human evolution may resemble a bush, with many different branches that grew but ended, and only one branch that continued on to found the AMH *homo sapiens*.

Some researchers claim that the early studies that support the African replacement model do not provide a complete picture of the events because the early studies relied on analysis of mtDNA and Y chromosome DNA sequences. The mtDNA is inherited solely from the mother, and the Y chromosome is inherited just from the father, so the mtDNA and Y chromosome DNAs cannot be used to analyze events as many years ago as is possible by analyzing autosomal chromosome DNA. Critics argue that the findings of the mtDNA and Y chromosome DNA studies must be confirmed with research on the autosomal chromosome sequences in order to provide the information necessary to resolve the debate. However, because nuclear genome DNA degrades much more rapidly than the mitochondrial DNA, it is difficult to accurately study nuclear DNA sequences from ancient specimens using current DNA testing technology. Even though some recently published studies support the African replacement model, others argue against it in favor of a model that postulates some level of interbreeding between the African migrants and the other hominids.

> Most experts support the model of African replacement, which concludes that little interbreeding took place between the African migrants and the other hominids in Europe and Asia, although other reports indicate that some degree of interbreeding occurred. Further studies on autosomal chromosome DNA sequences will help to resolve this debate.

## The Colonization of the Americas

The theories that describe the colonization of the Americas must account for the presence of three cultural and linguistic groups: Amerinds, NaDene, and Eskimo-Aleut. Before the advent of DNA testing, many theorists believed that each of these three groups came to the Americas in its own independent wave of migration, but more recent DNA studies strongly contradict this theory. Researchers discovered four common (and one rare) versions of the mtDNA sequence in Native Americans. All four of the common mtDNA sequences can be found in people who descended from the Amerinds, but only one of the common mtDNA sequences is found in the descendants of the NaDene, and only two mtDNA sequences are found in the descendants of the Eskimo-Aleuts. Most researchers conclude that the Americas were populated by a single wave of Amerind migrants from Northeast Asia and that the NaDene and Eskimo-Aleut peoples diverged from the Amerind people after the Amerinds arrived in the Americas. Although the experts generally agree on these points, they disagree on the specific timing of the migration, the size of the original founding population, and whether the founding population experienced extreme or mild **bottlenecks** as they went from one phase of the process to the next. A population bottleneck is an evolutionary event in which most of the individuals in a population are killed or are otherwise unable to reproduce. A bottleneck can increase inbreeding in the smaller population as a result of a reduced pool of possible mates. Different migration dates have been proposed that range from 40 kya to 13 kya, and estimates of the size of the effective founding population range from 70 to 5000 individuals.

Most authorities believe that the forerunners of the Amerinds (proto-Amerinds) expanded from East Central Asia into Northeast Asia sometime before 50 kya and experienced a period of gradual population growth around 43–36 kya (Figure 8.10). A subgroup of the proto-Amerinds is thought to have migrated into the regions on either side of what is now the Bering Strait, which was then a landmass that connected Northeast Asia to North America until rising water created the Bering Strait approximately 11–10 kya. The region was productive grassland, home to many species of plants and mammals. At approximately 36–16 kya, the proto-Amerind population grew in size and genetic diversity, but ice sheets blocked access to the Americas until about 17–14 kya when the ice sheets began to melt. Many believe that the Pacific coast may have been the first region to become passable, perhaps as early as 19 kya, and that the first proto-Amerinds to migrate into the Americas traveled via a Pacific coastal route as soon as the retreating ice offered the opportunity. As the ice retreated even farther, several more inland routes appeared, and other proto-Amerinds migrated using these paths. These theories are supported by the fact that human settlements at the Monte Verde site on the coast of Chile are believed to be 14,500 years old, which predates the more inland settlements such as the Clovis complex in the southwestern United States by as much as 2400 years. Once these early settlements were established, the human population in the Americas expanded rapidly, both in size and in geographic range, from approximately 16–9 kya.

Most experts agree that the first Amerinds to settle in the Americas migrated along the Pacific coast about 19–17 kya and settled as far south as the coast of Chile. Other Amerinds migrated to the Americas shortly thereafter, using more inland routes that were exposed by the receding glaciers that once covered what is now Canada.

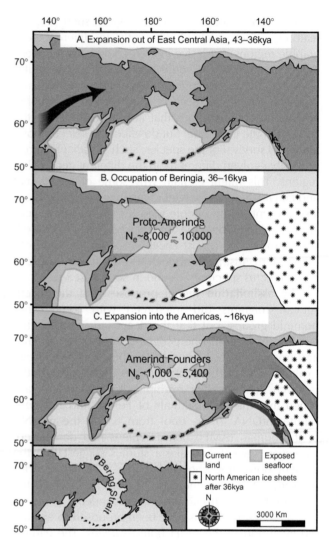

**FIGURE 8.10** Ancestors of the Amerinds (proto-Amerinds) expanded from East Central Asia into Northeast Asia. This diagram shows the most likely routes and timing for the colonization of the Americas. The ancestors of the Amerinds (proto-Amerinds) expanded from East Central Asia into Northeast Asia sometime before 50 kya and experienced a period of gradual population growth around 43–36 kya. Reprinted from Kitchen A, Miyamoto MM, Mulligan CJ (2008). A Three-Stage Colonization Model for the Peopling of the Americas. *PLoS ONE* 3(2):e1596. doi:10.1371/journal.pone.0001596.

## SUMMARY

Each individual has a genome DNA sequence that is unique, so that when someone leaves biological material at a crime scene, it is as if the person had left a calling card containing his or her name. Similarly, because we pass our chromosome DNA sequences to our children by inheritance, every child bears DNA sequences that unmistakably match DNA from both biological parents. When DNA testing is used to determine paternity, the analyst determines the DNA profile of the child and the mother, then subtracts the alleles that the child inherited from the mother from the child's DNA profile, which reveals the alleles that the child inherited from the father. To identify the source of crime scene evidence containing DNA, investigators look for someone whose DNA profile matches the profile of the DNA evidence. Everyone involved in this process must remain unbiased, however, because a DNA profile match by itself is not proof of guilt; it is one more piece of evidence for the jury to weigh.

The probability of a coincidental DNA match between the defendant and the evidence is referred to as the random match probability (RMP). This is the mathematical probability that a person who was randomly selected from the defendant's racial/ethnic group would have a DNA profile that would match the defendant. The higher the RMP, the less strongly the DNA match implicates the defendant as the source of the evidence. If the DNA in the sample has not degraded, forensic DNA analysts can usually get data from enough genome markers to produce an infinitesimal RMP, sometimes as low as 1 in 1 quintillion—way beyond reasonable doubt. To calculate the RMP for the DNA profiles for the suspect and the evidence, analysts must determine the frequency of the marker genotypes in the suspect's DNA and in the general population. Law enforcement agencies have collected DNA profiles from many people who are not criminals but who have volunteered to give their DNA information to build better DNA databases so that scientists and lawyers can determine the most accurate frequencies of common marker alleles and genotypes in the general population.

DNA testing has also proven to be a highly useful tool for investigating the evolution and migration of humans. Although most experts currently support the African replacement model, many also believe that little to no interbreeding occurred between the African migrants and the other European and Asian hominids. Some research points out flaws in the earlier studies in the light of findings that indicate traces of DNA sequences from some of these other hominid groups with modern human DNA.

Although there is still some debate regarding the timing of migration and the size of the founding population, DNA testing appears to have settled the main points of the debate regarding whether the Americas were settled by three peoples who migrated to the Americas independently of each other or whether a single wave of migrants gave rise to the three distinct cultural/linguistic groups believed to have been the Americas' earliest settlers. The DNA data suggest that the Americas were originally settled by Amerinds and that the NaDene and Eskimo-Aleut peoples evolved from the Amerinds. The Americas' first settlers appear to have traveled along a Pacific coastal route, whereas later settlers took advantage of the more inland routes, which developed as the glaciers that covered present-day Canada receded.

DNA testing can be performed on any life form, including recent as well as ancient specimens, and it has become a highly versatile and powerful molecular tool. DNA testing is useful for investigating violent crimes, not only those involving humans but also acts of poaching. DNA studies can help identify military casualties and victims of mass disasters and can identify the strains of microorganisms used in bioterror attacks. DNA testing has made it routine to use DNA to trace family lineages by genealogy and to identify the specific bacteria or viruses that cause various diseases. DNA has also offered scientists the power to reach far back in time and recreate the evolutionary events that created the diverse modern human population. DNA has a universal significance to all life forms as the molecule that carries genes, and as a result DNA testing can be applied to answer a variety of questions.

## REVIEW

This chapter focuses on two major areas of DNA applications in forensics. The first is the traditional DNA forensics used in law enforcement and by the criminal justice system. In addition, many examples were presented to demonstrate how forensic DNA testing is used in many fields, including paternity testing, identifying the perpetrators of violent crimes, and tracing the spread of infectious diseases. The second area of focus explores the use of DNA testing applied to the ongoing debate about the origins of humans and the colonization of the Americas. To test your understanding of these concepts and issues, answer the following review questions:

1. Describe the relationship between the following terms: "genotype," "DNA profile," "allele," and "polymorphic marker."
2. What is meant by the term "polymorphic tetranucleotide repeat"?
3. Imagine you are a DNA analyst assigned to a paternity case. You test the mother and child and produce the following genotypes:

|          | Mother | Child |
|----------|--------|-------|
| Marker 1 | 7,9    | 9,11  |
| Marker 2 | 27,31  | 25,31 |
| Marker 3 | 12,19  | 16,19 |
| Marker 4 | 6,12   | 3,6   |

What can you conclude about the father's DNA profile?

4. Imagine you are a forensic DNA analyst assigned to a murder case in which there is a match between the DNA profile of the evidence and that of the suspect. Why is it important for you to choose a reference database that has DNA profiles from people whose racial/ethnic heritage matches that of the defendant? Under what circumstances would the use of the wrong database prejudice the calculation against the defendant versus tilting the calculation in the defendant's favor?

5. Imagine you are a defense attorney and you are defending a murder suspect whose DNA profile matches that of some hairs that were found on the victim's shirt. How many different alternative explanations can you give for the presence of your client's hair at the crime scene, other than the explanation that your client committed the murder?

6. Describe one other application of forensic DNA testing, apart from the investigation of violent crimes.

7. Describe the limitations and advantages inherent in using mtDNA versus nDNA for forensic and anthropological investigations.

8. What three hypotheses have been offered as possible descriptions of the way in which the anatomically modern human arose on the Earth, and which of these three hypotheses do most authorities currently support?

9. What is the most important point of debate between authorities regarding the DNA pool from which the modern human DNA has descended?

10. How has the DNA evidence helped settle the debate as to whether the NaDene and Eskimo-Aleut peoples evolved from the Amerind people who first settled the Americas or whether all three groups coexisted in Northeast Asia and migrated independently to the Americas?

## ADDITIONAL READING

Bhattacharjee, Y., August 2008. The anthrax case: the trail of the spores. ScienceNOW Daily News.

Biello, D., December 10, 2007. Culture speeds up human evolution. Sci. Am.

Cartmill, M., September 1997. The third man. Discover.

Michaelis, R.C., Flanders, R.G. Jr., Wulff, P.H., 2008. A Litigator's Guide to DNA: From the Laboratory to the Courtroom. Elsevier/Academic Press.

Minke, J.R., August 2007. New fossils illustrate "Bushiness" of human evolution. Sci. Am.

National District Attorneys Association/American Prosecutors Research Institute. DNA forensics program web site. www.ndaa.org/apri/programs/dna/dna_home.html

Nemecek, S., September 2000. Who were the first Americans? Sci. Am.

President's DNA Initiative web site. Principles of forensic DNA for officers of the court. www.dna.gov/training/otc/

Scientific American. http://cogweb.ucla.edu/Chumash/EntryDate.html

Smithsonian Institute web site on human origins. http://anthropology.si.edu/humanorigins/

Sternberg, S., 1997. A Doughboy's lungs yield 1918 flu virus. Sci. News 151, 172.

Tattersall, I., April 1997. Out of Africa again ... and again. Sci. Am.

Tishkoff, S.A., Williams, S.M., 2002. Genetic analysis of African populations: human evolution and complex disease. Nat. Rev. Genet. 3 (8), 611–621.

United States Department of Energy web site. Human genome project information: DNA forensics. www.ornl.gov/sci/techresources/Human_Genome/elsi/forensics.shtml

Wade, N., December 11, 2007. Selection spurred recent evolution, researchers say. New York Times.

Wickenheiser, R.A., Jobin, R.M., 1999. Comparison of DNA recovered from a contact lens using PCR DNA typing. Can. Soc. Foren. Sci. J. 32, 67.

# Exploring Cell Fate

---

**"Patient Leads Fight for His Life"**

Josh Sommer: College Student and Cancer Researcher

"This is not an academic exercise at all. It is a matter of life and death"—Josh Sommer (Todayshow.com, February 2009).

Josh Sommer knows a lot about patience from working in a research lab; experiments cannot be rushed. Sommer's work is a small but vital part of an international effort to understand the genetic mechanisms behind cancer cells. So Sommer is very patient; he knows a mistake could set the project back a week, and thousands of lives are at stake, including his. In 2006, Sommer was a freshman at Duke University when an MRI showed a tumor pressing on his brain stem and growing around major arteries in his brain. Sommer's cancer is very rare and is known as an orphan disease because only about 300 people are diagnosed with chordoma tumors every year. Usually malignant, these tumors grow slowly in the spine or at the base of the skull and can spread to other organs. Chordoma tumors originate in the cells of the embryonic notochord, which is normally replaced by the bones of the spine early in fetal development (see Chapter 12). There are no effective treatments for chordoma, with a life expectancy of only seven years after diagnosis. Sommer and his mother were devastated by this horrible news, but they vowed not to give up. Sommer had the tumor removed and during his recovery, Sommer began to learn as much as possible about chordoma cancer, but there wasn't much to find. Like many other orphan diseases, rare cancers like chordomas are left out of the spotlight and out of the research funding system.

Then Sommer discovered that one of the world's leading chordoma researchers, Dr. Michael Kelley (MD), worked at Duke. Sommer met with Kelley and asked what he could do to help support chordoma cancer research. The next week, Sommer began doing research in Kelley's lab, searching for the genes that cause chordoma cancer (Figure 9.1). Sommer learned that one major obstacle to curing cancer was the lack of chordoma cancer cell lines, but unfortunately the tumors needed to generate the cancer cell lines were routinely discarded after surgery, and chordoma research was not well funded.

In 2007, Sommer and his mother began the Chordoma Foundation, created a chordoma database, and began to build an effective international scientific team to study the rare cancer. Sommer knows that his best chance for successful treatment requires a coordinated team effort by "scientists and doctors working hand-in-hand with patients toward this common goal." Sommer said, "I view every chordoma patient, family member, doctor and research as

Continued

(A)

(B)

**FIGURE 9.1** Josh Sommer: College student researches his own type of cancer. (A) Josh Sommer *(left)* with his research advisor at Duke University, Dr. Michael Kelley *(right)*. (B) Tumors can arise from tissues of the nose [1], sinuses (ethmoid sinuses [2], maxillary sinuses [3]), and the base of the skull [4].

teammates in the search for a cure.... Working together, we can turn our dreams for a cure into reality."

Josh Sommer is one of the first people to work in a research lab on his own disease, but he is not alone. Cystic fibrosis patient Jeff Pinard worked on the genetics of his disease, and Tulane medical student Andy Martin studied a cancer even rarer than chordoma called sinonasal undifferentiated carcinoma, which claimed his life in 2004. Sommer cautions people not to be distracted from the urgent reality of the lives that hang in the balance. "For me, this is a high-stakes race to outrun my disease," said Sommer, "I guess the way I look at it is that there will be a time for every disease when one can in essence outrun their disease.... For Andy his disease was too fast and the science too slow."

## LOOKING AHEAD

The human body is made up of trillions of tiny cells that work together to keep the body alive; each cell must also have the ability to act in ways that are completely independent of the other cells. In fact, each cell has the ability to commit to grow by cell division (mitosis) and make more cells of the same type or to commit to cell suicide (apoptosis), a normal process that rids the human body of unneeded cells. Both processes involve individual cells, but in the end they both benefit the entire cell population. These processes reflect steps toward a longer-term goal such as the development of fingers in the human embryo; the cells that make up the webbing between the digits die off as a result of apoptosis. Cells follow different developmental pathways as a result of interpreting the signals transmitted between cells. Cancer cells are of particular interest because they fail to follow the rules that control the cell cycle and begin to grow out of control. To understand these different cell fate decisions and responses, we need to know how genes and proteins control the eukaryotic cell cycle and what happens to trigger the development of deadly cancer cells.

On completing the chapter, you should be able to do the following:

- Understand how protein receptors on the surfaces of certain cells can participate in deciding cell fate.
- Explain the important differences between mitosis and meiosis in terms of maintaining or changing the total number of chromosomes in the cell.
- Explain how the four stages of the eukaryotic cell cycle are related to the part of the cell cycle called interphase.
- Describe how the checkpoint feedback system is used to avoid creating potential cancer cells carrying unstable genomes.
- Explain why cell suicide (apoptosis) is an important process for healthy cells even though it sounds like a strange choice for an individual cell to make.
- Understand that tumor cells need a blood supply to grow, and explain how scientists have taken advantage of this feature of cancer cells to develop potential anticancer drugs.
- Describe the types of genes and proteins that directly cause cancer cells to develop.

## INTRODUCTION

The cells in the human body do not exist alone, but are members of a network of diverse communities of cells that make up the different tissues and organs in the body. Some organs are made up of one type of specialized cells, whereas other organs contain different types of cells; both cell types contribute to the overall functions of the organs. For example, the kidney and heart contain different cell types and have different functions. The pancreas contains several different types of cells with different specialized functions, including the islet cells that produce and secrete insulin into the bloodstream. This chapter focuses on the biological processes that determine the developmental "fates" of different cells during the life span of the organism. The chapter explores the biochemical processes that cause cells to change fate, usually by altering the expression of certain genes that determine which pathways the cell will follow:

- Fate 1: cell division (reproduction to make more cells identical to the parent cells)
- Fate 2: cell differentiation (specialization changes cell structure and function)
- Fate 3: cell death (apoptosis to rid the body of selected cells)

It could be argued that fate 4 is the prolonged quiescent state adopted by highly specialized cells such as memory T and B cells and egg cells in females, but we will focus on the three major cell pathways. Cells in the human body are constantly lost and replaced. During a lifetime, about 100 trillion cells in the human body will age, die, and be replaced. The epithelial cells lining the intestines routinely slough off and are replaced with new epithelial cells. The human body normally changes all of its skin about once every month, shedding 30,000 dead skin cells every minute! Cell renewal is necessary so that the tissues and organs continue to function. The blood cells in the human body are replaced continuously, including the red blood cells that carry oxygen to the body tissues. The body's immune system must also routinely replace white blood cells (B and T lymphocytes), which have a very short life span. These immune system cells are replaced by the rapidly dividing blood precursor stem cells located in the marrow of the long bones.

Whether present in the fetus before birth or years later in the adult human body, cells routinely change developmental pathways in response to many different signals received from the environment and to interactions with neighboring cells. Cells also interpret information from distant parts of the body such as the brain. Signals can come in the form of protein hormones that are released into the circulating blood, an efficient process because the blood comes into contact with all cells and tissues. Hormones and other circulating factors can prompt cells to change gene expression patterns, which in turn causes cells to begin to develop into specialized cells (Figure 9.2).

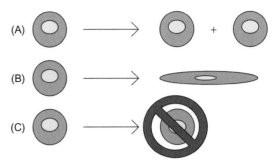

**FIGURE 9.2** Cell division can have more than one outcome. (A) Cell division by mitosis (reproduction to make more of the same types of cells). (B) Cell differentiation (cell specialization to change cell structure and function). (C) Cell death (apoptosis or programmed cell death to rid the body of unwanted cells).

Every cell in an individual human body contains identical DNA genomes regardless of cell type (except for red blood cells, sperm, and egg cells). It is not the differences in DNA sequences that allow specialized cells to perform different jobs in the human body. The 220 types of different cells in an individual's body all contain exactly the same genome DNA sequences. But different cells in the body do express different genes at different times. This fundamental concept explains how the cells with identical genomes can perform such diverse functions in the human body. Differential gene expression permits lung cells to absorb oxygen, stomach cells to express digestive enzymes, and brain cells to send nerve impulses; even though all of these cells carry identical DNA genomes, the specialized functions are dictated by the specific genes expressed in the individual cells involved. In human cells, this means that at any one time a myriad of different genes in the human genome are being turned on and off in response to complex external and internal cellular signals. These cellular events trigger many of the changes that occur during a human lifetime, from removing the skin cells between the fingers of a developing embryo to the explosion of sex hormones in the teen years and the slow growth of a cancerous tumor later in life.

This chapter focuses on cell fates, including cell division (reproduction) to make more body cells (or gametes), cell specialization to change the function performed by the cell, and programmed cell death (apoptosis) to rid the body of unneeded cells.

## FATE 1: CELL DIVISION AND REPRODUCTION

Healing a wound is a common human experience that offers a good opportunity to observe how the body responds to trauma involving the top layers of skin

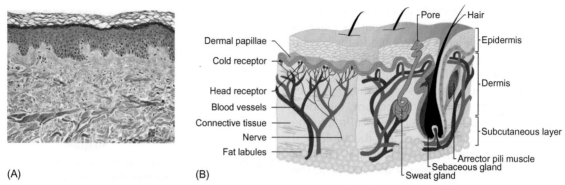

**FIGURE 9.3** The top layers of human skin cells are dead or dying. (A) Microscopic image of stained human skin cells (cross section). (B) Diagram indicating the different layers of the human skin.

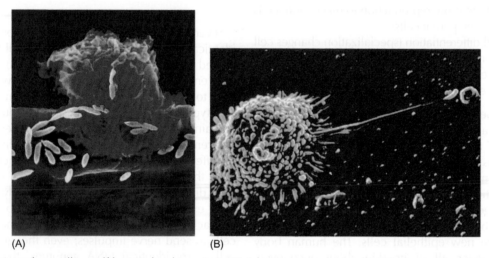

**FIGURE 9.4** Macrophage cells engulf bacteria by phagocytosis. (A) A human macrophage cell (pink) devours the much smaller bacteria (yellow rods) on the outer surface of a blood vessel. (B) A macrophage cell extends a long, thin projection to capture a nearby bacterial cell.

cells (Figure 9.3). The healing process starts when the undamaged cells near the site of the injury multiply by mitosis to replace the destroyed cells. The purpose of mitotic reproduction is to make new cells that are genetically identical to the parent cells, contain the same number of chromosomes, and can perform the same functions in the body. In the skin the process of **regeneration** involves three steps called the inflammatory, proliferation, and remodeling phases. Wound cleanup starts during the inflammatory phase, when cellular debris is engulfed by **macrophage** cells and eliminated by phagocytosis (Figure 9.4). Then signal proteins are released that instruct the cells near the edge of the wound to begin mitosis (cell division) and produce replacement cells. Cells migrate to new positions in the wound area during the remodeling phase, which is characterized by the growth of new blood vessels. The fibroblast cells secrete collagen and fibronectin into the wound area to build a new extracellular matrix (ECM) around the outside of the cells. Epithelial cells grow over the wound, myofibroblasts cause the wound site to contract, and collagen deposits are remodeled and realigned, all with the goal of healing the wound without forming scar tissue. Finally, programmed cell death (**apoptosis**) is activated in selected cells causing those specific cells to die, removing unnecessary and damaged cells left after cellular remodeling is complete.

The healing characteristics of skin cells are amazing, but the fact is that most organs and tissues in the human body do not respond to injury or disease by cellular regeneration. A well-known example is injury to the spinal cord. Even though some damaged nerve cells can regenerate, the nerves in the spine cannot regenerate a spinal cord, so the spinal cord damage remains and some level of paralysis is usually the result.

The human liver is unusual because it is one of the few organs in the body that can totally regenerate. Injury to the liver triggers the liver cells to undergo mitotic cell division to produce more liver cells that repair the damage to the liver.

## Chromosome Number is Controlled by Cell Division

Chromosome number (**ploidy**) is a key feature of every eukaryotic cell and it is carefully controlled during cell division. Most of the cells in the human body are **diploid** cells containing two copies of each chromosome (46 total) per cell. (see Chapter 6). Each **haploid** sperm or egg cell carries 23 chromosomes and contributes one human genome to the fertilized egg, producing a diploid zygote with 46 chromosomes (Figure 9.5). During each cell cycle, the entire DNA genome must be replicated and the duplicated chromosomes must be correctly transferred to the offspring cells. A high-fidelity chromosome segregation process is essential for the correct chromosomes (and genes) to be inherited by the progeny cells. A defect in this process of chromosome segregation can cause cells to inherit the wrong chromosomes (and genes) when the cells divide. Before a diploid cell can undergo cell division to make two new cells, the DNA in 46 chromosomes must be duplicated and packaged into 92 chromosomes, which are then segregated during cell division to produce two progeny cells with 46 chromosomes each. When cells divide by **mitosis**, chromosome number (ploidy) is *maintained* (Figure 9.6). When cells divide by **meiosis**, chromosome number is *decreased* by half (Figure 9.7).

Although eukaryotic cells are extremely diverse in form, function, and lifestyle, all eukaryotic cells reproduce by following the same general life plan. It is essential to understand the basic steps in cell division to be able to appreciate what occurs when cancer cells grow out of control. The eukaryotic **cell division cycle** contains four main stages, which are always executed in the same order: G1, S, G2, and M (Figure 9.8). Each stage of the cell cycle is dedicated to completing a series of specific events that must occur as the cell transits through each successive stage of the cell cycle. The overall plan makes sense: the cell must replicate its DNA (duplicate the chromosomes) before the cell can proceed to divide into two cells, with each progeny cell inheriting an equal number of chromosomes from the parent cell. Briefly, cells in G1 are preparing for DNA replication in S phase (the DNA synthesis phase). Cells in G2 have already duplicated their chromosomes and are preparing to segregate

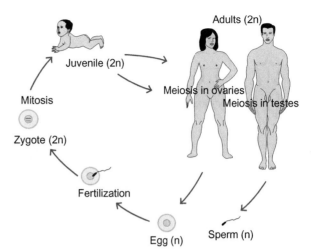

**FIGURE 9.5** The human life cycle. Meiotic cell divisions produce egg or sperm cells in ovaries and testes, respectively. Each sperm cell carries one copy of each chromosome; there are 23 chromosomes in each haploid germ cell (1n). When an egg is fertilized by a sperm cell, the result is a diploid zygote containing 46 chromosomes (2 copies of each chromosome; 2n). (see Chapters 6 and 10)

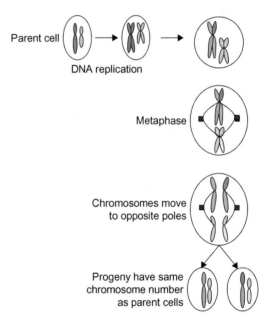

**FIGURE 9.6** Mitotic cell division maintains chromosome number. In mitosis and meiosis, the DNA in the chromosomes replicates and the duplicated chromosomes line up in metaphase to be distributed correctly to the progeny cells. Mitotic cell division maintains the parental number of chromosomes.

the chromosomes to the progeny cells in mitosis (M). Typically cells spend most (95%) time in G1, G2, and S phases (**interphase**). Mitosis (M), the part of the cell cycle where the important process of chromosome segregation (the actual molecular inheritance of genes)

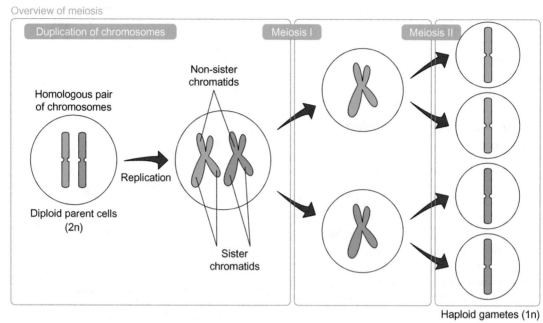

**FIGURE 9.7** Meiotic cell division reduces chromosome number. In mitosis and meiosis, the duplicated chromosomes line up in metaphase to be distributed correctly to the progeny cells at the end of the first of two nuclear divisions (meiosis I and meiosis II). The final cell products of meiosis are four haploid daughter cells, each cell containing one-half of the parental number of chromosomes (23 chromosomes in each human haploid cell). Meiotic cell division reduces the chromosome number from diploid (2n) to haploid (1n) cells.

occurs, is often the shortest part of a cell's life cycle (see Figure 9.8).

Cells in the G1 stage of the cell cycle are metabolically active, but DNA replication does not start until the cells transit into S phase. After the chromosomes duplicate, the cell enters G2 and continues to grow in size, synthesizing many different proteins in preparation for mitosis. For every cell division, the cell assembles an apparatus of special fibers (microtubules) that function to properly move (segregate) the duplicated chromosomes when cells divide in mitosis and meiosis. Once mitosis is finished, the process of cytokinesis physically creates the two progeny cells and the cell cycle is complete. The future health of any cell depends on inheriting a complete genome, so the process of chromosome segregation during cell division is arguably the most important stage in the cell cycle. Nonetheless a successful cell cycle also depends on the cell's ability to complete each stage of the cell cycle.

To control each cycle, eukaryotic cells use a checkpoint surveillance system to obtain feedback information about the progress of each stage of the cell cycle (Figure 9.9). The checkpoint system is an important protection against the formation of dangerous cancer cells. The checkpoint system significantly reduces the number of potential cancer cells with abnormal genomes that escape the safeguards provided by the genome repair and apoptosis pathways.

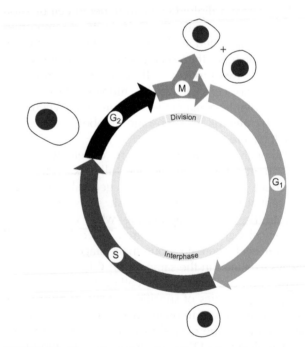

**FIGURE 9.8** The eukaryotic cell cycle consists of two unequal parts, mitosis (M) and interphase (I). The four stages of interphase always occur in the same order: G1, S, G2, M *(clockwise)*. During interphase (G1, S, and G2) the chromosome DNA is replicated and the chromosomes are duplicated. During mitosis the duplicated chromosomes are segregated (moved) in the dividing cell so that each daughter cell inherits one copy of each chromosome. G1: gap before S phase, S: DNA replication, G2: gap after S phase, M: mitosis.

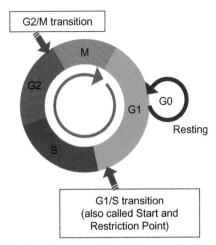

**FIGURE 9.9** The eukaryotic cell cycle is controlled by checkpoint genes. The cell cycle feedback system monitors the key transition points in the cell cycle, such as when a cell passes from G1 into S phase (called the G1/S transition; restriction point or "Start") or from G2 into M phase (G2/M transition).

## CANCER CELLS GO TO THE "DARK SIDE" AND EVADE CELL CYCLE CONTROL

The checkpoint surveillance system is a biochemical feedback mechanism used by eukaryotic cells to monitor the progress of each cell cycle. As the cell finishes each stage of the cycle, the checkpoint feedback system signals that it is safe for the cell to proceed to the next stage of the cycle. For example, if a cell does not finish replicating its chromosome DNA during S phase, it is dangerous for the cell to start mitosis (M phase) with partly replicated genome DNA. In this situation the checkpoint system detects the incomplete replication and sends a signal to temporarily delay the cell cycle at the S/G2 transition point until DNA replication is complete (Figure 9.9). The cell cycle delay is only a temporary solution however, because with time the cell will overcome the checkpoint delay and continue to divide. Cells that proceed with mitosis before genome replication is complete risk significant genome damage including the accumulation of multiple mutations and abnormal chromosome distribution during mitosis. An "unstable" genome can cause several forms of human cancer. Inheriting the wrong number of chromosomes, a condition called **aneuploidy**, is often lethal for cells missing one or more chromosomes.

A relatively small number of human genes encode proteins with important roles in the prevention, initiation, and progression of cancer. In many types of cancer, the genes that normally regulate the cell cycle are mutated, causing the cells to grow out of control. **Proto-oncogenes** are usually harmless genes in the human genome, but occasionally the accumulation of multiple DNA mutations will convert harmless proto-oncogenes into dangerous **oncogene** and mutant **tumor suppressor** genes (Figure 9.10).

> The cell cycle requires a series of highly regulated events that depend on the proper expression of specific cell division cycle proteins. Gene mutations can destroy normal cell cycle control and cause checkpoint failure. Without proper cell cycle surveillance, the cells begin to grow rapidly without control, the key hallmark of cancer cells.

Human genome DNA is wrapped by special proteins into chromosome packages that are further protected by residing inside the double-membrane cell nucleus. Despite these protections, human genome DNA can be easily damaged by environmental factors such as x-rays, ultraviolet radiation from the sun, physical trauma, high heat, and mutations can also accumulate due to DNA mistakes made during replication. A damaged genome might contain nicks and breaks in the DNA helix, creating regions of chromosome DNA that cannot be accurately copied and replicated during S phase or properly segregated in M phase, mitosis. The consequences of these damaging events can have a devastating impact on the cells and the organism (Figure 9.11).

Cells that divide while carrying damaged chromosome DNA are vulnerable to lethal events involving chromosome instability, including inheriting the wrong number of chromosomes or suffer broken or rearranged chromosomes. Multiple mutations leading to genome instability are key steps in the formation of cancer cells.

### Cell Division Cycle Genes Control Cell Growth

The biochemical feedback mechanism that controls the progression of the cell cycle is essential to avoid the devastating consequences that occur when cells attempt to divide with damaged genomes. The cell cycle feedback system monitors the status of the entire genome and identifies DNA damage and other genome abnormalities. In the event that DNA damage is detected, a feedback signal is sent that causes the cell to delay the forward progress of the cycle, giving the cell time to repair the DNA damage or to complete DNA replication. This feedback system is an important protection against the formation of cancer cells because it is an effective way for out-of-control cells to be identified and destroyed, sometimes by inducing programmed cell death (apoptosis) in the cancer cells.

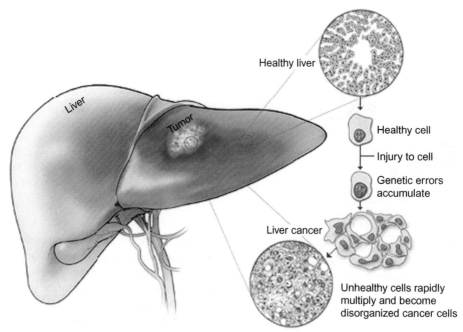

**FIGURE 9.10**   Multiple genome mutations can cause cancer cell development. When the cells in a tissue or organ suffer genome damage, the accumulation of genetic errors can cause cancer cells to develop.

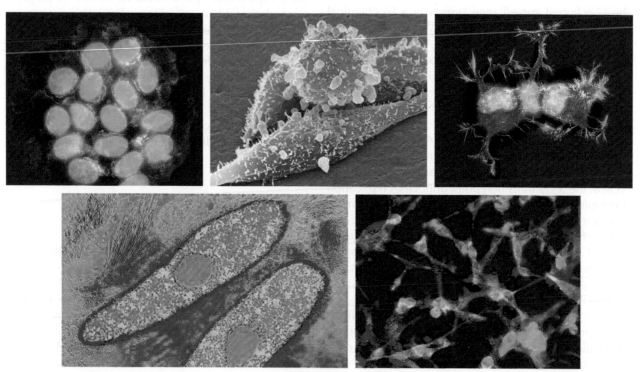

**FIGURE 9.11**   Cancer cells can develop from different types of cells. Cancer cells are shown. *(Left to right) Top:* breast, lung, neuroblastoma [microtubules *(purple)*, actin *(green)*] *Bottom:* Sarcoma (nucleus is orange), prostate cancer cells.

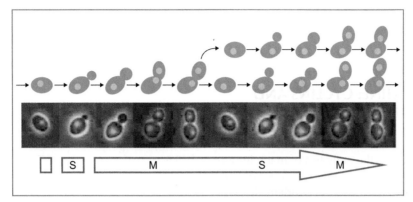

**FIGURE 9.12** Budding yeast cell cycle. The cell cycle of the budding yeast *Saccharomyces cerevisiae* proceeds from left to right. A single yeast cell at left has a nucleus and is beginning to make a bud. DNA replication (S) takes place followed by mitosis (M) to distribute chromosomes (and nuclei) to the new cell (growing bud), which separates from parent cell by cytokinesis.

The cell cycle feedback system monitors the key transition points in the cell cycle, when a cell passes from G1 into S phase (called the G1/S transition) or from G2 into M phase (G2/M transition). Special proteins search the genome for DNA damage and report on the status of the genome to the cell cycle control system, which is interpreted by the cell before the cell cycle is allowed to continue. In cases where the cell senses damage to the genome DNA at the G1/S transition, the DNA must be repaired before the genome can finish replication. Similarly, if the cell senses genome damage at the G2/M transition, the cell must repair the DNA before the chromosomes can be safely segregated during mitosis.

When the cell receives a "DNA damage" signal at either the G1/S or G2/M transitions, the checkpoint system responds by delaying the progress of the cell cycle, giving the DNA repair enzymes time to correct the damage or the DNA polymerase time to finish replicating the chromosome DNA. The cell cycle delay is designed to give the cell time to successfully correct the genome defects. In cases where the cell fails to send or receive the DNA damage signals, continued uncontrolled growth causes serious consequences such as genome instability, chromosome loss, and chromosome rearrangements. The failure of the cell checkpoint system is a characteristic feature exhibited by all cancer cells (see Figure 9.9).

The cell division cycle (CDC) genes that control the eukaryotic cell cycle were identified using genetic approaches in budding yeast (Figure 9.12). These single-celled eukaryotes use a relatively simple version of the same cell cycle control mechanism that operates in multicellular animals and mammals. In the 1980s, genetic studies on yeast helped scientists to understand how the key CDC genes and proteins control the cell cycle and execute a cell cycle checkpoint mechanism that is common to all eukaryotic cells.

The eukaryotic cell cycle is controlled by a series of biochemical reactions involving cell division cycle proteins that trigger, coordinate, and carry out the key events in the cell cycle. The cell cycle plays a key role in cancer development; loss of cell cycle control is a defect common to cancer cells. In normal cells, the feedback mechanisms work efficiently because components of the cell cycle machinery interact with cell cycle regulatory proteins in a highly interconnected network that continually responds to internal and external signals. The cell cycle is controlled by proteins that perform cell surveillance and transmit signals informing on the status of the genome each time the cell transitions from one stage of the cell cycle to the next.

The main checkpoints occur when the cell transitions between stages, such as between G1 and S (G1/S) and between G2 and M (G2/M). At the checkpoints, the cell gathers information and responds by regulating the progress of the cycle (see Figure 9.9). For example, at the G1/S checkpoint, also called the "restriction point" in mammalian cells and "Start" in budding yeast, the cell can either continue to grow in overall size or it can commit to a cell division pathway, depending on factors such as nutrient availability and lack of genome damage. If conditions at the G1/S checkpoint are satisfactory, then the cell division cycle will continue. At the G1/S transition, however, some cells have the option to exit the cell cycle and enter a nondividing (or resting) cell state called G0. Some highly developed cells in G0 can exist in a quiescent state in the human body for long periods of time. At the G2/M checkpoint, the cell must confirm that the chromosomes had replicated properly in S phase and are ready for the critically important step of chromosome segregation in mitosis. Incomplete chromosome duplication or mistakes in DNA replication cause genome damage that can trigger a cell cycle delay at the G2/M checkpoint. This system also monitors molecular events at the centromere checkpoint, confirming that each chromosome is physically attached to a spindle fiber before mitosis begins.

## CELL-CYCLE MACHINE: CYCLINS AND CYCLIN-DEPENDENT KINASES PROMOTE MITOSIS

Cell cycle control requires the activities of many important proteins with special functions. Two key types of proteins, the **cyclins** and **cyclin-dependent kinases (Cdk)**, play essential roles in cell cycle control and cancer prevention (suppression) (Figure 9.13). Cyclins are cell-cycle regulator proteins named because the number of cyclin proteins present in the cell cycles, increasing and decreasing as the cell cycle progresses (Figure 9.14). As the concentration of the cyclin proteins increases in the cell, cyclin binds to Cdk and makes a cyclin-Cdk complex called **MPF (maturation promoting factor or M-phase promoting factor)**. MPF is an active protein kinase enzyme complex, which regulates the activity of many key cellular proteins by adding or removing phosphates from target proteins. The cyclic nature of the cell cycle reflects periodic fluctuations in the cellular concentrations of cyclin proteins and the enzymatic activities of MPF (Cdk-cyclin) complexes, which determine the timing of the successive events of the cell cycle (Figure 9.14).

The term "cancer" actually represents many different, closely related diseases that can affect almost any type of human cell. Genetic studies indicate that no two cancers are exactly the same, even when the same types of cells develop into tumors. In most cases, cancers form as a result of multiple mutations in a cell's DNA genome, which cause defects that interfere with important biochemical pathways and protein functions in the cell. Acquiring multiple genetic mutations in the cell's genome is a key step in the development of cancer. Understanding the biochemical steps involved in controlling the cell division cycle is critical to understanding how cancer develops. Cancer cells undergo profound changes in gene expression to overcome the controls that normally restrict cell division, allowing the cancer cells to reproduce without limit and eventually migrate to other places in the body to establish new tumors (**metastasis**).

### Early Cancers Need Nutrients

Cells in the human body need a constant supply of oxygen and nutrients delivered by the circulatory system. The tissues are full of many tiny blood vessels called capillaries that come in close contact with most cells. When the cancer cells are dividing rapidly, they put additional demands on the supplies of oxygen and nutrients. The oxygen-starved tissues release signal molecules to promote **angiogenesis**, the growth of new blood vessels (Figure 9.15). The observation by Dr. Judah Folkman that cancer cells require access to

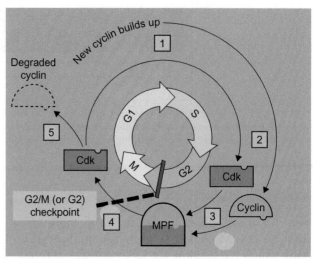

**FIGURE 9.13** The eukaryotic cell cycle is controlled by cyclin proteins, Cdk enzymes, and (Cdk-cyclin) kinase (MPF). (1) Cyclin proteins are synthesized during the G1, S, and G2 phases and accumulate in the cell. (2) A critical concentration of cyclin proteins is reached at the transition from G2 to M phase, called the G2/M checkpoint *(red bar)*. (3) At this point in the cell cycle the cyclin proteins bind to the inactive Cdk proteins to make active MPF (Cdk-cyclin) kinase. MPF is an enzyme that adds phosphates to key cell cycle control proteins. (4) This triggers a signal cascade mechanism that permits the cell to transit the G2/M checkpoint and begin mitosis (M). (4) Once mitosis is underway, the activity of the MPF (Cdk-cyclin) kinase enzyme is turned off when the cyclin proteins bound to Cdk are degraded. (5) In the beginning of the next cell cycle, newly synthesized cyclin proteins start to accumulate and bind to the Cdk proteins to form active MPF kinase enzymes at the G2/M checkpoint.

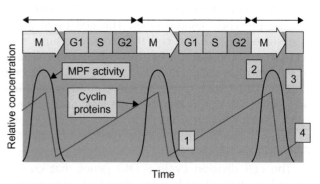

**FIGURE 9.14** The number of cyclin proteins changes during the cell cycle. (1) The concentration of cyclin proteins increases during G1 and S phases. (2) At the G2/M transition the abundant cyclin proteins bind to inactive Cdk proteins to make activate MPF kinase (Cdk-cyclin) enzymes. (3) At the metaphase-anaphase transition point in mitosis, the chromosomes segregate and the cyclin proteins are rapidly degraded, which inactivates the MPF kinase enzyme. (4) In the next cell cycle, the process begins again. At the G2/M transition, the cyclins and Cdk proteins form the active MPF kinase enzymes needed to initiate mitosis.

a blood supply was an important discovery because it raised the idea that anticancer drugs could be developed that would block angiogenesis and inhibit cancer cell growth.

## Metastasis is Inefficient but Deadly

The early diagnosis and treatment of cancer is often the best chance for a complete cure for most cancer patients; early cancer detection is essential. The majority of cancer deaths are caused by cells that have **metastasized;** these cancer cells are released from a primary tumor and travel through the blood and lymph system to begin new tumors at distant parts of the body. The much less dangerous **benign tumors,** also called neoplasms, grow in only one location in the body and do not metastasize. Benign tumors can cause illness and even death if they interfere with the functions of vital organs such as the brain. However, benign tumors are almost never as lethal as **malignant tumors** that have spread from a primary site to another location.

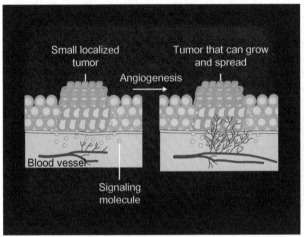

(A)

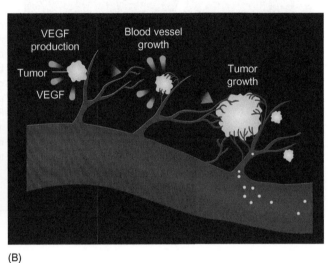

(B)

**FIGURE 9.15** The growth of new blood vessels is needed for cancer to spread in the body. (A) Small tumors emit signaling molecules that promote the growth of new blood vessels (angiogenesis) and provide the nutrients needed for the small tumor to grow into larger tumors. (B) Cancer cells shed from the growing tumor travel to other locations through the blood vessels, which spreads the cancer in the body.

### Box 9.1   Dr. Judah Folkman

World-Famous Cancer Researcher and Free Thinker

Dr. Judah Folkman, a world-famous cancer researcher whose insights led to whole new fields of medicine, died in 2008 at age 74. Folkman worked as a cancer doctor and researcher at Children's Hospital in Boston for 36 years. His free-thinking, persistent style often went against the grain of the conservative medical community, and Folkman became a frequent target of criticism in 1971 after he proposed that preventing angiogenesis could inhibit tumor growth by starving the cancer cells for nutrients. At the time, the dogma among cancer specialists overwhelmingly favored surgery and toxic chemotherapy drugs to stop cancer from spreading, so other approaches including Folkman's idea were largely discounted. When Folkman's research made national headlines in 1972, he was accused of offering people false hope for breakthroughs in future cancer treatments. "If your idea succeeds, everybody says you're persistent," Folkman liked to joke. "If it doesn't succeed, you're stubborn" (Figure 9.16). Folkman did not give up trying to prove his theory. He performed

experimental cancer treatments in mice that began to convince his critics. Folkman and many other researchers began to search for agents that could block the formation of tumor blood vessels in humans. Despite this interest and the potential of the discovery as an effective cancer treatment, the first natural angiogenesis inhibitor, thrombospondin, was not identified until 1989, followed by angiostatin in 1994 and endostatin in 1997. The vascular endothelial growth factor (VEGF) protein binds to the tyrosine kinase enzyme and regulates angiogenesis during biological processes such as wound repair, embryonic development, bone formation, and reproductive changes. VEGF expression is up-regulated (increased) by mutant oncogenes that permit the cancer cells to use VEGF to recruit new blood vessels.

In 2004, based on Folkman's pioneering research, Genentech developed the anti-VEGF drug, Avastin, which contains monoclonal antibodies that are directed against VEGF proteins. The Food and Drug Administration (FDA) approved Avastin to treat solid tumors, and it effectively

Box 9.1   Continued

**FIGURE 9.16**   Dr Judah Folkman's research led to the discovery of more than 10 new cancer treatment drugs and had profound scientific insight that led to new fields of medical research. Dr. Folkman's insights into the role of blood vessel growth in cancer development also gave us breakthrough treatments for age-related macular degeneration, a leading cause of blindness in humans. In addition, Dr. Folkman also played an important role in the development of a new form of birth control that is implanted under a woman's skin.

added to the life expectancy of patients with advanced colon cancer, even without chemotherapy. Unfortunately, Avastin is not the one-size-fits-all cancer cure that many hoped for. Even Folkman conceded that it is much harder to cut off the blood supply to a tumor than he once believed. "The ideas are simple, but getting them figured out is very complicated." In 2008, Avastin was approved to treat advanced breast cancers. Researchers around the world are studying at least 10 different antiangiogenesis drugs, hoping that treatments with combinations of angiogenesis inhibitors will keep tumors from growing for extended periods of time.

Folkman's genius extends beyond his inspired work on cancer to the development of effective treatments for blindness. Angiogenesis inhibitors can help patients to successfully avoid the serious vision problems and blindness caused by age-related macular degeneration in humans. In this eye disease, new blood vessels grow into and destroy the central part of the retina. VEGF protein expression is prevented by RNAi gene therapy treatments in people (see Chapter 11). The angiogenesis inhibitors restored vision in people blinded by macular degeneration; one grateful patient donated $100,000 to Children's Hospital.

Metastasis, the spread of cancer cells around the body, is more like a planned invasion and not a passive process. The cancer cells in the small primary tumor enter the bloodstream by secreting enzymes that digest the extracellular components surrounding the blood vessel cells, which allows the cancer cells to squeeze between cells and enter the bloodstream (Figure 9.15). The metastasizing cancer cells circulate through the body looking for a place to reenter the tissues and start tumor growth in the new location. Some metastasized cells develop into a tumor made of the same types of cells as the cells in the primary tumor. For example, when bone cancer cells metastasize to the liver, the secondary tumor in the liver is made up of bone cancer cells that grow in the liver. Overall the process of metastasis is inefficient, but it is a successful way to spread the disease because millions of cancer cells can be released from a tumor each day. If only a fraction of metastasized cells survive to form a new tumor, the odds favor the growth of a secondary tumor during an individual's lifetime. Cancer cells can also travel through the lymph system, which is an extensive network of fluid that flows throughout the body. The timing of the movement of cancer cells into the lymph nodes in patients is one measure used in the detection and staging of individual cancers.

## GENES CONTROLLING CANCER: TUMOR SUPPRESSOR GENES AND ONCOGENES

Two main types of human genes are directly involved in the development of human cancers, **oncogenes** and **tumor suppressor genes**. Oncogenes are the mutant forms of genes found in the genome that normally control how often a cell divides and also control genes that regulate the process of differentiation when cells acquire specialized functions. A mutation can change a harmless wildtype gene into a dangerous oncogene that drives uncontrolled cell division. Tumor suppression proteins function in cells by controlling processes that have the potential to cause cancer if left unregulated.

More than 100 oncogenes have been identified that encode proteins with five major functions (Figure 9.17):

- *Growth factors.* Specific growth factor proteins promote the growth of certain types of cells.
- *Growth factor receptors.* Growth factor receptor proteins are located on the surfaces of the cells where they detect and transmit signals between cells and into cells.
- *Signal transducers.* Signal transducers are signaling components that transmit signals between the growth factor receptors on the surface of the cell

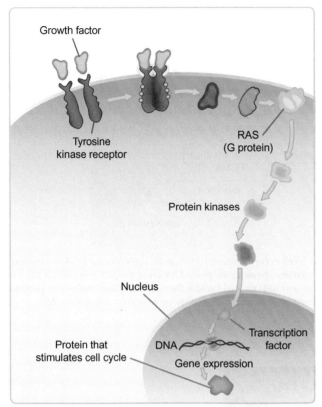

**FIGURE 9.17** Proto-oncogene proteins normally function in the pathway that transmits signals into the cell. Proto-oncogenes encode several different types of proteins including growth factors, growth factor receptors, signal transducers, transcription factors, and programmed cell death regulators. These proteins normally transmit signals from the outside surface of the cell to the genome in the nucleus. However, DNA mutations can convert the proto-oncogenes into dangerous oncogenes.

and the inside of the nucleus, where the genome can respond to the signal. Mutations in these components can prevent the cancer cell from controlling its own division.

- *Transcription factors.* Transcription factors turn gene expression on and off and regulate genes that control cell division. Mutations that overactivate the Myc oncogen can stimulate rapid cell division in lung cancer, leukemia, and lymphoma.

- *Programmed cell death regulators.* Programmed cell death (PCD) proteins control apoptosis like a master switch that commands certain cells to commit suicide.

## p53: A Tumor Suppressor Superstar

Cells normally benefit from the functions of the p53 tumor suppressor protein. Expression of the p53 gene, named for the mass of the encoded protein (53 kD), is essential to protect the body from cancer. The p53 protein is involved in the checkpoint feedback system that safeguards the genome and acts to control the cell cycle to avoid cancer cell development. Cells

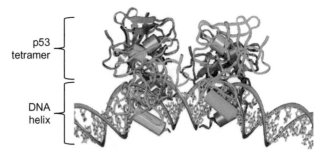

**FIGURE 9.18** The p53 protein binds to the DNA genome and turns on expression of the p21 gene. Four identical p53 proteins *(shown in gold, blue, green, and magenta)* make up the p53 tetramer that binds to DNA *(silver)* and turns on expression of the p21 gene.

with mutant p53 proteins continue to divide even with damaged genome DNA because they fail to trigger apoptosis in the absence of p53. The genomes in these mutant cells accumulate additional mutations and become unstable. p53 was chosen as the Molecule of the Year by *Science* magazine in 1993, which reflects the key role that the p53 tumor suppressor protein plays in these biological processes.

In normal cells, the wildtype p53 tumor suppressor protein functions in the nucleus where it acts as a transcription factor. p53 is one of a large number of transcription factors, which are different proteins that control the expression of genes by regulating how many RNA copies are made from each gene. The p53 protein induces the expression of selected target genes by binding directly to chromosome DNA (Figure 9.18). p53 is the master control protein that functions at the center of the large protein network that monitors the health and security of the cell and reports on the integrity of the cell's DNA genome. p53 function is essential to the proper function of the surveillance system that controls the cell cycle. The activity level of the p53 protein in the cell influences which developmental pathway the cell will follow next, the genome DNA repair pathway or apoptosis.

When damaged genome DNA is detected, the p53 gene is expressed and the p53 proteins are made. A tetramer of four identical p53 proteins binds directly to the control DNA at the start of the p21 gene, which turns on expression of the p21 gene (Figure 9.19). The p21 protein has two major functions in the cell. First, p21 inhibits activity of the MPF (Cdk-cyclin) kinase enzyme, which is necessary for the cells to pass from G1 into the S phase of the cell cycle. At G1/S, the p21 protein prevents the cells from progressing past G1 and into S phase. As a result of this G1/S block, the cells in G1 continue to grow until encountering G1/S, at which point the cells stop growing and adopt a uniform cellular morphology called a G1 arrest. Cells that have already passed the G1/S checkpoint and entered S phase also arrest because the p21 protein inhibits the

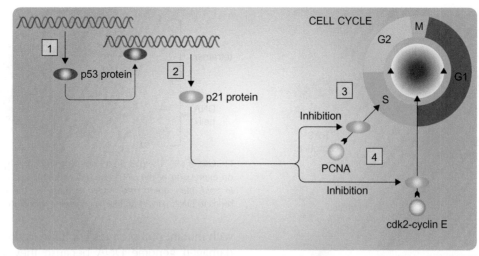

**FIGURE 9.19** The p53 and p21 proteins act to halt the cell cycle at the G1/S cell cycle transition. (1) When the cell's genome DNA is damaged, the p53 gene is turned on and the p53 protein is synthesized. (2) The p53 tetramer binds to the control region of the p21 gene, which triggers p21 protein production. (3) The p21 protein blocks Cdk-cyclin (MPF) activity and halts the cell cycle at the G1/S transition. (4) The p21 proteins also inhibit the activity of the PCNA protein and block DNA replication in S phase. A similar p53 pathway controls entry into mitosis (M).

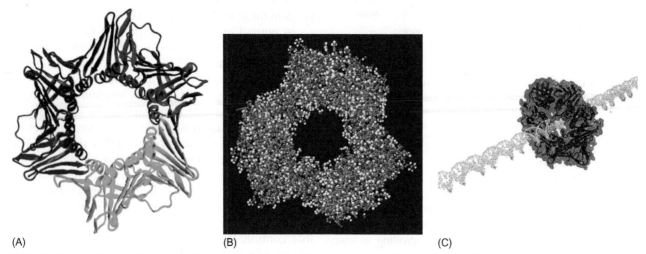

(A)                                               (B)                                               (C)

**FIGURE 9.20** Proliferating Cell Nuclear Antigen (PCNA) is a DNA clamp protein complex that is involved in replicating the cell DNA genome. (A) The structure of the PCNA DNA clamp protein is shown in ribbon form (A) and space-fill form (B). The shape of the PCNA protein complex is a clue to its function in the cell. The DNA clamp protein "clamps" around the DNA helix during the process of DNA replication (C).

function of the DNA clamp protein, proliferating cell nuclear antigen (PCNA), which is required for DNA replication (Figure 9.20).

The p53 gene plays a key role in the development of most types of cancers; over 50% of human cancers carry mutations in the p53 gene. If the mutant p53 protein cannot bind properly to DNA, the p53 protein fails to induce expression of the p21 gene. Without the p21 protein, the MPF kinase remains active and the cell continues through the cell cycle instead of halting at G1/S. The p53 protein master switch can trigger a cascade of biochemical events, mediated through the p21 protein, which shut down the cell cycle and initiate apoptosis.

Retinoblastoma (Rb) is a rare childhood cancer of the eye that affects a few hundred children in the United States annually and accounts for about 3% of all children's cancers (under age 15). If the cancer is detected early enough, the Rb tumors can be treated successfully by radiation therapy, laser surgery, or cryotreatment (freezing), and in many cases the child's vision can be preserved. However, undiagnosed, Rb cancers metastasize and spread to the brain and central nervous system (Figure 9.22).

Much like the p53 tumor suppressor gene, the Rb protein performs a checkpoint surveillance function that protects the genome by detecting damaged DNA and causing the cell cycle to delay to provide time for DNA

## Box 9.2   In the Eye of a Child: A Real-Life Retinoblastoma Story

NBA star Derek Fisher has a contract with the Lakers for millions, but even he couldn't avoid a parent's worst nightmare. He is new to the fight against childhood cancer, which for Fisher began during the playoff games between the Utah Jazz and the Golden State Warriors (Fisher played for Utah). Fisher arrived extremely late to game 2; he had a family emergency.

Fisher and his wife, Candace, have a daughter, Tatum, and at 10 months old Candace Fisher noticed an odd reflection in one of her daughter's eyes (Figure 9.21). Soon after, Tatum was diagnosed with an advanced retinoblastoma (Rb) tumor in her left eye. Rb is the most common childhood cancer of the eye, with 350 cases diagnosed in the United States each year. Immediate treatment usually results in an excellent prognosis for these children, but the doctors needed to decide if they should remove Tatum's left eye entirely or try a risky, cutting-edge procedure called intraarterial chemotherapy (IAC), developed by Drs. David Abramson and Pierre Gobin at New York Presbyterian Hospital. IAC involves introducing chemotherapy

drugs directly into the tumor, hoping to make the tumor shrink to prevent loss of the eye altogether. After careful consideration, Derek Fisher and his wife decided that the intraarterial chemotherapy procedure was the best chance they had to cure the advanced cancer and save Tatum's eye.

The IAC treatment involved injecting a very high dose of chemotherapy into the artery leading into Tatum's cancerous left eye, allowing the blood to carry the chemotherapy agents directly to the cancer cells. IAC has been performed only a few times before and was not yet reported in medical journals.

Derek Fisher arrived to game 2 of the playoffs under police escort during the third quarter and was greeted with a standing ovation from the fans. Once in the game, Fisher forced a crucial turnover and a three-point shot that propelled his team to win the game. Time will tell if the IAC treatment also performed for Tatum, helping her to win the most important game of her life. Four months after the treatment, Tatum was responding well to the chemosurgery and her prognosis is excellent.

 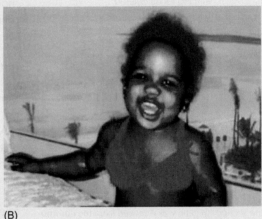

(A)                                        (B)

**FIGURE 9.21**   In the eyes of a child: a real-life retinoblastoma story. (A) Basketball star Derek Fisher fights for his daughter's life. (B) Tatum has a retinoblastoma tumor.

repair. The Rb protein also plays a key role in triggering apoptosis in cells. If the genome damage is not repaired, the cell can trigger its own destruction through apoptosis, providing a way for the body to eliminate cells that pose a greater risk if allowed to grow without controls. Alterations in cell functions that reduce or block apoptosis will significantly alter the cell development and fate in the body.

In healthy children, each retinal cell carries two copies of chromosome 13 with two wildtype Rb genes. Studies on the genetics of retinoblastoma tumors show that the genomes of afflicted children have a deletion mutation in human chromosome 13, which removes one of the two Rb genes (Figure 9.23). Retinal cells that inherit one wildtype and one mutant Rb gene are unaffected and remain healthy, but retinal cells that inherit two copies of

the mutant Rb genes are destined to develop into a retinoblastoma tumor. Genetic testing performed in infancy can detect any chromosome 13 mutations, which is a big step toward the early detection and treatment of Rb cancer.

## Proto-oncogenes Cause Human Cancers

Wildtype proto-oncogenes are normally harmless DNA elements carried in all human genomes, but a random mutation can convert a harmless proto-oncogene into a dangerous oncogene. These mutant oncogene proteins drive abnormal cell growth by increasing the transcription (expression) of selected genes. The mutant src and ras oncogenes are good examples of activated proto-oncogenes, which are continuously expressed (transcribed) in tumor cells, causing the cells to lose

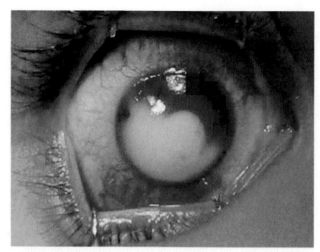

**FIGURE 9.22** Retinoblastoma tumor growing in the eye of a young child.

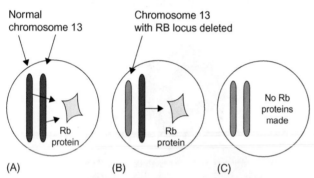

**FIGURE 9.23** Retinoblastoma tumor cells are missing both Rb genes. (A) Normal cells in the retina of the human eye contain two normal Rb genes, one on each copy of human chromosome 13. (B) Retinal cells at risk for developing cancer have only one normal Rb gene because the second Rb gene was removed from the chromosome due to a deletion in the q14 region of human chromosome 13. (C) Retinal cells that are missing both Rb genes grow out of control and develop into retinoblastoma tumor cells.

cell cycle control and divide much more rapidly than normal.

**Src** was the first oncogene to be identified. The src gene codes for a tyrosine kinase enzyme that catalyzes the transfer of phosphate groups onto tyrosine amino acids in certain target proteins. The addition and removal of phosphate groups on proteins can act like a master switch that controls the activities of the target proteins. Healthy cells make only low levels of the src protein, which functions to transmit signals to the nucleus. However, mutations altering src gene expression overproduce the protein and cause neuroblastoma, lung, colon, and breast cancer.

Receptor proteins sitting on the surfaces of healthy breast cells receive and transmit signals from neighboring cells to the cell nucleus. The human epidermal growth factor receptor-2 (Her-2) protein is also located on the surfaces of the breast cells. Her-2 binds to specific growth factor proteins and functions in cell-to-cell signaling. Some types of breast cancer cells overproduce the Her-2 protein, flooding the surfaces of the cells with receptor proteins, and triggering uncontrolled cell division. Certain breast cancer treatments are designed to block overexpression of the Her-2 protein in cancer cells, including Herceptin (trastuzumab), which was made by the biotechnology company Genentech and was approved by the FDA in 1998. Herceptin is another example of a drug that is actually a monoclonal antibody designed to bind to a specific protein, in this case, Her-2. The binding of the antibody to the Her-2 protein blocks the activity of the Her-2 protein and inhibits transmission of the signals telling the cell to divide. Herceptin treatment kills the rapidly dividing breast cancer cells that are making large amounts of Her-2 protein. Not all types of breast cancers overexpress Her-2, but the "Her-2 positive" breast cancers are usually more aggressive and have a high risk of cancer recurrence compared to the breast cancers that test negative for Her-2 overexpression.

The **ras** proto-oncogene and protein are also involved in the kinase-mediated signaling pathways that control the expression of certain genes in the cell, which, in turn, control cell division and cell development. The ras protein acts like a switch controlling a cellular pathway; when ras binds to guanosine triphosphate (GTP) in the cell, the pathway is turned on. The ras protein must release the GTP molecule to turn the pathway off. However, a mutation in the ras proto-oncogene causes the production of mutant ras oncoproteins that cannot release the GTP, and as a result the pathway becomes stuck in the "on" position, leading to uncontrolled cell growth and proliferation (Figure 9.24). The consequences of the mutant ras oncogenes are devastating to the cell because the ras protein is normally involved in so many critically important signaling pathways. In addition, ras mutations have been identified in many different cancers, including pancreas (90%), colon (50%), lung (30%), thyroid (50%), bladder (6%), ovarian (15%), breast, skin, liver, kidney, and some forms of the blood cancer, such as leukemia.

The **myc** proto-oncogenes encode a transcription factor protein that normally regulates the expression of several different genes (Figure 9.25). The myc proto-oncogenes are activated into myc oncogenes by gene rearrangements or by DNA amplification events involving the breakage, rearrangement, and duplication of regions of chromosome DNA. These types of chromosome changes have the potential to alter the normal expression of many genes. Myc oncogenes are linked to several types of cancer including Burkitt's lymphoma, B-cell leukemia, and lung cancers. A specific

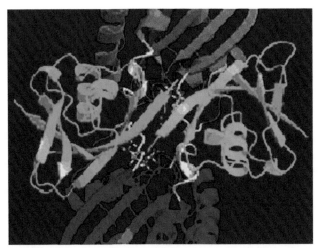

**FIGURE 9.24** Ras oncoprotein. The ras proto-oncogene and ras protein are also involved in the kinase-mediated signaling pathways that control the expression of certain genes.

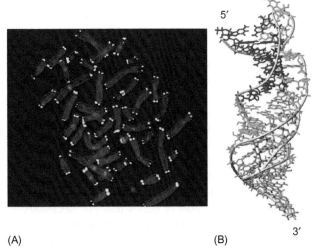

(A) (B)

**FIGURE 9.26** Telomeres are the ends of linear human chromosomes. (A) Telomere proteins *(yellow)* are located on the ends of the blue chromosomes. (B) The RNA component of the telomerase ribozyme, an enzyme made up of protein and RNA components.

**FIGURE 9.25** Myc protein binds to the DNA helix. The myc proto-oncogene encodes a transcription factor protein that normally regulates the expression of several different genes. The alpha helices in the Myc protein are shown in light purple; the DNA helix is shown multicolored.

DNA translocation from one chromosome to another chromosome causes overexpression of the myc gene (and protein), and eventually leads to a B-cell cancer.

The proteins involved in DNA replication can successfully copy the chromosome DNA until they reach the telomeres at the ends of the chromosomes, where technical difficulties prevent the DNA polymerase enzyme from copying the DNA strands located at the

extreme ends of the linear chromosomes (Figure 9.26). This complication would cause the progressive shortening of the chromosome ends with each cell division, with successive divisions potentially deleting essential genes and killing the cell. The problem of how to replicate the ends of the chromosomes is solved by a novel **ribozyme** called telomerase (hTERT), which is a complex composed of a protein bound to an RNA molecule. The hTERT gene encodes the human protein component of the telomerase ribozyme, and the gene for the short RNA component is encoded elsewhere in the genome. Telomerase replicates the DNA at the ends of linear chromosomes using a completely different molecular mechanism than that used by DNA polymerase. The telomerase enzyme is made in large amounts in rapidly dividing cells such as in a developing fetus, but most adult cells divide less often and as a result do not need much telomerase. Cells with limited amounts of telomerase can enter **senescence** (G0 in the cell cycle), which represents long-term growth arrest in G1.

Since cancer cells divide rapidly they also produce high levels of telomerase enzyme. Like myc and src, hTERT is a proto-oncogene that functions as an oncogene when mutated and promotes uncontrolled cell growth.

In normal cells, tumor suppressor proteins function to limit the frequency of cell division, and control processes such as gene expression, DNA repair, and cell-cell communication. Mutations that alter the function of the tumor suppressor proteins can lead to abnormal cell division and cancer development. An example is most colon cancers that develop slowly over a period of years, allowing time for multiple genetic changes to accumulate and convert the normal colon cells into tumor cells. About 5% of all colon cancers are the result of inherited genetic abnormalities that are characterized by the early appearance of polyps in the colon, growths that have the potential to become a colon cancer. Of the several kinds of inherited colon cancers, the most common are familial adenomatous polyposis (APC or FAP) and hereditary nonpolyposis colon cancer (HNPCC) (Figure 9.27). Nonhereditary colon cancers rarely develop before age 40 and are caused by sporadic genome mutations.

The APC protein, like many tumor suppressors, controls the expression of genes that are necessary to the cell-division process. A mutation in the APC gene on chromosome 5 prevents expression of the APC protein and causes increased cell division and the development of colon cancer. Individuals who inherit a mutation in only one of the two APC genes inherit a precancerous condition that causes many colon polyps to grow, with each polyp having the potential to develop into colon cancer. People who inherit preexisting mutations are at a much higher risk for developing

colon cancer because these individuals require fewer genetic changes to convert a cell into colon cancer.

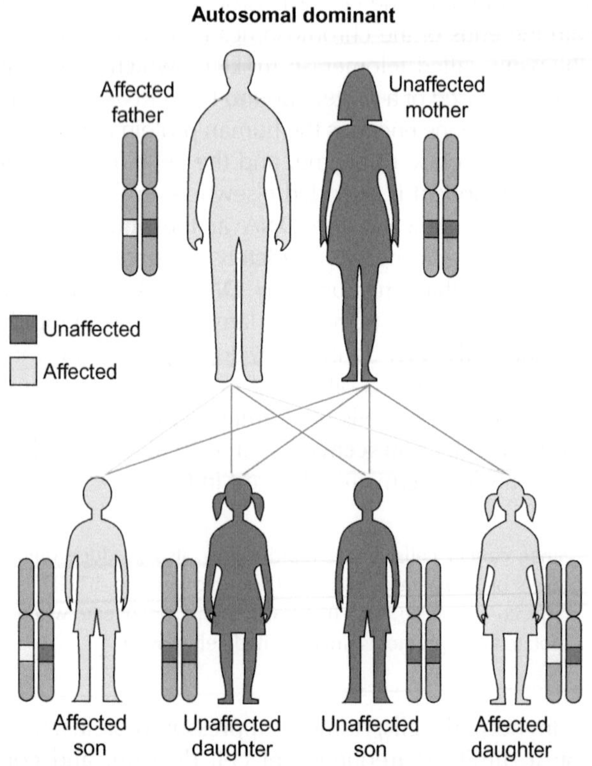

## Early Detection and Treatment of Cancer are Critically Important

The importance of early detection of colon cancer was brought to the public's attention when Katie Couric, then a host of NBC's *Today* show, had her colon screened by having a colonoscopy test on live television (Figure 9.28). Couric's husband had succumbed to colon cancer, a devastating loss that inspired her to start a campaign to help inform the public about the importance of early detection for successful cancer treatment. Couric surprised everyone by having her preventive colonoscopy performed for the first time on live morning television. A follow-up study revealed that Couric's efforts to spread the word about the early detection of colon cancer had saved many lives. Because genetic testing is available, people with a mutant gene can be monitored closely for the first signs of colon cancer, to take advantage of the benefits of early detection and treatment. The increased risk of developing colon cancer usually prompts these people to be monitored often for signs of cancer. More widespread genetic testing should increase the number of people who decide to seek early detection and treatment for many different types of cancers.

**Autosomal dominant**

**FIGURE 9.27** The genetics of hereditary colon cancer (HNPCC). If one parent carries a copy of the HNPCC gene mutation, there is a 50% chance that each child will inherit the HNPCC mutation and a 50% chance that each child will inherit two normal genes. Only one mutant form of the gene is necessary to cause disease because HNPCC is an autosomal dominant mutation.

## Human Breast Cancer Genes BRCA1 and BRCA2

According to the National Cancer Institute, more than 192,000 women in the United States are diagnosed with breast cancer each year (Figure 9.29), but only 5% to 10% of these women have an inherited form of breast cancer disease. These women have mutations in the BRCA1 or BRCA2 genes that make them more susceptible to developing breast cancer. The names of the BRCA1 and BRCA2 genes stand for *breast cancer 1* and *breast cancer 2*, respectively (Figure 9.30). The proteins encoded by the wildtype BRCA1 and BRCA2 genes are involved in regulating gene expression and repairing genome damage. The BRCA1 and BRCA2 genes contain highly repeated DNA sequences that are prone to acquiring mutations. The BRCA proteins interact with transcription factor proteins that control the transcription of several genes including the p53 and p21 genes. When the BRCA proteins do not function properly, the failure to correct the DNA damage in the genome generates cells with potential chromosome rearrangements and abnormal chromosome number that promotes the development of malignant cancer cells (Figure 9.31). Scientists estimate that about 13% of American women have a lifetime risk of developing breast cancer, compared to 36% to 85% of women

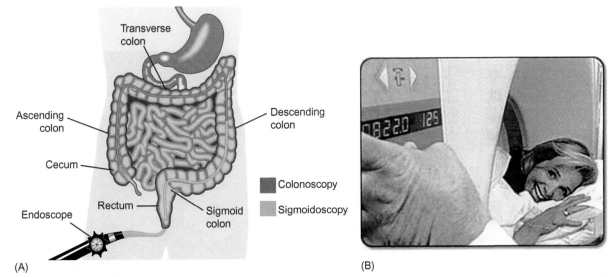

(A)                                                                (B)

**FIGURE 9.28**  Early detection of colon cancer. (A) The colon is part of the large intestine. Sigmoidoscopy and colonoscopy are two methods doctors use to look for tumors. (B) Then NBC anchor woman Katie Couric has her colon screened for cancer on live TV. Katie Couric hosted a series on the NBC *Today* show called "Confronting Colon Cancer" to increase public awareness of colorectal cancer. (B) She stressed the importance of early detection of cancer, including the critical step of undergoing a colon exam (colonoscopy) after age 50.

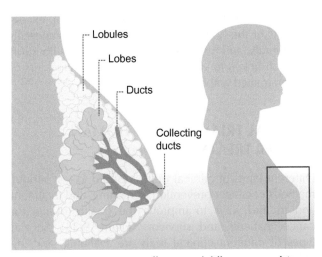

**FIGURE 9.29**  Breast cancers affect several different types of tissues.

who inherit the mutant BRCA1 or BRCA2 alleles. These women have an increased risk of developing breast or ovarian cancers at a young age (before menopause), often have family members with the disease, and may also face an increased risk for colon cancer. BRCA2 mutations are also associated with an increased risk of lymphoma, melanoma, and cancers of the pancreas, gallbladder, bile duct, and stomach.

The lifetime risk of ovarian cancer in the general population indicates that 1.7% of women will get ovarian cancer, compared to 16% to 60% of women with mutant BRCA1 or BRCA2 genes. Genetic research on BRCA1 and BRCA2 involved studies on large families with members affected by cancer and provided estimates

of the cancer risks associated with inheriting the BRCA1 or BRCA2 mutant genes. However, because family members typically share common environments as well as genes, it is possible that the increased number of cases of cancer in these families is due at least in part to genetic or environmental factors unrelated to mutations in the BRCA1 or BRCA2 genes, and the increased risk in these families might not accurately reflect the levels of risk in the general population.

> Acquiring multiple spontaneous mutations in the same DNA genome is a rare event, which explains why cancers caused by multiple mutations take time to develop and tend to occur later in life. The total cancer risk for any individual depends on personal exposure to genetic and environmental risk factors and the individual's unique genetic makeup.

## Environmental Factors Contribute to Cancer Development

Many environmental factors contribute to the development of cancer cells, including some viruses carrying oncogenes that cause human cells to become cancerous. Epstein-Barr virus, for example, not only causes infectious mononucleosis ("kissing disease"), but it is also implicated in the development of Burkitt's lymphoma and nasopharyngeal cancers. Exposure to ultraviolet light (UV), x-rays, and ionizing radiation also cause damage to the genome DNA and alter the genetic makeup of an organism (Figure 9.32). Many environmental agents are known to damage DNA and induce

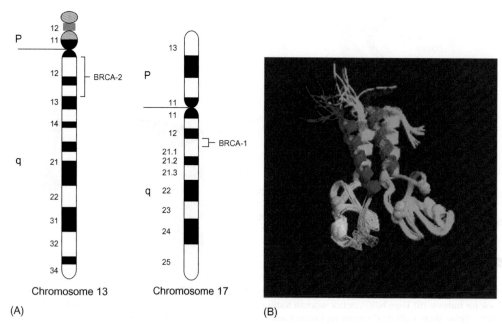

FIGURE 9.30   BRCA genes and proteins. (A) The BRCA1 gene is carried on chromosome 17 and BRCA2 on chromosome 13. (B) The breast cancer protein (BRCA1).

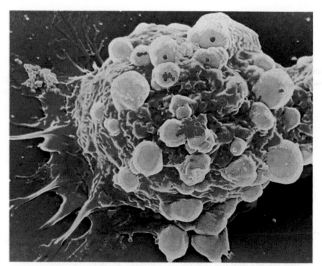

FIGURE 9.31   Malignant breast cancer cell. This breast cancer cell is capable of metastasis, spreading to other locations in the body.

the development of cancer (carcinogenesis), including DNA alkylating agents (leukemia), asbestos (mesothelioma of the lung), aromatic hydrocarbons and benzopyrene from air pollution (lung cancer), tobacco smoke (lung cancer, oral cavity and upper airway cancer, pancreatic cancer, esophagus cancer, bladder and kidney cancer), and vinyl chloride (angiosarcoma of the liver) (Figure 9.33). The human diet also plays an important role in the development of cancers of the gastrointestinal tract; the consumption of high levels of animal fat, food additives containing nitrates, some types of sugar substitutes, and chemicals associated with charbroiled meat is linked to gastrointestinal cancers. The role of hormones

in cancer development is not yet clear, but excess amounts of the hormone estrogen can cause cancer in test animals, and a synthetic estrogen (diethylstilbestrol) has been linked to vaginal cancer in some daughters of women treated with the synthetic hormone.

## CLINICAL TRIALS TO TEST HUMAN CANCER TREATMENTS

Different kinds of clinical trials are used to test various methods of cancer prevention, screening, and treatment and to find ways to improve the quality of life for cancer patients and survivors (Figure 9.34). Several types of different clinical trials are available, and the public is allowed to participate:

- **Prevention trials** test new ways to reduce the risk of developing certain types of cancer, including the use of medications, vitamins, and other supplements. These trials can include cancer survivors who want to prevent a recurrence or reduce the risk of developing a different type of cancer.
- **Screening trials** are early detection studies and treatments designed to work before the cancer metastisizes and is much less treatable.
- **Diagnostic trials** study the effectiveness of different detection protocols to identify early cancers more easily.
- **Quality-of-life (also called supportive care) trials** are designed to study approaches to improve the quality of life for current cancer patients and cancer

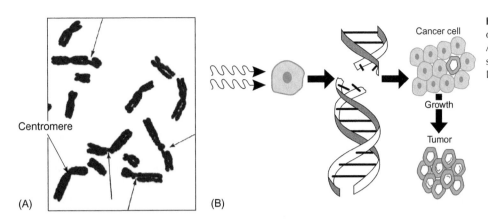

**FIGURE 9.32** Ionizing radiation damages the DNA genome. (A) Arrows indicate broken chromosomes. (B) Radiation damages the DNA in the chromosome.

Centromere

(A)                                 (B)

Cancer cell

Growth

Tumor

**FIGURE 9.33** Environmental sources of cancer-causing agents. The environmental sources of cancer-causing agents (carcinogens) that harm DNA include water and air pollution, nuclear power plants, and cigarette smoke.

**FIGURE 9.34** Clinical trials to test new treatments for cancer and other serious diseases.

survivors. The issues of concern include nausea, vomiting, depression, and sleep disorders as well as many other negative side effects of cancer and cancer treatments.

- **Genetic clinical studies** focus on the genes involved in specific cancers to determine how the genetics of the cancer cells affect the way that the cells respond to cancer treatments.
- The clinical trials that study the effectiveness of new cancer treatments, including drugs, vaccines, surgery, and chemotherapy, are organized into three phases. Phase I trials are designed to test the safety of drugs in human patients. Phase II trials extend the safety information and test different doses of drug

treatments. Phase III trials involve patients with cancer who are treated with a new drug or therapy with the goal of comparing the effectiveness of a new approach to the currently accepted treatment for a given cancer.

## FATE 2: DEVELOPMENT OF SPECIALIZED CELLS

There are about 220 different types of cells in the adult human body that perform many thousands of diverse cellular functions. Highly specialized cells perform many important jobs in the human body: islet cells in the pancreas produce insulin, nerve cells send impulses in the brain, and stomach cells secrete digestive enzymes, to name just a few (see Chapter 12). These biological activities require a cast of thousands; the average human body contains about 60 trillion to 100 trillion cells. Highly specialized cells usually develop from the precursor cells through the processes of differential gene expression and cellular differentiation. Under normal circumstances, most cells in the human body are fully differentiated into a specialized shape and are permanently committed to performing a certain function. Under normal circumstances highly differentiated cells are unable to revert back to an undifferentiated, nonspecialized state. Many types of cells have the capacity to undergo developmental changes, for example, in response to a threat. When fighting an infection, the body's immune system produces highly specialized cells that recognize and kill the invading pathogen. In response to the infection, immune precursor cells in the bone marrow differentiate and form the mature immune cells that produce antibodies against the invading microbe.

## Specialized Cells are Generated by Stem Cells

To better understand the complex processes involved in developing various specialized types of human cells, it is useful to start at the beginning and review how cells develop and change during embryogenesis. It starts when a sperm fertilizes an egg cell, and the resulting zygote contains two genomes. The zygote divides by mitosis and makes a ball containing a small number of genetically identical embryonic cells. Four to five days after fertilization, the growing human embryo has taken the form of a blastocyst, a hollow ball that consists of about 200 cells and contains a handful of (the amazing) embryonic stem cells (ESC) growing inside (Figure 9.35).

Embryonic stem cells are special because they give rise to all of the different types of cells that will eventually make a human body. The ESCs have the genetic potential to perform all of the different biochemical reactions required in the body and can develop into every cell type needed. Embryonic stem cells have the most developmental potential compared to other types of stem cells. For example, adult blood stem cells, which are found in the bone marrow of adults, are capable of developing into all the different types of cells found in the circulating blood (Figure 9.36). But the adult stem cells in the bone marrow cannot develop into nerve or muscle cells. The unlimited developmental potential of embryonic stem cells have made them the focus of biomedical research across the globe (see Chapter 12). During normal human embryogenesis, the ESCs divide, differentiate, and migrate to form an embryo with three distinct layers

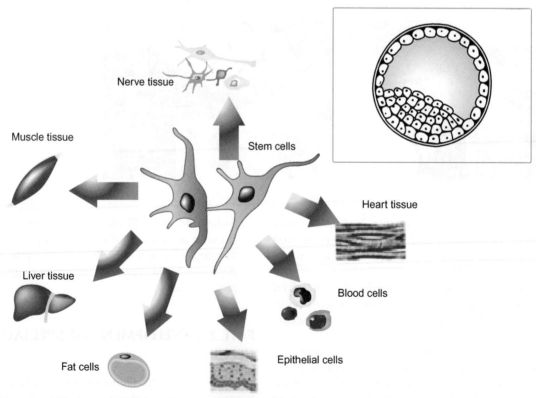

Nerve tissue

Muscle tissue

Stem cells

Heart tissue

Liver tissue

Blood cells

Fat cells

Epithelial cells

**FIGURE 9.35** ESCs develop into different types of cells. During embryonic development, the embryonic stem cells (ESCs) develop into a large variety of diverse cells that make up the brain, muscles, stomach, bone, and all other tissues in the human body. Inset shows a blastocyst (left) and cross-section of a blastocyst (right) showing ESCs inside.

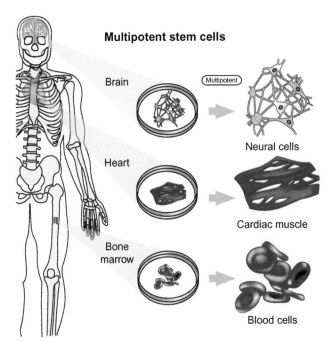

**Multipotent stem cells**

Brain → Multipotent → Neural cells

Heart → Cardiac muscle

Bone marrow → Blood cells

**FIGURE 9.36**  The adult human body has multipotent stem cells. The adult human body contains many examples of precursor cells that each develop into certain subsets of cells. However, these cells are different from the ESCs because the multipotent cells cannot develop into all of the cells in the body.

of embryonic tissue, the ectoderm, endoderm, and mesoderm. Each of these three tissue layers contains stem cells with the potential to develop into some but not all of the cell types needed in the human body. The ectoderm tissue gives rise to the central and peripheral nervous systems, the mammary glands, the pituitary gland, and tooth enamel, whereas the endoderm forms the gastrointestinal tract and the mesoderm generates bone, cartilage, connective tissue, and muscle cells.

## FATE 3: APOPTOSIS IS PROGRAMMED CELL DEATH

During embryo development and throughout adult life, body cells are selected to die at appropriate times and are sometimes replaced by other more specialized cells. Precursor cells give rise to different types of specialized cells that perform specific functions and have life spans measured in days or weeks and not years. Apoptosis or cell suicide occurs routinely in the human body, sometimes to remove unneeded cells and other times as a result of injury to tissues and cells (Figure 9.37). As suggested by the alternative name "programmed cell death" (PCD), apoptosis proceeds

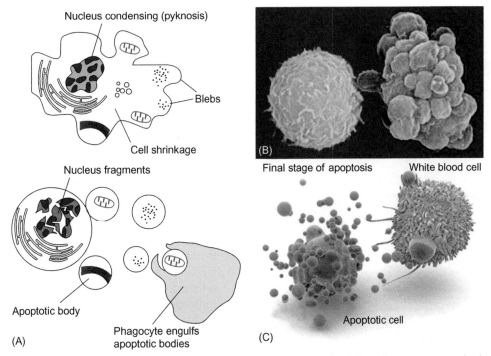

**FIGURE 9.37**  Apoptosis (cell suicide) occurs when genome damage cannot be repaired and the cell triggers its own death. (A) Once apoptosis is triggered (top), the cell begins to shrink. The chromosomes condense, the DNA is cut into small fragments (bottom) and the nucleus collapses. The plasma membrane begins to form bubbles and blebs. Apoptotic bodies form and lyse and cell debris is engulfed by macrophage cells that are attracted to the dying cell by the blebbing membranes. (B) A white blood cell (right) dying by apoptosis (programmed cell death). (C) A cancer cell (right) undergoing apoptosis.

through a series of genetically controlled biochemical events that cause characteristic changes and trigger cell death. When a cell follows this pathway toward self-destruction, and the apoptosis genes are expressed, the cell undergoes a series of morphological changes, including blebbing of the cell membrane, cell shrinkage, and chromosome fragmentation. As the cells die, they form apoptotic fragments of cell debris that are subsequently removed by the macrophage cells.

## Apoptosis Makes Embryos with 10 Fingers and 10 Toes

Cell death is carefully controlled during the intricate cellular processes that occur during embryogenesis of plants and animals. In fact, a defect in apoptosis can cause a common birth defect that affects the fingers of newborn babies. Human embryos develop with tissue growing between their fingers and toes that normally disappears before birth. However, sometimes the webbing persists after an infant is born. This birth defect, called **syndactyly**, is caused by a mutation in a specific gene that prevents apoptosis in the webbing between the fingers of the embryo (Figure 9.38). Depending on the severity of this condition, the fused fingers are often surgically corrected in a newborn. The gene that causes this disorder is inherited as an autosomal dominant genetic mutation (see Chapter 10).

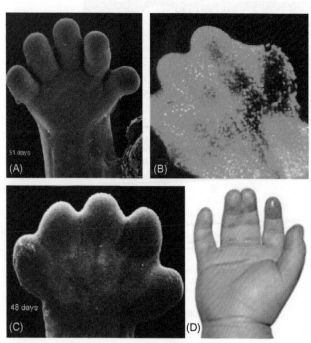

**FIGURE 9.38** Apoptosis gives babies their fingers and toes! (A) Day 51 human hand. (B) Growing embryo shows webbing. (C) Day 48 human hand. (D) Hand with syndactyly.

Apoptosis is an important focus for the discovery of novel cancer treatments, such as new anticancer drugs that induce apoptotic death in the cancer cells without affecting the surrounding healthy cells.

## SUMMARY

The purpose of this chapter is to better understand the genetic and biochemical mechanisms used by eukaryotic cells to determine cell fate. The term "cell fate" refers to processes such as cell division and reproduction, which act to maintain or change chromosome number (ploidy) during the cell cycle. This chapter discusses the origin of more than 200 different cell types required by the adult human body. The embryonic stem cells differentiate and generate the new shapes, structures, and functions of the highly specialized cells that perform functions such as transmitting nerve impulses or secreting digestive enzymes. Human cells are not designed to last a lifetime. When highly specialized cells wear out and can no longer function adequately, the cells are replaced using a natural recycling system. The human body also needs to replace cells damaged by disease or trauma, which often involves apoptosis (programmed cell death).

The human genes involved in making cell fate decisions control the cell cycle, regulate gene expression, and convert normal cells into cancer cells. Proto-oncogenes, oncogenes, and tumor suppressor genes all play key roles in the prevention and development of cancer cells.

## REVIEW

In this chapter we discussed how eukaryotic cells execute gene expression patterns that determine cell fate: to divide by mitosis or meiosis, to develop into a specialized cell type, or to commit to apoptosis. To test your comprehension of the chapter's contents, answer the following questions:

1. What types of information (signals) does a cell use to decide its fate?
2. What types of cell fates can a cell choose?
3. Cyclin proteins get their name from what characteristic feature?
4. Indicate the order that the four phases of the cell cycle occur in normal cells.
5. What happens to the checkpoint feedback mechanism in cancer cells?
6. Explain the relationship between proto-oncogenes and oncogenes.

7. Describe the most important features of one of the tumor suppressor genes (and proteins) described in the chapter.
8. Explain the important role that apoptosis plays in embryogenesis and fetal development.
9. Explain the structure and function of MPF in the cell cycle.
10. Explain the role of metastasis in cancer disease.

## ADDITIONAL READING

Cooper, G.M., Hausman, R.E., 2007. The Cell: A Molecular Approach, fourth ed. Boston University, Sinauer.

Fisher, J.P., Staff Writer, 2007. Patient leads fight for his life. Raleigh News Observ.

Cooke, R., author 2001 biography, "Dr. Folkman's War."

Varmus, H., Weinberg, R., 1993. Genes and the Biology of Cancer. Scientific American Library, New York.

## WEB SITES

American Academy of Orthopedic Surgeons (bone cancers). http://orthoinfo.aaos.org/topic.cfm?topic=A00084.

Allen, S., Boston Globe, Research giant Judah Folkman dies January 15, 2008. www.boston.com/news/health/blog/2008/01/research_giant.html.

BRCA1 and BRCA2 Genetic Testing. www.cdc.gov/genomics/training/perspectives/factshts/breastcancer.htm.

Breast cancer tissues. allenneighborhoodcenter.org/.../breastcancer.htm.

Cancer genes. http://genome.wellcome.ac.uk/doc_WTD020846.html.

Cancer Quest site. www.cancerquest.org.

Cell cycle checkpoint. homepage.mac.com/enognog/checkpoint.htm.

Embryogenesis. http://gallery.elon.edu/poet/dayesjourney/pag14.html.

Entertainment Industry Foundation's National Colorectal Cancer Research Alliance (EIF's NCCRA). www.eif.nccra.org; www.eifoundation.org; www.cdc.gov/cancer/screenforlife.

Gene Testing. www.geneclinics.org.

Jay Monahan Center for Gastrointestinal Health. www.monahancenter.org.

Josh Sommer interview. Todayshow.com 022008.

National Cancer Institute (NCI). www.cancer.gov.

National Institutes of Health. www.nlm.nih.gov, p53 tumor suppressor, www.cancerwatch.org/index.cfm?d=3005&c=4038&p=10661&do=detail.

Pope, J., AP Education Writer. News and Observer, College student fights his own cancer. www.newsobserver.com/2187/story/953939.html.

Stem cell development. www.stemcure.com/images/pic_sc_stemcell.gif.

# Human Genetic Diseases

## Autism Symptoms Reversed in Lab

BBC News, June 27, 2007

Reversing autism, even in lab mice, is really amazing news. For the first time ever the symptoms of mental retardation have been reversed in laboratory mice. Autism has risen to strike 1 in every 150 people born in the United States in 2007. Mutations in the fragile-X mental retardation gene (FMR1) gene are currently the leading genetic cause of mental retardation and autism in humans. To create a model animal with autism symptoms to study in the lab, the scientists made "knockout (KO) mice", which no longer carry the FMR1 gene in their genomes. The FMR1 KO mice showed evidence of cognitive disorders and exhibited hyperactive, purposeless, and repetitive actions, all behaviors that are quite different from normal mice.

From human studies, scientists knew that an enzyme called PAK3 (p21-activated kinase 3) was implicated in mental disorders and might be a good target for developing new drugs to treat autism. Scientists tried to reverse the symptoms of autism in the FMR1 KO mice by blocking the action of the PAK enzyme in the mouse brain. Not only was the treatment effective, it even worked in mice that exhibited pronounced symptoms associated with autism. Microscopic analysis of tissue from the brains of FMR1 KO mice treated to block PAK activity showed that the treatment resulted in the repair of damaged nerve cells and rebuilt the connections between neighboring nerves, restoring proper electrical communication between the cells in the mouse brain.

"This is very exciting because it suggests that PAK inhibitors could be used for therapeutic purposes to reverse already established mental impairments in fragile X children."—Professor Eric Klann, New York University Center for Neural Science.

When scientists reported that they had successfully "cured" fragile X syndrome and autism in mice for the first time, it might seem like modest progress. But genetic disorders affecting nerve and brain function in humans and other mammals are caused by complicated interactions among many genes. Defective nerve cell development can result in a variety of mental deficits and physical limitations that are difficult to study in humans. The results of testing a PAK3 inhibitor drug on the FMR1 KO mice were very encouraging and have prompted scientists to search for drugs that are safe to use in humans and effectively block PAK3 activity in people with autism or fragile X syndrome.

## LOOKING AHEAD

This chapter describes some of the many thousands of normal (wildtype) genes and mutant genes that are implicated in causing some common genetic diseases and disorders in humans. On completing the chapter, you should be able to do the following:

- Describe the differences between simple and complex genetic diseases in humans.
- Explain how a mutation in a gene affects the product of the mutant gene.
- Appreciate the role of the environment in damaging DNA and in the development of human genetic disease.
- Make the connection between biomedical research involving animals and the essential treatments for human diseases that directly result from research on animals.
- Explain how the biological families of patients with genetic diseases contribute to our understanding of human gene function.
- Understand the structure of a DNA probe, and explain how it is used to identify the location of a target gene in a chromosome.
- Describe the role that the DNA differences between individual human genomes have played in searching for human genes involved in genetic diseases.
- Have a better understanding of the functions of the macromolecular machines and how the machines are assembled from many protein parts in the cell.
- Access the most accurate and up-to-date information about human genetic diseases available using a computer and the Internet.

## INTRODUCTION

The past 50 years have seen amazing progress in our understanding of genes and how genes function in the human body. Scientists from all over the world collaborated to determine the entire DNA sequence (the exact order of the bases in the DNA) of all 3.2 billion bases that comprise the human genome (see Chapter 6). Ongoing genome studies have focused on learning about the functions of the proteins encoded by approximately 20,000 human genes. However, even though we know the entire human genome DNA sequence, the identities and functions of all 20,000 individual genes are not yet known. Unfortunately the function

of a gene product, usually a protein, is not usually obvious just from DNA sequence analysis alone. Additional studies are usually required to understand how different proteins function in the cells. The technical ability to explore the structure and function of genes at the level of the DNA molecule has dramatically changed and revolutionized our concept of how genes work and substantially increased our understanding of how a mutation alters the structure and function of a protein and sometimes changes the fate of the entire organism.

## GENETIC DISEASES ARE CAUSED BY MUTANT GENES

Humans all start life by inheriting one version of each gene from their mother and one version of each gene from their father. These two different versions of the same gene are called **alleles**. The maternal and paternal alleles of a gene can be slightly different from each other in DNA sequence; one allele might be the normal (wildtype) version of the gene, whereas the other allele might be a mutant version of the gene. The consequence to the person who inherits these alleles depends on the specific gene involved, the nature of the mutation, and to some extent on the environment in which the person lives. The interaction of inherited normal and mutant genes with the many factors in an individual's environment is so complex that, at present, we do not have all the information needed to entirely understand the impact of environment on genetics. However, since we can usually determine the genes involved and the effects of mutations in genes, this genetic information is often the starting point for unraveling the complex networks of genes that dramatically influence our health, personality, and, in fact, every aspect of our lives.

Many different kinds of gene mutations are discussed in this chapter, and they all represent changes in the linear sequence or order of the DNA bases. In the case of mutations, a very small change in the DNA sequence can either have drastic consequences, or have no effect at all. The impact of small DNA changes is evident from studies on hemoglobin, an essential protein in the red blood cells that carry oxygen in the bloodstream. Adult hemoglobin (HbA) contains the β-globin protein, which binds to oxygen. A specific single base pair change (called a point mutation) in the β-globin gene sequence alters one amino acid in the globin protein (HbS) and causes the disease sickle cell anemia (Figure 10.1A).

The mutant HbS proteins are made inside the red blood cells of a person with sickle cell anemia clump together and form long, insoluble rod-shaped polymers, which distort the red blood cells into characteristic

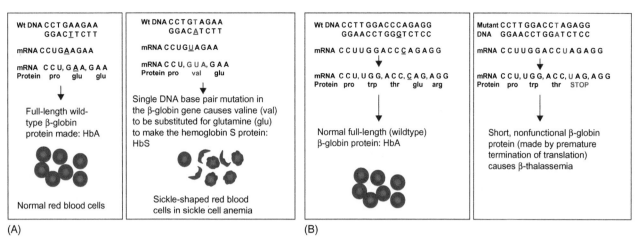

**FIGURE 10.1** Mutations in β-globin cause blood cell diseases. (A) A point mutation in β-globin makes sickle cell protein. The wildtype β-globin gene makes the normal protein, HbA, which is found in healthy red blood cells *(left)*. This point mutation in the β-globin gene causes a valine (val) amino acid to be substituted for a glutamine (glu) amino acid to make the hemoglobin S protein (HbS); the mutant HbS protein makes the red blood cells become sickle-shaped and causes sickle cell anemia *(right)*. (B) Point mutation in β-globin makes the short protein that causes β-thalassemia. The wildtype β-globin gene makes the normal protein, HbA, which is found in healthy red blood cells. *(left)* The C to T mutation in the DNA encodes a UAG stop codon in the mRNA, which in turn causes premature termination of the growing β-globin amino acid chain. The result is a short, nonfunctional mutant β-globin protein.

sickle or banana shapes. The sickle-shaped red blood cells tend to clog the smallest blood vessels, the capillaries, and interfere with the blood flow in the body, causing severe pain and other disabling symptoms.

A different point mutation in the β-globin gene creates a new protein synthesis stop signal, UAG, in the β-globin mRNA. This change replaces a glutamate (glu) amino acid in the β-globin protein with a stop codon that forces the production of a prematurely shortened β-globin protein chain containing only 39 amino acids, instead of the usual 146 amino acids (Figure 10.1B). The short β-globin protein is functionally useless and causes the blood disorder, β-thalassemia.

Humans inherit one copy of each gene from each biological parent, so it is possible to inherit two normal (wildtype) versions of the gene, two mutant versions of the gene, or one wildtype copy and one mutant copy of the gene. When a cell inherits two different versions (alleles) of the same gene, does the trait controlled by one allele "win out" over the other? The answer is sometimes yes, sometimes no. For our purposes, there are two possibilities. First, it is possible that both gene products are made in the cell and both proteins have an effect, a situation called **codominance**. A second possibility is that one version of the gene makes a functional protein product and is the **dominant allele**, whereas the other version of the gene makes a nonfunctional product and is the **recessive allele**. The dominant allele is the form of the gene that has an observable effect on the organism, the trait. This observable change is called the **phenotype** (blue eyes, black hair, an inherited disease, or other features).

A "point mutation" changes only one DNA base in an entire gene sequence, yet the consequences to the organism can be very serious; witness how a point mutation can damage the structure and function of hemoglobin.

Genes in human chromosomes send instructions to the cell by the process of **gene expression** (Figure 10.2A). Two copies of each gene are present in the genome of each cell, but at any one point in time, some genes are expressed (turned on) while others are silent (turned off). When a gene is turned on, the DNA sequence encoding the gene is copied into a long precursor RNA, which is then processed by RNA splicing to make the final messenger RNA (mRNA). The mRNA is exported out of the nucleus (through a pore in the membrane) to the cytoplasm where it delivers a copy of the specific genetic information to the ribosome, which produces proteins in the cell according to the specific instructions encoded in the mRNA. A gene without a mutation usually produces a protein that functions correctly. However, a mutant gene can encode a protein that is defective and is unable to function correctly in the cell (Figure 10.2B).

## 10,000 HUMAN GENES POTENTIALLY CAUSE GENETIC DISEASES

Scientists have known for years that inherited diseases are caused by mutations in genes, which in turn produce

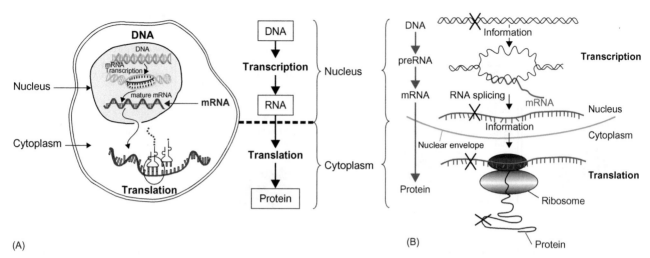

(A)                                                                                              (B)

**FIGURE 10.2**   Wildtype and mutant genes are expressed using similar pathways. (A) Overview of the major gene expression pathway used by eukaryotic cells. A gene is copied into many long RNA transcripts in the nucleus. Most human genes require splicing; the interrupted coding regions of the gene are copied into long precursor RNAs that contain both introns and exons. These long RNA transcripts are processed by RNA splicing, which removes the introns from the precursor RNAs to make "mature" mRNAs containing exons in the nucleus. RNA splicing is required before the mRNAs can be transported to ribosomes in the cytoplasm, where each mRNA is translated into a chain of amino acids specific for that protein. (B) Closeup of the expression of wildtype and mutant genes and proteins. The gene DNA sequence is copied into a precursor RNA (transcription) that is spliced to make the mRNA in the nucleus. (The X indicates a mutation in the DNA that changes the DNA sequence of the gene.) The mRNA is transported out of the nucleus and into the cytoplasm where the mRNA is translated by the ribosome into a specific protein. The mutation in the DNA sequence is transferred into the RNA sequence by transcription and is then incorporated into the mutant protein product in the form of incorrect amino acids.

defective proteins that fail to function properly in the cell. When the human genome DNA sequence was completed, it became clear that in the past we had overestimated the total number of human genes and probably underestimated the number of genes involved in human diseases. In 1966, the noted founder of the science of medical genetics, Victor McKusick at Johns Hopkins University, had cataloged the 1500 genetic diseases known in humans at that time in his classic book, *Mendelian Inheritance in Man*. Now, more than 40 years later, McKusick is lead editor of the Mendelian Inheritance in Man (OMIM) web site, with a database of more than 10,000 human genes associated with human diseases (Figure 10.3).

> The flow of genetic information in the cell starts with the gene's DNA, is transferred to the messenger mRNA, and then to the protein product: DNA to mRNA to protein.

All human genetic diseases pose complex scientific and medical questions, whether the disease involves a single base pair change in a single gene or is caused by changes in many mutant genes. The impact of even a single defective gene product (protein) on cellular functions can be very complex. Many genes are part of a network of genes encoding proteins that control the expression of other genes (Figure 10.4). Mutant genes

**FIGURE 10.3**   Victor McKusick is the "Father of Genetic Medicine." McKusick is the lead editor of the Mendelian Inheritance in Man (OMIM) web site. OMIM has information on more than 10,000 human genes that are associated with human diseases.

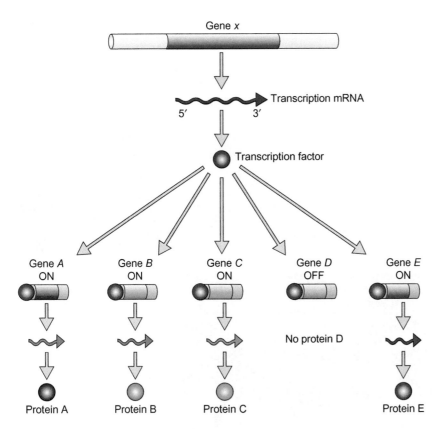

Gene x

Transcription mRNA

5'          3'

Transcription factor

Gene A ON    Gene B ON    Gene C ON    Gene D OFF    Gene E ON

No protein D

Protein A    Protein B    Protein C                  Protein E

**FIGURE 10.4** Expression of some genes can affect the expression of other genes. The expression of genes in human cells is complicated and highly regulated. Genes are turned on or turned off, and the amount of gene product is attenuated (increased or decreased) depending on the needs of the cell. A gene coding for a transcription factor protein can in turn control the expression of many other genes through the function of the transcription factor.

often, but not always, produce defective proteins, which fail to perform a job in the cell. To determine how a mutant gene might cause a particular disease, scientists study how the gene and the protein function in normal cells.

With so many human genes implicated in genetic diseases, it is not possible to discuss them all in this chapter, or even in this book. Instead we have decided to focus on certain human genes and diseases that are particularly instructive, showing key concepts in gene expression and protein function, as well as those with especially interesting science stories to tell. A section at the end of this chapter provides advice about how to find accurate and reliable online information about the thousands of genes and diseases that we will not be able to cover here.

> At least one-half of the 20,000 genes in the human genome have been implicated as causing genetic diseases in people.

## Environment Can Cause Mutations in Human Genome DNA

The expression of our genes controls who and what we all are, to a great extent, but the impact of our genes is also strongly influenced by various environmental factors. Heart disease is a good example of how an individual's environment can impact the overall risk of a disease. Factors known to increase the risk for coronary disease include obesity, lack of exercise, and a family history of heart attacks. The risk of heart disease for any one individual depends on that person's genes, the effects of genetic mutations, and the cumulative results of a lifetime of environmental exposure.

Inheriting a mutant gene (or genes) is one way to acquire a genetic disease, but it is not the only way. A genetic disease can be caused by a gene mutation that was not inherited but instead arose as a result of a **spontaneous mutation** in the DNA sequence encoding the gene. Damage to a person's DNA can be caused by factors in the environment, such as exposure to sunlight, hazardous chemicals, and radiation (not including medical or dental x-rays) (see Chapter 9). Sun exposure causes damage to DNA because sun light is the primary environmental source of highly **mutagenic** UV (ultraviolet) radiation. If the damage to the DNA in the genome is not repaired, the mutant gene will potentially produce defective proteins in the cell.

A spontaneous mutation will persist in the genome DNA of the original cell and in all cells derived from the original cell. However, because the spontaneous mutation did not alter the genome of a sex cell (egg or sperm), the mutation will not be passed on to the offspring. The blood-clotting disorder hemophilia is usually caused by inheriting mutant genes that encode defective blood-clotting factors. Surprisingly, 30% of the

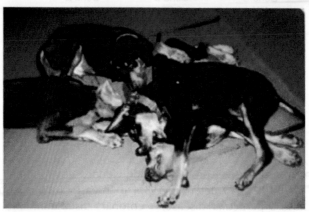

**FIGURE 10.5**   Man's best friend sleeping on the job? These dogs with the sleep disorder narcolepsy are playing actively (top left) until suddenly they reach a threshold of excitement and they collapse together in a tangled heap, unable to move a muscle (top right, bottom). Frozen in a state of cataplexy, the dogs are still and silent. A few seconds or minutes later, the dogs jump up, shake it off, and carry on as before, until the next exciting event, such as dinnertime!

people diagnosed with hemophilia had acquired the disease due to spontaneous mutations in the clotting factor genes.

> A combination of genetic and environmental factors influences who we are and who we become, from personality to body type. The environment also influences the impact of many inherited diseases, although no known vitamins or amount of exercise can avoid the most serious consequences of even a "simple" genetic mutation such as sickle cell anemia.

A relatively small number of **monogenic diseases** are caused by mutations in single human genes, including Huntington's disease, cystic fibrosis, alpha-1 antitrypsin deficiency, adenosine deaminase (ADA) deficiency, neurofibromatosis 1, phenylketonuria (PKU), severe combined immunodeficiency syndrome (SCID), and sickle cell disease. Most inherited diseases (and human

traits) are **multigenic** and are caused by many genes. Multigenic disorders include heart disease, hypothyroidism, colon cancer, other cancers, Alzheimer's disease, and diabetes (Table 10.1).

## INCONSISTENT GENETIC TESTING LAWS

The information obtained from identifying the mutant gene involved in a disease can often be used to develop a genetic test to detect people who might be at risk for a particular inherited disease. The development of **gene-specific DNA probes** has made it routine to test for a specific mutant or wildtype gene allele in an individual's genome. Increased genetic testing has brought about the need for more consistent access to reliable genetic counseling. The genetic screening tests usually given to newborn infants in the hospital shortly after birth have proven to be an extremely effective way to avoid the consequences of inheriting certain

**TABLE 10.1** Human diseases caused by single gene mutations

| Monogenic diseases | Description | Symptoms |
|---|---|---|
| Adenosine deaminase (ADA) deficiency | Immune disease makes the body open to severe infections | Growth retardation, opportunistic infections, poor/little immune system |
| Alpha-1 antitrypsin deficiency | Lack of a liver protein that normally blocks destructive enzymes; deficiency leads to emphysema and liver disease | Shortness of breath with wheezing, unintended weight loss, fatigue, respiratory infections, rapid heartbeat on standing, vision problems |
| Cystic fibrosis (CF) | Chronic illness affects lungs, digestive system, and other body organs | Very salty skin, shortness of breath, wheezing, coughing phlegm, frequent lung infections, poor growth and weight gain, difficulty with bowel movements |
| Neurofibromatosis | Progressive neurological disorder; abnormal embryonic neural cells | Café-au-lait spots on the skin and freckles, neurofibromas, bone defects, bilateral acoustic tumors (type 2) |
| Phenylketonuria (PKU) | Genetic disorder prevents body from utilizing the amino acid phenylalanine | Light skin color (relative to biological family), eczema, possible mental retardation |
| Severe combined immunodeficiency syndrome (SCID) | Immune deficiency; abnormal T and B lymphocytes | Severe infections in first several months of life; pneumonia and meningitis |

devastating genetic diseases. Currently, these screening tests identify 29 treatable genetic disorders using just a few drops of blood taken from a pin prick in a newborn's heel.

The genetic screening tests are relatively inexpensive and are essential to identify newborns at risk. Testing immediately after birth assures that the babies will get early treatment to avoid lifelong disability or death. One of the earliest tests available detected the inherited metabolic disorder phenylketonuria (PKU), which causes severe mental retardation if not rapidly diagnosed; early treatment avoids permanent damage to the brain. The importance of these newborn screening tests is clear, but progress toward guaranteeing testing for all infants born in the United States is very slow. Surprisingly, in 2005 only 38% of American infants were born in states that screen for at most 21 of the 29 treatable genetic diseases. Many groups have recommended for years that states adopt federal guidelines requiring that every infant born in the United States be tested for the same 29 genetic disorders. Federal standards are needed to correct the widespread inconsistencies that exist between state-run testing programs across the country.

## GENETIC DISEASES ARE FREQUENTLY CAUSED BY MORE THAN ONE GENE

Most human genetic diseases are caused by the actions of more than one gene, together with influence from various environmental factors. To find the multiple genes involved in a complex condition such as heart disease, it is necessary to conduct large genetic studies involving many people. The Wellcome Trust sponsored a large study testing DNA samples from 17,000 British residents, and processing the equivalent of almost 10 billion pieces of genetic information. (The Wellcome Trust is the world's largest medical research charity funding research into human and animal health.) The scientists used the most recent gene mapping techniques based on SNP markers (DNA polymorphisms) to successfully identify several new genes associated with bipolar disorder, Crohn's disease, coronary heart disease, hypertension, rheumatoid arthritis, and diabetes. The study is also pursuing the human genes involved in tuberculosis, breast cancer, autoimmune thyroid disease, multiple sclerosis, and ankylosing spondylitis (Table 10.2).

A genetic mutation that alters the three-dimensional shape of the protein product prevents the mutant protein from functioning properly, often because the misshapen protein cannot fit together with other proteins to assemble a functional molecular machine.

The human brain is the most amazing, intricate, and complicated organ in the body, so it makes sense that diseases affecting the brain are also very complex. Alzheimer's disease often brings early memory loss and dementia even in middle-aged people. Named in the early 1900s for Alois Alzheimer, a German neurologist, the disease causes a progressive loss of intellectual function, followed by the inability to speak, walk,

**TABLE 10.2** Human diseases caused by mutations in more than one gene

| Multigenetic diseases | Description | Symptoms |
|---|---|---|
| Heart disease | Several diseases affect the heart and the blood vessels; coronary artery disease (CAD) and heart failure | Chest pain, shortness of breath, fatigue; different symptoms for each heart disease in men and women |
| Hypothyroidism | Deficiency in thyroid hormone causes slow metabolism; at birth causes cretinism | Dry skin, puffy face, hair loss, slow speech and heart rate, mental retardation |
| Colon cancer | Colon cancer can metastasize; early detection is essentials; a leading type of cancer in both males and females | Fatigue, weakness, shortness of breath, change in bowel movements, abdominal pain, cramps, and bloating |
| Alzheimer's disease | Impaired higher intellectual and cognitive brain function; progression to dementia over a 5- to 10-year period | Memory loss, confusion, hallucinations, emotional instability, inability to concentrate, perform daily activities and personal care |
| Diabetes mellitus | Chronic disease; very high glucose levels in the blood (hyperglycemia); insulin missing or not functional; type I and type II diabetes | Thirst, blurry vision, hunger, fatigue, dry itchy skin, weight loss, excess urination, tingling in the hands and/or feet, sores that take a long time to heal |

or perform even basic skills. The cause of Alzheimer's disease was completely unknown until 1987 when researchers identified the first human gene linked to the development of the abnormal brain tissues that characterize the disorder (Figure 10.6). Brain tissues analyzed from people with Alzheimer's disease typically show fibrous plaques made by the accumulation of short amyloid proteins. *Familial Alzheimer's disease* (FAD) is caused by a mutation in the gene located on human chromosome 21 that codes for the amyloid precursor protein (APP). The large APP protein is located in nerve cell membranes at the junctions between nerve cells (synapses). Abnormal cleavage of the precursor APP protein produces the short amyloid proteins that form the ubiquitous amyloid plaques in Alzheimer's brains. The protein complex that cleaves APP requires the function of presenilin, a gene that when mutated has been identified as a major genetic risk factor for Alzheimer's disease.

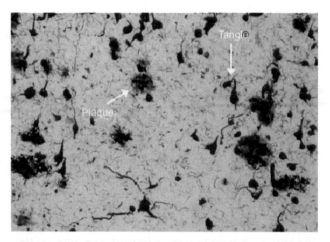

**FIGURE 10.6** Plaques and tangles form in the brains of Alzheimer's patients. Alois Alzheimer studied the brain tissue and identified the characteristic changes that are now known to be hallmarks of Alzheimer's disease.

> Diseases like Alzheimer's are especially hard to study because diseases of the brain are often caused by defects in the complicated interplay of the many genes involved in the specialized development and function of nerve cells.

Amyotrophic lateral sclerosis (ALS) is commonly known as Lou Gehrig's disease in honor of the New York Yankees first baseman who died from ALS in 1941 (Figure 10.7). ALS is a progressive disease that destroys motor nerves in the brain and the spinal cord, causing muscle weakness, paralysis, and death. Surprisingly,

only 5% to 10% of ALS cases are inherited; the remaining cases result from spontaneous mutations. An international team of scientists identified a mutant form of the superoxide dismutase (SOD1) gene by studying people with ALS from 13 families. In healthy cells, the SOD1 enzyme eliminates toxic **free radicals** from the cells, an important function because free radicals attack and mutate DNA and proteins. The defective SOD1 enzyme also fails to control the function of an important transporter protein that normally removes excess glutamate from nerve cells. The accumulation of high levels of glutamate is especially toxic to nerve cells, so the brain relies on the SOD1 enzyme to eliminate both free radicals and glutamate from nerve cells.

**FIGURE 10.8** Astrophysicist Stephen Hawking. This amazing scientist, author, and father lives with the devastating disease, ALS.

**FIGURE 10.7** Lou Gehrig, renowned New York Yankees first baseman. Lou Gehrig died of ALS in 1941.

Many healthy people find it difficult to imagine living every day with the challenges of a chronic, disabling disease like ALS. Stephen Hawking is a man who has accomplished more than most people in his lifetime of 60 or so years. Born in England, where he lives and works as an award-winning astrophysicist, Hawking has three children and one grandchild. In addition to teaching and traveling worldwide giving seminars on his work, Hawking wrote several books including the all-time best-selling book *A Brief History of Time*. And Stephen Hawking has ALS disease (Figure 10.8).

On his web site, in plain and often humorous language, Stephen Hawking answers many of the questions that people have about living with a serious disease. Hawking describes his childhood and his education in England. He explains how he was diagnosed with ALS, and he discusses finding a place to live and a job. Also in plain language, Stephen Hawking describes his stellar research that changed our understanding of the basic laws of the universe, beginning with the Big Bang and ending in what Hawking calls, the "not-quite-black" holes. About ALS, Hawking says, "I am quite often asked: How do you feel about having ALS? The answer is, not a lot. I try to lead as normal a life as possible, and not think about my condition, or regret the things it prevents me from doing, which are not that many."

Free radicals damage DNA and proteins, but a diet rich in colored vegetables is the best source of antioxidants to fight the free radicals in your cells. Recent studies show that the dietary supplements marketed as antioxidants do not work.

## Diabetes and Obesity Epidemics in the United States

Type I and type II diabetes are complicated serious metabolic disorders affecting animals and humans. Interactions among at least 15 gene mutations have been linked to type I diabetes in humans. The characteristic feature of type I diabetes is the failure of the pancreas to produce **insulin**, a **hormone** protein, which is required for cells to take up sugar from the blood for use in energy metabolism (Figure 10.9). People with type I (or juvenile-onset) diabetes often become insulin-dependent as children, requiring daily injections of animal-purified or recombinant DNA-produced insulin (Humulin) to survive. People who develop type II diabetes (also called adult, maturity, or late-onset diabetes) are often overweight adults with sedentary lifestyles, both important risk factors for type II diabetes. Often type II diabetics can rely on oral medications to

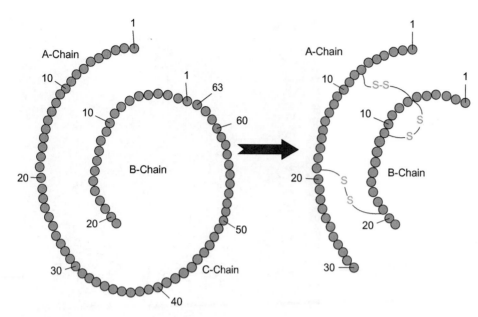

FIGURE 10.9 Insulin is a protein hormone used to treat diabetes. Insulin is made in the pancreas as a precursor protein containing three amino acid chains synthesized as one protein. In the blood, the insulin precursor protein is processed, which removes the C peptide and produces the active insulin hormone containing the A and B peptides, linked together by these disulfide bonds (–S–S–).

control their blood sugar levels, but neither oral drugs nor insulin injections are a cure for type I or type II diabetes. These treatments delay the physiological damage to the body that occurs when blood sugar levels are not controlled. For example, the tiny capillary blood vessels and nerves in the diabetic's hands and feet and the retina in the eyes are particularly sensitive to the detrimental effects of high blood sugar levels, making blindness and loss of feeling in the extremities just two of the many complications of this disease.

The Diabetes epidemic is international. Diabetes (type I and type II) will kill about 3.8 million people worldwide in 2007, the same number killed by AIDS. In the next 20 years, diabetes is projected to increase by 80% in Africa, 100% in Latin America, and by 43% in the United States.

For many people, **obesity** is an intractable medical problem that significantly increases the risk of having many medical problems in addition to diabetes, including high blood pressure, heart disease, and breathing problems. In some cases, excess body weight is inherited and has a genetic and biochemical basis. Obesity is increasing among U.S. children, accompanied by a staggering rise in the number of children diagnosed with type II diabetes.

Scientists study the role of genes in weight control and obesity in humans using mice. In the early 1990s a new gene was discovered that causes a severe form of inherited obesity when mutated in mice (*ob* gene) (Figure 10.10). Similar protein hormones were found in both the mouse and human cells called leptin. Studies show that the leptin hormone is made only in the fat cells of the mouse that functions in weight control and helps the mouse to maintain its ideal body weight. Defects in

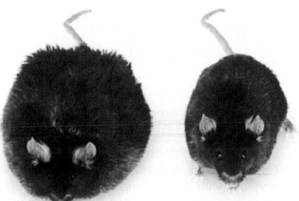

FIGURE 10.10 The ob mouse gene encodes a weight control protein. The obesity gene (ob gene) in mice codes for a hormone protein called leptin. The mouse carrying a mutant ob gene gains weight (left), while the mouse with the normal ob gene makes functional leptin proteins and exhibits appropriate weight control.

the leptin protein cause a severe form of inherited obesity in mice and humans, another indication that mice and man use similar proteins and biochemical processes to control body weight.

The leptin hormone is secreted into the blood by the fat cells and delivers a signal to the hypothalamus, the part of the brain that regulates food intake and energy expenditure in both the mouse and the human. Leptin has enormous potential as a drug if it can safely and effectively promote weight control in humans.

The large number of obese children diagnosed with diabetes in the United States is now recognized as a public health emergency. The direct solution is to take immediate action—decrease caloric intake in the diet and increase body activity.

# HUMAN CHROMOSOME KARYOTYPES REVEAL GENETIC DISEASES

The analysis of whole chromosomes plays a critical role in diagnosing genetic diseases and is equally important in routine prenatal genetic testing. Normally each cell in the human body contains 46 chromosomes in the nucleus. When a cell divides to make two cells, each new cell also contains 46 human chromosomes. The exceptions are the sex cells. The sperm and egg cells each contain 23 chromosomes, half of the 46 chromosomes found in every body cell. This difference in chromosome number is extremely important because when the sperm and egg cells come together to make a fertilized egg, the egg receives 23 chromosomes from the mother (unfertilized egg) and 23 chromosomes from the father (sperm) for a total of 46 chromosomes (see Chapter 6). An analysis that reveals the number and structures of the human chromosomes can determine the gender of an individual as well as diagnose certain genetic disorders.

Human chromosomes dramatically change structure as the cell proceeds through the different stages of the cell cycle (see Chapter 9). Cells that are in different stages of mitosis can be identified by staining the cells that are involved in cell division. In preparation for mitosis, the chromosomes condense as the cell begins to divide into two cells (see Chapter 9). After DNA replication is finished, each chromosome contains two duplicated DNA molecules that each extend from one end of a **chromatid** to the other end of the same chromatid. In order to be able to physically move the long strands of duplicated DNA, the DNA helix becomes tightly packaged with special **histone** proteins into compact structures called **mitotic chromosomes** (Figure 10.11, A, B, and C). Once the chromosomes have been distributed to the offspring cells, **cytokinesis** occurs and the two new cells physically separate. The mitotic spindle apparatus is then disassembled until the next cell division cycle. The compact chromosomes decondense and become threadlike chromatin fibers, completing the chromosome cycle.

> Cells must protect, duplicate, and segregate their genomes during each cell cycle. The cell starts by duplicating its chromosome DNA and then assembles a spindle fiber apparatus that attaches to the chromosomes, and moving half of the chromosomes into each of the progeny cells to ensure that each cell receives the proper number of chromosomes.

Human mitotic chromosomes are prepared for morphological studies by **cytological staining** methods so that they are visible when magnified in a light microscope (Figure 10.12). The condensed chromosomes are taken from cells in mitosis and are stained with Giemsa dye to create a distinct banding pattern on each chromosome (see Chapter 9). The 46 chromosomes from individual cells were sorted, identified, and arranged in a typical human **karyotype** display in black and white. Currently, amazing spectral technology allows scientists to "paint" human mitotic chromosomes in a rainbow of colors and generate biologically accurate computer images of karyotypes with multicolored chromosomes (Figure 10.13). To make a spectral image, the researchers treat the chromosomes with several different fluorescent tags attached to different DNA probes that base pair to different sequences on specific chromosomes. The light emitted from the bound probes is collected and sent to a camera to create the digital image. Although several of the fluorescent dyes appear to be the same color to the naked eye, subtle differences in the wavelengths of the dyes can be detected by an interferometer, which uses the interference of light waves to determine distance (wavelength). The computer then assigns colors to the different chromosomes to display the digital "painted" karyotype. These spectral karyotypes are not only beautiful, but they also offer scientists a much better look at chromosome structure at a higher resolution than is possible using a standard karyotype test (Figure 10.14).

## Extra Chromosome Copies Upset Gene Balance

Genetic destiny is set at conception when the sperm and egg both contribute chromosomes. Some genetic disorders are caused by inheriting an extra copy of an entire chromosome, a condition called **trisomy**. Inheriting three copies of one chromosome and genes (usually present in only two copies) can only be tolerated in rare cases such as trisomy 21, Down syndrome (Figure 10.15). However, depending on which chromosome is involved, inheriting an extra chromosome copy, or losing a chromosome copy, can be a lethal event. Down syndrome is the most common chromosome imbalance in humans (1 in 800 live births), affecting more than 350,000 people in the United States. Down syndrome occurs when an individual inherits an extra copy of chromosome 21 (trisomy 21), which causes mental retardation, distinctive facial features like slanted eyes, and heart problems. In 90% of the Down syndrome cases, the extra chromosome copy was inherited from the mother's egg, explaining the observation that the incidence of Down syndrome increases with the increasing age of the biological mother. Chromosome mis-segregation events are much more frequent as we age, and the eggs produced in older women are more likely to have a chromosome imbalance even before fertilization occurs.

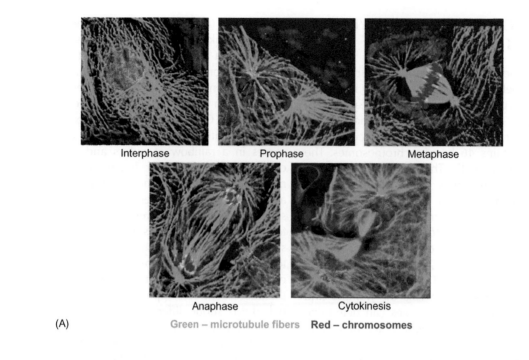

Interphase    Prophase    Metaphase

Anaphase    Cytokinesis

(A)

Green – microtubule fibers    **Red – chromosomes**

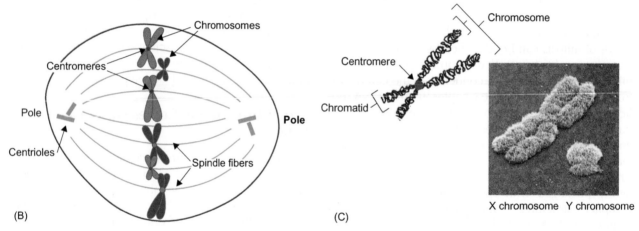

(B)

(C)

X chromosome  Y chromosome

**FIGURE 10.11**  Mitosis is the process of cell division. (A) Chromosomes in cells in different stages of mitosis. Chromosomes change shape and position during the different stages of mitosis in the cell cycle (mitosis). Tubulin proteins in fibers in the cells are stained green by antitubulin antibodies; the chromosomes are stained red. (B) Chromosomes line up in metaphase before they segregate. Chromosomes are moved (segregate) using spindle fibers that attach to the centromere DNA in each chromosome. Accurate chromosome segregation ensures that each daughter cell will inherit the correct number of chromosomes. (C) Condensed mitotic chromosomes are narrow at the centromere. Each chromatid contains one linear (unbranched) double-stranded DNA molecule extending from one end to the other end. In preparation for mitosis, the chromatin fibers are packed into condensed chromosomes. Shown are scanning electron micrographs of the human X and Y chromosomes.

The power of the sex chromosomes to influence trait development is demonstrated by people who inherit an imbalance of the X or Y chromosomes, normally X, X for human females and X, Y for males. A female who inherits one X chromosome instead of two X chromosomes has Turner syndrome, which affects more than 60,000 women in the United States. Individuals who inherit one Y chromosome and two X chromosomes have Klinefelter syndrome (1 in 500 to 1000 live births); they develop as males but lack male secondary sex characteristics (facial, underarm, or pubic hair).

## Mixed-Up Chromosome Sequences Cause Genetic Disorders and Cancer

Human chromosome **translocations** involve breaking the DNA helix in one chromosome and attaching it to a different DNA helix in a completely different chromosome. A **reciprocal translocation** involves an even swap of DNA sequences between two chromosomes, and occurs without the loss of any genetic information (Figure 10.16). This is because genes located completely within the boundaries of the translocated

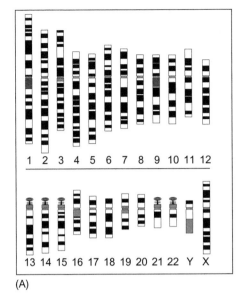
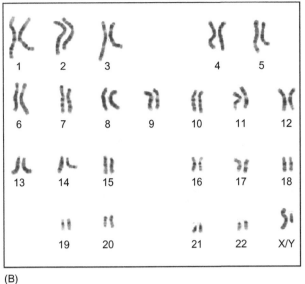

(A)                                                                                  (B)

**FIGURE 10.12** Human chromosomes have distinct banding patterns. (A) The drawing shows the banding patterns that are characteristic of human chromosomes. (B) This typical karyotype display shows 46 chromosomes (22 pairs plus X, Y) from a male human.

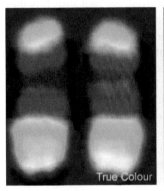

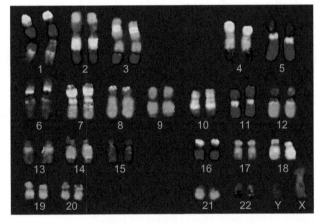

**FIGURE 10.13** "Painted" human chromosomes make a spectral karyotype. (A) Human chromosomes are "painted" with different DNA-specific fluorescent dyes that allow each chromosome and different regions on each chromosome to be identified by color (left: true color; right: pseudo color). (B) Mitotic chromosomes from a human male are arranged into a spectral karyotype, emphasizing the similar colored banding patterns exhibited by homologous chromosomes from the mother and father.

region of DNA are not affected by moving the DNA to a different chromosome. However, the genes encoded by DNA sequences that bridge the chromosome junctions will be disrupted by the DNA translocation event, sometimes with devastating consequences. A good example is chronic myelogenous leukemia (CML), a type of blood cancer that causes the uncontrolled growth of white blood cells (leukocytes). About 5000 people are diagnosed with CML each year, and more than 20,000 people in the United States live with the disease every day.

CML begins to develop when a chromosome translocation occurs in a single blood stem cell in the bone marrow (Figure 10.17A). In the first step of the translocation,

the DNA helices in chromosomes 9 and 22 are broken, within disrupting the ABL and BCR genes. In the next step, the DNA in chromosome 9 becomes attached to the end of chromosome 22, which effectively fuses the sequences encoding the beginning of the BCR gene to the end of the ABL gene, and creating a new fusion gene, BCR-ABL, which is carried on the Philadelphia chromosome, and is easily identified by karyotype analysis (Figure 10.17B).

In the bone marrow, the single leukocyte stem cell carrying the Philadelphia chromosome expresses the BCR-ABL fusion gene, which acts as an **oncogene** and promotes the development of cancer. The BCR-ABL fusion protein functions as a kinase enzyme that attaches phosphate groups to target proteins in the cell.

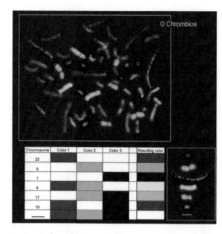

FIGURE 10.14   Colored human chromosomes are visible in a spectral array (top). The colors exhibited by each chromosome can result from combinations of different fluorescent colors (bottom).

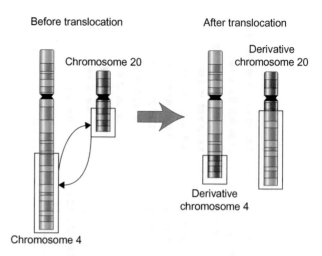

FIGURE 10.16   Reciprocal chromosome translocation. Reciprocal translocation of DNA between chromosomes 4 and 20 occurs without the loss of essential genes.

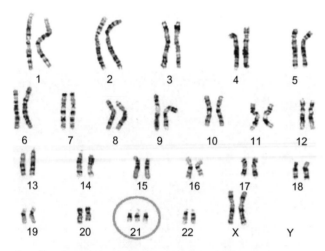

FIGURE 10.15   Colored human chromosomes are visible in a spectral array (top). The colors exhibited by each chromosome can result from combinations of different fluorescent colors (bottom).

A Robertsonian translocation occurs when the long arms of two chromosomes break off and fuse together at a centromere, causing the loss of the genetic material on the two short arms of the chromosome. Surprisingly, there are no obvious negative consequences for people who inherit Robertsonian translocations involving chromosomes 13, 14, 15, 21, or 22, even though all the genes on the short chromosome arms are lost. Karyotype analysis revealed that people with Robertsonian translocations have only 45 chromosomes in each cell (instead of 46), yet these individuals function normally. However, the biological children of people with Robertsonian translocations are at high risk of inheriting a chromosome imbalance, the wrong number of chromosomes per cell.

## COMPARISON OF HUMAN GENOMES REVEALS IMPORTANT DNA DIFFERENCES

The ability to make connections between certain human genes and specific human diseases yields important information that often accelerates the development of new medical treatments. It is accurate to say that we inherit two copies of each chromosome, one from our mother and one from our father, but the two chromosome copies are not completely identical in DNA sequence. Studies that compared the genome DNA sequences from many different people reveal that individual human genomes have single base pair differences located at many sites along the genome DNAs. By 2007, scientists had identified and cataloged more than 1.8 million **single nucleotide polymorphisms (SNPs)** on the 46 human chromosomes.

The proteins with the phosphates become permanently active, even though they normally function only when the cell receives a signal to start cell division. The mutant BCR-ABL enzyme not only promotes rapid cell division, but it also protects the new cancer cells from committing suicide by programmed cell death (apoptosis).

At this stage the CML disease is chronic. The cancer cells are still subject to partial cell cycle control, and occasionally they differentiate into mature cells that perform normal functions in the blood. However, the cancer cell carrying the Philadelphia chromosome can acquire a second mutation that activates another oncogene (ras) or destroys the function of the p53 tumor-suppressor gene, resulting in a large increase in reproduction of the cancer cells. Cells that carry both mutations fail to differentiate and cannot function properly in the blood, bringing on the crisis phase of CML disease.

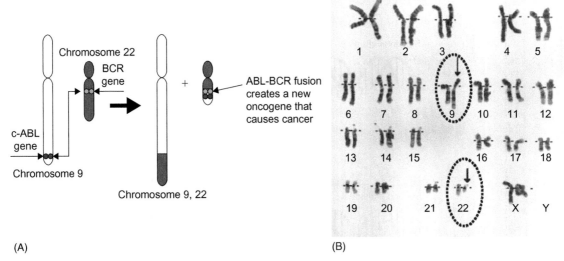

**FIGURE 10.17** A chromosome translocation causes a human blood cancer. (A) The Philadelphia chromosome results from translocation events that attach part of chromosome 22 onto the long arm of chromosome 10 and move a small part of the short arm of chromosome 10 onto chromosome 22. This chromosome DNA rearrangement creates a fused gene that acts as an oncogene and causes a blood cancer called chronic myelogenous leukemia (CML). The Philadelphia chromosome 10, 22 (Ph1) carries a new gene fusion (ABL-BCR) that functions as an oncogene and causes cancer; (B) Philadelphia chromosome karyotype. The reciprocal translocation between chromosome 10 and chromosome 22, designated t(10;22), results in one chromosome 10 that is longer than normal and one chromosome 22 (Philadelphia chromosome: Ph1) that is shorter than normal.

Before the human genome sequence became available, genes were identified and mapped on the chromosomes using genetic information in combination with chromosome markers, usually restriction fragment length polymorphisms (RFLP) (see Chapters 6 and 8) (Figure 10.18). Finding human genes by RFLP mapping is much more difficult than using SNPs because the RFLP markers occur much less frequently in the human genome than SNPs, and therefore provide much less information for the gene search. The RFLP and SNP markers can be used to identify and locate genes on human chromosome because the positions of many RFLP and SNP markers are known in the human genome DNA. The RFLP and SNP markers are used to identify which chromosomes and genes are reproducibly inherited by individuals who also inherit the particular disease under study. In other words, if a chromosome marker is inherited by the same family members who also inherit a genetic disease, then a possible link exists between that chromosome marker (or gene mutation) and a specific genetic disease. When an SNP marker and a particular gene are inherited together generation after generation, then the SNP marker and the gene are considered to be **genetically linked**; the marker and the gene are located on the same chromosome within a region containing about 5 million base pairs of DNA.

## Family Connections Are Important to the Search for Disease Genes

This chapter describes the genetic mapping methods used to identify genes involved in human genetic diseases. Behind the scenes in these studies are the people and their families who live every day with genetic diseases and disorders. Several groundbreaking genetic studies would not have been possible without the willing cooperation of many devoted patients and their families who continue to work to promote research to develop future treatments. Gene mapping studies often involve making a pedigree, a kind of biological family tree that illustrates a pattern of gene inheritance and the diseases exhibited by the family under study. This approach was used to identify many human genes including the gene associated with a devastating neurological disorder called Huntington's disease, named for George Huntington, the American physician who studied the disease in the early 1900s. Probably best known as the disease that killed folksinger Woody Guthrie in 1967, Huntington's disease is an invariably fatal illness that slowly destroys brain function (Figure 10.19). People with Huntington's disease suffer progressive deterioration of the nervous system, starting with minor neurological symptoms but inevitably leading to involuntary thrashing and loss of motor control. These writhing movements earned the disease the name of Huntington's "chorea," from the Greek word for dance, *chorea*.

Mapping human genes depends on studying the incidence of the disease in members of a biological family because the pattern of gene inheritance will reflect any genetic link between a specific mutant gene and the disease in question.

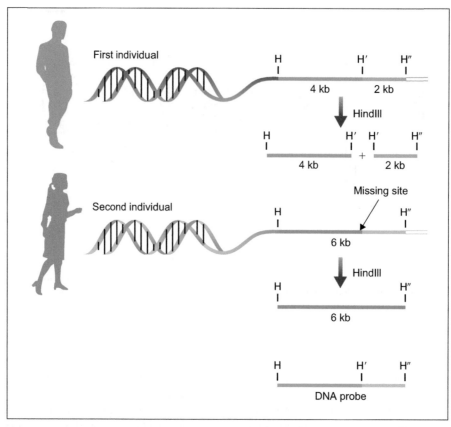

**FIGURE 10.18**   Detecting an RFLP in genome DNA. The first individual has a region of DNA with three cleavage sites for the HindIII restriction enzyme extending over about 6 (kilobase pairs) kb (or 6000bp) in the genome. If we cut just that region of the DNA with HindIII, the result would be two DNA fragments 4kb and 2kb in length, which add up to 6kb in total. Analysis of the same region of DNA in the genome of a second individual shows that one of the three HindIII sites has been altered by a mutation in the second genome and as a result the DNA at that site is no longer recognized or cut by the HindIII restriction enzyme. If we cut just that region of the DNA with HindIII, the result would be one DNA fragment that is 6kb in length. The DNA probe in this experiment allows the scientist to visualize only the desired DNA bands without interference from all of the other DNA fragments generated by cutting all 46 chromosomes with HindIII.

**FIGURE 10.19**   Folk singer Woody Guthrie died from Huntington's disease.

**FIGURE 10.20**   The Gusella research group worked on Huntington's disease.

In the 1970s, geneticist James Gusella at Harvard University and neuropsychologist Nancy Wexler at Columbia University and colleagues began to look for the human Huntington's disease gene, a quest that led them to Venezuela (Figure 10.20). They found a family descended from a European woman who migrated to Venezuela in the early 1800s and brought with her the mutant gene that causes Huntington's disease.

The research team led by Nancy Wexler traveled to Venezuela to collect information and biological samples from the villagers, which were then taken to the Gusella lab at Harvard for analysis.

When this study began, the chromosomal location of the Huntington's gene was not known, so the first goal of the research team was to find the specific DNA probe that could reproducibly detect the Huntington's gene on human chromosomes. This study predated SNP gene mapping by many years, but at the time Gusella's team had several human RFLP DNA probes available to test the dozens of DNA samples gathered from the people in Venezuela. One by one the team screened the DNA samples from Venezuela with each RFLP DNA probe to search for a link between the family members who inherit the mutant gene and the incidence of Huntington's disease in the Venezuelan family. After the first several RFLP candidate probes failed the linkage test, Gusella decided that the next candidate probe would be named G8 after his lab technician, Ginger Weeks. The G8 DNA probe worked so well that it was used to accurately track the mutant Huntington gene through seven generations of the Venezuelan family!

These experiments finally revealed the chromosome location of the Huntington's disease gene, but it wasn't until 1993 that an international team of research scientists including Gusella's group announced that they had finally isolated (cloned) and sequenced the Huntington's disease gene. They found that the wildtype Huntington's gene encodes the huntingtin protein, which normally functions in nerve cells in the brain. However, despite the successful identification and cloning of the gene for the huntingtin protein, over a decade later there was still no effective treatment or cure for people who inherit this devastating disease. Nancy Wexler returns frequently to Venezuela to add more data to the growing pedigree chart of the family with Huntington's disease that covers both walls of the hallway outside her office. This work holds a personal interest for Wexler who is also at risk for inheriting the mutant Huntington's disease gene. More about the Huntington's disease gene is presented later in this chapter.

## CARRIERS OF GENETIC DISEASES HAVE A MUTANT GENE BUT DO NOT GET SICK

Every year more than 30,000 children in the United States become chronically ill with a persistent cough, constant lung infections, shortness of breath, and poor growth because they have inherited cystic fibrosis. There is no cure for cystic fibrosis disease, but due to advances in medical treatments many people with cystic fibrosis now live well into their forties (median age 37 years). The gene that causes cystic fibrosis disease is called the cystic fibrosis transmembrane conductance regulator (CFTR) gene (Figure 10.22A). The wildtype CFTR gene is written in all capital letters and the mutant gene allele written in lowercase letters (cftr). To develop the symptoms of cystic fibrosis disease, a person must inherit two copies of a mutant cystic fibrosis gene (cftr/cftr).

Healthy human cells must constantly regulate the internal concentrations of salts and many other potentially harmful compounds. As you might guess from the name, cystic fibrosis transmembrane conductance regulator, the CFTR gene codes for a key protein that builds channels in the cell membrane to export excess salt from the cell. In people with cystic fibrosis disease, the defective proteins cannot build functional membrane channels and as a result the cells fail to properly regulate internal salt concentrations.

More than 10 million Americans carry one wildtype copy (CFTR) and one mutant (cftr) copy of the cystic fibrosis gene and are genetic carriers of cystic fibrosis disease. However, people who are carriers of cystic fibrosis do not suffer from disease symptoms themselves, but although they are at risk of passing the mutant cftr gene to their children. A child who inherits a mutant cftr allele from each parent (cftr/cftr) will become sick with cystic fibrosis disease because the child's cells lack the normal gene and can make only the defective cftr proteins. People who are carriers of cystic fibrosis disease do not become sick because the normal (CFTR) and mutant (cftr) forms of the CF proteins are both made in the cells of a CF carrier. As a result, the normal CFTR proteins can build functional membrane channels and properly regulate internal salt concentrations, even in the presence of mutant, non-functional cftr proteins (Figure 10.22B). In genetic terminology, the wildtype CFTR allele is **dominant** over the **recessive** mutant cftr allele.

> Carriers of cystic fibrosis do not become sick because their cells carry both the mutant and normal alleles of the CFTR gene and make both the mutant and normal forms of the proteins; the wildtype CFTR proteins are sufficient to compensate for the lack of function by the defective cftr proteins.

Muscular dystrophies are not a single disease called Muscular Dystrophy but instead represent a group of 30 genetic diseases that cause progressive degeneration of the muscles that control body movements. Duchenne muscular dystrophy (DMD) is the most prevalent type of muscular dystrophy disease in humans and was named for Guillaume Duchenne, the French neurologist who described the condition in 1868. DMD causes the rapid degeneration of muscle tissues early in life, striking

## Box 10.1    Get to Know Nancy Wexler: Huntington Gene Hunter

If you are a baby boomer, you might remember seeing Nancy Wexler years ago when she appeared on television on *60 Minutes* in "Gene Hunter," a show that described Nancy's search for the gene that causes Huntington's disease. It took 10 years of research and the assistance of the entire Venezuelan village of Laguneta, but finally, in 1983, Nancy Wexler and her colleagues discovered the Huntington's disease gene (Figure 10.21).

What led Nancy Wexler to become interested in searching for the genes that cause genetic diseases? Growing up, Nancy developed a strong interest in science, especially psychology. She attended Harvard University's Radcliffe College, where she studied literature, anthropology, and psychology. Later, Nancy studied in London under Anna Freud, the daughter of the famous Dr. Sigmund Freud.

Huntington's disease has been a part of Nancy's life as long as she can remember. Nancy's three uncles died from Huntington's disease and her mother, who also developed this devastating disease, died when Nancy was 32. Nancy still faces this personal challenge; she and her older sister Alice are at risk of inheriting the disease. In 1969 when Nancy first began studying Huntington's disease, there was no way to know which children born to parents with Huntington's disease might inherit the gene that causes the disease. Nancy joined other scientists dedicated to finding the gene that causes Huntington disease, with the hope that finding the gene might lead to effective treatments and eventually a cure.

Nancy Wexler led a team of scientists who traveled to Venezuela to collect dozens of biological samples from people in the village of Laguneta, people who potentially carry the Huntington's disease gene. The team kept careful records on the people who donated the samples, including medical histories and exams, gathering information that was critically important for the scientists to correlate candidate genes with the occurrence of Huntington's disease. The dedicated efforts of many people finally paid off when the chromosomal

**FIGURE 10.21**    Nancy Wexler Gene Hunter helped to identify and isolate the Huntington's disease gene.

location of the mutant Huntington's gene was finally determined in 1983 using the genetic studies and biological samples provided by Nancy Wexler, her team of scientists, and the people of the Venezuelan village.

Unfortunately there is as yet no effective treatment or cure for this devastating disease, even though researchers were able to isolate and characterize the Huntington's disease gene. Now, because of scientists like Nancy, people at risk for Huntington's disease can take a blood test to find out if they have inherited the mutant gene. Whether or not to take this test is a difficult decision for the person at risk to make, because a positive test result means you will develop a terrible disease with no treatment and no cure. What would you do?

---

children at the age of 3 years, gradually causing unsteady movements as they lose muscle strength and control. Often confined to a wheelchair by age 10, children with DMD usually die in their teens or early twenties. There is no cure for DMD.

Scientists have known for years that people with DMD make mutant dystrophin proteins, so hopes for a cure were raised when the human dystrophin gene was first identified in 1986. Unfortunately, DMD is a good example of why identifying the gene responsible for a genetic disease is not always enough to lead to a rapid cure for the disease. One reason is that the dystrophin protein is encoded by the largest known human gene, which extends over more than 2.6 Mb (megabases) (2.6 million bases) of human genome

DNA and contains 97 exon sequences (Figure 10.23). Different mutations in the dystrophin gene are responsible for either the most severe form of the disease, Duchenne muscular dystrophy, or a much milder form of the disease, Becker muscular dystrophy. At first scientists could not explain why the mutation that removed the largest amount of DNA actually causes the mild disease, while, of DNA a mutation removing a much smaller region causes the severe form of the disease (Figure 10.24A). Further study of the expression of the mutant alleles revealed that the mutant dystrophin RNA transcripts were processed by RNA splicing to make different mutant mRNAs, containing the appropriate coding sequences to be translated into the two mutant dystrophin proteins associated with the

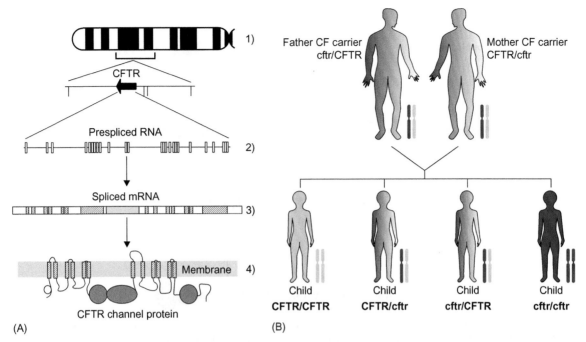

**FIGURE 10.22** (A) Mutation in CFTR gene causes cystic fibrosis. (1) The normal cystic fibrosis gene CFTR is on human chromosome 7. (2) CFTR gene DNA is copied into prespliced RNA. (3) CFTR prespliced RNA is spliced into mRNA. (4) mRNA is translated into the CFTR protein that weaves back and forth through the cell membrane to build the CFTR protein channel needed to regulate the amount of salt inside the cell. (B) Genetic carriers of cystic fibrosis inherit one mutant cftr gene. Cystic fibrosis is caused by a recessive mutation in the CFTR gene; it is recessive because two copies of the mutant cftr gene must be present to cause the disease to develop (cftr/cftr). The mother (CFTR/cftr) and father (cftr/CFTR) are both carriers of cystic fibrosis but are normal because each carries a normal dominant CFTR gene to make up for the single copy of the mutant cftr gene. The children inherit genes in pairs, one copy from each parent. There are four possible gene outcomes for the children of these parents. Three children are normal because they carry at least one copy of the dominant CFTR gene. Two of these three children are carriers because they carry one normal and one mutant copy of the gene (CFTR/cftr). One child inherits two copies of the mutant gene (cftr/cftr) and becomes ill with cystic fibrosis. Key: CFTR (wildtype gene); cftr (mutant gene); CFTR/CFTR (no cystic fibrosis); cftr/cftr (cystic fibrosis disease); CFTR/cftr (CF carrier; no symptoms).

mild and severe forms of muscular dystrophy disease (Figure 10.24B).

> Most human genes are copied into precursor RNAs containing both introns and exons. Successful RNA processing (splicing) removes the introns from each precursor RNA, and links the exons together, in the correct order, to produce a mature mRNA containing just exon protein coding regions.

DMD, the most common form of childhood muscular dystrophy, affects 1 in 3500 males worldwide but rarely affects females. Why? The answer lies in the fact that dystrophin is an **X-linked gene**; it is located on the X chromosome. Human males inherit one X and one Y chromosome, whereas females inherit two X chromosomes. In the case of X-linked genetic mutations, females often have an advantage over males. A female who inherits a mutant gene carried on one copy of the X chromosome has a 50% chance of also inheriting a normal (wildtype) version of that gene on her second copy of the X chromosome; one normal copy of the gene might be sufficient to compensate for the deficit caused by the mutant gene. Of course, she has an equal chance of inheriting two copies of the X chromosome carrying the mutant genes; lack of a wildtype allele in XY males means that the detrimental consequences of the mutant genes cannot be rescued.

Human males inherit only one copy of the X chromosome, so they carry only one version of each gene on the X chromosome, either mutant or wildtype. As a consequence of human genetics, when a male inherits an X-linked disease such as DMD, there are no backup copies of the X chromosome available in his male cells, which could potentially provide a normal version of the dystrophin gene and protein. For this reason, boys with the X-linked disease cannot make normal dystrophin proteins.

Having no second, backup copy of the X chromosome means that disease genes located on the X chromosome (X linked) will always predominantly affect boys over girls.

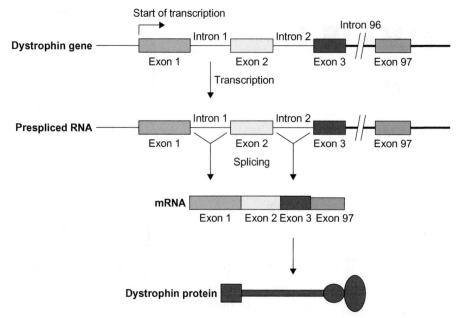

**FIGURE 10.23**  The dystrophin gene has 97 exons. It is transcribed into precursor RNAs, which are spliced to remove the introns. The numerous exons are retained in the spliced mRNA. Dystrophin is a rod-shaped protein containing 3684 amino acids.

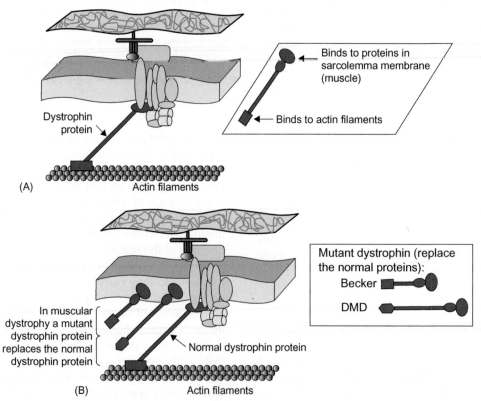

**FIGURE 10.24**  (A) Dystrophin protein connects a complex in muscle membrane to actin filaments. The dystrophin protein connects the actin filaments to a complex anchored in the muscle membrane. One end of the normal dystrophin protein binds to the actin filaments, and the other end attaches to the sarcolemma membrane proteins. (B) Dystrophin mutations cause the molecular machine in the muscles to malfunction. Mutations that shorten the dystrophin helix cause the milder form, Becker muscular dystrophy, because the mutant protein is not long enough to reach from the actin filaments to the sarcolemma membrane. Mutations that alter the actin-binding region of the dystrophin protein prevent the complex from becoming anchored to the actin filaments, which causes the most severe form, Duchenne muscular dystrophy.

## Triplet Repeat Mutations Cause Many Different Genetic Diseases

The triplet repeat diseases (TRD) are grouped together even though they involve a variety of different genes and diseases. This is because the mutant forms of the TRD genes all have the same type of unusual DNA mutation; triplets of three DNA bases (e.g., CCG), are repeated in tandem often hundreds of times in the coding region of the mutant gene (e.g., CCG CCG CCG CCG CCG). The normal (wildtype) allele of the gene usually encodes only a few triplet repeats, but in the TRD mutants the few triplet repeats have expanded to create a mutant gene containing very large numbers of tandem triplet repeats. In most cases, the TRD mutation is located in the coding region of the gene, which means that the mutant triple repeat sequence is translated into a long stretch of the same amino acid repeated many times in the nonfunctional protein.

Huntington's disease, discussed earlier in this chapter, is caused by a TRD mutation in the coding region of the huntingtin gene. As a result, mutant huntingtin proteins contain regions where the amino acid glutamine is repeated hundreds of times. It is not yet clear why this extremely glutamine-rich region destroys the function of the huntingtin protein and causes a degenerative and fatal nerve disease. However, there is a correlation between the severity of an individual's Huntington's disease symptoms and the number of TRD repeats contained in that individual's huntingtin genes. The number of gene repeats might also explain the age of onset of symptoms and the rate of disease progression, which both vary considerably in Huntington's disease. Studies show clearly that the patients who decline most rapidly have invariably inherited huntingtin alleles that encode the largest numbers of triplet repeats, and their cells produce mutant huntingtin proteins containing very long stretches of glutamine residues.

Preliminary studies suggest that the mutant TRD proteins might bind directly to a protein enzyme called glyceraldehyde-3 phosphate dehydrogenase (GAPDH), which normally functions to metabolize glucose (sugar) and oxygen in all cells in the body. Researchers propose that abnormally tight binding between the mutant TRD proteins and the GAPDH enzymes in brain cells might block GAPDH enzyme activity, which would inhibit energy production in nerve cells and cause brain tissue to atrophy. Interestingly, preliminary lab experiments show that the mutant proteins containing the longest stretch of a single amino acid bind most tightly to the GAPDH enzyme compared to the normal protein. This result potentially has wide reaching consequences because a similar TRD mechanism could potentially be the underlying cause of several different but related degenerative neurological disorders including myotonic dystrophy, spinobulbar muscular atrophy, Friedreich's ataxia, spinocerebellar ataxia, and some forms of muscular dystrophy (Table 10.3).

> TRD mutations alter the protein coding regions of genes, which cause changes in the structure and properties of the mutant protein, which can lead to degenerative neurological diseases such as Huntington's disease.

During the 1990s, autism became the fastest-growing developmental disability in the United States, increasing 172% at a time when the U.S. population increased by only 13%, and occurring in 1 in 150 births by 2007. The emotional and financial cost to patients and families is enormous and the cost to society every year is about $90 billion and is expected to grow to $200 billion to $400 billion by 2017. The reason for the startling increase in children diagnosed with autism is not clear, but it is not explained solely by the increase in public awareness of the disease. In addition, scientific evidence does not support a link between autism and childhood vaccines.

Concern about the increase in autism has helped to raise funding and support for new research on autism including studies aimed at identifying the genes involved in autism and related disorders. A consortium of 11 scientific institutions began a comprehensive gene linkage study designed to identify all of the human genes involved in autism. Early results from this research suggest that two proteins in the brain called neuroligin-1 and neuroligin-2 are involved in autism in humans. In rats, the same proteins play a critical role in the development of nerve cell connections in the brain. The neuroligin proteins function at the junction where two nerve cells meet, the **synapse**, and enable nerves to make connections with other nerves. One neuroligin protein increases the excitability of the nerve cells, while the other neuroligin protein inhibits nerve cell activity. This important balance in nerve cell activity is disrupted in autism.

A TRD mutation in the FMR1 gene is the leading cause of mental retardation and autism in humans. Unlike mutations that alter the coding region of a gene, the FMR1 mutation contains 200 tandem repeats of CGG base pairs located in the **control region** of the gene. This mutation prevents the FMR1 gene from being transcribed into RNA or expressed as protein, causing the degenerative mental disorder called fragile X syndrome. The FMR1 protein normally regulates the translation of certain mRNAs, whose protein products are needed for nerve cell development. The extremely long stretch of CGG DNA repeats in the mutant FMR1 gene distorts the physical conformation of the DNA double helix in the chromosome, blocking gene expression and

**TABLE 10.3** Human diseases caused by triplet repeat mutations

| Triplet repeat diseases | Description | Symptoms |
|---|---|---|
| Fragile X syndrome | Inherited mutant gene on X chromosome; most common form of genetic mental disease and autism; males typically affected more than females | Mental retardation, attention deficit hyperactivity, anxiety, moodiness, large chin, ears |
| Freidrich's ataxia | Inherited progressive disease damages nervous system; often scoliosis; heart disease | Difficulty walking: clubfoot, hammer toes, muscle loss in feet, legs, and hands; loss of knee and tendon reflexes |
| Muscular dystrophy | Family of several inherited muscle-destroying disorders | Progressive weakness, loss of muscle strength, movement; specific muscles involved and disease progression depend on type of muscular dystrophy |
| Myotonic dystrophy | Most common form of adult muscular dystrophy; type 1 and type 2 | Weakness and loss of muscle control in lower legs, hands, neck, face; muscles fail to relax after use, cardiac defects |
| Spinobulbar muscular dystrophy | Inherited X-linked recessive disorder; neuromuscular disease affects spinal and bulbar neurons; primarily affects males | Tremors, cramps, muscular atrophy, decreased motor and primary sensory neuropathy (nerve degeneration) |
| Spinocerebellar ataxia | Describes many inherited genetic disorders that are characterized by loss of coordination | Progressive uncoordinated gait, hands, speech, and eye movements; atrophy of the cerebellum; symptom progression depends on disease type |
| Huntington's disease | Degenerative, fatal nerve and brain disease; monogenic | Chorea (uncontrollable jerky movements), balance and coordination problems, trouble shifting eyes without moving head, dementia, seizures; Parkinson-like symptoms, muscle rigidity, and tremors |

preventing the DNA from being copied into mRNA. Because the mutant FMR1 gene is not transcribed, the mutant FMR1 protein is not made in the fragile X cells. The cells carrying the FMR1 mutation also exhibit broken X chromosomes because the long stretch of CGG base pairs in the FMR1 DNA weakens the chromosome structure at the site of the mutation (Figure 10.25).

> Hundreds of tandem CGG repeats in the FMR1 gene control region distort the local chromosome structure, which not only prevents FMR1 gene expression, but also creates a "fragile site" in the X chromosome. The X chromosome DNA helix breaks easily at the position of the FMR1 mutant gene.

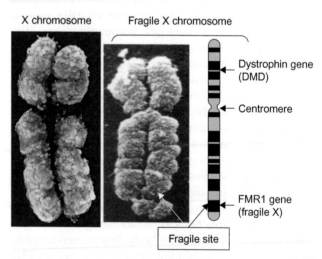

**FIGURE 10.25** The fragile X mutation causes the X chromosome to physically break at the fragile site in the DNA.

## AMERICANS SEEK INFORMATION ON GENES, HEALTH, AND BIOMEDICAL RESEARCH

Recently the American public has become more interested in gaining a better understanding of the human body, how it works, and how genes and environment contribute to health, happiness, and longevity. Public interest increased in the 1990s with media coverage of the Human Genome Project, including an ongoing debate over the total number of human genes needed to design and assemble a person. By the early 2000s, new technologies such as nuclear transfer (animal cloning; see Chapter 14) and embryonic stem cell research (see Chapter 12) continue to fuel often contentious public discussions about the scientific and ethical issues raised by biomedical research.

It is important that the public appreciate the direct connection between human genes and human health so that people will understand the relationship between basic medical research, clinical testing, and the development of new treatments. In addition, support services such as health education and genetic counseling are increasingly important to help patients and their families deal with the emotional and financial impact that a genetic disease places on an entire family. Many people now use their computers to access information on medicine and health; the World Wide Web (Internet) has dramatically changed how we do home research on everything, including health and medicine. Many excellent Internet web sites provide reliable information about common and rare genetic diseases (see the Additional Reading section at the end of this chapter). Information about web sites dedicated to searching databases and manipulating DNA sequences are included in bioinformatics (Chapter 7), and online resources for visualizing biomolecules in three dimensions are described in Chapter 2.

> Americans now use the Internet as a primary source of information on health topics. But buyer beware, it is often difficult to decide between online fact and fiction and it is easy to make poor health choices based on faulty or outdated medical information available online.

The Dolan Learning Center at Cold Spring Harbor Laboratory (CSHL) (Cold Spring Harbor, New York) is an excellent resource on genetic diseases that offers clear explanations and animations covering everything from the basics on DNA to detailed information on new advances in biotechnology and medicine. *Your Genes Your Health* provides up-to-date, reliable information about 15 genetic diseases including some discussed in this chapter. The Cold Spring Harbor web site also offers clear basic information about genes in *DNA from the Beginning*. Cold Spring Harbor Laboratory is an acclaimed biomedical research center that also offers excellent science education programs for children. There is free online access to information about the CSHL scientists and their research, including the international SNP consortium, which is responsible for cataloging the human genome SNP markers used to construct the haplotype map or "HapMap" of the human genome.

The HapMap will be an important genetic tool for researchers to use DNA genome variations to find genes involved in the complex diseases such as diabetes, which are caused by multiple genes. The DNA sequences of any two people are almost identical, but the tiny variations in the DNA add up to what makes us different and unique and explain why some people are at risk for heart disease while others are more likely to get cancer. When the SNPs are positioned near each other on the same chromosome DNA, they are inherited together and comprise a haplotype block. When complete, the HapMap will show the locations of all the haplotype blocks in the human genome. Researchers will scan the HapMap to find genes that cause complex human diseases, to identify genetic factors that cause susceptibility to infection, and to predict in advance the risk of adverse reactions to drugs and vaccines, all because scientists mapped the tiny differences among us to benefit everyone.

The University of Utah is a top-notch center for human genetics that maintains an excellent web site with readable excellent educational information on the nuts and bolts of human genetics. At the University of Utah Genetics Science Learning Center, the public can find accurate information about many topics such as calculating genetic risk, genetic counseling, personalized medicine, genetics and the brain, and much more. The web sites for foundations such as the Muscular Dystrophy Association, the National Fragile X Foundation, the Cystic Fibrosis Foundation, and the March of Dimes are all excellent sources of current, accurate information. These sites are often maintained by people with personal connections to the disease and can be the best sources of information about new treatments, organization activities, and patient and family support groups.

The Internet provides information on just about everything, but it is important to realize that the information on the Internet is not necessarily accurate (even if it says so in Wikipedia). In terms of scientific accuracy, the public can access an international database of all published scientific research articles from most scientific and medical fields through PubMed, a huge database of published research papers (see Chapter 7). Some articles require journal subscriptions for access to the entire paper, but many scientists pay to allow free access to their published reports, so the public can find reports on basic biomedical research from around the world. Be forewarned that the majority of the research papers include highly technical descriptions of experiments and results, but they often include references for review papers that offer approachable summaries of the research field. The references listed at the end of each paper can be directly accessed online.

> The U.S. government maintains excellent web sites to help the public navigate some complicated medical issues and to connect people with experts in every area of medicine and disease. Resources feature human genes and diseases, updates on genetic counseling and clinical human trials, reports about new drugs in the pharmaceutical pipeline, and more.

## SUMMARY

In this chapter we discussed the important roles that mutant human genes play in the inheritance of genetic diseases, whether caused by a mutation in a single gene or by the actions of many defective genes. A mutation in a gene can alter the product of the gene, usually a protein, which often prevents the mutant protein from functioning in the cell. Mutations in genes can be inherited or can be the result of a spontaneous DNA change caused by environmental insults to the genome. For scientists to understand diseases caused by mutant genes, they need to understand how the gene and the gene product function in normal cells. DNA differences between human genomes have been identified and are used in RFLP or SNP studies to identify the locations of genes on the chromosomes and to trace the inheritance of genetic diseases from generation to generation.

Scientists can tell a lot about genetic diseases by examining condensed human chromosomes in a karyotype display. Some disorders are caused by inheriting extra chromosome copies, whereas others occur because of large DNA rearrangements that alter the entire chromosome structure. Many genetic disorders result from tiny changes in the genome DNA that can only be detected by cytological methods using specific DNA probes. DNA probes are essential tools used in many methods including gene searches and RFLP/SNP mapping. A DNA probe is typically a short, single-stranded DNA molecule containing a specific DNA sequence that is designed to bind to (base pair with) its complementary DNA target sequence in the chromosome. When the probe DNA base pairs with the target DNA, it generates a signal to indicate the physical location in the chromosome of the DNA probe bound to the target DNA.

Genetically linked markers (genes) are almost always inherited together and are used to make correlations between a mutant gene on a specific chromosome and the inheritance of a specific genetic disease. Over the years, thousands of dedicated people with genetic diseases and their families have participated in studies designed to trace genetic diseases through family generations. New information and new genes have come from these studies, but progress toward novel treatments for these human genetic diseases lags behind.

## REVIEW

To test your knowledge of the concepts in this chapter, answer the following questions:

1. What is the meaning of the statement "Two genes are genetically linked"?
2. Why do DNA probes bind specifically to target DNA sequences in a chromosome?
3. Summarize how RFLP and SNP markers are used to find mutant human genes.
4. Explain why people who are genetic carriers of cystic fibrosis do not become sick with cystic fibrosis disease.
5. Explain the difference between an inherited gene mutation and a spontaneous mutation in a gene.
6. What does the term "X-linked gene" mean in terms of human chromosomes?
7. Explain why testing newborn infants to uncover genetic diseases right after birth is important.
8. Explain how the Internet has created free, unlimited access to science and medicine for the public.
9. Explain why it is advantageous to the cell to inherit two copies of every chromosome, even though the chromosomes might carry both normal and mutant alleles of the same gene.
10. Describe the evidence in laboratory mice that indicates how body weight is controlled at least in part by genes.

## ADDITIONAL READING

Alpert, M., April 2007. The autism diet. Sci. Am. 296 (4), 19–20 PMID: 17479620.

Autism Symptoms reversed in lab. BBC News June 27, 2007.

Siegel, J.M., August 20, 1999. Narcolepsy: a key role for hypocretins (orexins). Cell 98 (4), 409–412 PMID: 10481905.

Siegel, J.M., November 2003. Why we sleep. Sci. Am. 289 (5), 92–97 PMID: 14564818.

## WEB SITES

American Diabetes Association, www.diabetes.org. Provides information on the prevention and treatments for diabetes and helps to improve the lives of all people affected by this disease.

Autism Society of America, www.autism-society.org. Works to raise funds to address the many unanswered questions about autism.

Autism Speaks, www.autismspeaks.org. Provides accurate information on new research on autism.

Cold Spring Harbor Lab (CSHL), Dolan Learning Center (Cold Spring, NY), www.cshl.edu. Maintains excellent online resources about DNA, genes, and related research with clearly worded explanations, animations, and videos.

Cystic Fibrosis Foundation, www.cff.org. Site has information for people living with CF; testing for genetic carriers of CF.

DNA from the Beginning, www.dnaftb.org. A step-by-step description of basic information on DNA and genes.

Genetics Science Learning Center (GSLC), University of Utah, www.gslc.genetics.utah.edu. Site has excellent educational information on human genetics.

March of Dimes, www.marchofdimes.com.

Mendelian Inheritance in Man (OMIM), www.ncbi.nlm.nih.gov/Omim/mimstats.html. Learn about human chromosomes and genes; includes direct links to information on specific human genes.

Muscular Dystrophy Association, www.mdausa.org.

National Center for Biotechnology Information (NCBI), www.ncbi. nlm.nih.gov/Omim/mimstats.html. Site provides many useful resources and Fact Sheets describing human genes and associated genetic diseases.

National Fragile X Foundation, http://www.fragilex.org.

Nancy Wexler Gene Hunter, www.iwaswondering.org/nancy_ homepage.html. Site features information about Nancy Wexler's life, including pictures of the village in Venezuela and a comic book on the Hunt for Huntington's disease gene.

Stanford University: Narcolepsy, www.med.stanford.edu/school/ Psychiatry/narcolepsy/moviedog.html. Site provides information on narcolepsy research and includes links to videos on dogs and other animals with narcolepsy.

Your Genes Your Health, www.ygyh.org.

# Gene Therapy

## Could Gene Therapy Help Alcoholics to Stay on the Wagon?

*New Scientist*, June 2007

Many people of East Asian descent get very sick when drinking even a small amount of alcohol. These individuals carry a mutation in the gene encoding the enzyme aldehyde dehydrogenase (ADA), which not only causes the bad reaction to alcohol but also reduces the risk of becoming an alcoholic by more than two-thirds. Disulfiram (or Antabuse) is a common drug used to help alcoholics quit their addiction to alcohol. This drug blocks the activity of the same enzyme, aldehyde dehydrogenase, and it seems to help alcoholics stop drinking. However, for this treatment to be effective, the drug must be taken every day, and this is a very difficult challenge for the addicts to successfully meet.

A gene therapy approach might solve this problem because it provides a way to prevent expression of the aldehyde dehydrogenase gene for the long term. Scientists tested this idea by constructing a virus vector carrying an "antisense" DNA copy of the ADH gene. Once inside the cell, the antisense DNA base-paired to the ADH messenger RNA (mRNA) and formed an RNA:DNA duplex. This action selectively blocked translation of the ADH mRNA and prevented production of the ADH protein (enzyme). This therapy has great potential because it blocks enzyme activity just as if the patient had taken the disulfiram drug every day.

Scientists tested this ADH therapy on "addicted" lab rats bred to crave alcohol. One injection of the vector carrying an "antisense" DNA copy of the ADH gene into the "addicted" rats decreased the ADH enzyme activity in the liver by 80%. The rats that had previously craved alcohol drank 50% less for more than a month after the gene therapy treatment. (American Society of Gene Therapy, 2007)

## LOOKING AHEAD

Gene therapy is an innovative use of recombinant DNA technology that offers tremendous potential for the widespread treatment of many human genetic diseases and disorders. For many years, gene therapy has been the center of public controversy, and the early gene therapy trials with human patients suffered very serious setbacks, raising the distinct possibility that the sunny promises offered by the proponents of gene therapy will never be realized. Progress in gene therapy will always have technological challenges and will have to answer ethical, legal, and social questions. However, on the positive side, the gene therapy field has made some amazing advances not only in the research lab with animals but also in some clinical cases involving human patients. This chapter examines the history and current status of gene therapy and explores

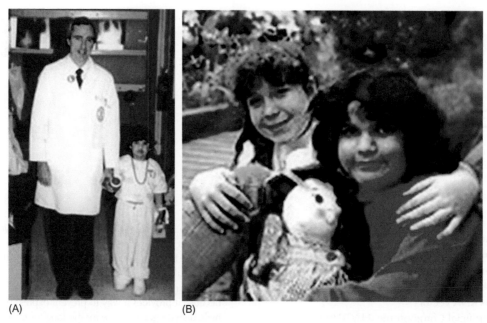

**FIGURE 11.1**  Ashanti and Cynthia were the first human gene therapy patients. (A) Four-year-old Ashanti DeSilva is holding the hand of her gene therapy doctor, Dr. W. French Anderson in 1999. (B) Ashanti (right) and Cynthia (left) visit the Cleveland Zoo together, doing well three years after receiving gene therapy treatment.

the applications of the principles of gene therapy to the successful treatment of human diseases.

On completing the chapter, you should be able to do the following:

- Understand how "normal" (wildtype) genes can be introduced into the cells of a patient through the use of vectors or other delivery methods.
- Explain what goes wrong in individuals who are deficient in a specific protein (enzyme), and outline a possible gene therapy treatment for that specific disease or disorder.
- Understand the characteristics of the different vectors used in gene transfer, and appreciate the mechanisms used by the different vectors during the gene therapy process.
- Describe how RNA interference (RNAi) technology is used as a gene therapy application to treat disease.
- Describe the reasons that different gene delivery methods were chosen to treat a brain disorder like Parkinson's compared to treating an eye disease that destroys the retina and causes blindness.
- Explain the medical risks and potential complications of gene therapy.

## INTRODUCTION

In September 1990, DNA history was made when two little girls named Ashanti and Cynthia became the first two people to receive gene therapy treatments for a genetic disease (Figure 11.1). The girls each inherited a mutant gene that causes a serious immune deficiency disease, leaving them unable to fight infections. This historic gene therapy treatment provided Ashanti and Cynthia with billions of treated cells carrying the normal (wildtype) gene and directing the synthesis of wildtype proteins. In this case the function of the wild-type proteins compensated for the nonfunctional mutant proteins produced in the disease cells.

The fundamental idea behind gene therapy seems almost too easy: just replace an altered (mutant) gene with the corresponding normal (wildtype) gene. In theory, when a disease is caused by a single gene mutation it should be straightforward to treat the disease with a wildtype gene. The defective, mutant gene in the patient's cells can potentially be rescued by introducing the wildtype gene, which makes the functional wildtype proteins that function properly in the cells. The ability to plan a successful gene therapy strategy depends on knowing the identity of the mutant gene that causes the disease. The scientists must learn about the specific gene expression patterns and biochemical processes in the healthy and disease cells in order to design the best approach for the gene therapy treatment, including the best way to deliver the therapeutic wildtype gene to the target cells in the body.

Despite the initial successful use of gene therapy to treat Ashanti and Cynthia (discussed later), much controversy surrounded gene therapy issues in the

1990s and continues today, with plenty of critics and proponents on both sides. Gene therapy still faces many challenges before it can become a widespread treatment for genetic diseases, but specific gene therapy approaches have been used as successful treatments for certain diseases and disorders. The potential of gene therapy should not be underestimated, especially in light of the exciting research advances in the field. This chapter covers the facts and controversies surrounding gene therapy and reviews the government regulations for the oversight of human gene therapy trials.

Advances in recombinant DNA technology solved many of the technical obstacles that have inhibited the clinical implementation of gene therapy treatments. The complete DNA sequence of the entire human genome helped to launch the fields of **genomics** and **bioinformatics** (Chapters 6 and 7) and flooded the scientific (and public) databases with extensive information about novel human genes. These studies have revealed that many new human genes are implicated in genetic diseases and are potential targets for gene therapy treatments. Different types of gene therapies are used to treat different genetic diseases and some gene therapy treatments have advanced to clinical trials with human patients (see Chapter 10).

In some ways, gene therapy is a relatively new field, but even so it has already had a large impact on many areas of science, medicine, and society. This chapter describes how clinical gene therapy trials are conducted to test some of the many candidates for gene therapy treatments. It is beyond the scope of this chapter to comprehensively report on all of the gene therapy activities currently ongoing around the world. Instead, the goal of this chapter is to learn about some of the most interesting and exciting gene therapy trials, including the different vectors used, and the cell targeting and delivery system used for successful gene therapy.

## GENE THERAPY: A METHOD TO RESCUE MUTANT GENES

Gene therapy is a medical process where the wildtype (normal) version of a gene is introduced into a patient's cells to treat the disease caused by the mutant form of the gene, which failed to function properly. A disease that is caused by a mutation in a "single gene" is a good candidate for successful gene therapy treatment because it is sometimes possible to replace the single mutant gene in the patient's cells with a normal (wildtype) copy of the same gene. When the function of normal proteins can compensate for the defective function of the mutant proteins, then the gene therapy strategy

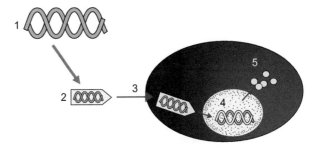

**FIGURE 11.2** Key steps in a gene therapy treatment. (1) Therapeutic (wildtype) gene (blue DNA helix). (2) Therapeutic gene (blue DNA) is inserted into a vector (yellow arrow). (3) The vector carrying the therapeutic gene is delivered into the host cell. (4) The vector DNA travels into the host cell nucleus. (5) The therapeutic gene is transcribed into mRNAs (not shown), and the mRNAs are translated to produce the therapeutic proteins.

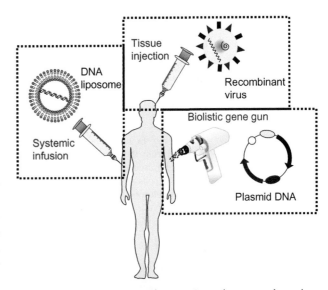

**FIGURE 11.3** *In vivo* gene therapy. Gene therapy performed on cells inside a patient's body using therapeutic wildtype gene DNA.

of transplanting a wildtype gene into disease cells is likely to succeed (Figure 11.2).

There are two main approaches used in gene therapy depending on whether the diseased cells are treated inside the body (**in vivo**) (Figure 11.3) or are removed from the body for treatment (**ex vivo**) and then returned to the patient (Figure 11.4). *In vivo* gene therapy methods must use gene delivery systems that accurately target the therapeutic genes into specific cells in the patient's body (Figure 11.3). If the diseased cells can be removed from the patient, they will be treated with therapeutic DNA in the lab and then the genetically altered cells are returned into the body of the same patient (Figure 11.4).

One of the biggest challenges in gene therapy is the need for gene delivery vehicles called vectors that can accurately target the therapeutic gene into the correct cells in the patient's body. Gene therapy vectors are

Gene therapy treatments are designed to rescue the defect in the cells carrying the mutant gene by introducing the wildtype gene into the cell to make functional proteins that can rescue the defective function of the mutant protein.

usually made of functional DNA molecules that deliver the therapeutic gene cargo into the disease cells in the body. Vectors have been derived from many sources including RNA and DNA viral genomes, circular double-stranded DNA bacterial plasmids, and artificial eukaryotic chromosomes. Recently a new delivery system was developed involving carbon nanotubes that carry a DNA or drug cargo to the target cells (see Chapter 13).

## Virus DNA Vectors

Successful gene therapy requires a vector or other delivery system to carry the therapeutic gene into the diseased cells. Many early gene therapy experiments depended on vectors made from the linear adenovirus double-stranded DNA genome. Adenoviruses have linear DNA genomes covered in a capsid made of virus proteins (Figure 11.5). Adenoviruses cause upper respiratory tract infections in people with low immunity. However, in the absence of the viral capsid, the adenovirus DNA genome alone is not infectious to humans. To introduce an adenovirus vector into cells, the vector DNA is packaged into a capsid made of viral proteins, so it can be delivered into the nucleus with high efficiency (Figure 11.6). Different vectors have been derived from adenovirus genomes and

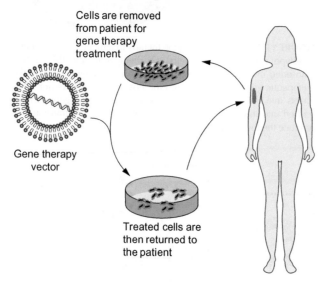

**FIGURE 11.4** *Ex vivo* gene therapy. *Ex vivo* gene therapy is performed on disease cells removed from the patient. Different gene therapy vectors and methods can be used to treat the cells depending on the genetic disease. After the cells are treated with the appropriate gene treatment, the cells are returned to the patient.

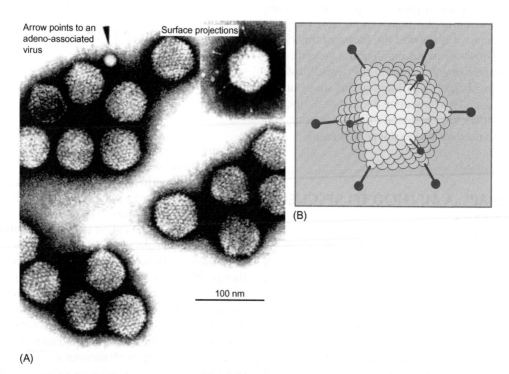

**FIGURE 11.5** Virus genomes and viral vectors can be packaged into capsids made of viral proteins. (A) The electron micrograph (EM) shows the protein structures of adenovirus capsids and the much smaller adeno-associated virus (arrow). (B) The protein projections on the surfaces of the virus capsids are barely visible in the EM but are shown on the diagram of the adenovirus capsid structure.

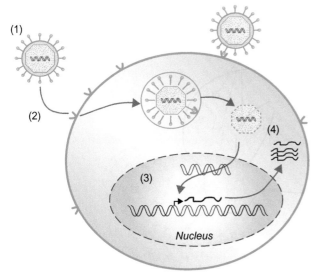

**FIGURE 11.6** Therapeutic genes in viral DNA vectors are delivered into cells. (1) The wildtype (normal) gene is carried by the viral vector, which is packaged with virus capsid proteins into recombinant gene delivery system. (2) The recombinant virus carrying the wildtype gene is used to infect the diseased cells. (3) The viral vector and wildtype gene are inserted (integrated) into the chromosome DNA in the host cell (patient's) genome. (4) The therapeutic wildtype gene in the cell is expressed as mRNAs and wildtype (normal) proteins.

technological improvements over time have made the modern adenovirus vectors much more efficient and safer to use. The earliest adenovirus vectors carried the risk of serious side effects in animals and people.

Once the DNA vectors and the genes have entered the target cells in the body and penetrated the nucleus, the therapeutic gene DNA is copied into mRNAs by the host cell enzymes. Then the mRNAs are processed and transported into the cytoplasm where they are translated into proteins that provide therapeutic benefits (see Figure 11.6). Scientists must consider many factors when deciding which disease treatments might benefit from using a gene therapy approach. Of course, the disease must result from a DNA mutation in a known gene, and a wildtype, unmutated copy of the gene must be identified and available for cloning (insertion) into an appropriate vector. Each gene therapy treatment includes a strategy to deliver the wildtype gene(s) into the diseased cells.

Inside the nucleus, the fate of the vector DNA (and the therapeutic gene) depends on the specific properties of the vector used. In some cases, successful treatment requires that the therapeutic genes be actively expressed for prolonged periods of time in the host cell nucleus. To accomplish this, scientists often use a type of vector that **integrates** (inserts) directly into the DNA in the chromosome and becomes a permanent part of the host genome. The vector DNA integrates

into the genome through homologous recombination between DNA sequences in the vector and DNA in the host chromosome (see Chapter 6). Once integrated, the vector and the therapeutic gene replicate when the genome duplicates and are distributed to offspring cells along with the native chromosomes during cell division.

Many different virus genomes in addition to adenoviruses have been modified for use as animal vectors in human gene therapy trials. Once inside the cells the vectors derived from circular double-stranded DNA plasmids do not persist for long periods in the nuclei. Even though the circular DNA vectors replicate (duplicate) inside the cell, they are not equally distributed (**segregated**) when the cells divide and make more cells. As a result, DNA vectors that do not insert into the genome are chosen when it is appropriate to provide only transient expression of the therapeutic gene in the host cells.

> Choosing the best gene therapy vector to use for each treatment depends on understanding the basic biological mechanisms responsible for the disease process, including the tissues and cells involved and the timeline of disease progression.

## Retrovirus RNA Vectors

Some vectors designed for use in human gene therapy have been derived from **retrovirus RNA genomes.** These viruses get their name from the clever strategy that they use to infect and take over the biosynthetic machinery in the cell ("retro-" = reversal). The retrovirus infection depends on reversing the usual flow of genetic information in the cell, from RNA to DNA to protein instead of the traditional flow of information: DNA genes are copied into RNA transcripts and translated into proteins (see Chapter 3). This feat is accomplished by a special enzyme called reverse transcriptase, which has the unusual ability to copy the retrovirus RNA into double-stranded DNA (Figure 11.7). In the cell nucleus the reverse transcriptase enzyme makes a double-stranded DNA copy of the retrovirus RNA genome, which is then inserted (integrated) into the host cell chromosome to make a **provirus** genome. The genes encoded by the provirus genome are expressed as mRNAs and are translated into the proteins that package the new retrovirus RNA genome. The virions also contain reverse transcriptase enzyme proteins ready for use in the next infected cell.

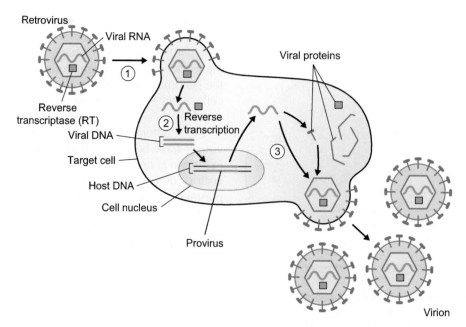

**FIGURE 11.7** Retrovirus life cycle. (1) The retrovirus gains access to a cell when a glycoprotein on the surface of the virus capsid binds to a receptor protein on the surface of the cell. The envelope surrounding the virus particle fuses with the cellular membrane, and empties the retrovirus RNA genome into the cytoplasm of the cell. The retrovirus virion carries reverse transcriptase enzyme proteins in addition to the RNA virus genome. Reverse transcriptase enzymes have the unusual activity of copying RNA into DNA, instead of DNA into RNA, the normal gene express ion pathway in the cell nucleus. (2) In the nucleus the viral reverse transcriptase enzyme copies the retrovirus RNA genome into double-stranded DNA, which is then inserted (integrated) into the host cell genome. (3) The virus genes encoded by the integrated retrovirus genome (the provirus) express viral mRNAs that are translated into viral proteins. The capsid proteins package the new retrovirus RNA genomes into virions, which acquire a membrane envelope with viral proteins as they exit through the cell's plasma membrane.

To make the retrovirus vectors as safe as possible, researchers used standard recombinant DNA methods (see Chapters 4 and 5) to remove the dangerous genes from the retrovirus genome and create vectors that lack the three retroviral genes (pol, gag, env) (Figure 11.8). These virus proteins package the genome (or vector) into viral capsids (see Figure 11.7). Without these capsid proteins the vector cannot be packaged into a virion and cannot spread to other cells.

The development of safe and effective gene delivery options has been a major obstacle to the routine use of gene therapy in medical treatments. Most of the early gene delivery systems rely on DNA vectors that were derived from viral genomes. Viral delivery systems are often more efficient than nonviral methods, because the viral vectors can be packaged into protein capsids that enter the cells using the efficient pathway normally used by the invading virus. Viral vectors are modified viral genomes, which can sometimes cause serious adverse reactions in patients. Or the viral vector might insert accidentally into an essential gene in the chromosome, causing a mutation that might be lethal to the patient. There is also the potential risk that viral vectors might

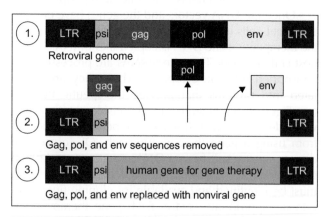

**FIGURE 11.8** Building a retroviral vector. (1) The wildtype retrovirus genome contains viral genes (gag, pol, env) and has long-terminal repeat (LTR) regulatory sequences located at the ends of the linear RNA genome. (2) To make the retroviral genome safe for use as a gene therapy vector, the genes that the virus uses to make more viruses during an infection were removed (deleted) from the retroviral genome (gag, pol, env). Without these three genes and their proteins, the retrovirus genome is not infectious. (3) The therapeutic gene needed for the gene therapy was inserted between the LTR sequences at each end of the retrovirus vector. Retrovirus gene therapy vectors retain the ability to integrate into the host cell chromosome, where the vectors can remain integrated in the genome for years without becoming infectious.

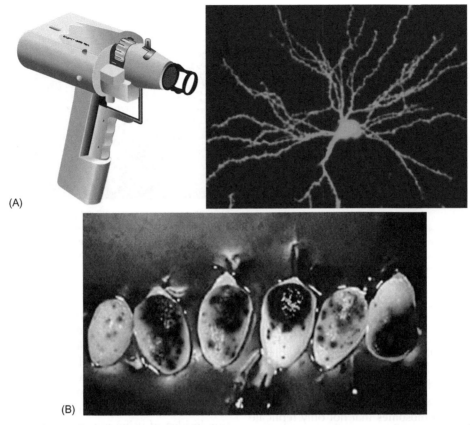

(A)

(B)

**FIGURE 11.9** Cells can be transformed using DNA fired by a gene gun. (A) DNA encoding a green fluorescent protein (GFP) gene was introduced into a nerve cell in the brain causing the body of the nerve cell and its extended axons to fluoresce with a green color. (B) Rice embryos were genetically modified by DNA from a gene gun. The transferred DNA carried with it a "marker" gene that was expressed in the cells, and caused the rice embryos to have a blue color (see Chapter 15).

spread to parts of the body outside the gene therapy treatment zone. Many scientists and biotechnology companies have developed nonviral gene delivery systems to use as possible alternatives to viral vectors for gene therapy, including **electroporation**, which uses an electric field to drive DNA into the cells, and **gene guns**, designed to shoot DNA projectiles directly into the nuclei of the cells to be treated. Gene guns have also been used to successfully propel genes into the subcellular organelles, including mitochondria and chloroplasts (plant cells) (Figure 11.9) (see Chapter 15). **Liposome fusion** is another approach to nonviral gene transfer. Liposomes are hollow spheres surrounded with membranes made up of fatlike molecules called phospholipids (Figure 11.10). The vector DNA and therapeutic gene is inserted into the liposome spheres (vesicles), which are then mixed together with the diseased cells to be treated. The liposomes fuse with the plasma membrane of the cell, releasing the DNA into the cytoplasm. By an unknown mechanism, the vector DNA is transported into the nucleus where the therapeutic gene is expressed (Figure 11.11).

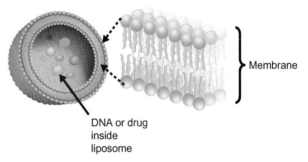

Membrane

DNA or drug
inside
liposome

**FIGURE 11.10** Liposomes can carry DNA genes. Like eukaryotic cells, liposomes are surrounded by phospholipid bilayer membranes. Liposomes carrying therapeutic DNA genes or drugs can fuse with the plasma membrane of a eukaryotic cell to deliver the contents inside the cell.

Viral vectors are the most common vehicles used to deliver therapeutic genes, but new approaches are under development using nonviral methods designed to improve delivery and accuracy, and that also offer ways to control expression of therapeutic genes in the disease cells.

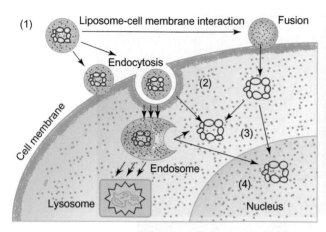

FIGURE 11.11 Liposomes deliver genes into the cell nucleus. (1) Liposome with DNA inside (DNA-lipid complex). (2) Cell takes up the liposome-DNA complex by endocytosis. (3) DNA-lipid complex enters cell nucleus. (4) Therapeutic gene is expressed in the nucleus.

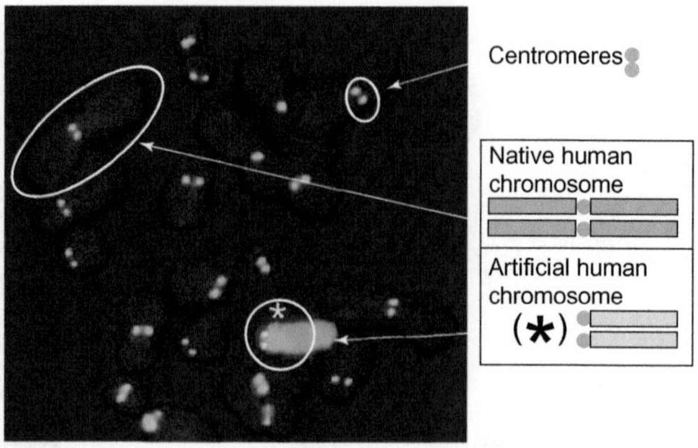

FIGURE 11.12 Human artificial chromosome. An artificial human chromosome (*) is shown among many native human chromosomes. The centromere regions of the chromosomes, including the artificial chromosome, show up as two bright dots and the artificial chromosome is a light green color.

## Human Artificial Chromosome (HAC) Vectors

Artificial chromosome vectors are actually derived from native chromosome DNA sequences and they replicate in the nucleus alongside the native chromosomes. These linear double-stranded dsDNA vectors mimic the cell's chromosomes and assemble with histone proteins to make linear **mini-chromosomes** that not only replicate (duplicate) just like the native chromosomes, but they also segregate properly during cell division, insuring proper gene inheritance by offspring (progeny cells).

Perhaps the most sophisticated gene therapy vectors ever developed are human artificial chromosomes (HAC). These linear vectors contain all of the DNA elements necessary to function as a native human chromosome, only are much shorter. In 1997, researchers at Case Western Reserve University assembled an entire human chromosome from DNA fragments made in the lab. Each chromosome vector included a **centromere** to ensure that the vector DNA can attach to spindle fibers and be inherited properly, a **DNA replication origin** to start DNA replication during cell division, and **telomere DNA** at both chromosome ends. HAC vectors have no upper limit on the length of DNA that can be inserted between the two telomere ends, because the artificial chromosomes are not restricted by the need to be packaged into virus capsids. These vectors also carry the DNA regulatory elements necessary to control expression of the therapeutic gene once inside the disease cells. Imaging techniques offer one way to monitor the artificial chromosome vectors inside the cells (Figure 11.12).

## POSSIBLE RISKS: THE HUMAN SIDE OF GENE THERAPY

All medical treatments have inherent risks, and the development of new drugs and new treatments such as gene therapy carries special risks for the patient. Processes such as **informed consent** were put in place to protect the many people who volunteer to participate in clinical trials of all kinds. Jesse Gelsinger had just turned 18 years old when he volunteered for a gene therapy trial in 1999 (Figure 11.13). A look at the circumstances of Gelsinger's case shows how the simple idea of using a good gene to replace a mutant gene is actually a difficult goal, not only from the perspective of the science involved, but also from the standpoint of human motives. Gelsinger died as a result of gene therapy, but Gelsinger's case also changed the way that federally supported human gene therapy trials are reviewed and regulated for the safety of the patient.

Gelsinger was fortunate to have the support of his family when he was diagnosed with a nonfatal form of a rare genetic disorder called ornithine transcarbamylase (OTC) deficiency in 1983. Lack of the OTC enzyme interferes with the ability of the liver to metabolize ammonia, causing toxic buildup of ammonia in the body. Babies born with the severe form of OTC deficiency rarely survive. As Jesse Gelsinger grew up, he struggled with the restrictions of his disease, but for the most part, Gelsinger did well. He had times when OTC made him sick, but Gelsinger persisted. He reached his teen years by managing his disease with a strict low-protein diet (to cut down on the amount of ammonia in his body) and by taking more than 30 medications every day.

In 1998, Gelsinger's specialist told his family about a clinical OTC gene therapy trial under way at the University of Pennsylvania in Philadelphia. Gelsinger was very interested, but he could not participate until age 18. That year he again became very ill with OTC and was hospitalized in a coma. Gelsinger finally recovered, and in May of 1999 he graduated from high

**FIGURE 11.13** Meet Jesse Gelsinger. Jesse Gelsinger had just turned 18 years old when he volunteered to participate in a gene therapy clinical trial for a disease called OTC. Gelsinger died a few days after the start of the gene therapy trial, but he also helped to change the way human gene therapy trials are regulated and supervised in the United States.

school. That summer was Gelsinger's 18th birthday, and with the support of his family, he decided to volunteer for the OTC gene therapy trial.

The Gelsinger family flew to Philadelphia to be with Jesse when he was screened in the hospital for acceptance into the OTC gene therapy trial (called the Batshaw-Wilson study). Gelsinger was very excited when he learned that he had qualified for the gene therapy trial to be headed by top physician Dr. James Wilson and performed at one of the premier institutions doing gene therapy at the time, the Institute for Human Gene Therapy at the University of Pennsylvania. One of Wilson's colleagues explained how the gene therapy technique would work. While he was under sedation, two catheters would be placed into Gelsinger's liver; one in the hepatic artery to inject the viral vector into the liver and another to monitor the blood leaving the liver to be sure that the vector was being absorbed by the cells in the organ. In addition to the risk involved, the gene therapy treatment would not provide a long-term benefit to Gelsinger and the other volunteers. Even if the therapeutic genes worked properly in this trial, the effect would be transient because the body's immune system would attack and kill the virus vector in four to six weeks after treatment.

On September 13, 1999, Gelsinger's hepatic artery was injected with the genetically altered adenovirus. Twenty hours into the therapy, Gelsinger developed jaundice, a yellow tinge in the eyes and skin that is often a sign of liver failure. Some of the animals had exhibited the same complication in earlier testing of this gene therapy vector protocol. Soon Gelsinger entered a coma, then he suffered multiple organ system failure and was placed on life support until he died four days later.

Seven years before Gelsinger volunteered for the OTC clinical trial, the head of the trial, Dr. James M. Wilson, had started a for-profit, private company named Genovo, Inc., with Wilson as the major shareholder. During the 1990s, Genovo, Inc. gave almost $5 million each year to the Institute for Human Gene Therapy at the University of Pennsylvania, at a time when Wilson was its director. In 1993, Mark Batshaw and Wilson conducted lab experiments on OTC-deficient mice using an adenovirus vector to carry the wildtype OTC gene. These animal studies predicted future success with the gene therapy treatments in humans, but the data also revealed significant safety problems with the OTC gene therapy protocol. In a separate experiment, three monkeys died from a treatment using an adenovirus vector similar to the vector used in the human trials, and additional animals used in studies suffered severe hepatitis and liver failure after exposure to adenovirus vectors. Genovo, Inc. produced the adenovirus vector used in the human OTC trial.

At least five years before Gelsinger's death, the gene therapy community appeared to be aware of possible dangers posed by the adenovirus vector. In 1995, when the Recombinant DNA Advisory Committee of the National Institutes of Health approved the Batshaw-Wilson OTC gene therapy protocol for use on human subjects, but two dissenting experts warned that the trial was too risky to include volunteers with the non-fatal form of OTC. Eventually 19 OTC-deficient adults, including Gelsinger, enrolled in this gene therapy trial, and Gelsinger received the highest possible dosage of the adenovirus vector. Since Gelsinger's death in 1999, there have been several additional gene therapy fatalities in the United States. Questions about the safety of gene therapy rose again in the summer of 2007 after the death of 36-year-old Jolee Mohr, who took part in an experimental gene therapy trial to treat rheumatoid arthritis.

Independent investigators concluded that Jesse Gelsinger's death was partly due to the failure of the scientific team to get appropriate informed consent from the patient and his family. Informed consent is designed to ensure that patients have all of the information that they need to make the best possible decision about whether or not to participate in a specific clinical trial. In January 2008, a top scientific journal in the field, *Human Gene Therapy*, published articles emphasizing the importance of informed consent in gene therapy trials, stressing that it is imperative for patients to understand the risks and benefits of a clinical trial. Ironically, the authors of these articles included some of the key doctors involved in Gelsinger's gene therapy trial; the editor-in-chief of *Human Gene Therapy* was Dr. James Wilson and author bioethicist Arthur Caplan was also a member of Gelsinger's medical team at the University of Pennsylvania. Both Wilson and Caplan were defendants in the lawsuit brought by the Gelsinger family, which was settled out of court in 2000.

## SUCCESSFUL GENE THERAPY TREATMENTS FOR HUMAN GENETIC DISEASES

Some of the gene therapy methods used to successfully treat human diseases are described here to help illustrate the different scientific approaches, technologies, and people involved in the field of gene therapy. Many human diseases not mentioned here are good candidates for gene therapy treatments, and many research and clinical studies are currently under way.

A different type of gene therapy approach was used to treat melanoma, a virulent form of skin cancer. Special cells called tumor-infiltrating lymphocyte cells (lymphocytes that attack tumors) (TILs) were isolated from the patients and treated in the lab with a gene encoding, the tumor necrosis factor (TNF) protein. The cells begin to make this anticancer protein and are then returned into the patient. The production of TNF will inhibit the growth of the cancer cells. More about cancer cells and the new approaches to treating cancer in Chapter 9.

## Cardiac Gene Therapy Improves Heart Function

The gene encoding the vascular endothelial growth factor (VEGF) protein has been used in successful cardiac gene therapies. VEGF is a naturally occurring protein that can trigger the growth of new blood vessels in the body (see Chapter 9). The VEGF DNA can be injected directly into the heart safely and was highly effective in treating advanced coronary heart disease. The VEGF protein is expressed in the sick heart muscle cells, triggering the growth of new blood vessels in the regions of the heart containing the oxygen-starved cardiac cells. Instead of using a virus genome as a vector to carry the gene DNA into cells, this form of cardiac gene therapy uses a DNA plasmid to carry the VEGF gene into the heart tissues where it temporarily stimulates the development of new blood vessels.

## SCID and Adenosine Deaminase (ADA) Deficiency: Bubble Boy Disease

In the United States, about 1 in 100,000 babies are born with some form of severe combined immunodeficiency disease (SCID), which is caused by a mutant gene that makes a faulty adenosine deaminase (ADA) enzyme. Normally the ADA enzyme degrades DNA and RNA molecules into components that are recycled by the cell for other uses. The ADA enzyme is produced in specialized T lymphocytes (T cells) in the immune system. T cells are essential to the body's immune system because they control the activity of B lymphocytes, the cells that make the antibodies that fight infectious agents. Individuals without the ADA enzyme cannot mount an

---

**Box 11.1   The Impact of Jesse Gelsinger's Life**

Jesse Gelsinger's death was the first report of a human fatality known to be directly linked to a gene therapy trial. His death served as an alert to warn scientists everywhere that serious and possibly fatal side effects can accompany the use of some vectors in gene therapy treatments.

Gelsinger's case helped bring attention to the potential dangers when the scientists, doctors, and staff fail to follow safety standards when doing human studies. Formal investigation into the circumstances of Jesse Gelsinger's death revealed serious problems with the OTC gene therapy trial that included reports of sloppy practices used to select volunteers, failure to confirm that the volunteers met the necessary medical criteria for enrollment, and soliciting volunteers using inadequate informed consent. Investigation officials found that the doctors had removed language from the consent forms describing the animal deaths and sickness earlier in the research, including previous studies

on the adenovirus vector that noted cases of liver failure in animal subjects. They also failed to report that two volunteers in a previous OTC study had suffered severe reactions at vector dosages lower than the dose Gelsinger received. The doctors apparently continued the OTC trial using higher vector dosages without consulting the Food and Drug Administration (FDA).

Sanctions were brought against the doctors and researchers involved in Gelsinger's case, prohibiting them from conducting human gene therapy trials in the future. In 2005, when Wilson settled with the U.S. Department of Justice, he agreed not to conduct human clinical trials for five years. As a direct result of Gelsinger's case, the U.S. Department of Health and Human Services created the Gene Therapy Clinical Trial Monitoring Plan, which dramatically increased the level of oversight and scrutiny for the entire gene therapy testing process in humans.

effective immune defense against germs and will always be at risk for life-threatening infections.

The initial ADA gene therapy performed on Ashanti and Cynthia in 1990 were *ex vivo* treatments, which means that the lymphocyte cells were removed from each patient and treated in the lab with billions of retrovirus vectors carrying wildtype ADA genes. Then the lymphocyte cells treated with the normal ADA genes were returned to the patient's body (Figure 11.14). The entire gene therapy procedure was repeated several weeks later because T lymphocyte cells normally survive for only a short time in the human body (whether or not they have undergone gene therapy). The ADA

gene therapy administered to the two girls in 1990 was successful, although there is still some controversy because the girls received an ADA enzyme replacement treatment in addition to the gene therapy.

This type of immune deficiency disorder first came to the public's attention in the 1970s through the story of David Vetter, a boy born with SCID. Vetter lived in a sterile plastic bubble for most of his 12 years, until he died in 1984 after an unsuccessful bone marrow transplant (Figure 11.15). The lives of David Vetter and Ted DeVita sparked public interest and prompted a TV movie made in 1976 called *The Boy in the Plastic Bubble* starring John Travolta. Since that time,

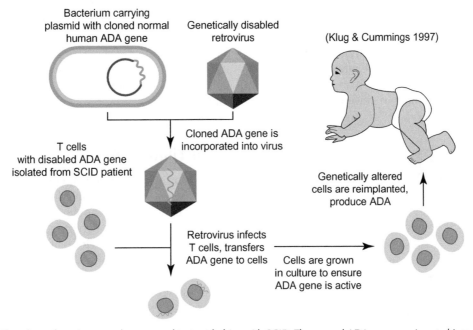

**FIGURE 11.14** Flowchart of *ex vivo* gene therapy used to treat babies with SCID. The normal ADA gene was inserted into a retroviral vector and packaged into a viral capsid. T cells with the mutant ADA gene were taken from the babies with SCID and treated with the ADA retrovirus vector, which transferred the normal ADA gene into the T cells from the babies. The T cells expressed the wildtype ADA proteins and were transplanted back into the babies where they continued to produce functional ADA enzymes.

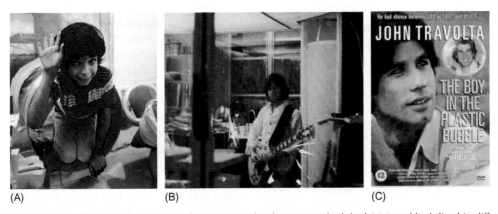

**FIGURE 11.15** David was the first "Bubble Boy." David Vetter (A) and Ted DeVita (B) both had SCID and both lived in different versions of "sterile bubbles" all their lives, waiting for the development of an effective treatment for SCID. (C) The amazing lives led by David and Ted in their respective sterile environments inspired the movie *The Boy in the Plastic Bubble* starring John Travolta (1976).

the progress made in organ transplant technology has greatly improved the effectiveness of bone marrow transplants as treatments for SCID and for other medical conditions caused by the failure of a patient's bone marrow to produce various types of blood cells (Figure 11.16). To be successful, the donated bone marrow must come from a living genetic relative of the patient (such as a sibling) or from an unrelated donor who is "matched" to the patient through special blood tests using HLA (Human Leukocyte Antigen) genetic typing. The risk of **rejection** (host versus graft disease) occurs whenever a patient receives a transplant of cells or tissues from a genetically unrelated donor. Rejection is a serious complication that can sometimes be suppressed by medications. These powerful drugs come with dangers as well, because they significantly reduce the body's ability to fight infections.

Almost 15 years after David Vetter died, a baby named Owen was living in the United Kingdom in a sterile bubble, waiting for a bone marrow transplant to cure his SCID. Owen finally received a bone marrow transplant and even though the bone marrow came from an unrelated donor, the donor was still a close genetic match, and the transplant was successful. Owen survived and remarkably, at age 3 years, he was at center stage once again when he became the first child to donate his own bone marrow to cure his infant brother, Niall, also born with SCID.

In 1995, infants diagnosed with ADA were treated for the first time with stem cells collected after birth from their umbilical cords. In the lab the umbilical cord cells were treated with vectors carrying the wildtype ADA genes. A few days later, each ADA infant received an infusion of his or her own treated umbilical cord cells, which now carried both the mutant and normal ADA genes. In this case the long-range treatment goal was to create a permanent population of ADA-producing cells in the infant's bodies. Surprisingly, after only a single gene treatment, the immune cells in both children began to produce wildtype ADA enzymes, and have continued to do so for several years.

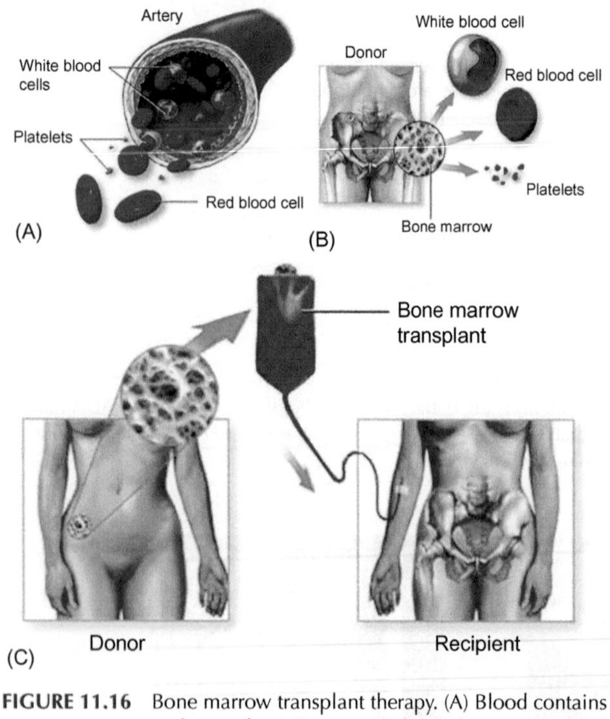

FIGURE 11.16 Bone marrow transplant therapy. (A) Blood contains many components that perform important jobs such as transporting oxygen and nutrients to body tissues and removing carbon dioxide and producing cells essential to the immune system. Blood factors are required for clot formation and help to prevent and fight infection. (B) Bone marrow is a tissue found inside the long bones in the skeleton, which produces blood cells (red blood cells, white blood cells, and platelets). Red blood cells carry oxygen in the blood, white blood cells fight infection, and platelets function in blood clotting. (C) A bone marrow transplant is performed to treat patients who have serious forms of their disease.

## Cystic Fibrosis Disease

Cystic fibrosis is the most common life-threatening genetic disease in the United States, affecting about 30,000 children and adults. Cystic fibrosis occurs when a person inherits two mutant cystic fibrosis transmembrane receptor (CFTR) genes (one from Mom and one from Dad). The normal CFTR is a large membrane protein that regulates the concentrations of salt ions in the cells (see Chapter 10). People with cystic fibrosis disease can make only the defective mutant cftr proteins in their cells, causing chloride ions to build up in the cells lining the organs and vessels in the body. About 10 million Americans are unknowing carriers of a mutant cystic fibrosis gene; they have both the wildtype (CFTR) and the mutant (cftr) genes but do not suffer disease symptoms. Genetic carriers of cystic fibrosis risk having a biological child who inherits two copies of the defective cftr gene and becomes sick with cystic fibrosis disease, including the risk of serious respiratory tract infections and reduced life span (see Chapter 10.)

Although the cftr mutations affect all cells in the body, the defect is most damaging to the lung cells because the elevated level of the salt ions causes water to enter the cells, creating sticky mucus in the airways that makes it difficult to breathe. Similar problems in cystic fibrosis disease cause severe damage to the pancreas cells. Scientists first focused on developing cystic fibrosis gene therapy treatments that would work on lung and pancreas tissues.

In initial gene therapy tests, scientists took cells from the lungs of a cystic fibrosis patient and grew them in tissue culture in the lab. They used a viral vector to insert the wildtype CFTR gene into the cultured human lung and pancreas cells from the cystic fibrosis patient (cftr/cftr). Before treatment, the lung cells contained high levels of salt as expected for tissues from a cystic fibrosis patient. After the gene therapy treatment, the cultured lung and pancreatic cells expressed the wild-type CFTR gene and proteins, and both types of cells showed a large decrease in internal salt levels.

New drugs and therapies for use in treating humans are evaluated in a federal system of clinical trials with human patients. The approval process involves four phases of clinical trials that can take several years to complete. Drugs or therapies that successfully pass the tests in phases I, II, and III are usually approved for use by the general public. Extensive preclinical studies must be completed before the clinical trials start. Phase I trials begin to test human subjects, usually a group of 50 to 80 healthy volunteers. The phase I trials are not designed to test how well the drug works as a treatment for a disease. Phase I trials are designed to test the safety of the drug and also to determine the appropriate therapeutic dose of the drug. After the drug passes the safety tests in phase I trials, the phase II trials involve larger groups (20 to 300) of volunteers and patients and are designed to assess how well the drug works at the therapeutic dose(s). Phase III involves multicenter trials with large groups of patients (300 to 3000 or more) that are designed to provide a clear assessment of the drug's effectiveness compared to the current standard treatment. In preparation for the large-scale phase III cystic fibrosis gene therapy trial scheduled to begin in 2009, scientists wanted to extend the length of time that the therapeutic CFTR gene was expressed in the cystic fibrosis cells to increase the overall effectiveness of the treatment (see Chapter 10).

---

### Box 11.2 Restoring Sight to Blind Dogs and Humans

For decades researchers have worked toward the goal of treating inherited diseases that cause childhood blindness. It took longer than 15 years for scientists to identify the mutant gene involved in Leber's congenital amaurosis (LCA), which causes vision to rapidly deteriorate early in life. Interestingly, this form of inherited blindness is caused in humans and dogs by inheriting the mutant form of either the human or canine RPE65 gene (Figure 11.17). The RPE65 proteins made in humans and dogs have similar structures and perform similar functions in their respective retinal cells.

In humans and dogs, the mutations that alter the RPE65 proteins invariably damage vision and can cause complete blindness. This reflects the critically important role of the RPE65 protein in recycling retinol, a molecule that is required for the retinal cells to capture light. The image seen by the eye is interpreted by the visual center of the brain not by the cells in the retina. The discovery that mutations in the RPE65 gene causes dogs and humans to go blind was a very important step because it allowed researchers to test potential treatments on the blind dogs before using the therapy on humans with LCA.

Scientists at the University of Pennsylvania used gene therapy to treat blind dogs with LCA that had inherited the mutant canine RPE65 genes. The eye was ideal for the first gene therapy treatments because the retina is easily accessible for injections to deliver vectors and genes to the target cells (Figure 11.18). For the LCA gene therapy test on dogs, the researchers injected the normal, wildtype RPE65 gene DNA directly into the eyes of dogs that had been blind since birth. Amazingly, after the treatment with the wildtype RPE65 gene the dogs could successfully navigate their surroundings without difficulty.

(A)                              (B)

**FIGURE 11.17** Mutations in the RPE65 gene cause loss of vision in humans and dogs. (A) A retinal fundus photograph from a Briard breed of dog with an inherited eye disease. (B) Lancelot, a Briard-mix dog, was born blind but his vision was restored by gene therapy that provided wildtype (normal) RPE65 gene and protein. Gene therapy involving dogs holds out hope for curing blindness in people with LCA.

The dogs' behavior implied that transfer of the wildtype RPE65 gene into the retina cells had restored the dog's vision, but it was important for the scientists to confirm this result by finding with evidence that the dogs' brains had actually responded to having restored sight. Using functional magnetic resonance imaging (MRI) the scientists discovered that the RPE65 gene therapy treatment dramatically changed how the dogs' brains responded to light and the gene treatment had restored function to the visual center of the brain in a dog that was blind since birth (Figure 11.19). The recovery of visual brain function persisted in one dog for at least two-and-a-half years

**Box 11.2    continued**

**RETINAL GENE THERAPY**

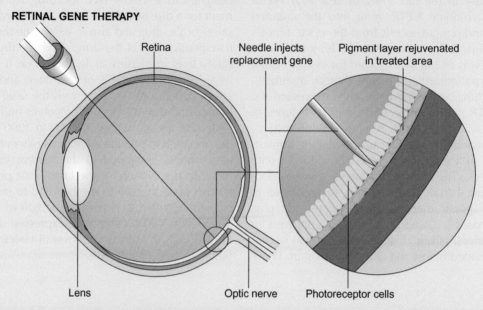

**FIGURE 11.18**   Gene therapy to treat inherited blindness. Eye diseases are good candidates for gene therapy treatments because doctors have relatively easy access to the eye for direct application of the therapeutic gene DNA to the target cells that require the gene therapy treatment.

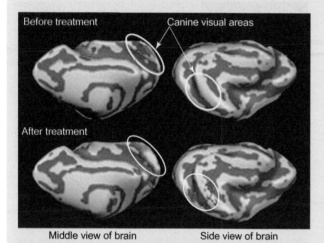

**FIGURE 11.19**   MRI shows activity in visual region of blind dog's brain. Activity in a dog's brain is compared before and after gene therapy treatment for the eye disease LCA. Functional MRI measures the activity of the brain cells in the part of the dog's brain involved in vision. After gene therapy treatment for LCA, the previously blind dogs could maneuver through a maze. Evidence from functional MRI studies supported the conclusion that dogs blind since birth could see after the LCA gene therapy treatment.

after gene therapy. These amazing studies indicate that blindness in infancy does not permanently alter the structure and function of the blind brain. When the retina can again detect light, the brain can properly process the information and restore vision.

Based on the success of animal studies on Leber's congenital amaurosis (LCA), the first human gene therapy trial to treat inherited blindness due to LCA is now under way. A team in London injected the wildtype RPE65 gene (carried on a viral vector) directly into the retinal cells of 12 people who were losing their vision due to inherited LCA. The doctors injected the normal RPE65 gene DNA directly into the retinal cells, which produced wildtype RPE65 proteins to compensate for the defective protein caused by the inherited mutant RPE65 gene. A similar gene therapy strategy could potentially be used to treat about 100 different inherited diseases that affect vision. Based on the success of LCA animal studies, scientists are hopeful that this therapy will also work, but it will be some time before the results of this human gene therapy trial are known.

## Parkinson's Disease

More than 500,000 Americans have Parkinson's disease, a nerve disorder in the brain that is caused by a decrease in dopamine, a key **neurotransmitter** in the brain. Neurotransmitters propagate nerve signals across the open spaces between two adjacent neurons called

synapses (Figure 11.20). Without sufficient dopamine, nerve signal transmission is interrupted at the synapses and the neurological symptoms characteristic of Parkinson's disease appear, including tremors and rigidity in the limbs, slow movements, impaired balance, and poor physical coordination. Medications

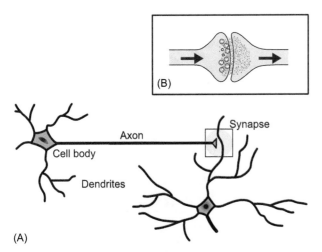

(B)

Synapse

Axon

Cell body

Dendrites

(A)

**FIGURE 11.20** Neurotransmitters carry impulses across the gap between adjacent nerves. (A) The diagram shows two nerve cells that send signals to each other over the synapse or gap between the two nerves. (B) When the nerve impulse reaches the end of the first nerve, neurotransmitters form and are released into the gap between the nerves. The neurotransmitters carry the signal across the synapse gap to the next nerve cell (nerve impulse is traveling left to right).

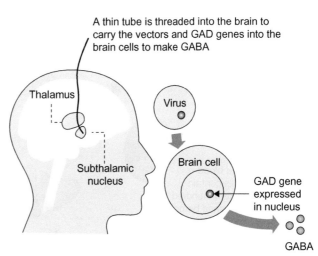

A thin tube is threaded into the brain to carry the vectors and GAD genes into the brain cells to make GABA

Thalamus

Virus

Subthalamic nucleus

Brain cell

GAD gene expressed in nucleus

GABA

**FIGURE 11.21** First human gene therapy to treat Parkinson's disease. A thin tube is threaded through a hole in the skull, deep into the brain to contact the subthalamic nucleus in this Parkinson's patient. A virus vector carrying the GAD gene is injected into the brain through the tube. The treated brain cells express the GAD protein and produce GABA, which calms the excess nerve activity in Parkinson's patients.

that replace or mimic dopamine are available to treat Parkinson's disease, but relief is temporary because the medications lose effectiveness and the Parkinson's disease symptoms get progressively worse.

Individuals with Parkinson's disease have extremely overactive nerve cells in the walnut-size subthalamic nucleus in the brain, causing further decrease in the production of dopamine. The first human gene therapy trial to treat Parkinson's disease was conducted by a scientific team from Cornell University, which constructed a virus vector carrying a gene for glutamic acid decarboxylase (GAD). In normal brain cells, the GAD enzyme produces gamma-amino butyric acid (GABA), a neurotransmitter that naturally inhibits overactive nerves in the subthalamic nucleus in the brain.

Nathan Klein was a 59-year-old TV producer living in New York City who suffered from debilitating Parkinson's symptoms; his voice was weak, his gait was unsteady, and his hands trembled badly. In 2003, Klein entered the first human gene therapy trial for Parkinson's disease and a year later his neurological symptoms had greatly improved. Gene therapy for Parkinson's disease takes advantage of the ability to directly access and treat the region of the brain affected in Parkinson's patients, the subthalamic nucleus. The scientists inserted thin tubing through a quarter-sized hole in the top of Klein's skull and threaded it down into the interior of his brain (Figure 11.21). A vector carrying the GAD gene DNA was administered through the tubing directly into the cells in the subthalamic nucleus in Klein's brain. Expression of the GAD gene

produced the GABA neurotransmitter, which inhibited the overactive nerves in the subthalamic nucleus and relieves Klein's symptoms. Following the gene therapy treatment, Klein's neurological symptoms gradually improved and the 12 patients who also participated in the first human Parkinson's gene therapy trial reported similar results.

## Sickle Cell Disease

Sickle cell disease is a serious genetic blood disorder that affects predominantly 1 in 500 African Americans and Hispanic Americans. Sickle cell was the first human disease for which the gene mutation, a single DNA base pair change in the $\beta$-globin gene, was characterized at the molecular level. This seminal discovery was published in 1957, yet the goal of establishing a routine gene therapy approach for the treatment of sickle cell disease remains elusive even today.

The "sickle cell" mutation in the $\beta$-globin gene dramatically changes the structure of hemoglobin, a critical component of red blood cells that is required to transport oxygen from the lungs to other parts of the body (see Chapter 10). The sickle cell mutation alters an amino acid in the $\beta$-globin protein, which in turn causes the mutant hemoglobin complexes to stick together and form stiff fibers that distort the physical shape of the cell into the sickle-shaped red blood cells characteristic of sickle cell disease (Figure 11.22). The painful, often debilitating symptoms of sickle cell anemia occur because the sickle cells get stuck in the

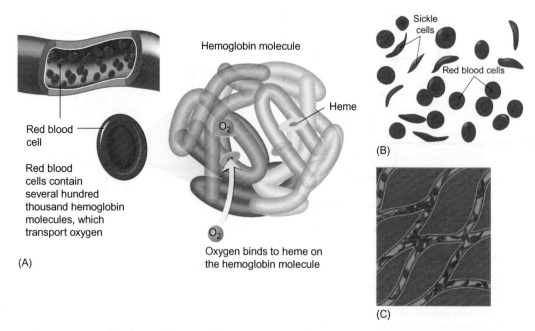

**FIGURE 11.22** Oxygen is carried by hemoglobin in red blood cells. (A) Normal red blood cells have a characteristic frisbee shape, which allows the cells to easily navigate even the smallest blood vessels, called capillaries. The red blood cells contain hemoglobin, which binds to oxygen, permitting the red blood cells to carry oxygen throughout the body. (B) Sickle red blood cells exhibit a dramatic sickle shape caused by a single base pair mutation in the β-globin protein gene. (C) Blood cells must squeeze through tiny capillaries to reach extremities in the body, a process that is much more difficult for sickle-shaped red blood cells.

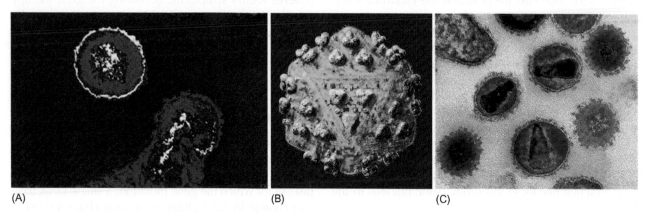

**FIGURE 11.23** Lentivirus vectors carry genes into mouse cells. Lentiviruses are unusual among retroviruses because lentiviruses can infect cells that are not dividing (noncycling), whereas most other retroviruses can only infect cells that are actively undergoing cell division. (A) Lentiviruses attach to cell surface. (B) Diagram of Lentivirus protein capsid structure. (C) Lentivirus virions.

smallest blood vessels and block the normal flow of blood to tissues and organs (see Chapter 10).

In 2001, scientists developed a gene therapy strategy that successfully corrected sickle cell disease in mice, suggesting that gene therapy might also successfully treat sickle cell disease in humans (Pawliuk et al. 2001). Scientists used a lentiviral vector to deliver wildtype β-globin genes into the mouse cells expressing the mutant β-globin protein (Figure 11.23). The lentivirus family of retroviruses includes the human immunodeficiency virus (HIV), which causes AIDS. But the lentiviral vectors have been carefully modified for safety; they cannot replicate in human cells or cause disease, but they can efficiently transfer genes to a broad range of different human host cells and tissues. Lentiviruses are unusual among retroviruses because they can infect cells that are not dividing (noncycling cells), whereas other retroviruses can infect only cells that are actively undergoing cell division.

Over the past three decades, biotechnology companies have developed many commercially available "kits" designed to help scientists produce appropriate vectors for use in biomedical research. One example is the Lentiviral Expression System kit, which is available to help scientists propagate lentivirus vector particles carrying the therapeutic gene of choice for a given gene therapy treatment or clinical trial. The scientist provides the DNA encoding the therapeutic gene, but the kit provides the vector and the additional components needed to help the cells grow and package the vector into viral particles. Excellent kits are available to construct plasmid or viral vectors, to sequence DNA by chain termination reactions, to amplify DNA using polymerase chain reaction (PCR), and for many other techniques in molecular biology (see Chapter 5).

## RNAi: THE FUTURE OF GENE THERAPY?

In the early years of the twenty-first century, scientists developed a powerful new approach called RNA interference (RNAi), which became an essential, ubiquitous tool in biomedical research. One cancer researcher called RNAi a transforming technology. "You can't do a [genetic] experiment without doing RNAi." In a remarkably short time, RNAi grew beyond being one of the top-10 basic science breakthroughs in 2002 and 2003 to become the focus of clinical trials to test RNAi-based drugs in treating genetic disease.

Before scientists can develop a treatment for a specific genetic disease, they must first investigate the wildtype and mutant forms of the gene and understand the structural characteristics and functions of the protein that causes the disease symptoms in question. To determine the function of an unknown gene, scientists often observe cells that have had a specific gene removed from the genome (a **gene knockout**) or have had transcription of that specific gene turned off (gene is not expressed as RNA). The ability to silence genes using gene knockouts or RNAi is a useful research approach that permits scientists to turn off expression of a specific gene at will. Then the scientists can observe what happens to the cells when the protein encoded by the silenced gene is absent.

An essential gene encodes a protein that performs a function required for the cell to live. A knockout of an essential gene is usually fatal because the protein with an essential function is not produced. RNAi technology allows scientists to attenuate or fine-tune the decrease in expression of a particular gene at the mRNA level.

## RNAi Security: Slicing and Dicing RNA in Cells

Research teams led by Andrew Fire and Craig Mello conducting research on nematode worms (C. elegans)

**FIGURE 11.24** RNAi scientists win a Nobel Prize. Dr. Andrew Fire and Dr. Craig Mello were awarded the Nobel Prize in 2006 for their research on RNA interference (RNAi).

discovered a novel cellular process called RNA interference that "silences" or "turns off" the expression of selected genes in living cells. **RNAi gene silencing** is an intriguing and important biological mechanism for which Fire and Mello won the Nobel Prize in 2006 (Figure 11.24). The natural RNAi process has been adapted for use in many applications of basic research, biotechnology, and potential medical treatments. At first the researchers thought that they were studying a phenomenon in worms, but then they discovered that RNAi is a biological process essential to all eukaryotic cells.

In wildtype cells, RNA interference (RNAi) shuts off the expression of selected genes, whether the RNAi mechanism is part of a normal cellular defense process or co-opted as a gene therapy strategy (Figure 11.25). The natural process of RNA interference (RNAi) in cells is triggered by the presence of short interfering RNA (siRNA) molecules in the cells, usually produced by the destruction of invading double-stranded viral RNA genomes. Fire and Mello designed siRNAs that when added to cells will target the mRNA from a specific gene for cleavage. The siRNAs bind together with cellular proteins to make the RNA-induced silencing complex (RISC) (see Figure 11.25). RISC guides the siRNA to its target messenger RNA (mRNA) in the cell and the RISC-siRNA complex base pairs to the target mRNA. This is a key step in the degradation of the selected target mRNA, because it immediately shuts down translation of the target mRNA and expression of the protein encoded by the mRNA.

Researchers used the RNAi strategy to design specific siRNA molecules that can be used as therapeutic

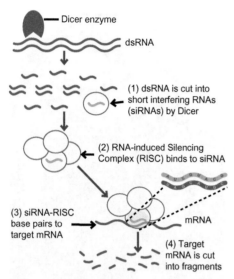

**FIGURE 11.25** RNA interference mechanism (RNAi) (1) dsRNA is cut into short interfering RNAs (siRNAs) by Dicer enzyme, (2) RNA-induced Silencing Complex (RISC) binds to siRNA, (3) siRNA-RISC base pairs to target mRNA, and (4) the target mRNA is cut into non-functional fragments.

"drugs" and are engineered to destroy only certain targeted mRNAs, for example, the mRNAs produced by expressing a harmful mutant gene. The ability of RNAi to target the mRNA transcripts made from a specific gene without affecting the RNAs transcribed from other genes is a very powerful feature of RNAi technology.

Researchers engineer the siRNAs to base pair to a short sequence in the target mRNA, shutting down expression of the proteins encoded by the target mRNAs; RNAi can target and degrade invading viral RNAs, or it can be used to target mutant genes that cause genetic diseases.

## Human siRNA Therapy: Treatment for Macular Degeneration in the Human Eye

The ability of scientists to use RNAi technology to silence selected genes in living cells has made RNAi an increasingly valuable tool for the treatment of human diseases. Some genetic diseases are caused by the inappropriate expression of genes that are normally held silent (are not transcribed) by control mechanisms in the cell. RNAi suppression of these harmful genes has the potential to have a large impact on the future treatment of human diseases. RNAi technology promises to permit doctors to unambiguously target a specific

disease-causing gene and silence that gene in the human body, all without interfering with the expression of other human genes.

Cancer cells have the ability to activate the biological process of **angiogenesis**, the growth of nearby blood vessels, by releasing the vascular endothelial growth factor (VEGF) protein (see Chapter 9). Cancer cells require access to new blood vessels in order for them to continue to divide and grow into a tumor. New RNAi-based cancer drugs were developed with the goal of using RNA interference to shut off expression of the VEGF gene in cancer cells and, as a result, cut off the blood supply, starving the cancer cells.

Recently RNAi was used to treat age-related macular degeneration (AMD), which impairs the vision of more than 1.5 million adults over the age of 50 in the United States. AMD is an eye disease that destroys central vision by damaging the macula, the part of the retina that lines the inside of the eye. Macular degeneration causes the excessive growth of blood vessels in the retina, resulting in swelling and inflammation that progressively interfere with central vision. In 2004, Sirna Therapeutics began a clinical human trial to test an RNAi-based treatment for AMD. Scientists used RNAi technology to turn off expression of the VEGF gene in the retinal cells of people with AMD. The siRNA "drug" used in the AMD trial was a short RNA (siRNA) molecule containing an RNA sequence designed to base pair specifically with a short sequence of the mRNA encoding the VEGF protein.

The phase I AMD gene therapy trial tested, specific siRNAs that base pair to the VEGF mRNA sequence as RNAi drugs designed to destroy only the targeted VEGF mRNAs in the cells. The siRNA drug was injected into the patients' eyes to deliver the siRNA drug directly to the cells in the retina. Although this phase I trial was designed to look for possible side effects from the use of the siRNA drugs and was not meant to measure the effectiveness of the RNAi treatment, as many as 25% of the AMD patients in the phase I trial reported improvement in vision, and the vision of the remaining patients stabilized and did not worsen (Figure 11.26).

Many additional RNAi-based human gene therapy trials are currently in the early stages of human trials. Other human diseases are candidates for siRNA clinical trials, including a new approach to genetically reverse sickle cell disease using stem cell-based gene therapy combined with RNA interference. Sickle cell disease can be cured by a bone marrow transplant to transfer the healthy blood-forming stem cells from a biological relative or close genetic match to the patient. However, this option is not available to most patients because of the difficulty in finding a compatible donor, especially for minority patients due to the smaller number of available tested donors.

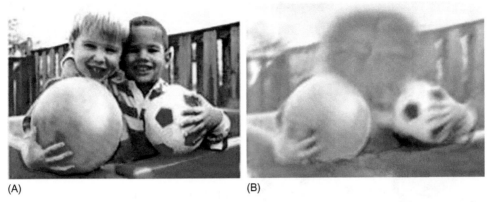

**FIGURE 11.26** RNAi gene therapy for macular degeneration. Age-related macular degeneration is a common cause of vision problems in older people, often beginning with loss of central vision. (B) Normal vision is shown in (A). The VEGF protein promotes the growth of new blood vessels in the retina cells in the back of the eye. The gene therapy strategy to treat age-related macular degeneration is based on the ability of siRNA technology to silence specific genes, in this case the VEGF gene in the retina cells.

## ENZYME REPLACEMENT THERAPY: AN ALTERNATIVE TO GENE THERAPY?

Some genetic diseases can be treated successfully with appropriate doses of the wildtype enzyme protein (not the gene). This enzyme replacement therapy (ERT) is a strategy that would potentially bypass the need for gene therapy treatment in some cases. For patients with adenosine deaminase deficiency (ADA, SCID), ERT is a possible option because PEG-ADA is an effective treatment for SCID. The PEG-ADA drug was created by linking purified ADA enzyme proteins to a nonmetabolizable carrier made from polyethylene glycol (PEG). The PEG-ADA drug provides the patient's cells with the active ADA enzyme, without many side effects. Although this is a promising treatment, the PEG-ADA drug is extremely expensive and at this time it must be taken by injection for life.

Another candidate for ERT is Gaucher's disease (GD), a genetic lysosomal storage disease that causes a genetic defect in the cellular lysosomal pathway. In healthy cells, fatty molecules called glucocerebrosides are degraded by special glucocerebrosidase enzymes carried inside the lysosome vacuoles (Figure 11.27). People with Gaucher's disease have a mutation in the gene for the glucocerebrosidase enzyme, so that the glucocerebrosides are not degraded properly but instead these toxins accumulate in the spleen, liver, and bone marrow, causing serious problems such as anemia; bone, liver, and spleen damage; and neurological deficits. In samples taken from Gaucher's disease patients, the macrophages stain dark blue because of the accumulation of excess glucocerebrosides (Figure 11.28).

ERT requires a commercial source of the specific enzyme proteins to be used in the treatments. In initial tests on ERT, 12 Gaucher's disease patients received

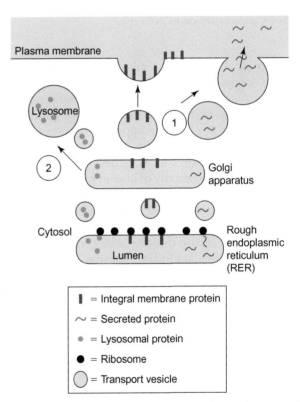

**FIGURE 11.27** Lysosome pathway in the cell. (1) The RER and Golgi are part of the membrane system used by cells to secrete newly made proteins out of the cell, or to incorporate the proteins into the plasma membrane. (2) Lysosome vacuoles (vesicles) contain many different degradation enzymes (proteases, lipases, nucleases, and polysaccharidases).

doses of wildtype glucocerebrosidase enzyme purified from rare human placental tissues, and all the patients showed dramatic improvements in disease symptoms. The success of this approach prompted scientists to

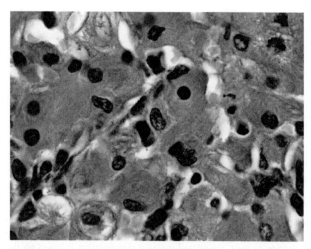

**FIGURE 11.28**  Lysosomes from cells with Gaucher's disease. Gaucher's disease is caused by a defective glucocerebrosidase gene and enzyme. In lysosomes, glucocerebrosidase enzymes normally catalyze the breakdown of sphingolipids. Macrophages from this Gaucher's disease patient are swollen and stained dark blue because of the accumulation of glucocerebrosides in the cells.

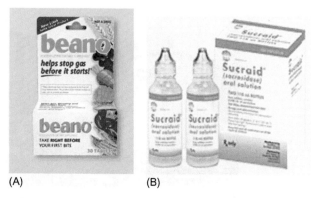

(A)                                            (B)

**FIGURE 11.29**  Over-the-counter enzyme replacement therapy (ERT). (A) People who are intolerant of complex carbohydrates (CCI) and cannot digest sugars properly suffer abdominal pain and gas. Fortunately, CCI symptoms can be prevented by Beano, which provides the cells in the body with the missing enzyme and allows people with CCI to include complex carbohydrates in their diet. (B) The over-the-counter ERT product Sucraid treats sucrase enzyme deficiency, which is caused by inheriting a sucrase-isomaltase genetic disease.

develop an alternative, reliable source of human glucocerebrosidase enzyme by cloning the glucocerebrosidase enzyme gene into a DNA vector and producing large amounts of the glucocerebrosidase protein in the bacterial host cells. This approach ensures a future source for the large amounts of enzyme needed for lifetime ERT treatments for Gaucher's patients without relying on a limited supply of human placental tissue.

Beano is a commercially available enzyme replacement therapy used to treat individuals with the gastrointestinal disorder called complex carbohydrate intolerance (CCI). This disorder affects individuals who lack the alpha-galactosidase enzyme that degrades complex carbohydrates into sugar building blocks (Figure 11.29). Sugars that are not properly digested by people with CCI will produce gas and cause pain in the abdomen. Beano prevents the symptoms of CCI by providing the cells in the body with the missing enzyme. This treatment also permits people with CCI to receive the health benefits available from foods rich in complex carbohydrates. Another over-the-counter ERT is Sucraid, which is used to treat sucrase deficiency in people who inherit a sucrase-isomaltase genetic disease and lack the sucrase enzyme.

## SUMMARY

Gene therapy is a remarkable approach to curing genetic diseases. In a gene therapy treatment, the patient receives the wildtype gene that produces the necessary functional protein, and as a result the disease symptoms are alleviated. The goal of gene therapy is to rescue the genetic defect caused by the mutation by using the wildtype gene rather than just treating the disease symptoms. Gene therapy offers great promise for the future development of life-saving treatments for patients with many different genetic diseases.

A key part of any gene therapy strategy is the choice of vector to carry the therapeutic gene to the cells in need of treatment. Many virus genomes have been modified for use in gene therapy treatments based on the characteristics of the vector and the specific requirements of the treatment protocol. Some vectors used for gene therapy are derived from retrovirus genomes, but the vectors have been altered to be sure that the vector does not harm human cells. Once the therapeutic gene has been inserted into the retrovirus vector, the vector enters the nucleus of the cell and becomes inserted into host cell chromosome, one way to achieve long-term gene therapy for the patient.

Nonviral vectors are becoming more prevalent as scientists have combined DNA technology with nanotechnology to better approach the challenges of gene therapy (see Chapter 13). Researchers continue to test other ways to deliver genes into cells, including liposomes, nanoparticles, and inhalation aerosols. Safe and effective gene therapy treatments are based on the large amount of information available about the mechanism of the disease in question. They must decide which disease cells are accessible for the delivery of therapeutic DNA, without harming other cells in the body. Scientists often find that the technologies developed to treat one disease can also be useful for the treatment of different diseases.

The field of gene therapy is controversial, and ethical questions continue to be raised. To guard against the accidental transmission of altered human genomes to the children in the next generation, only nonreproductive (somatic) human cells are legal to use in gene therapy in the United States. A multilayered review system was established to ensure the safety and efficacy of future gene therapy treatments that are reviewed and funded through the National Institutes of Health (NIH) and the Food and Drug Administration (FDA). Gene therapy opponents point out the availability of alternative treatments for some diseases, including enzyme replacement and bone marrow transplants for some blood diseases. Proponents argue that the high cost of alternative treatments is not usually covered by health insurance, and the success of any bone marrow transplant still requires a closely matched donor.

In this chapter we discussed only a few of the human diseases treated with gene therapy approaches, but many other diseases, including cancer, are also good candidates for gene therapy treatments. Updated information on human genetic studies including the ongoing and planned human gene therapy trials (phases I, II, III) is available online. The completion of the Human Genome Project provided the scientific community—in fact, the entire world community—with access to the 20,000 (or so) gene instructions about how to make humans, human. Thousands of these human genes, possibly as many as 10,000 single-gene mutations, can potentially cause human diseases (see Chapter 10). Clearly, many people who inherit these genetic mutations and the associated diseases are likely to benefit from the development of safe, effective gene therapy treatments.

The future potential of gene therapy technology lies in a combination of ingenious ideas and new technologies that are sufficiently flexible to meet the needs of the patient and beat the disease. The explosion of RNA technologies, many based on applications of the RNAi (siRNA) treatments, has inspired and excited the entire scientific community. Our understanding of the way genes work (combined with the huge wealth of information offered by the human genome sequences and genomic research) is finally poised to benefit everyone.

## REVIEW

In this chapter we reviewed the current and future prospects for different human gene therapy treatments. To test your comprehension of the chapter's contents, answer the following questions:

1. What types of vectors are used for gene therapy experiments, and how do scientists decide which vector to use in a particular gene therapy case?

2. Describe the first successful experiment involving gene therapy for individuals with ADA deficiency.

3. What are the main differences between *in vivo* and *ex vivo* gene therapy methods?

4. Summarize how gene therapy can be used to treat cancer.

5. What are the major advantages and disadvantages associated with using RNAi for gene therapy?

6. What is the genetic basis of cystic fibrosis, and what intervention can be made with gene therapy?

7. Describe the main characteristics of a nonviral gene delivery system.

8. Explain why "Kill the messenger RNA" is an appropriate nickname for the RNA interference (RNAi) process.

9. How do scientists decide what vector to use in a particular gene therapy?

10. Explain the risks associated with gene therapy treatment.

## ADDITIONAL READING

Anderson, W.F., September 1995. Gene therapy. Sci. Am., 124–128.

Blaese, M., June 1997. Gene therapy for cancer. Sci. Am., 111–115.

Blue, L., 2007. A gene to cure blindness. Time. www.time.com/time/health/article/0,8599,1623086,00.html

Could gene therapy help alcoholics stay on the wagon?, 23 June 2007. New Sci. www.newscientist.com/article/dn12117-could-gene-therapy-help-alcoholics-stay-on-the-wagon.html

Culver, K.W., January-February 1993. Splice of life: genetic therapy comes of age. Sciences (New York) 33 (1), 18–24.

Friedman, T., 1997. Overcoming the obstacles to gene therapy. Sci. Am. http://www.lsic.ucla.edu/classes/winter09/syllabi/hnrs70a_syl09w

Gene therapy alleviates symptoms of Parkinson's disease, 30 June 2007. New Sci. (2610), 20.

Gene therapy restores sight in blind dogs, Philadelphia, June 2007 (UPI), Journal PLoS Medicine online. www.upi.com/Science_News/2007/06/26/UPI-NewsTrack-Health-and-Science-News/UPI-69941182894240/

Glausiusz, J., January 1996. Gene therapy: special delivery. Discover.

Howard, K., July/August 1995. New arteries for old. Sciences.

Hsiau, TH.-C., Diaconu, C., Myers, C.A., Lee, J, Cepko, C.L., Corbo, J.C. 2007. The cis-regulatory logic of the mammalian photoreceptor transcriptional network. Public Library of Science One, (Washington University School of Medicine, St. Louis). Gene therapy has passed an important milestone with the first successful treatment of sickle cell anemia in a mouse model of the disease.

Jaroff, L., Fall 1996. Keys to the Kingdom. Time (special issue).

Pawliuk, R., Westerman, K.A., Fabry, M.E., Payen, E., Tighe, R., Bouhassira, E.E., Acharya, S.A., Ellis, J., London, I.M., Eaves, C.J., Humphries, R.K., Beuzard, Y., Nagel, R.L., Leboulch, P., 2001. Correction of sickle cell disease in transgenic mouse models by gene therapy. Science 294, 2368–2371.

New research may lead to future gene therapies for patients with sickle cell anemia and beta-thalassemia, Medical Research News, August 2007, Prepublished online Aug. 3; Journal Blood

(American Association for Hematology); www.vcu.edu; www. news-medical.net/print_article.asp?id=28544

Thompson, L., September-October 2000. Human gene therapy: harsh lessons, high hopes. FDA Consum. Mag.

Travis, J., 1998. Inner strength: gene therapy for AIDS. Sci. News 153, 174–178.

Mejía, J.E., Willmott, A., Levy, E., Earnshaw, W.C., Larin, Z., August 1, 2001. Functional complementation of a genetic deficiency with human artificial chromosomes. Am. J. Hum. Genet. 69 (2), 315–326 doi:10.1086/321977.

Zoia Larin, A., Mejíab, J.E., 1 June 2002. Advances in human artificial chromosome technology. Trends Genet. 18 (6), 313–319 doi:10.1016/S0168-9525(02)02679-3.

## WEB SITES

American Society of Gene Therapy, www.asgt.org/index.php.

Copernicus Therapeutics, January 2008. www.cgsys.com was redirected to http://168.144.36.118/index/index.asp.

Cystic fibrosis: genetic testing for human cystic fibrosis carriers, www. acog.org/from_home/wellness/cf001.htm; www.sciencedaily.com/ releases/2003/04/030430082133.htm.

Cystic fibrosis gene therapy trial results encouraging, University Hospitals of Cleveland Science Daily, 2003.

Comprehensive source of information on worldwide gene therapy clinical trials, http://cmbi.bjmu.edu.cn/news/0304/212.htm

NIH gene therapy update: http://history.nih.gov/exhibits/genetics/ sect4.htm. Sickle cell treatment marks gene therapy milestone 14 December 2001 by Philip Cohen New Scientist, 2001. www. newscientist.com/article/dn1690-sickle-cell-treatment-marks-gene-therapy-milestone.html

Howard Hughes Medical Institute, www.hhmi.org/biointeractive/rna/ rnai. A good overview of RNAi mechanisms including some history of how the technology was developed.

NOVA Science Now, www.pbs.org/wgbh/nova/sciencenow/3210/02. html. A NOVA site available free online, all about RNAi plus additional resources.

J. Gene Ther.-Prog. Prospects www.nature.com/gt/progress_and_ prospects.html. Nature suggests current review articles on various aspects of gene therapy, including RNAi.

RNAi-net, www.rnai.net. For the latest RNAi news and updates.

Top 10 scientific advances of 2003, by Gareth Cook, Boston Globe December 23, 2003, http://www.boston.com/news/specials/year_ in_review/2003/articles/science/

# Stem Cell Research

## Windpipe Transplant Breakthrough

BBC News, http://news.bbc.co.uk/2/hi/health/7735696.stm
By Michelle Roberts

Surgeons in Spain have carried out the world's first tissue-engineered whole organ transplant—using a windpipe made with the patient's own stem cells. The groundbreaking technology also means for the first time tissue transplants can be carried out without the need for antirejection drugs. The patient, 30-year-old mother-of-two Claudia Castillo, needed the transplant to save a lung after contracting tuberculosis. The Colombian woman's airways had been damaged by the disease. Scientists from Bristol helped grow the cells for the transplant and the European team believes such tailor-made organs could become the norm. To make the new airway, the doctors took a donor windpipe, or trachea, from a patient who had recently died. Then they used strong chemicals and enzymes to wash away all of the cells from the donor trachea, leaving only a tissue scaffold made of the fibrous protein collagen. This gave them a structure to repopulate with cells from Ms Castillo herself, which could then be used in an operation to repair her damaged left bronchus—a branch of the windpipe. By using Ms Castillo's own cells the doctors were able to trick her body into thinking the donated trachea was part of it, thus avoiding rejection.

Two types of cell were taken from Ms Castillo: cells lining her windpipe, and **adult stem cells**—very immature cells from the bone marrow—which could be encouraged to grow into the cells that normally surround the windpipe. After four days of growth in the lab in a special rotating bioreactor, the newly-coated donor windpipe was ready to be transplanted into Ms Castillo. Her surgeon, Professor Paolo Macchiarini of the Hospital Clínic of Barcelona, Spain, carried out the operation in June. Five months on the patient, 30-year-old mother-of-two Claudia Castillo, is in perfect health, *The Lancet* reports.

## LOOKING AHEAD

This chapter gives an overview of the exciting and dynamic study of stem cell biology and introduces the key concepts necessary to understand this fast-moving field while providing the reader with a working knowledge of this interesting subject. All stem cells have the same defining characteristics, but there are many different kinds of stem cells, perhaps as many types of stem cells as there are types of tissues. This chapter reviews the basic concepts about these amazing cells and updates the status of work on the most widely studied stem cells.

On completing this chapter, you should be able to do the following:

- Provide the functional definition of a stem cell listing the two most important features.
- Know the major differences between multipotent stem cells and pluripotent stem cells and how they are related to each other.
- Explain the essential functions of embryonic stem cells during embryo development.
- Understand the most important differences among adult skin cells, adult stem cells, and embryonic stem cells.
- Describe the advantages and disadvantages of using adult stem cells or embryonic stem cells for the development of stem cell therapies.
- Describe the advantages of someday having access to patient-specific blastocyst embryos.
- Explain the origin of induced pluripotent stem (iPS) cells and why they represent a very important advance in the field of regenerative medicine.
- Describe the basic features of genome reprogramming and the role of epigenetic changes in the cell.

## INTRODUCTION

Stem cells are an amazing type of cell that have captured the fascination of scientists and the public alike. Stem cells have been studied for longer than 50 years. Treatments such as bone marrow transplants have become routine for some diseases, but it was not until 1998, when scientists described the origin of human embryonic stem cells, that stem cells became a major topic of discussion and debate in the research community, the halls of Congress, and at the kitchen table.

Human embryonic stem cells hold great promise for advances in basic research and have enormous potential for the development of novel therapies to treat many degenerative diseases.

Despite their enormous potential, the biological source of embryonic stem cells (ESCs), human embryos, has made research and the development of human treatments extremely controversial (Figure 12.1). President H. W. Bush entered the debate in 2001 at the start of the explosion of stem cell research when he banned all federally funded research on human embryonic stem cells in the United States. This stalled public research on stem cells but promoted an increase in private, corporate sources of support for stem cell research in the United States and abroad. It also increased research to develop new ways to generate embryonic stem cells without using human embryos.

> Human embryonic stem cells hold great promise for the development of novel therapies to treat many diseases. Despite such amazing potential, the fact that the ESCs cells come from human embryos has made this research extremely controversial and led to the development of new methods to avoid the use of human embryos.

## STEM CELLS GENERATE NEW TYPES OF SPECIALIZED CELLS

The field of stem cell research was launched in the early 1960s by scientists James Till and Ernest McCulloch who, for the first time, discovered that stem cells in mice give rise to the many different specialized mouse cells in the blood and immune systems. In the early stages of development, mammalian embryos, including mouse and human embryos, form

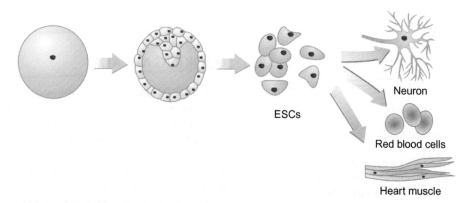

**FIGURE 12.1** Embryonic stem cells (ESCs) have the most developmental potential. Embryonic stem cells (ESCs) grow in the inner cell mass (ICM) inside the hollow blastocyst embryo (visible cross-section of blastocyst embryo). ESCs are vitally important because they have the potential to generate all of the different types of specialized cells needed in the embryo and to develop an adult human. When the ESCs are used to generate specialized cells for tissue replacement therapy, the blastocyst embryo is destroyed.

hollow balls of cells called blastocysts (Figure 12.2). Till and McCulloch were studying the bone marrow cells in mice, looking for components in the bone marrow that might rescue mice that were previously exposed to lethal doses of radiation. The radiation destroyed the cells in the bone marrow of the irradiated mice, but these scientists found that the doomed mice could be saved by a bone marrow transplant from nonirradiated mice.

Till and McCulloch analyzed the bone marrow cells from the nonirradiated donor mice to look for specific factors that could account for the regenerative potential of the nonirradiated blood. They discovered that only one type of transplanted cell could regenerate healthy cells in the irradiated mice, the stem cells that also give rise to the blood cells. For their fundamental work on stem cells, Ernest McCulloch and James Till won the Lasker Award for Basic Medical Research in 2005. Even today, scientists judge the developmental potential of putative stem cells using the standards of self-renewal and multi-lineage differentiation set by Till and McCulloch.

## Stem Cells Have Developmental Potential

Stem cells are categorized as either **pluripotent** or **multipotent** cells based on their ability to grow and reproduce (self-renewal) and their potential to generate different types of cells by cell differentiation (Figure 12.3). The stem cells with the highest potential to generate all types of specialized cells grow inside the blastocyst. These **pluripotent stem cells** are called **embryonic stem cells (ESCs)** (Figure 12.4). ESCs exist temporarily in mammalian embryos at a certain time during very early development when the embryo begins to make cells with more specific functions. Eventually the ESCs in a human embryo will generate all of the 200 or so different types of cells required to grow a human body. The underlying promise of stem cell research is the potential to use the amazing ESCs to develop new and powerful cell replacement treatments for many human diseases and disorders (Figure 12.5). Cell and tissue replacement therapies involve replacing the damaged or diseased tissues much like organ transplantation is designed to replace diseased or damaged organs.

ESCs are pluripotent cells that can generate multipotent stem cells, which have far less developmental potential than the ESCs and develop into far fewer types of specialized cells. Multipotent stem cells are partially specialized or partly **differentiated** stem cells, which are produced in both adult and embryonic

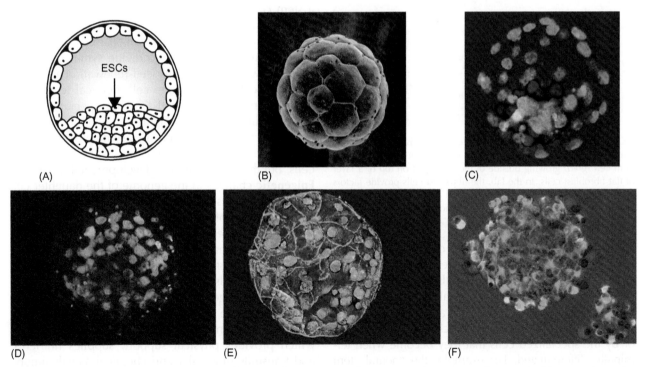

(A)    (B)    (C)

(D)    (E)    (F)

**FIGURE 12.2**  Blastocyst embryos contain the infamous embryonic stem cells (ESCs). (A) Cross section diagram of a blastocyst shows that the embryo is a hollow ball with the ESCs growing inside. (B) Image shows the surface topology of an intact blastocyst embryo, stained to indicate the individual cells in the embryo. (C) The small number of human ESCs *(green cells)* can be seen growing inside a hybrid developing mouse embryo *(blue cells)* (D) Blastocyst embryo from a Cheetah; (E) Human blastocyst shows each cell nucleus stained pink; (F) This blastocyst was made from a new approach using four transcription factor genes to reprogram ordinary adult skin cells and convert them into induced pluripotent stem (iPS) cells.

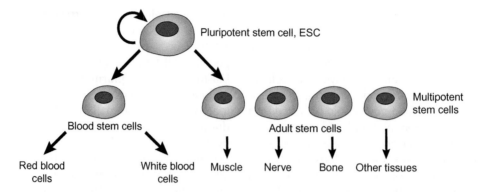

**FIGURE 12.3** ESCs generate other important stem cells. During embryo development the ESCs give rise to other stem cells, which are the precursors that produce the more specialized cells required by the adult body. For example, the ESCs in the embryo generate the hematopoietic stem cells in bone marrow that give rise to the red and white blood cells in the adult human. Other progenitor stem cells give rise to muscle, nerve, bone and other specialized tissues.

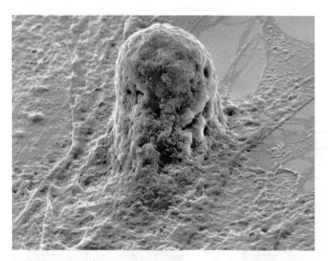

**FIGURE 12.4** A human embryonic stem cell. This color-enhanced scanning electron micrograph (SEM) shows a human embryonic stem cell (gold) growing outside of the blastocyst on top of a layer of flat fibroblast cells. In the lab, the fibroblast cells provide factors that support the growth of undifferentiated ESCs. ESCs can generate any type of specialized cell in the human body.

tissues. Multipotent adult stem cells are generated when pluripotent embryonic stem cells differentiate to form more specialized cells (see Figure 12.2). Multipotent stem cells are also called "somatic" stem cells or "tissue-derived" stem cells and are further identified either by the tissue from which they were isolated or by the specialized cells that they become when terminally differentiated. For example, the "neural stem cells" give rise to highly specialized nerve cells in specific areas of the adult human brain. The primary difference between pluripotent and multipotent stem cells is that pluripotent stem cells have a broader developmental repertoire and can generate cells that contribute

to all three major germ layers required for embryo development: the ectoderm, endoderm, and mesoderm (Chapter 9). Multipotent stem cells have a much more limited potential to differentiate into specialized kinds of cells.

The ability of stem cells to execute asymmetric cell division is fundamental to the mechanism that permits a stem cell to acquire new and specialized functions by differentiation (see Chapter 9). When a stem cell begins to divide by cell division, it must follow one of two major paths, either to divide symmetrically to produce two identical cells (self-renewal) or to divide asymmetrically to yield one cell identical to the parent cell and a second cell that has specialized and exhibits different traits (Figure 12.6). The daughter cells that result from such an asymmetric cell division event contain the same genes as the parent stem cells, but they have different developmental potentials. **Epigenetic** changes in the genome of the daughter cell alter gene expression patterns but do not change the DNA sequence of the genome.

Embryonic stem cells have the highest potential to generate all types of specialized cells in the body. To develop successful regenerative therapies involving human stem cells, it is imperative to understand how stem cells are generated, sequestered, and nurtured in the body. The **stem cell niche** refers to the immediate environment surrounding stem cells in the body, which provides physical support, supplies nutrients, and transmits molecular and chemical signals among the cells. These signals control the development of the stem cells and influence the differentiation of stem cells into the required types of cells at the appropriate time. The stem cell niche is a dynamic environment supported by components of the extracellular matrix,

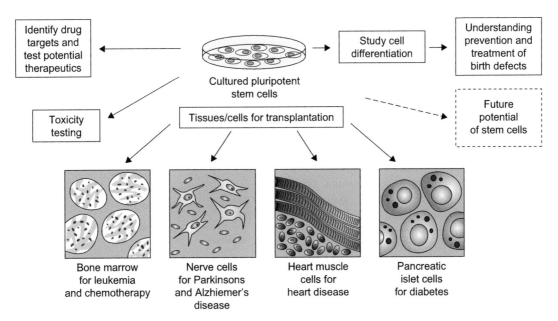

FIGURE 12.5 The promise of stem cell research. The underlying promise of stem cell research is based on the amazing potential of ESCs not only to develop new and powerful cell replacement treatments but also for use testing drugs and toxins, to better treat and prevent birth defects.

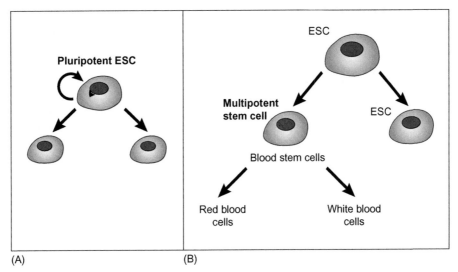

FIGURE 12.6 Asymmetric cell division generates differentiated cells. (A) When stem cells undergo mitotic cell division, they can divide symmetrically so that each dividing cell produces two identical progeny cells (self-renewal). (B) Alternatively the cells can divide asymmetrically in which case the dividing cell will produce one identical stem cell and one differentiated cell that has undertaken a different developmental pathway.

which includes various physiological metabolites, growth factors, and other hormones made by the stem cells and their progeny cells.

## Diminishing Potential with Increasing Specialization

At some point during development, stem cells commit to becoming a specific type of specialized cell in the adult body. As stem cells move through the differentiation spectrum from the undifferentiated fertilized egg toward a more fully differentiated and mature state, each cell can generate fewer and fewer different types of cells. When the cell becomes terminally differentiated it will adopt a final form and function.

Two biological characteristics define all stem cells: (1) stem cells can produce identical cells by cell division (self-renewal), and (2) stem cells can cease cell division and differentiate into multiple types of cells depending on developmental potential.

## THE POTENTIAL AND THE PROBLEMS OF EMBRYONIC STEM CELLS

In the developing embryo, ESCs grow in the inner cell mass (ICM) of very early, blastocyst-stage embryos (see Figure 12.2). As the embryo continues to develop, the ESCs differentiate into the cells that will eventually give rise to all of the cell types in the human embryo. The first report of ESCs isolated from human embryos sparked a great deal of discussion and debate worldwide. This research caused huge excitement in the scientific community and fueled speculation about the future potential for human ESCs to play a major role in treating diseases, injuries, and the effects of aging in humans.

Embryonic stem cells were first isolated from mice and not humans, but since then, the ESCs from lab mice have laid the groundwork for research on human ESCs. Much later the mouse ESCs were used to produce gene-targeted knockout mice, an advance that profoundly changed the nature of biomedical research (discussed later).

### Embryonic Stem Cells Have the Most Developmental Potential

In early development at the blastocyst-stage, a human embryo resembles a hollow ball containing two main types of cells, the outer layer **trophectoderm**, which gives rise to the **placenta**, and the inner cell mass (ICM), which gives rise to the cells in the embryo itself (Figure 12.7). Inside the blastocyst, a small number of ESCs are generated from cells in the ICM. These pluripotent ECSs differentiate into multipotent stem cells, which in turn develop into the tissue layers and the germ cells of the embryo itself.

Although the ICM is only transiently present in the developing early embryo, once the ESCs have been established in the lab culture as **embryonic stem cell lines**, they can often be maintained and reproduced for years without losing developmental potential (pluripotency). Human ESC lines are used for developmental research, for toxicology screening assays on new drugs, and to generate new cell replacement therapies.

Clearly the human ESCs have amazing potential for the successful development of cell replacement and regeneration therapies for human patients. However, the fact that these stem cells originate in human embryos poses serious ethical and moral concerns that threaten to destroy the potential benefits that ECSs offer. As a result, intense efforts were initiated to find less-controversial alternative sources of pluripotent human ESCs. ESCs are the most widely studied type of pluripotent stem cell, but to date ESCs have been derived successfully from the embryos of only three mammals; mice, nonhuman primates, and humans. The ban in 2001 on federal support for stem cell research using human ESCs helped to

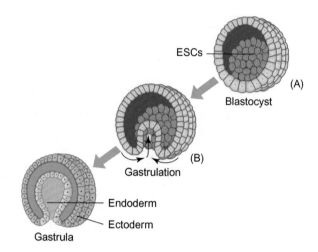

**FIGURE 12.7** Embryo cells move during development. (A) The blastocyst embryo is a hollow-ball of cells (shown in cross-section) that contains the inner cell mass, which gives rise to the pluripotent ESCs inside the blastocyst. (B) Gastrulation begins with cell migration into the hollow ball (arrows) and the formation of the three major cell layers in the embryo: the ectoderm, endoderm, and mesoderm.

encourage efforts to develop alternate sources of human stem cells for research and medicine, with some surprising successes.

> The first embryonic stem cells were isolated from mice not humans. Since then, work on ESCs from mice laid the fundamental groundwork for research on human ESCs and promoted the investigation of possible applications of ESCs to treat human diseases and disorders.

### Human Adult Stem Cells Have Limited Developmental Potential

Most cells in the adult human body are not stem cells but are fully (terminally) differentiated into over 200 types of specialized cells that perform many important functions in the body (Figure 12.8). Examples of fully differentiated cells include red blood cells that carry oxygen, epithelial cells that line the walls of the small intestine, skin cells covering the body, cardiac muscle cells that allow the heart to beat, and nerve cells that develop long extensions that transmit impulses to the brain.

Although the majority of adult cells are fully differentiated and specialized, the human body also contains **adult stem cells** (somatic stem cells) that have the power to generate specialized cells to replace damaged tissues following injury or disease. Adult stem cells are part of a highly regulated system that produces new cells in the right tissue at the right time and in the right amount to meet the needs of the body without causing uncontrolled cell growth or disease.

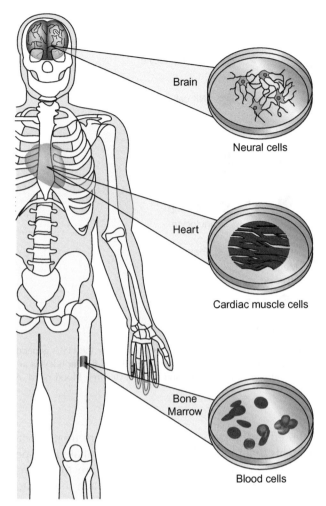

**FIGURE 12.8** Adult cells have specialized functions and are fully differentiated. Most cells in the human body have differentiated into specialized cells that do not divide but perform many specialized functions in the body. Examples of fully differentiated somatic cells include the red blood cells that carry oxygen, the epithelial cells that line the walls of the small intestine, cardiac muscle cells that make the heart beat and nerve cells that transmit impulses to the brain.

Adult stem cells have now been isolated from tissues such as brain, bone marrow, peripheral blood, blood vessels, skeletal muscle, epithelia cells in skin and digestive tract, cornea, dental pulp, retina, and liver. The fact that adult stem cells have been found in so many different adult tissues suggests that most tissues have a specific adult stem cell population that is available for limited repair and regeneration of that tissue.

> Adult stem cells are relatively rare multipotent somatic stem cells that reside in adult organs and tissues. In the human body, the adult stem cells give rise to a small number of specialized cell types in the tissue from which they were derived.

## Adult Stem Cells Generate Specialized Human Cells

Adult stem cells are multipotent, which means that they have limited developmental potential. Regardless of the source, adult stem cells can differentiate only into cells of the same type as the tissue from which they originated. For example, hematopoietic stem cells can generate only blood and immune system cells and neural stem cells can generate only specialized cells found in the nervous system. Nevertheless, even though adult stem cells have limited developmental potential, in many cases they present a possible alternative to human ESCs for use in cell and tissue transplantation therapies.

Here we will focus on three well-studied types of adult stem cells: the hematopoietic stem cells (HSCs), which give rise to the different types of specialized cells in the blood and immune systems (Figure 12.9), the mesenchymal stem cells (MSCs), which produce bone and cartilage, and the neural stem cells (NSCs), which generate nerves and brain tissues. Although adult stem cells reside in the adult, they retain the main features of all stem cells, the ability to reproduce as undifferentiated cells and the capacity to change developmental direction and generate different kinds of specialized cell types.

> Adult stem cells are multipotent and have limited developmental potential compared to the pluripotent ESCs. Still adult stem cells are capable of generating certain types of specialized cells in the body and can potentially be used in cell transplantation therapies.

## Hematopoietic Stem Cells Generate Blood Cells

**Hematopoietic stem cells (HSCs)** have been a research focus because HSCs generate all of the different types of specialized cells in the human blood and immune systems. The HSCs are familiar to the public because they provide the cells most commonly used in bone marrow transplants to treat cancer and other diseases of the blood. The traditional sources of HSCs for cell transplant therapies include the bone marrow in the adult long bones and the peripheral blood. More recently, the HSCs for transplantation have been obtained from umbilical cord blood, which is easily collected at birth and can be cryopreserved for long-term storage. Once thawed, the cord blood stem cells can be readily grown in the lab to provide HSCs with no risk to the donor.

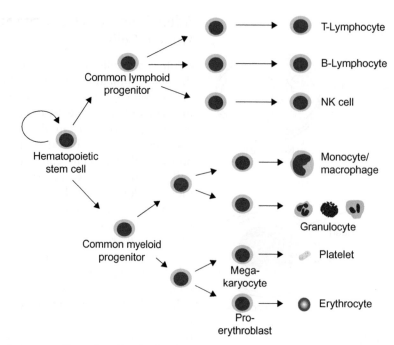

**FIGURE 12.9** Hematopoietic stem cells produce blood cells. The hematopoietic stem cell (HSC) family tree shows how the HSCs generate different types of cells in the blood. The HSCs also generate the lymphoid progenitor cells, which give rise to the T and B lymphocytes and NK cells, and the myeloid progenitor cells, which give rise to macrophages, granulocytes, platelets, and erythrocytes (red blood cells). As cells become more specialized, they lose some developmental potential.

> Adult stem cells are multipotent somatic stem cells that reside in very small numbers in adult organs and tissues. In the human body, adult stem cells give rise to a small number of specialized cell types in the tissue from which they were derived.

## Mesenchymal Stem Cells Generate Bone, Cartilage, and Muscle

The **mesenchymal stem cells (MSCs)** in the adult bone marrow are necessary for the body to generate tissues such as bone, cartilage, muscle, ligament, tendon, adipose, and bone marrow. The MSCs represent an ideal population of adult stem cells for stem cell therapies because they are comparatively easy to isolate from the body, can be grown easily in lab cultures, and can be induced to differentiate into lineage-specific cell types without becoming contaminated with other cell types (Figure 12.10). MSC stem cells can be grown for many generations in the laboratory and still retain a stable morphology and normal chromosome complement. These properties explain why the MSCs are well suited for **autologous stem cell transplants**, in which the patient's own stem cells are used to replace damaged or diseased tissues in the patient's own body. MSCs taken directly from a patient are

cultured in the laboratory where they are treated with reagents that induce the MSCs to differentiate into specific types of specialized cells, which are then transplanted into the patient. This approach eliminates the risk that the patient will reject the transplanted cells because the same individual is both donor and patient.

Both the HSCs and MSCs develop in the bone marrow, but these two cell types can be easily separated from each other in laboratory preparations of bone marrow cells because the two cell types have different physical properties on the cell surfaces. As a result the MSCs attach to the bottom of the plastic culture plates, whereas the HSCs remain unattached and float in the medium. Proliferating MSCs (stromal stem cells) have a fibroblast-like morphology, but they retain the ability to differentiate into multiple mesenchymal cell lineages (see Figure 12.11). MSCs have been used in regenerative stem cell therapies to treat degenerative diseases such as arthritis.

> MSC stem cells can be grown for many generations in laboratory and maintain a stable chromosome number, and retain the ability to differentiate into pure populations of specific types of bone, cartilage, and muscle cells.

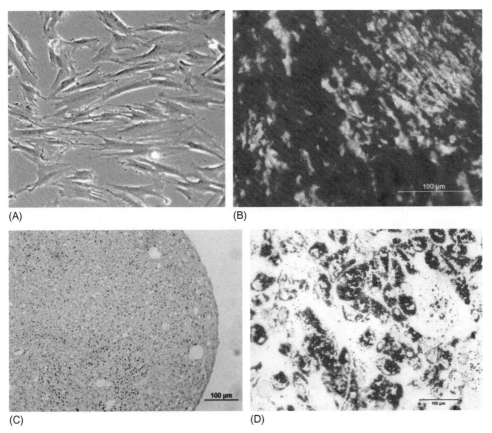

(A)    (B)

(C)    (D)

**FIGURE 12.10** Mesenchymal stem cells make bone and cartilage. (A) Mesenchymal stem cells (MSCs) give rise to three main types of cells: (B) osteocytes in bone (the calcium deposits are stained with alizarin red). (C) cartilage cells (proteoglycans are stained with toluidine blue), and (D) adipocytes (fat cells are stained with oil red).

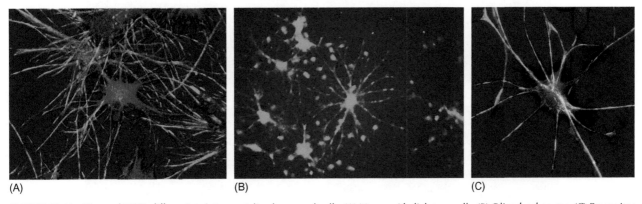

(A)    (B)    (C)

**FIGURE 12.11** Human hNSCs differentiate into specialized neuronal cells. (A) Neuroepithelial stem cells. (B) Oligodendrocytes. (C) Progenitor neuroglial cells.

## Neural Stem Cells Generate Specialized Nerve Cells

Except for severe damage to the spinal cord, research shows that it is not correct to conclude that injured nerve cells never regrow. In 1992, for the first time, Reynolds and Weiss isolated **neural stem cells (NSCs)** from the brains of adult mice and successfully grew the dividing NSCs in lab cultures for long periods of time. This work challenged the "no new neurons" dogma and proved to be a major breakthrough in the fields of neurobiology and cell regeneration. Remodeling the neural network in the human brain and the related process of nerve growth (neurogenesis) continue throughout our lives. Nerves die off and new nerve connections form in response to learning, injury, disease, the environment, and much more.

Two main regions of the human brain actively grow new nerve cells: the hippocampus, which is important for memory formation, and the ventricular and subventricular zones, both areas of actively dividing cells in the fetal and adult brain. In adult rodents, neurogenesis also occurs in the olfactory bulb, the part of the brain that receives odor signals from the nose. Scientists still do not understand how the body controls the growth of neural stem cells or the precise signals needed to initiate differentiation into different types of nerve cells.

Neural stem cells are also called **neural progenitor cells** because they give rise to three lines of specialized nerve cells: (1) neurons, of which there are hundreds of different types, (2) astrocytes, which support the neurons, and (3) oligodendrocytes, which make myelin, a lipid-rich fatty material that surrounds the neurons. Myelin is vital for the proper transmission of electrical signals along the nerves and is defective in degenerative nerve diseases such as multiple sclerosis (Figure 12.11). Remodeling the neural network in the human brain includes the death of some nerves and the growth of others throughout a lifetime, and new nerve connections form in response to learning, injury, and disease.

To successfully grow human nerve cells in culture, it is necessary to provide an environment that mimics the physiological conditions of the human body in the lab, including the appropriate temperature, atmosphere, and nutrient medium. Cell culture methods usually require **serum**, which is blood plasma without the clotting factors, as a nutrient source, but scientists found that prolonged exposure to serum prevented the NSCs from differentiating into specialized nerve cells. A huge advance came with the development of a serum-free culture medium for NSC growth, which also contained large amounts of epidermal growth factor (EGF) and bovine fibroblast growth factor (bFGF). The large quantities of hormones needed to propagate NSCs in culture are produced by recombinant DNA methods. Growth factors are necessary for the NSCs to proliferate (divide) and grow as a single layer of cells on a surface substrate such as fibronectin (Figure 12.12).

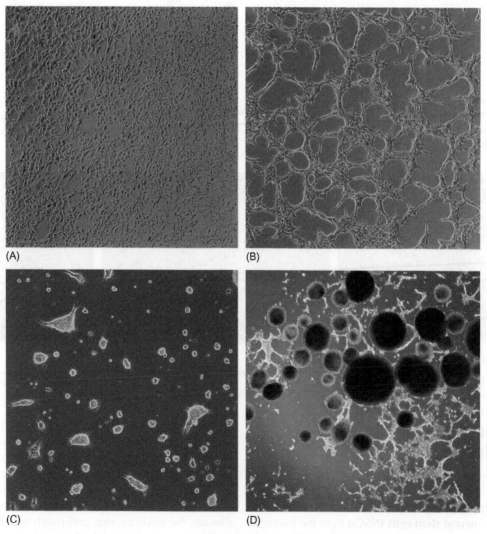

**FIGURE 12.12** Human neural stem cells growing in culture. (A) The NSCs are growing attached to fibronectin substrate. (B) Cells proliferate and begin to aggregate. (C) Further aggregation of cells. (D) Neurosphere induction by growing cells in nutrient restricted medium.

When induced to differentiate, some of the NSCs detach from the surface of the plate and form multicellular structures called **neurospheres** (see Figure 12.13). This type of NSC culture can be propagated indefinitely in the lab by periodically breaking up the neurospheres and dispersing the cells into plates with fresh medium for further growth. The NSCs can be induced to develop into specific types of neural cells by adding or removing specific proteins such as growth factors. The specific conditions needed to induce stem cells to differentiate are derived experimentally by the scientists in the research lab and depend on the type of stem cells and the type of desired specialized cells.

The ability to get NSCs to generate specialized neurons in the lab is an amazing achievement, but better methods are needed to prepare large quantities of nerve stem cells for routine use in the treatment of degenerative nervous system diseases. Specialized nerve cells are under intense study, and scientists continue to develop better methods to prepare large quantities of nerve stem cells for use treating degenerative nervous system diseases.

The types of specialized cells generated by differentiation of NSCs are routinely identified in the lab using fluorescent immunocytological staining (see Figure 12.13). In this method, the cells are treated with specific antibodies that have been raised against the different proteins (antigens) that are found on the cell surfaces of either neurons, astrocytes, or oligodendrocytes. Because each antibody can recognize and bind to a specific protein on the cell surface, the different types of nerve cells in a population of cells can be easily identified by color (see Figure 12.11).

Highly purified preparations of neural stem cells can be derived from both cultured mouse and human ESCs by following established research protocols. Both mouse and human ESCs have been used to generate specialized neural cells that were then used to successfully restore partial function in test animals with spinal cord injuries.

> ESC-derived NSCs represent a promising future for the development of stem cell treatments for diseases such as multiple sclerosis, spinal cord injury, and Parkinson's. These three diseases are prime targets for stem cell replacement therapies because they are each caused by the loss of one type of mature, differentiated cell.

## REPROGRAMMED ADULT CELLS REGAIN POTENTIAL

People have long wanted to turn back the clock and regain their youth. For specialized adult cells, this would require differentiated cells to be recalibrated to a younger, less specialized state through **genome reprogramming**. The biological process of genome reprogramming is thought to reset the cell's DNA genome to a pluripotent state, which then permits the reprogrammed cell to change developmental fate and become a different type of specialized cell. An example of this type of dramatic change occurs just after fertilization when the adult sperm and egg genomes come together to create a new reprogrammed embryonic genome. The reprogrammed embryonic genome will direct the development of all the cells in the embryo, including the ESCs that give rise to the entire human body. The reprogramming mechanism does not change or mutate the genome DNA sequence, instead the **epigenetic** information is carried in the DNA modification patterns and the proteins bound to the chromosome.

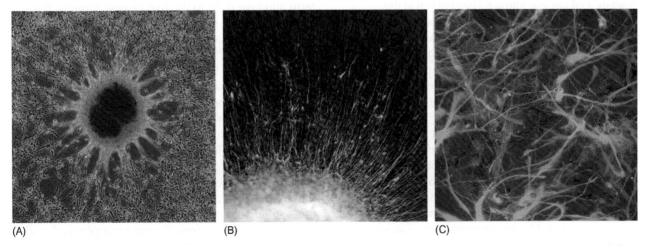

(A)                                    (B)                                    (C)

**FIGURE 12.13** Human neural stem cells develop in culture. (A) Neurosphere structure grows attached to the substrate (100× magnification). (B) Developing neuronal outgrowth contains DNA stained with DAPI *(blue)*, as well as the Nestin, *(red)*, Neuron Tuj1 *(green)* proteins (200× magnification). (C) Budding neurosphere shows expression of the Sox-2 *(red)*, and Nestin *(green)* proteins (400× magnification).

Genome reprogramming is an active area of research, driven in large part by the need to understand the molecular mechanisms that drive cell differentiation and the desire to develop patient-specific stem cells. Patient-specific stem cells are derived from and genetically related to a specific patient, making these cells especially useful in treating that particular patient's disease.

Most cell replacement therapies suffer from the disadvantage that the patient is usually treated with cells that are foreign to the patient's body. This situation triggers tissue rejection, and the patient's own immune system attacks the transplanted foreign cells. Tissue rejection requires treatment with strong immunosuppressive drugs to decrease the effectiveness of the patient's immune system with the hope of halting the rejection process. Unfortunately, many complications and side effects can accompany immunosuppressive treatments, making it even more important to develop cell replacement therapies that do not cause this immune response. Patient-specific stem cells or donor-specific stem cells will eliminate the need for immunosuppressive drugs because the stem cells in question were originally derived from the patient and are therefore a genetic match to the patient.

> Genome reprogramming is an epigenetic process that can turn back the clock in adult skin cells, and cause fully differentiated cells to adopt a much younger less specialized state.

## Reprogrammed Cells Avoid Human Embryos

Reprogrammed genomes have the power to dramatically alter the fate of specialized adult cells, as demonstrated in the following examples:

*Somatic cell nuclear transfer.* The genome in the nucleus of a specialized adult cell is reprogrammed when the somatic cell nucleus is transferred into an empty egg cell that has already had its nucleus removed. Genome reprogramming permits the somatic cell nuclear genome to direct the development of an entire embryo generated from an unfertilized egg cell.

*Cell fusion.* When a specialized adult cell (containing an adult nucleus and genome) and an embryonic stem cell (containing an embryonic nucleus and genome) are fused together, the contents of the two cells co-mingle, but the two nuclei remain separate. The adult nuclear genome is reprogrammed by key proteins in the cytoplasm of the embryonic stem cell.

*Induced pluripotent stem (iPS) cells.* Adult human skin cells grown in the lab were treated with DNA encoding four different transcription factor proteins. When expressed in the adult skin cells, the transcription factor proteins altered gene expression in the cell, and produced proteins that reprogram the adult genome into an embryonic genome. As a result the adult skin cells are converted into undifferentiated cells called induced pluripotent stem (iPS) cells that resemble normal embryonic stem cells. Most exciting, scientists have successfully triggered the iPS cells to differentiate into specialized nerve cells.

## Cloning by Somatic Cell Nuclear Transfer

Somatic cell nuclear transfer (SCNT) involves replacing the nucleus of an oocyte (an unfertilized egg cell) with a donated nucleus from a differentiated adult somatic cell (Figure 12.14) (see Chapter 14). The cytoplasm in a mammalian oocyte contains all of the molecules needed to reprogram the adult genome to direct the "cloned" egg to develop into a blastocyst embryo in the lab. Depending on what happens next, the "cloned" blastocyst embryos (generated by SCNT) are used for either **reproductive cloning** or **therapeutic cloning**, depending on the specific goal of the work, and the organism involved.

**Reproductive cloning** is used to generate genetic copies of an entire living organism. This approach gave us the first cloned animal, Dolly the sheep (see Chapter 14). Reproductive cloning first uses the process of somatic cell nuclear transfer (SCNT) to generate cloned blastocyst embryos carrying new genes. The blastocyst embryos are then implanted into surrogate females, which carry the developing embryos to term and give birth to the live cloned animals. Born in 1997, Dolly the sheep was the result of somatic cell nuclear transfer and reproductive cloning. Dolly was a major scientific accomplishment and a very photogenic sheep that introduced the public to the prospect of cloning animals almost overnight. Of course, for many reasons, the reproductive cloning of humans is banned in the United States.

Dolly was cloned from the single nucleus of a mammary epithelium skin cell obtained from a 6-year-old adult sheep (see Figure 12.14). The nucleus was removed from the sheep skin cell and introduced into the empty cytoplasm of an unfertilized sheep egg cell (the egg nucleus had already been removed). Under the influence of proteins from the egg's cytoplasm, the adult genome is reprogrammed and directs the cell to start embryo development. Once the embryos have reached the blastocyst stage of development, they are implanted into surrogate females to nurture the growing embryos to birth.

There is no genetic relationship between the implanted embryos and the surrogate females that carry the embryos to term. The genomes of the cloned animals are identical to the genomes in the donor nuclei; genetically identical copies are called clones. It is important to note that cloning animals is a very inefficient process overall, and it took persistent efforts to produce Dolly and other cloned animals. However, 17 species of cloned animals were generated by reproductive cloning methods in the 10 years between 1997 and 2007, including sheep, mouse, bull, pig, goat, Guar wild ox, Mouflon sheep, rabbit, cat, mule, rat, African wildcat, dog, water buffalo, horse, ferret, and wolf.

The reproductive cloning process that produced Dolly is quite different from the use of somatic cell nuclear transfer (SCNT) in the process of **therapeutic cloning** (Figure 12.15). Unlike in reproductive cloning, the blastocyst embryos generated by therapeutic cloning methods are never implanted in surrogate females and are never used to produce pregnancy. The therapeutic cloned blastocysts are most often used as the source of donor-specific embryonic stem cells used to generate specific cells to treat the patient's disease, and avoid the risk of transplant rejection (see Figure 12.15). To date, scientists have not been able to successfully generate cloned human blastocyst embryos by SCNT. Cloned human embryonic cells have been observed to undergo the early stages of cell division

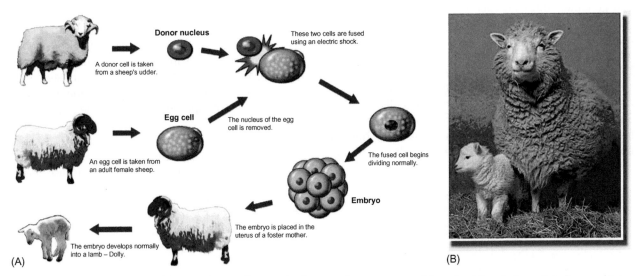

**FIGURE 12.14**   Dolly was created by reproductive cloning. The donor nucleus was removed from the skin cell of an adult sheep and introduced into the empty cytoplasm of an empty sheep egg to make a cloned embryo. Proteins in the egg cytoplasm reprogram the donated adult sheep genome, and the reprogrammed genome then directs the cell to become less specialized. The resulting cloned blastocyst embryos were implanted into surrogate females, and Dolly was born. (B) Dolly, the first cloned animal, is shown here with her biological daughter conceived the normal way!

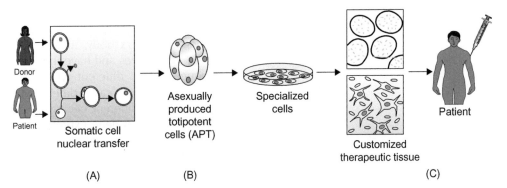

**FIGURE 12.15**   Therapeutic cloning by SCNT. (A) The nucleus of an egg cell oocyte is removed and replaced by the nucleus of an adult (somatic) cell donated by a specific patient. (B) This 'cloned egg' develops into a cloned blastocyst embryo. (C) The ESCs from cloned blastocyst embryos have the potential to develop into any type of adult cells and tissues, which are all are genetically matched to the specific human donor.

and development, but so far they have not achieved the blastocyst stage.

## Genome Reprogramming by Cell Fusion

When two cells are fused together the two nuclei are suddenly housed in the same cytoplasm influence each other. In many cases the genomes of one or both nuclei undergo reprogramming. When ESCs are fused to adult skin cells, the genome of the adult skin cell is exposed to the cytoplasm of the pluripotent ESC. Molecular components and proteins in the cytoplasm of the ESC reprogram the adult cell genome, causing the adult skin cell to convert to a pluripotent state.

Researchers fused human ESCs with human fibroblast skin cells and generated "cybrid" cells, and factors in the cytoplasm of the ESC reprogram the adult cell genome. Although each cybrid cell contains two nuclei, one nucleus from the ESC and one nucleus from the fibroblast, the cybrid cell maintains a stable tetraploid genome containing twice the normal diploid DNA content. In addition, the cybrids have the same cellular morphology, growth rate, and cell surface proteins as do human embryonic stem cells (hESCs). Surprisingly, cybrid cells can differentiate into the different specialized cell types, much like the human embryonic stem cells that give rise to the three cell layers in the developing embryo (Figure 12.16).

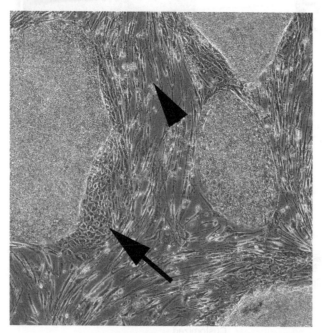

**FIGURE 12.16** Human embryonic stem cells in culture. Human ESCs grow in tightly clustered colonies of cells that are in very close contact with neighboring cells in the colony. The hESCs are commonly co-cultured with embryonic fibroblast cells, which provide factors that support the growth of the undifferentiated hESCs. Arrowhead shows MEF fibroblast. Arrow shows differentiating cells at the edge of the hESC colony. (100× magnification).

## INDUCED PLURIPOTENT STEM CELLS

The studies on induced pluripotent stem (iPS) cells will help scientists to better understand the molecular mechanisms responsible for genome reprogramming. In the future, scientists plan to use the iPS approach to generate pluripotent stem cells with greater diversity and to generate hESCs with known genetic mutations or possible predisposition to develop certain diseases. The ability to generate iPS from genetically diverse populations should help us to better understand the causes of these diseases and to develop better drugs designed to treat specific genetic diseases.

The discovery that adult somatic (body) cells can be reprogrammed to behave like embryonic stem cells is exciting, but scientists still need to better understand the similarities and differences between the iPS cells and normal ESCs and it is not yet clear how well the iPS cells will function in the laboratory and in the clinic. Nevertheless, the discovery of iPS cells represents a huge leap forward for the field of regenerative medicine. The potential to make pluripotent stem cells without using human embryos brings the reality of patient-specific pluripotent stem cells one step closer.

> Genome reprogramming occurred when the Oct4, Sox-2, and Nanog transcription factor genes were introduced into the adult somatic skin cells. The change in gene expression patterns caused by these transcription factors caused the production of proteins that reprogrammed the genome and triggered the formation of induced pluripotent stem cells (iPS).

## EPIGENETIC CHANGES IN GENOME REPROGRAMMING

The need to understand the genes involved in cell differentiation has created a strong motive for researchers to dissect the molecular mechanisms involved in reprogramming the DNA genome of a cell. Genome sequences differ slightly between people, but in an individual person's body, all of the cells contain exactly the same genes and have the identical genome DNA sequence. The different types of cells are not different because they carry different genes in their genomes, they are different because they express different genes. Different gene expression patterns direct the synthesis of different proteins in different types of cells, and provide the specialized cells with tissue-specific functions. For example, the genes expressed and the mRNAs made in a liver cell are very different from the mRNAs expressed in a muscle cell, because the liver cell needs proteins that are different than the proteins needed by the muscle cells. The liver and muscle

## Box 12.1 Human Skin Cells Are Converted into Pluripotent Stem Cells

Scientists are studying the gene expression patterns in the pluripotent embryonic stem cells growing in the lab to find out how the cells control and maintain the pluripotent (undifferentiated) and differentiated states. Since the pluripotent state of the ESCs is induced by the expression of specific genes the scientists decided to mimic this specific gene expression pattern in fully differentiated adult skin cells.

First the researchers needed to identify which transcription factor proteins are actively involved in maintaining the pluripotent state of the cells. They discovered that a small number of transcription factor genes including Oct4, Sox2, Kif4, and Nanog, among others, have essential roles in maintaining the cell's ability to remain pluripotent. These transcription factor proteins control the expression (transcription) of a series of other human genes required to establish and maintain pluripotency (Figure 12.17). These genes are vital to the production and maintenance of the inner cell mass in the blastocyst and to maintain the pluripotent state of ESCs.

Scientists reasoned that the transcription factor proteins might be able to influence the state of a differentiated adult cell by turning on expression of the genes in the skin cells that are normally expressed in ESCs, resulting in a change to pluripotency.

In 2007, two independent research groups, James Thomson (University of Wisconsin, Madison) (*Science*) and Shinya Yamanaka (Kyoto University) (*Cell*), tested the idea that certain transcription factor proteins might influence the state of a differentiated adult skin cell by inducing expression of genes that are normally expressed only in ESCs. The scientists introduced the transcription factor gene DNA into the adult skin cells in the lab (Figure 12.18). To their surprise both teams found that the transcription factor genes had converted the adult skin cells into apparently normal embryonic stem cells, called **induced pluripotent stem cells (iPS cells)**, to distinguish the iPS cells from the ESCs (Figure 12.19).

Both research teams then showed that the iPS cells could be induced to generate different specialized cells in culture,

including nerve cell tissue. The iPS cells meet all the criteria of embryonic stem cells, including self-renewal and pluripotency, except that the iPS cells are not derived from human embryos. ESCs that naturally express the Oct4, Sox-2, and Nanog proteins can be grown for extended periods in the laboratory and remain pluripotent and do not acquire chromosomal abnormalities that lead to cancer. It is significant that the iPS cells also exhibit high developmental potential and are capable of unlimited cell division without genome damage. Like the normal ECSs, the iPS cells respond to normal cell cycle regulation and avoid DNA chromosome damage. Potentially, the iPS cells can differentiate into any of the more specialized cells in the ectoderm, mesoderm, and endoderm germ layers. To date iPS cells have formed nerve tissue, but further research will tell whether iPS cells live up to this prediction.

The ability to induce pluripotent stem (iPS) cells was a major breakthrough in the field of stem cell science, which will result in the routine use of **patient-specific** and **disease-specific**

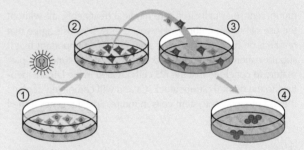

**FIGURE 12.18** The method used to generate induced pluripotent stem (iPS) cells: (1) Isolate and culture host cells (fibroblasts). (2) Introduce the ES specific genes (iPS factors) into the cells carried in a retrovirus vector. Red cells indicate the cells expressing the added iPS genes. (3) Harvest and culture the cells using feeder cells *(gray)*. (4) A subset of the cells generates ES-like colonies, the iPS cells.

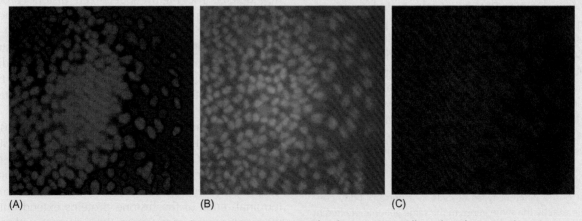

(A)  (B)  (C)

**FIGURE 12.17** Pluripotency transcription factor proteins are expressed in human ESCs. Cells in these human embryonic stem cell colonies (hESC) have been stained with antibodies to show the expression of three key pluripotency proteins: (A) Nanog, (B) Sox2, and (C) Oct4 (400× magnification).

Box 12.1   Continued

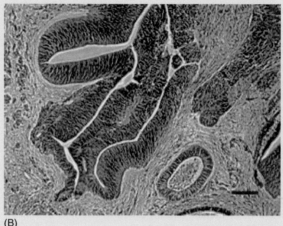

(A)

(B)

**FIGURE 12.19**   Human skin cells mimic ESCs and generate nerve cells. (A) Human skin cells in culture before the cells were genetically modified and converted into embryonic stem cells. (B) This photomicrograph shows nerve tissue generated from human skin cells. The transcription factors cause reprogramming of the skin cell genome, inducing the skin cell to become an embryonic stem cell (iPS). The black scale bar represents 0.1 millimeters, (about the width of a human hair).

pluripotent stem cells in future medical treatments, all without the use of human embryos. However, most scientists agree that research on human embryonic stem cells is so important that it should continue using donated embryos. At this time it is premature to conclude that the iPS cells created in the lab are identical to the normal pluripotent ESCs and will completely replace the use of embryonic stem cells in future. Specific genes turned on and expressed after fertilization trigger the growth of the inner cell mass within the developing blastocyst embryo. Other genes are required to direct the development of the embryonic stem cells (ESCs) inside the blastocyst, and to maintain the pluripotent state. This includes the transcription factor genes that make iPS cells, such as Oct4, Sox 2, and Nanog.

cells are very different in form and functions but they carry exactly the same DNA genomes.

Transcription factors and other specific proteins function to control tissue-specific gene expression, which permits only certain genes to be turned on in certain types of cells at certain times. Eukaryotic genes are regulated at many points along the gene expression pathway from DNA to RNA to protein, including processes such as transcription initiation, RNA processing, and transport, to protein translation and folding (see Chapter 3). Still, none of these processes is sufficient to explain what happens during genome reprogramming, when an adult nucleus becomes "young" again and adopts a pluripotent state. Recent evidence shows that **chromatin**, the specific combination of DNA and histone proteins that make up the structures of the chromosomes, plays a key role in the genome reprogramming process.

> All of the different types of cells in a person's body contain exactly the same genes in the same genome.. Different types of specialized cells are distinguished from each other not by carrying different genes in their genomes, but by turning on and expressing different genes from the same human genome in the different cell types.

## Reprogramming Involves Chromatin Modification

Chromatin (DNA and histone proteins) plays a central role in packaging a very long double-stranded DNA helix into a typical chromosome in a eukaryotic nucleus. The highly specialized histone proteins assemble into protein-DNA complexes called nucleosomes that are spaced at intervals of every 220–250 base pairs along the DNA helix, like beads on a string. Each nucleosome core contains an octamer of 4 different histone proteins, and about 200 base pairs of DNA helix wrapped twice around the outside of the octamer (Figure 12.20).

The DNA helix and the core histone proteins are the majority of components that make up the structure of human chromosomes. Histone proteins have highly globular middle regions flanked by the ends of the proteins (see Figure 12.20). The N-terminal (amino-terminal) ends of the histone proteins extend outside the nucleosome core where they interact with regulatory proteins in the nucleus. As a result the histone proteins play an important role in gene expression. Enzyme proteins in the nucleus modify the histones by adding specific chemical groups to certain amino acids

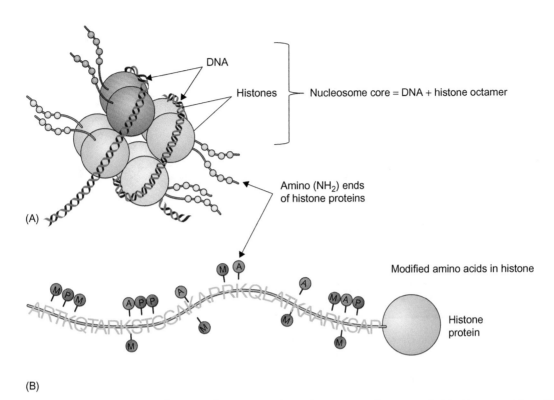

**FIGURE 12.20**  DNA and histone proteins form a nucleosome. (A) The nucleosome contains a core of eight histone proteins with the histone N-terminal ends projecting away from the core. Epigenetic modification of the protein tails allows the histones to transmit signals and influence gene expression. (B) The first 30 amino acids in the N-terminus of the human histone H3 protein are shown. Several amino acids in the N-terminus of histone H3 proteins are targets for epigenetic changes (acetylation, phosphorylation, and methylation). The collection of epigenetic modifications, called the histone code, causes a finely tuned transcriptional response in the cells, which alters the expression of a subset of genes in the cell.

in the N-terminal ends of the proteins. This **covalent modification** of the histone proteins in chromosomes is a good example of an **epigenetic modification** because the genetic signals transmitted by the histone protein "code" are not encoded in the primary sequence of the DNA genome. Epigenetic modification potentially impacts the entire genome, changes gene expression, and profoundly alters the state of the cell. **Epigenetic memory** is a process that allows cells to maintain their undifferentiated state, even while the cells are exposed to the conditions that usually induce differentiation. Epigenetic memory is an important function that helps to maintain stable cultures of pluripotent stem cells over time and helps prevent the formation of cancer cells derived from the pluripotent cells.

> Research shows that modification of the histone proteins in chromosomes turns on the expression of genes that direct the processes of cellular differentiation and genome reprogramming. Histone modification is an epigenetic change that occurs without changing the DNA sequence of the reprogrammed genome.

Genome reprogramming occurs after an egg cell (oocyte) containing 23 chromosomes is fertilized by a sperm, which also carries 23 chromosomes. In sperm the genome DNA is packaged into compact structures using protamine instead of the usual histone proteins (Figure 12.21). After the egg is fertilized by a sperm the protamine is removed and the sperm genome DNA is packaged into 23 traditional chromosomes by acetylated histone proteins (modified with acetyl-groups). The acetyl groups on the histone proteins help to maintain the chromatin structure in an "open" conformation, which offers access to the DNA genome and helps to prepare the DNA for transcription. Genome reprogramming continues when methyl groups are transferred from the genome DNA and are added to the histone proteins. In addition, the structure of the egg chromosomes change as the oocyte-specific linker histone H1oo protein is removed from the chromatin (see Figure 12.21).

In early embryo development, the embryo's genome, called the **epigenome**, is in a highly dynamic state. During blastocyst formation, the epigenome undergoes changes that set the stage for further embryo development. As the cells transition from pluripotent to

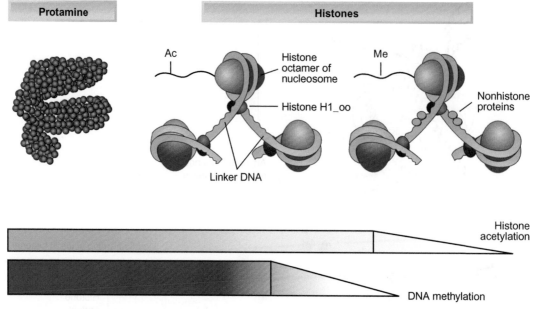

**FIGURE 12.21** Epigenetic changes in chromatin structure are important for genome reprogramming. The chromosome DNA in sperm is highly compacted with protamines (left), but after fertilization, the protamine is removed and the DNA is packaged into normal chromosome structures containing acetylated histones that maintain an "open" chromatin conformation that is ready for transcription (right). The demethylation of DNA and the methylation of the histone proteins is part of the genome reprogramming process that prepares the chromosome DNA for transcription.

more differentiated and restricted states, large changes in gene expression take place.

## The Future of Stem Cell Research and Therapeutics

Understanding the genes and underlying molecular mechanisms that maintain a cell in a pluripotent state or trigger cell differentiation is absolutely critical if stem cell therapies are to become available as routine treatments for disease. Scientists need to have a detailed understanding of the events in the cell that influence the function of the genome, but the small numbers of genes that can be monitored using conventional techniques has limited the search for answers to questions about the human genome. However, the recent development of genome-wide screening methods and systems-level evaluations now permit scientists to assess gene expression across entire populations of cells.

The ESC system is particularly well suited for the application of genomic and systems-level genetic screens. The earliest stages of embryo development and commitment to differentiation can be evaluated using a combination of genomic methods including genetic and pharmacological manipulation of the cells. Stem cells are temporary in the developing embryo and exist in very small numbers, but the ability to grow ESCs in culture now provides sources of progenitor cells in quantities that are sufficient for use in clinical therapies.

The systems biology approach to studying stem cells provides a dynamic overview of the molecular pathways at work in a given cell as the cell responds to different environmental signals. This work could lead to the development of more methods or agents that will reliably induce pluripotent cells in the lab to successfully differentiate into specific differentiated cell lineages.

The huge potential of human ESCs to generate specialized cells for cell replacement therapies has led to widespread proposals for the use of stem cells as therapeutic agents, including ways to overcome the limitations imposed by adult stem cells in therapeutic treatments. The very nature of an embryonic stem cell is to generate all of the different types of cells in the body, but this presents real challenges for scientists as well, because they must determine how to trigger the ESCs to differentiate into the type of cell needed for a particular cell replacement therapy. We know little about the factors required to induce ESCs to differentiate into nerve cells, which are probably different from those required to induce muscle cell development.

In addition to learning how to induce ESCs to differentiate into specialized cells, scientists also needed methods to detect the presence of undifferentiated cells that are mixed in with the differentiated cells. The potential genome instability of the undifferentiated stem cells carries the risk of developing cancer cells in the transplant recipient under some conditions. The progress in stem cell research has continued despite the controversy, technical difficulties, and funding problems precisely because of the substantial long-term benefits of stem cell work. Human studies and research on model animal systems will continue to be at the forefront of the efforts to develop effective regenerative medical treatments based on advances in stem cell science.

> Cells must accurately control the expression of certain genes that are turned on in early progenitor cells and are gradually turned off and then silenced at later developmental stages. Meanwhile, subsets of cell type-specific genes are turned on to make proteins that are required for the next stage of embryo development.

The ability to insert genes into specific sites in the genomes of human ESCs by DNA transformation is a major future goal for scientists in the field of regenerative medicine. The plan is to introduce DNA into the cells by transformation and then to grow the transformed cells containing the inserted DNA. When the cells are transplanted into the patient, the cells will all carry the repaired therapeutic DNA in their chromosomes. Using this approach, scientists will be able to repair or replace a defective or mutant gene in the genome of the stem cell. This gene repair will be inherited by the progeny cells and will also effectively repair the defect in all the future progeny cells. Achieving

this goal is still years away and there are many technical obstacles to be overcome and ethical issues to be discussed (see Chapter 11). Successful gene therapy relies on developing highly effective gene targeting techniques that will accurately insert DNA into target sites in the genome DNA of human ESCs. When appropriate gene targeting technologies for hESCs have been developed, the hESCs will become an even more important resource for the study of human development and disease and will also greatly promote the application of genetically engineered cellular therapeutics to the physician's arsenal for the treatment of disease or injury.

In the 17 years between the first report of the existence of mouse embryonic stem cells and the subsequent identification of human embryonic stem cells, scientists have constructed knockout mutations in the genomes of many different organisms. This is in addition to the hundreds of knockout strains of mice constructed with each strain having a different gene deleted or removed (knockout) from the mouse genome. Work with mouse ESCs has greatly aided the study of human ESCs and continues to provide important information to complement and enhance research on human ESCs.

> Scientists constructed a collection of mouse knockout strains that represent mutations in about half of the genes predicted in the mouse genome. Knockout mice with special mutations are custom-designed and available by special order, providing scientists with many research options.

---

### Box 12.2 Knockout Mice Are Right on Target

In 2007, the Nobel Prize in medicine was awarded to Mario R. Capecchi, Martin J. Evans, and Oliver Smithies for their work on **knockout mice**. They developed a landmark method that permits scientists to specifically inactivate a single mouse gene at will. This method permits scientists to target a single gene in the mouse genome for inactivation through the use of mouse embryonic stem cells. The different strains of knockout mice were analyzed to determine how the gene knockout and the loss of the normal mouse protein affected the normal traits exhibited by the mice.

Capecchi and Smithies knew how to genetically modify specific genes in the mammalian genome but neither scientist had ever worked with pluripotent stem cells, so they began to collaborate with Evans, who had extensive stem cell experience. This team successfully produced live animals carrying genetically modified genomes accurately inherited from the knockout parent mouse. Scientists constructed a collection of pluripotent mouse embryonic stem cell lines with each cell line carrying a different knockout gene mutation. This collection of knockout

mice has provided the scientific research community with a very valuable research tool used to study many different human diseases. The knockout mutations carried by these mouse strains are well characterized and the highly similar genetic backgrounds cause the knockout mice to exhibit traits (phenotypes) that are predominantly caused by the knockout mutation and the lack of an important protein product and are not the non-specific result of a spurious mutation elsewhere in the mouse genome. Research involving knockout mice and other organisms carrying knockout mutations or other forms of transgenic genomes continues to have a significant impact on scientific research. Gene targeting is a powerful tool to study many biological processes including mammalian cell, tissue, organ and body development, physiology, aging, and human disease. In addition to the Nobel Prize, in 2001 Capecchi, Smithies, and Evans were presented with the Lasker Award for Basic Scientific Research for their innovative research on knockout mice.

## SUMMARY

The field of regenerative medicine has made real progress since scientists discovered the amazing developmental potential of embryonic stem cells. Regenerative medicine is a highly dynamic area of research and clinical medicine in which new discoveries continue to challenge long-held beliefs about the success of cell regeneration. Stem cells are often in the national news and seem to hold a certain fascination for laypersons and scientists alike. The discovery of human ESCs had an impact on the general public that was equaled by few other scientific discoveries. The 1998 report on human ESCs generated wide-ranging excitement because of the probability that these cells will fundamentally change the way we study cell and tissue development, how we produce and test new drugs, and how we treat numerous degenerative diseases. Amid all this excitement and potential, however, is the big drawback that embryonic stem cells are typically isolated from human embryos, making them the subject of a great deal of controversy.

Although ESCs are a highly controversial area of research, the potential that stem cell therapy would dramatically improve treatments for degenerative diseases continues to put emphasis on stem cell research. Pluripotent stem cells can undergo cell division continuously while undifferentiated and have the potential to differentiate into all cell types in the body. Multipotent stem cells include adult stem cells such as the hematopoietic stem cells, which were derived from the bone marrow, and have been used to treat diseases and illnesses since the late 1950s. Few ethical concerns have been raised about the use of adult or somatic stem cells in research or transplant therapies, as they have long been used to treat diseases. The main controversy in stem cell research lies in the research using pluripotent embryonic stem cells derived from human embryos. Scientists, laypersons, and politicians have voiced views on the scientific, social, and ethical impact of this science.

Fortunately, recent research has revealed new ways to generate pluripotent human stem cells including the exciting discovery that adult skin cells can be reprogrammed to behave like embryonic stem cells and can even differentiate into nerve tissue. Of course, much work remains to find out how well these induced pluripotent (iPS) cells will work under clinical treatment conditions. Nevertheless, the ability to make pluripotent stem cells without using human embryos represents a huge leap forward for the field of regenerative medicine.

Scientists do not yet know which type of stem cells will be most effective in a specific area of research or cell therapy. Research on different types of stem cells has led to a better understanding of stem cell biology and the power of pluripotent developmental potential. Biotechnology researchers and pharmaceutical companies are developing stem cell lines for applications in toxicology assays and in wide variety of disease treatments. The ability to make induced pluripotent (iPS) cells has brought us closer to the reality of patient-specific stem cells suitable for cell transplant therapies that are predicted to alleviate many degenerative diseases and address the problem of the limited number of human organs donated for human transplant therapy.

## REVIEW

This chapter describes both the great potential offered by stem cells as well as the drawbacks associated with the development of embryonic stem cell technology. This chapter explains the science behind human embryonic and adult stem cells, and describes the production of stem-cell-like induced pluripotent cells (iPS) derived from adult human skin cells. To assess your comprehension of stem cell biology and technology in these topic areas, answer the following review questions:

1. Explain the role played by the embryonic stem cells in the development of the late-stage embryo.
2. Explain the fundamental differences between the developmental potential of embryonic stem cells and that of adult stem cells.
3. What is the difference between a multipotent stem cell and a fully differentiated somatic cell?
4. Describe the different characteristics between multipotent stem cells and pluripotent stem cells and explain the biological relationship between these two types of cells.
5. Describe the advantages and disadvantages of medical cell replacement treatments that use cells from patient-specific blastocyst embryos.
6. Explain the importance of the genome reprogramming that occurs after fertilization or when a somatic cell nucleus is transferred into an "empty" egg (oocyte without a nucleus).
7. Explain the fundamental difference between a DNA mutation in the genome and epigenetic changes affecting the genome.
8. Describe the objections raised by the public against the use of embryonic stem cells for the development of stem cell therapies for medical treatments.
9. To make induced pluripotent stem (iPS) cells in the lab, scientists treated adult skin cells with genes that encode what types of proteins?
10. Describe the general structure of the human blastocyst embryo and explain what part of the blastocyst structure plays a key role in generating embryonic stem cells.

## ADDITIONAL READING

Barinaga, M., 1992. Challenging the "No New Neurons" Dogma. Science 255, 1646.

Becher, A., McCulloch, E., 1963. Till "Targeting Gene Therapy to Neuroinflammatory Lesions in the field of hematopoiesis". Nature 197, 452–454.

Evans, M., Kaufman, M., 1981. Establishment in culture of pluripotential cells from mouse embryos. Nature 292, 154–156.

Hall, V.J., Stojkovic, P., Stojkovic, M., 2006. Using therapeutic cloning to fight human disease: a conundrum or reality?. Stem Cells 24, 1628–1637.

Loring, J.F., Wesselschmidt, R.L., Schwartz, P.S. (Eds.), 2007. Human Stem Cells: A Laboratory Manual. Elsevier, Inc, Amsterdam, The Netherlands.

Ruth Kirschstein, M.D. Acting Director., Lana, R., Skriboll, PhD, Director. June 2001. Stem Cells: Scientific Progress and Future Research Directions. Report prepared by the National Institutes of Health.

Takahshi, K., Tanabe, K., Ohnuki, M., Narita, M., Ichisaka, T., Tomoda, K., Yamanaka, S., 2007. Induction of pluripotent stem cells from adult human fibroblasts by defined factors. Cell 131, 1–12.

Thomson, J.A., Itskovitz-Eldor, J., Shapiro, S.S., Waknitz, M.A., Swiergiel, J.J., Marshall, V.S., Jones, J.M., 1998. Embryonic stem cell lines derived from human blastocysts. Science 282, 1145–1147.

Till, J., McCulloch, E., 1961. Cell fate determination from stem cells. Radiation Research 14, 1419–1430.

Windpipe transplant breakthrough, Michelle Roberts BBC News. http://news.bbc.co.uk/2/hi/health/7735696.stm/.

## WEB SITES

California Institute for Regenerative Medicine. www.cirm.org.
Canadian Stem Cell Network. www.stemcellnetwork.ca/.
International Society of Stem Cell Research (ISSCR). www.isscr.org.
Lasker Prize. www.lasker.org.
National Stem Cell Bank. www.wicell.org.
Nature Stem Cell Project. www.nature.com/stemcells.
NIH glossary of stem cells. www.nih.gov/stemcell/glossary.
NIH Stem Cell. www.nih.gov/stemcell.
Nobel Prize. www.nobel.org/medicine, 2007.
Online Mendelian Inheritance. www.ncbi.nlm.nih.gov/sites/entrez?db=OMIM.
Science Magazine's Subject Collections: Cell Biology. www.sciencemag.org/cgi/collection/cell_biol.
Science Magazine's Subject Collections: Development. www.sciencemag.org/cgi/collection/development.
Stem Cell Community. www.stemcellcommunity.org.
Stem Cell Genomics Project. www.StemBase.ca.
UK National Stem Cell Network. www.uknscn.org.

# Pharmaceutical Biotechnology

**A Cancer Genome: Scientists Decode Set of Cancer Genes**

*New York Times,* November 5, 2008
    www.nytimes.com/2008/11/06/health/research/06
cancer.html?pagewanted=2&fta=y
    By Denise Grady
    For the first time, researchers have decoded all the genes of a person with cancer and found a set of mutations that may have caused the disease or aided its progression.

Using cells donated by a woman in her 50s who died of leukemia, the scientists sequenced the DNA from her cancer cells and compared it to the DNA from her own normal, healthy skin cells. Then, they zeroed in on 10 mutations, DNA differences that occurred only in the cancer cells, apparently spurring abnormal growth, preventing the cells from suppressing that growth and enabling them to fight off chemotherapy.

Mutations are genetic mistakes, and the ones found in this research were not inborn, but developed later in life, like most mutations that cause cancer. (Only 5 to 10 percent of all cancers are thought to be hereditary.)

The new research, by looking at the entire genome—all the DNA—and aiming to find all the mutations involved in a particular cancer, differs markedly from earlier studies, which have searched fewer genes for individual mutations.

## LOOKING AHEAD

This chapter describes the relatively new and rapidly growing field of pharmaceutical biotechnology. This research focuses on using advances in biotechnology to improve many aspects of drug development and action. A better understanding of the biological targets of drugs in the body will improve the chances of designing new, medically effective drugs. Designing drugs for treating humans is an extremely important aspect of pharmaceutical research. Over the past decade, the science of drug design grew in response to new information from the human genome sequence. New studies revealing the three-dimensional structures of the proteins that cause different diseases were a key step in drug design. Pharmaceutical biotechnology includes the amazing progress made in some of the most intriguing new scientific fields including carbon nanotechnology, lab-on-a-chip technology, and DNA computers.

After completing this chapter, you should be able to do the following:

- Discuss the process of drug design in the context of personalized medicine.
- Distinguish between the fields of genomics and proteomics and describe an example of research performed in each field.
- Understand the challenges posed by the process of drug delivery in the human body.
- Better understand the basic approaches used to design new drugs.
- Describe the unusual properties of carbon-based nanostructures built using nanotechnogy.
- Explain the technology behind lab-on-a-chip applications and suggest possible future applications of this technology.
- Explain why nanoparticles might have advantages as vehicles to deliver drugs to tumor cells.
- Describe the development and possible applications of DNA computers.
- Understand the basic risks and security issues associated with genome privacy.

## INTRODUCTION

The field of pharmaceutical biotechnology represents the marriage of many scientific specialties including genomics, proteomics, personalized medicine, drug discovery and development, lab-on-a-chip microtechnology, nanomedicine, and more. Scientists doing research in pharmaceutical biotechnology are focused on increasing the effectiveness of the drug treatments while minimizing potentially serious side effects. Pharmaceutical biotechnology research also focuses on developing new medicines and therapies to treat diseases in humans. Access to the complete DNA sequence of the human genome, including the continual sequence corrections and updates, provides a powerful resource that offers scientists a much more detailed picture of how human cells work and what goes wrong when disease strikes.

Scientists must have a detailed understanding of the intricate processes inside the cell in order to design drugs that can function along with normal bodily activities and also kills pathogens, often in the same cells. This information is essential for research in pharmaceutical biotechnology. Inside the human body, pharmaceutical drugs must maintain a fine balance between offering an effective treatment for a disease while at the same time avoiding potentially serious side effects. The most effective medications are drugs that do not interact with most cellular components while successfully attacking the intended target molecules that cause the disease.

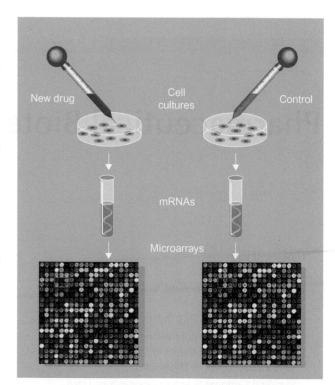

**FIGURE 13.1** Microarrays are used to detect genes that are expressed in some cells and not other cells. Scientists use microarray technology to study the expression of entire genomes to determine how gene expression is altered under different conditions. The two plates shown contain the same types of cells; the cells on the left were treated with the new drug, whereas the cells on the right were not treated (control cells). The mRNAs made in both cell samples were isolated separately, copied into cDNA probes, and analyzed by microarray technology.

## Genomics Is an Important Tool in Pharmaceutical Biotechnology

When the Human Genome Project revealed the entire DNA sequence of the human genome, for the first time scientists had access to all of the human genes, which had a positive impact on the growth of pharmaceutical biotechnology. Genomics research encompasses much more than just the study of the human genome, it also includes analysis of non-human genomes, coding and noncoding sequences, from both prokaryotic and eukaryotic organisms. Researchers study the structure and function of the encoded genes to explore how the gene products interact with each other and with factors in the environment. In pharmacogenomics research, scientists often use **microarray technology** to analyze changes in the expression of hundreds or thousands of genes in cells either treated with a drug or induced to undergo differentiation (Figure 13.1 and Figure 13.2).

Scientists use microarray technology to find out which genes are expressed in healthy cells compared to diseased cells or to determine how an administered

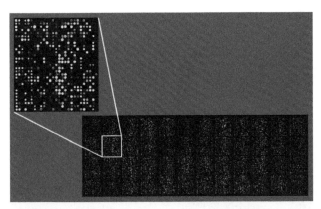

**FIGURE 13.2** Microarray DNA base pairs with cDNA probes. The genome DNA fragments are immobilized onto a microarray grid. The mRNA molecules isolated from the cells are copied into labeled, complementary single-stranded DNA s (cDNAs) and used as hybridization probes to search the microarray grid for complementary DNA sequences. *(Inset)* A closeup of the signals emitted by probes hybridized to DNA on the microarray. When a gene is expressed as mRNAs in a cell, the DNA spot representing that gene on the microarray will base pair to the cDNA probe and give off a detectable signal.

drug might affect the expression of specific genes in healthy and diseased cells. For this type of experiment, a scientist starts with two identical samples of the same cells but treats just one of the cell samples with a new drug (experimental sample). The remaining cell sample is not treated and is used as the control cell sample. The genes expressed as mRNAs in the two different cell samples are isolated and analyzed separately. The mRNA molecules are very fragile and cannot be easily manipulated or studied. It is preferable to convert the mRNA molecules into DNA strands by copying the mRNAs into cDNA (complementary DNA) molecules using reverse transcriptase enzyme. The single-stranded cDNA molecules are synthesized with incorporated labels that permit the DNA strands to be used as hybridization probes to detect and base pair to the complementary DNA sequences displayed on the microarray grid. The microarray grid contains a large array of DNA spots distributed in a uniform pattern that permits each spot to be located. The DNA on the microarray represents genome DNA isolated from the control cells, and contained DNA from all of the genes in the cell genome. The results of this analysis are interpreted from the spots of DNA on the microarray that give a positive signal indicating the presence of complementary DNA. When a particular gene is expressed as RNA in the cell, then the specific DNA spot(s) representing that gene on the microarray grid will indicate that fact by emitting a visual signal when the chromosome DNA base pairs with the cDNA probes copied from that gene's mRNAs. Microarray analyses allow scientists to monitor changes in the expression of hundreds or thousands of genes in many

different types of cells under a wide variety of different growth conditions and drug treatments.

Genomics involves studying the cell from the standpoint of the genome and the nucleus, with a focus on the genes, whereas **proteomics** is the study of the huge collection of proteins made in a typical cell. The term "proteome" was first used in 1994 to describe all the proteins that are expressed in a given cell at a given time. The genes in the human genome provide the basic instructions for the proteins that make the molecular-machines in the cells. The proteins are the cellular components that perform so many different jobs in the body. For example, proteins make muscles move, they digest food into the nutrients absorbed by the cells, and they help nerve cells to send signals in the brain. The field of proteomics includes the study of interactions between and among proteins, protein modifications, and other changes that affect protein functions in the cell.

Proteomics researchers analyze the structures and functions of the thousands of different proteins expressed in the trillions of cells in the human body. Genomics researchers focus on the DNA genome and the inherited genes that direct the synthesis of all the proteins in the cell.

## Protein Profiles Change in Diseased Cells

The field of proteomics includes the various modified and processed proteins produced in different cell types. Cells have three main types of **posttranslational modifications**: **phosphorylation** (phosphates added to proteins), **methylation** (methyl groups added to proteins), and **glycosylation** (sugar units added to proteins). Proteomics researchers also compare the different proteins made in diseased cells and healthy cells, searching for clues about how the disease process might affect the proteins made in the cell.

Studies on the spectrum of proteins made in a specific cell type, a protein profile, have successfully identified **biomarkers**, specific proteins that invariably change in some way with the onset of a disease. Biomarker proteins can play an important role as a characteristic indicator of specific diseases, which can be easily detected in the patient's blood or urine. "Early" biomarker proteins are produced in cells during the early stages of a disease, before full-blown disease symptoms occur. A commonly known example is prostate-specific antigen (PSA), a protein that is made by the prostate gland and measured in the blood by a PSA test. In many cases increased PSA levels result from the uncontrolled growth of prostate cells, which is a possible indication of early prostate cancer. This allows doctors to use PSA as an early biomarker indicator for this cancer. Unfortunately, the PSA test does not provide enough information to distinguish between real prostate cancer and benign conditions of the prostate that also raise PSA levels. But an increased PSA level should not be ignored and usually prompts both the doctor and patient to follow up on the possibility of prostate cancer.

The huge numbers of proteins visualized in the protein profiles of normal and diseased cells are compared by computer to identify the key proteins produced only in the diseased cells. These proteins might be essential to the disease process and therefore are potential targets for drug development.

## PERSONALIZED MEDICINE: DREAM OR REALITY?

Human genome research continues to have a large impact on our understanding of human genetic diseases and the large role that genes play in conferring biological traits. Many scientists predict that in the future the use of genetic information taken from an individual's genome, called personalized medicine, will greatly improve the diagnosis and treatment of human diseases. If personalized medicine becomes standard, it will be routine for the nurse in the doctor's office to put a drop of fingertip blood from a patient into a handheld lab device for immediate analysis. Once the tests are complete, a personal genetic profile will appear on the handheld computer screen. If medication is needed, the doctor will include an analysis of your genome information when deciding the appropriate drug to prescribe. The long-term success of personalized medicine will depend in part on the public's willingness to have their genes tested, the willingness of doctors to change how they prescribe drugs, and the level of genome security that will be available to patients.

### Unusual Drug Reactions Linked to Genetic Variation

The normal variations found in the DNA sequences of different human genomes such as RFLPs and SNPs are linked to the inheritance of some specific human diseases and disorders (see Chapters 6 and 10). These genetic variations in the human genome can influence the way that individuals respond to treatment with different drugs. For example, certain painkiller medications are effective in the body only after specific activator proteins convert the drug from an inactive to an active form. The idea that genes are involved in the

**FIGURE 13.3** Cheek swab is used to collect cells and DNA. The cell samples for DNA testing are collected using a safe, inexpensive, and painless method such as gently scraping cells from the inside of the cheek using a swab. The cells on the swab are then processed for DNA testing.

different responses that some people have to the same drug is not new. About 40 years ago, scientists found that drugs were removed (cleared) from the body at different rates in identical and fraternal twins, showing that genes do contribute to variable drug responses. These experiments contributed to the formation of the modern fields of pharmacogenetics and pharmacogenomics.

Scientists are studying the genomes of people who react differently to common medicines by searching for specific variations in genome DNA sequences that correlate with an individual's reactions to certain drugs. Samples for DNA testing are easily collected from blood or more routinely by scraping the cheek cells from inside the mouth (Figure 13.3). These studies showed that certain proteins responded to drugs differently in different individuals as a direct result of the genetic variation between human genomes. Studies on the genetic makeup of individual people provides important information for scientists and health care professionals, but also pose potential privacy risks for patients, even if the personal genetic information is stored in a secure computer database to protect the privacy of the patients.

Proponents say that personalized medicine will help patients to avoid the unnecessary and potentially dangerous side effects that can result from the routine practice of one-size-fits-all medicine. The research innovations offered by personalized medical care will also enhance our understanding of the human genes that contribute to cancer and many other human diseases. The widespread use of personalized medicine will depend in part on continued rapid advances in the science of genetic testing. Researchers can now perform relatively inexpensive, and very accurate genotype analyses using high-throughput genetic screening tests that are already available to the public. Despite such advances, few physicians actually practice personalized medicine and few customized drugs have been developed for use in humans.

A patient's personal genetic information will help doctors to prescribe the right medicine for that individual and avoid side effects from the medications. This would represent a big step toward making medicines much more effective and less expensive due to reduced risk.

## Personalized Cancer Treatments

In the mid-2000s, the United States government invested $100 million in the Cancer Genome Atlas project and gave much smaller amounts to other research programs designed to study the genomes of patients with different types of cancer. The goal was to identify the collections of genome DNA differences that occur in cancer cell genomes but are not present in the genomes of healthy cells from the same individual. Different types of cancers develop in different tissues, but the process always starts with a single cell that acquires a new DNA mutation in its genome DNA. This new DNA mutation is not usually sufficient to cause the cell to become transformed into a cancer cell, but over time the cell genome acquires additional DNA mutations. The development of the cancer-like characteristics in the cell does not occur until the cell genome has acquired a certain set of DNA mutations that act together to convert the normal cell into a cancer cell. This set of mutations is often called the **cancer genome**.

Scientists working on the Cancer Genome Project (CGP) (Washington University) focused on identifying specific DNA genome mutations that are involved in developing acute myelogenous leukemia (AML), an aggressive cancer that affects about 13,000 adult Americans each year and kills 8,800. The scientists studied the genome of an AML patient who volunteered for the Washington University study, and they found eight mutations linked specifically to cancer development. Unfortunately, AML took this patient's life just two years after the cancer diagnosis despite chemotherapy and two bone marrow transplant treatments. In the future researchers hope to avoid this outcome by using information from the cancer genome study. Sometimes if doctors know early enough that a specific patient's cancer is likely to be unusually aggressive, they can decide to prescribe more powerful treatments much earlier in the course of the disease.

Further studies on the AML-specific genome variations will help to determine which cellular function is affected by each genome mutation. For example, changes in the genome DNA might increase the functions of genes that promote the growth of cancer cells (oncogenes) or destroy the genes that code for the tumor suppressor proteins that usually protect healthy cells from cancer (see Chapter 9). Other genome mutations give the tumor cells the advantage by efficiently removing the chemotherapeutic drugs from the tissue before the cancer cells are damaged or destroyed. Scientists continue to search for the specific locations in the human genome of the specific DNA mutations that finally tip the cell's biochemical balance toward becoming a cancer cell. It is important to understand the signal or set of signals or biochemical events that finally change and cause the cell to abandon normal growth controls and become cancerous.

## Personal Genome Science: Google Your DNA

Over the years, the cost of DNA sequence analysis decreased substantially because of improved DNA sequencing technology and the widespread use of automatic DNA sequencing machines (see Chapter 6). Ongoing innovations in DNA sequencing technology will continue to provide faster and even more accurate DNA sequencing methods, and further important information will come from comparing the genome DNA sequences from many different individuals. The members of the Personal Genome Project (PGP) volunteer to share their personal genome DNA sequences and personal genetic information with other people. Despite the potential security issues, they even post their DNA sequences on the Internet to share with the entire world. An important goal of the PGP is to make personal genome sequencing more affordable and accessible and to help people to better understand how genes and the environment influence human traits.

Scientists no longer need to actually determine the complete DNA sequence of an entire genome in order to glean a large amount of genetic information about an individual. Sometimes instead of sequencing an entire genome, scientists analyze only those regions of the genome that vary in DNA sequence between individuals. This approach forms the scientific basis for the DNA fingerprinting technology used in forensic DNA testing (see Chapter 8). Variable regions of the genome often include single base pair differences (SNPs) in individual human genomes. Many of these SNPs and other specific variations in DNA sequences have been genetically linked to an increased risk for developing a specific human disease.

The science of personal genomics is in its infancy but has already had a large impact on the fields of medicine, health insurance, and economics. Personal genomics has immediate applications in health care because of the wealth of information in an individual's genome that can be accessed and interpreted by genetic testing. In the past, the ability to explore the human genome has been the exclusive domain of scientists working in sophisticated research labs. This situation changed dramatically in 2007 when biotechnology companies began to offer commercial genome testing for the first time. As part of their genetic testing services, the companies send clients their genotype results as well as an interpretation of the test results, including the potential lifetime risk of developing certain genetic diseases (Figure 13.4).

When a reporter from the *New York Times* became one of the first people to have her DNA genome tested,

23andMe analyzed 550,000 SNPs in her DNA and identified the sequence differences or genotypes at those locations in her genome (Figure 13.5). For example, the reporter's genotype of the alleles for adult lactose intolerance is "GG," in agreement with the reporter's experience that she is indeed unable to tolerate dairy foods containing lactose. A different genotype DNA sequence at that SNP location in the genome confers the predominant trait of easily digesting lactose. The three different biotechnology companies (23andMe, deCODE Genetics, and Navigenics) used similar methods to analyze the genotypes of different SNPs in each genome. At that time all of the genome testing companies voluntarily offer some type of educational assistance to help people to better understand the information in their own genomes. However, this assistance varies from professional genetic counselors available for consultation, to offers of referrals to genetic counselors. The cost of this

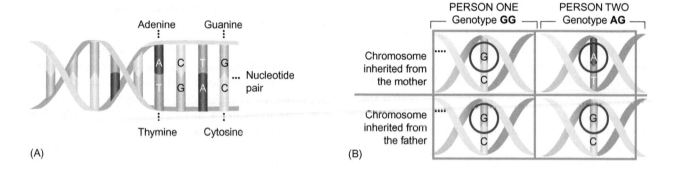

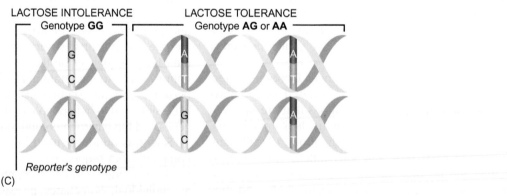

**FIGURE 13.4** Tiny variations in the human genome can have a large impact on the person. (A) The DNA sequence of the genome of a New York Times (NYT) reporter was tested by the 23andMe company in 2007. The company determined the specific order of the DNA base pairs at many regions of the DNA helix in the reporter's genome. (B) The order (sequence) of the DNA bases are almost identical in different human genomes, but the sequences can vary and there are millions of genome locations where the human genome sequences differ by single base pairs called single nucleotide polymorphisms (SNPs) (see Chapter 8). In a specific SNP location in a DNA genome, person 1 inherited the same base pair from each parent, whereas person 2 inherited different base pairs at that site. In other words, at this specific SNP location, person 1 has a genotype of "GG," and person 2 has the genotype "AG." (red circles) (C) To understand the biological consequences of inheriting different SNP genotypes, scientists studied the genotypes of people with various traits and established correlations between certain SNP genotypes and inheriting certain traits. In the example offered by the reporter's genome, an SNP variant located near a gene (LCT) that encodes a lactase enzyme is correlated with whether or not the lactase gene is expressed during adulthood. People with the genotype "GG" at this SNP, like the reporter, are more likely to experience lactose intolerance than people who inherited the "AG" or "AA" genotypes at that SNP genome location.

genome analysis service in 2007 ranged from the most expensive at $2500 for Navigenics (but this price also included genetic counseling) to about $1000 dollars for 23andMe ($999) and deCODE Genetics ($985).

Reporting genetic results without information or support from a genetic counselor or other healthcare professional is not ideal and raises the possibility that people will easily misinterpret the genetic predictions. For example, in this case it should be made clear to customers that inheriting any combination of ACTN3 alleles does not guarantee that a child will develop the physical attributes and talents of a natural athlete and compete in the Olympics. People should be reminded that human traits, including personality and behavior, result not just from the expression of inherited genes but are also influenced by the environment, as well as lifetime experiences and diet.

The availability of these personal genomics testing services has raised concerns about standardizing access to genetic counselors and increasing educational support services to help people better understand their own personal genome information. The implications suggest that personal genomics will affect a wide audience including doctors, medical educators, medical economists, insurers, and policy makers as well as the general public. Access to educational resources for individuals who have their genomes analyzed will be very important to help people to recognize misinformation and to be able to sort among the possible solutions to a problem according to the quality of the supporting evidence.

The new industry of personal genomics has already begun to create a new industry to respond to the demand for commercial products and services that are

| SNP | Location | Genotype | Genotype associated with |
|---|---|---|---|
| rs662799 | APOA5 | AA | Tendency to gain weight when eating fatty foods |
| rs174575 | FADS2 | CC | Higher I.Q. if breast fed for nine months as infant |
| rs6920220 | 6q23 | GG | Low risk of rheumatoid arthritis |
| rs17070145 | KIBRA | CC | Relatively poor verbal memory |
| rs1801260 | CLOCK | AA | Early rising |
| rs1953558 | OR11H7P | CC | Sensitivity to smell of sweat |
| rs17822931 | ABCC11 | CC | Wet earwax |
| rs4613903 | TAS2R38 | CG | Ability to taste certain bitter flavors |
| rs3751812 | FTO | GG | Lower body-mass index |

**FIGURE 13.5** Different SNP genotypes are linked to different human traits. When the DNA genome of the NYT reporter was tested by the 23andMe company in 2007 there were 550,000 SNPs identified in the human genome. The company screened many human genomes and determined the specific DNA sequences or genotypes at specific SNP genome locations. Specific SNP genotypes are correlated with inheriting certain human traits such as lower body-mass index.

---

**Box 13.1    Born to Run?**

**Little Ones Get Test for Sports Gene**

*New York Times*, November 29, 2008
    www.nytimes.com/2008/11/30/sports/30genetics.html?_r=1&emc=eta1
    By Juliet Macur
    A new DNA test recently became available, which boasts that it offers a way to test future athletic potential. Many parents were attracted to the possibility that DNA testing might be able to predict the athletic potential of their toddler. The "sports gene" test actually searches the human genome DNA for the human ACTN3 gene, which encodes the alpha-actinin-3 protein (Figure 13.6). The R gene variant (allele) of the ACTN3 gene produces the alpha-actinin-3 protein in the fast-twitch muscle fibers that perform the powerful, rapid contractions needed for speed, power, and endurance. The X allele of the ACTN3 gene is not expressed and the alpha-actinin-3 proteins are not made in these cells.

    To find out how the different forms of the human ACTN3 gene might influence the inheritance of athletic traits, the scientists determined which combination of ACTN3 alleles (R, R or R, X or X, X) were inherited by 429 top athletes, including 50 Olympians (see Figure 13.6). Then the scientists looked for correlations between the athlete's ACTN3 allele genotype and proven athletic performance. The results of this study showed that people who inherit either the (R, R) or (R, X) combinations of alleles have an advantage in terms of natural power and endurance, but those inheriting the X, X allele apparently do not exhibit enhanced athletic traits.

    The ACTN3 sports gene test was first marketed in 2004 in Australia, Europe and Japan by an Australian company, Genetic Technologies, and became available in the United States through Atlas Sports Genetics a few years later. The DNA sample is processed by the Atlas Sports Genetics lab for genotype analysis at a cost of $149 per test (2008). In a few weeks the DNA test results are sent to the customer by mail, along with information suggesting which sports are most appropriate for the child to pursue to help the child reach his or her full athletic potential.

Box 13.1    Continued

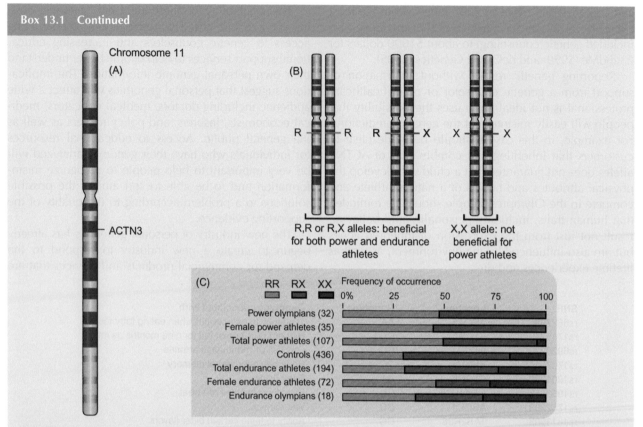

**FIGURE 13.6**  Human "sports gene" ACTN3 encodes the alpha-actinin-3 protein in muscle cells. (A) DNA variations in the ACTN3 gene on chromosome 11 affect inherited athletic traits. (B) The R allele (version) of the ACTN3 gene produces a muscle protein that has an important role in the function of fast-twitch muscle fibers. The R,R or R,X genotypes are beneficial for both power and endurance athletes, but the X,X alleles of the ACTN3 gene do not confer an athletic advantage. (C) The frequencies of the three ACTN3 genotypes (RR, RX or XX) in the population were determined to see if there is a correlation between specific ACTN3 alleles and people who are athletes with proven endurance or power skills.

designed to enhance the use of genomic information by individuals and healthcare professionals. As commercial genome sequencing and related DNA services become more popular, quality control will become increasingly important; imagine the damage caused by a mistaken genetic diagnosis of a terrible, fatal disease like Huntington's (see Chapter 10). The FDA could potentially be responsible for approving products associated with genome analysis services and specific medical applications. Genome privacy issues must be balanced with the ever present need to increase corporate profits.

## DRUG DISCOVERY AND DEVELOPMENT

Research in many areas of pharmaceutical biotechnology provides critically important information to help scientists to design and produce new and better medicines to treat disease. This research greatly contributes to our understanding of the human genes that are involved in cancers, heart disease, and many other diseases, just as research in these areas contributes to advances in pharmaceutical biotechnology. The definition of a drug is any chemical that is used for the diagnosis, cure, treatment, or prevention of diseases in humans and animals. There are many types of drugs including some drugs that act by physically binding to many different targets in the cells. The first step in developing an effective new drug often begins with studying the biochemical mechanisms involved in the specific disease process the drug will be used to treat. Most of the biochemical reactions in the cell, including disease processes, involve the actions of protein enzymes. Many protein-protein interactions can result in the assembly of large protein complexes containing many smaller proteins (Figure 13.7). When a protein binds to a drug, the activity of the protein can be enhanced or inhibited by the interaction, which can block the disease process (Figure 13.8). The goal of this research is to find out how a given disease affects the cell's mechanisms, looking for key proteins that might be possible targets for the action of the drug. A drug

(A)

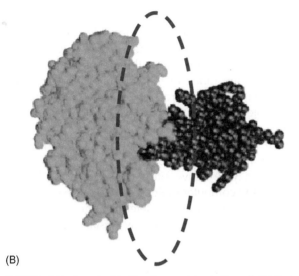

(B)

**FIGURE 13.7** Proteins bind to other proteins in the cell. (A) Proteins bind to other proteins to form everything from dimers with two proteins to the assembly of multi-protein complexes that often contain many protein subunits. Proteins bind to each other if the surfaces of the proteins have the appropriate chemical structures and shapes to fit together in three dimensional space. (B) The trypsin enzyme protein *(green)* is shown bound to the bovine trypsin inhibitor protein *(blue)*. These two proteins bind to each other by interacting through special regions of both proteins (red circle).

**FIGURE 13.8** Chemical is docked in the protein active site. The chemical compound (molecular stick structure) interacts with the protein by forming chemical bonds with specific amino acid side chains in the active site of the enzyme, blocking enzyme activity (see Chapter 3).

generally works best when it has a high binding affinity for its target protein because this avoids nonspecific binding of the drug to random cellular components that are not part of the disease process.

Receptor proteins are often used as the targets for drugs because receptors are exposed on the surfaces of the cells where they bind to specific proteins in the cell's environment. This means that is it important for scientists to identify the functions of proteins involved in the disease being treated. An unknown receptor protein can often be identified by comparing the amino acid sequence of the unknown receptor protein with databases of proteins with known functions (see Chapter 7). These structure-based approaches to drug design involve combining sequence information from bioinformatics with structural evidence from x-ray crystallography and nuclear magnetic resonance (NMR) spectroscopy studies. These results determine the molecular composition and three-dimensional structure of the unidentified protein. The information about the biological target molecule including the DNA and protein sequences, genetic variation, genetic maps, gene and protein expression data, and the protein structure results are used to devise biochemical and genetic screens to find effective new drugs.

The structural approach to drug design is based on understanding the molecular structures of the biological drug targets, with special attention paid to regions that interact with the active sites on the enzymes. Other drugs are specifically designed to inhibit key physical interactions between a protein and its target. This approach has potential, but engineered drugs sometimes do not bind to their targets as predicted when tested, probably because the engineered molecules are designed to bind to the static structure of the target protein, and not to the moving, "living" protein that changes shape in the cell. How a protein changes conformation in the cell is an important factor that must be considered by scientists working to improve drug design.

High throughput screening for new drugs starts with special computer software that creates structural models of a drug molecule using the information gained from studying interactions between the drug and established biological targets. When the structure and function of the active site of the receptor protein are well understood, scientists can use computer programs to design new drug molecules predicted to bind tightly to the target protein (Figure 13.9). Computer docking programs are used to test large libraries of chemicals for the ability to bind to the biological target, followed by other tests to assess the specificity and strength of the drug binding to the target protein. This method is used to identify molecules that interact accurately with the chosen target and do not bind nonspecifically to

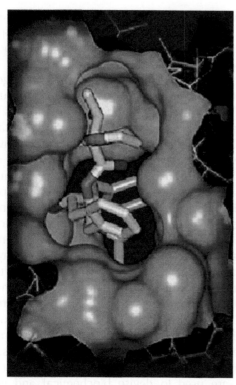

**FIGURE 13.9** The HIV protease enzyme is essential for an AIDs infection to develop. The active site of the HIV protease enzyme looks like a three-dimensional pocket in the protein structure. A drug designed to fit inside the active site has the potential to block the activity of the enzyme, stop the HIV virus and treat the disease. The x-ray crystallography studies show the molecular interactions between the HIV protein and the atoms of the drug (Kaletra): carbon atoms (gray), nitrogen (blue), and oxygen (red).

the "active site" of a specific enzyme. The computer searches through many thousands of different molecules, checking each one to look for the single molecule that fits tightly into the three-dimensional shape of the empty "active site" in the enzyme. Earlier computer approaches to find specific enzyme substrates were often unsuccessful because they relied on only stable protein structures. However this new research accomplished an amazing computational feat by producing simulated candidate substrates that accurately mimic the molecular structures of the unstable "intermediate" protein conformations. These transient protein structures exist only very briefly during the biochemical reaction as the enzyme binds to the substrate and transforms the substrate into the product molecule. The intermediate molecular structures are very unstable and are not detected by tests designed to find out how protein structures fit into the active site of the enzyme. In some cases the fit between the candidate substrate molecule and the enzyme active site can be predicted by the results of the computer molecular docking experiments, and confirmed by studies to determine the atomic structure of the substrate through x-ray crystallography.

> Some new drugs are synthetic molecules designed to fit into the active site of the key enzyme. This blocks the action of the enzyme and prevents the disease process from continuing to damage the cells.

## Drug Design and Personalized Medicine

The National Institutes of Health (NIH) sponsors research in the United States to determine why individuals have very different physical reactions to certain medicines. For example, some people get side effects but no pain relief from common painkiller medications, and certain allergy and asthma drugs work well for some people but are not at all effective for others. In 2008, millions of people in the United States risked an overdose reaction to taking standard amounts of a medicine that is commonly used to prevent blood clots. This NIH-supported research is focused on improving the health of all Americans by studying human cells and finding out what goes wrong when the human body experiences disease or injury. This research is directed at optimizing the positive effects of common, often essential, medications while preventing allergic reactions and other serious side effects.

similar molecules. One example is an AIDS drug that was designed with the right shape and other chemical characteristics so that it fits inside the active site of the HIV protease enzyme, much like a key fits into a lock. X-ray crystallography studies reveal the snug molecular interactions between the HIV protein and the atoms of the drug (see Figure 13.9).

Research is ongoing to determine how amino acid chains fold into three-dimensional protein structures that confer specific functions. Additional research is necessary for scientists to better understand the relationship between a three-dimensional protein structure and the biochemical activity of the protein. In 2007 for the first time a team of scientists (University of California, San Francisco, Albert Einstein College of Medicine, Texas A&M) determined the biological function of an enzyme based only on the three-dimensional structure of the protein. The scientists used computer-aided modeling and molecular docking to search a database of protein sequences for potential drug-protein interactions that block the action of the enzyme.

It is possible to use computer-aided molecular docking when the researchers know the atomic structure of

Different people often respond to common drugs differently, including medicines that lower cholesterol levels, or treat cancer, or anti-AIDS medications. In many cases these diverse reactions can be linked to naturally occurring variations in individual human genome

DNA sequences. These genome DNA differences influence how individual people respond to different drugs. Eventually results from this research will be used routinely by health care providers to prescribe the most effective medicines and treatment for individual patients. They focus on the liver, the key organ responsible for metabolizing the wide array of drugs and toxins that enter the human body. But liver function varies among people. For example, some liver enzymes act on painkiller drugs to convert the drug into an active form inside the body. The proteins that activate painkillers exhibit varying levels of function in different individuals, which correlates with DNA variation in individual human genomes.

Scientists have analyzed the genomes of people who have unusual reactions to common drugs, in order to identify variations in the DNA sequences of genes that are potentially involved in these atypical reactions. Studies designed to correlate diseases with sequence variations between individual human genomes will provide important information to help doctors prescribe the right medicine for each patient. In the future, personalized medicine will make drugs more effective and less expensive, and will decrease the unnecessary side effects that often result from one-size-fits-all medical treatment.

Research in drug development requires a detailed understanding of the biochemical mechanisms involved in the disease process. Protein-protein interactions play a central role in most diseases and often reveal successful targets for drug intervention and disease treatment.

## Modular Drugs: Leukemia Drug Shows Early Promise

A new type of bioengineered medicine called a modular immune-pharmaceutical drug was recently shown to be effective for the treatment of chronic lymphocytic leukemia (CLL) in lab animals. The new antileukemia drug, called CD37-SMIP (Trubion Pharmaceuticals, Inc.) binds to the CD37 proteins located on the surfaces of the leukemia cells. The binding of the CD37-SMIP drug to the CD37 protein sends a signal into the leukemia cell, which triggers programmed cell death (apoptosis). This process of cell suicide is used by the body for injury repair and to eliminate unnecessary cells during embryo development (see Chapter 9). Treatments with the new leukemia drug, CD37-SMIP, proved to be as effective as the standard anticancer drug, rituximab, which binds to a different target protein on the surface of the leukemia cells (Figure 13.10).

This new drug might be effective to treat other types of cancers that also express the CD37 protein on the

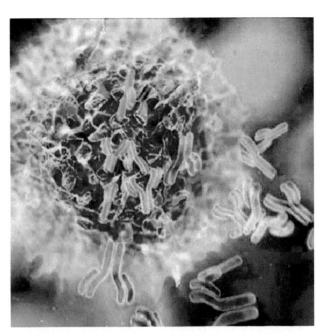

**FIGURE 13.10** Some anticancer drugs bind to target proteins on the surface of the cancer cells. Rituximab is an anticancer drug developed from a monoclonal antibody ("Y") that binds to a specific target protein on the surface of leukemia cells. In combination with chemotherapy, Rituxan has helped to treat patients with non-Hodgkin's lymphoma, a cancer that occurs in about 20 out of 100,000 people.

cell surface, including non-Hodgkin's lymphoma and acute lymphoblastic leukemia. The cell suicide mechanism triggered by the new anticancer CD37-SMIP drug differs from other anticancer drugs that induce apoptosis by triggering a cascade of caspase enzymes. This new type of anticancer agent could be an effective treatment for patients who have become resistant to other anticancer drugs, because the function of the CD37-SMIP drug is independent of the caspase enzyme pathway.

## Teaching an Old Drug New Tricks

Scientists also design effective new drugs by analyzing the mechanisms of less effective medications, which is one alternative approach to screening cells with thousands of small molecules to look for a specific biochemical response or protein activity. They decided that it might be possible to uncover new drug activities hidden within the structure of a drug commonly used to treat a different malady. A research team of computational structural biologists at Scripps Research Institute (La Jolla, California) focused on prostate cancer treatments using drugs that block the function of the androgen hormone receptor, a protein that plays an important role in the development of many cancers. The researchers started with a well-known and commonly used antipsychotic drug, which they transformed into a potential treatment for prostate cancer. Researchers studied 3D computer models of the structure of the androgen hormone receptor and used advanced computer algorithms to predict the 3D structure and properties of the binding pocket in the receptor protein. Information gathered from a large collection of binding pocket models, enabled the computer docking program and virtual screening functions to detect molecules that bind inside the active site pocket and specifically block the function of the receptor protein. This approach recently revealed inhibitors of other hormone receptors and proteins including the retinoic acid receptor, the epidermal growth factor receptor (EGFR), the anthrax lethal factor, dynamin, and alpha-1-antitrypsin.

Advances in the use of computers to analyze protein structure and predict accurate three-dimensional structures have greatly enhanced the ability of researchers to design new drugs effective for the treatment of many different human diseases.

## LAB ON A CHIP TECHNOLOGY

Tiny handheld lab on a chip devices are under development in many labs, and will soon permit physicians, crime scene investigators, pharmacists, and the general public to quickly conduct inexpensive tests on DNA and other biological compounds, anywhere, anytime. Scientists predict that within a decade the public will be able to purchase DNA test kits at pharmacies that are marketed for use at home, allowing people to self-test for a variety of diseases, even before the patient has developed any symptoms. The handheld lab-on-a-chip devices are about the size of a microscope slide, and will house a variety of miniature analytical laboratory tools (Figure 13.11). The handheld lab requires tiny reaction volumes that are thousands of times smaller than those used in a normal lab, which allows the reactions to proceed at rates that are 100 times faster than in a traditional lab.

In 1998 Hewlett-Packard Company and Caliper Technologies Corp. announced a collaboration to further develop lab-on-chip technology based on the original studies done at Oak Ridge National Laboratories, the University of Alberta, and Ciba Geigy. Caliper Technologies developed practical applications for microfluidics technology that represents a major advance in this technology. The Caliper scientists etched tiny integrated biochemical processing circuits into glass, silicon, quartz, or plastic supports to create channels about 80 micrometers wide and 10 micrometers deep. These lab chips perform the same protocols as conventional instruments, but the lab chip uses very tiny amounts of sample and performs at speeds that are much faster than usual because the fluids move at very high rates (Figure 13.12).

Voltages are applied to various channel intersections to direct the sample in the fluid channels throughout the chip, adjusting concentrations across three orders of magnitude, separating components, adding fluorescent tags, and sending the sample past detection devices for digital output. The microchip lab devices are designed to be flexible and in the future will offer an array of multipurpose workstations, with chips on hand to convert an instrument from one function in the morning to a different one in the afternoon. In combination with DNA on a chip and microarray technology, the

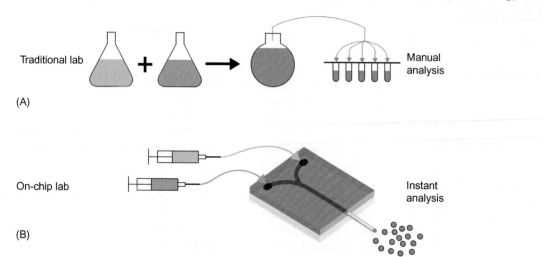

**FIGURE 13.11** Lab on a chip technology. (A) Scientists in a traditional molecular biology lab studying DNA and proteins will set up experiments using micro-pipets and involving tiny reaction volumes that take place in small plastic microfuge tubes. The idea behind the lab-on-a-chip devices currently under production is to miniaturize the traditional experimental protocols and reproduce them on a micro chip. The lab-on-a-chip devices will rapidly determine the chemical composition of substances as well as perform DNA testing and more.

science of microfluidics has the potential to dramatically improve the accuracy and speed of current diagnostic tests (see Figure 13.12). In the future, handheld lab devices will be used by health care professionals to screen for serious diseases and for rapid testing for an array of infectious diseases, including HIV, anthrax, the avian flu, the swine flu (H1N1 flu), and some cancers and genetic defects. One lab on a chip is being designed to travel to Mars to test for proteins that signal possible past life forms on the red planet (Figure 13.13).

> Lab on a chip technology allows the microchip to perform the same protocols as conventional instruments, except much faster and using much smaller samples.

Lab-on-a-chip technology is part of the trend toward "personalized medicine," in which health care is increasingly tailored to the specific genetic profile of each patient. Highly specialized personalized care allows physicians to design therapies and prescribe specific medications for patients while also considering the genetic profile of each patient. Together with simplified genetic testing, the lab-on-a-chip technology will reduce the overall cost of DNA testing

and will help to spread the idea of personalized care based on individual genetics. As this microtechnology is improved and refined, the costs of producing the handheld devices will go down. More people and health care groups will be able to afford the purchase price and the handheld labs will be adopted for use in a wide variety of clinical settings.

The diagnosis of human disease is complicated. For example, a magnetic resonance imaging (MRI) scan of the brain is one way to diagnose brain health, but even an MRI cannot reveal the health of individual nerve cells (Figure 13.14). In many cases, early diagnosis of a disease significantly improves the patient's chances of recovery, especially in the case of early cancer diagnosis. Eventually the technology of "proteomic fingerprinting" will be used to routinely reveal the protein expression patterns in healthy and diseased cells and will provide specific tests to rapidly assess the health of tissues and organs in an individual patient. Similar lab on a chip technology will be customized for use by people in law enforcement and forensic investigators who need to gather information quickly and often need to use small biological samples. The handheld nanolab will analyze tiny samples of blood or semen

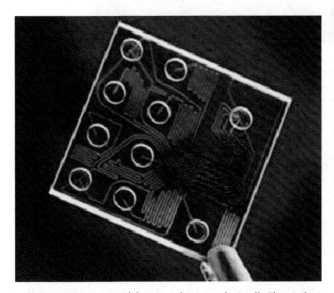

**FIGURE 13.12** A mini-lab on a chip can do it all. The Agilent Bioanalyzer (Agilent Technologies) carries out highly reproducible chemical reactions and DNA manipulations by utilizing a network of channels and wells etched into glass or polymer chips that accommodate very small sample and reagent volumes. Tiny changes in pressure or electro-kinetic forces propel the tiny volumes of fluids through the channels in a highly controlled process that depends on the desired test. The lab-on-a-chip can handle multiple samples without making an error while it performs tasks such as mixing or diluting reagents, resolving DNA, RNA and proteins by electrophoresis or chromatographic separation, and tags to provide visual detection.

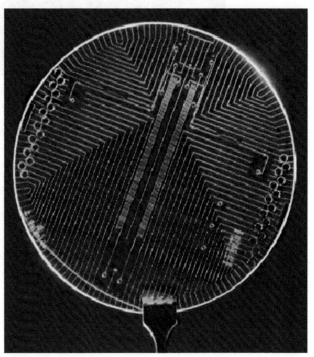

**FIGURE 13.13** Lab on a chip to visit Mars. This lab on a chip device is designed to fly to Mars in several years to look for signs of life (UC Berkeley). The chip has a microcapillary electrophoresis system that can determine the composition and "handedness" of any amino acids found on Mars. As chemicals, amino acids can exist as either left or right handed molecules, but on Earth living organisms contain only left handed amino acids. For this reason scientists need to determine if any amino acids found in space are left- or right-handed.

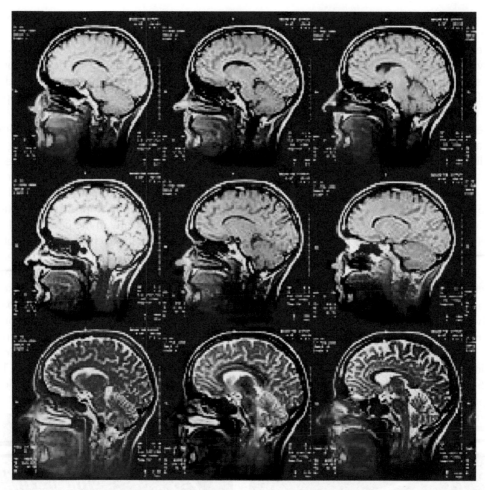

**FIGURE 13.14** MRI scans are one way to diagnose brain health and nerve function. These colorful pictures are magnetic resonance imaging (MRI) scans of a healthy human brain. MRI is a common method used to diagnose certain illnesses but requires a trip to the hospital or clinic. Scientists are developing a large array of tests for people to take at home in the future that would potentially check the health of every organ in the human body including the brain. A similar device could be used to screen people for infection during a flu epidemic or other infectious health emergency.

---

### Box 13.2  Nanotubes Help Advance Brain Tumor Research

*Science Daily,* January 2008

Nanotechnology promises to revolutionize the future of certain cancer therapies. Researchers at City of Hope have delivered cancer-fighting agents into the cells using a new type of carbon nanotube carrier. The carbon nanotubes are hollow cylinders made by rolling sheets of graphite-like carbon into tubes that are 50,000 times narrower than a human hair, and can be several centimeters long [Nano and Micro Systems Groups, Jet Propulsion Lab (JPL)]. These new carbon nanotubes efficiently carried drugs, DNAs, and siRNAs into cells, were not toxic to the brain cells of mice, and did not inhibit or promote cell reproduction.

JPL scientists and City of Hope doctors decided to collaborate on developing nanostructures to better diagnose and treat brain cancer. This nanomedical research collaboration focuses on attaching inhibitory RNA molecules to the nanotubes, which will be delivered into specific target regions of the brain. If the nanotube delivery technology can be effectively used to treat brain tumors, the same approach might also work for treating stroke, trauma, nerve disorders and other diseases that affect the brain.

Carbon nanotubes are extremely strong, flexible, and heat-resistant structures, making them excellent for use as field-emission cathodes, which help produce electrons, and for various aerospace applications involving x-ray and mass spectroscopy, vacuum microelectronics and high-frequency communications. "Nanotubes are important for miniaturizing spectroscopic instruments for space applications, developing extreme environment electronics, as well as for remote sensing," said Harish Manohara, a JPL supervisor. Nanotubes are not usually part of current NASA space missions, but they could be used in gas-analysis or mineralogical instruments for future missions to Mars, and other planets.

right at a crime scene and the results will be immediately compared to a genetic database, with the intent of identifying a suspect soon after the crime. Applications of nanolab technology would also impact many areas of agricultural biotechnology, including rapid genetic tests performed on thousands of hybrid plants to screen for the desirable traits such as resistance to disease and drought.

## NANOTECHNOLOGY

Nanotechnology research takes advantage of the ability to manipulate atoms and molecules to create new materials and nano-devices. Certain metallic and carbon-based nanometer-sized structures exhibit novel physical and chemical properties when they are produced in miniature. Examples of these dramatic property changes include opaque substances that are transparent nanostructures (copper), inert materials that become potent chemical catalysts (gold) and insulating materials that transform into efficient nanoconductors (silicon). The novel nanoscale properties of these tiny structures have the potential to produce a wide variety of useful nanoscale products with many potential applications, including pharmaceutical drug design and miniature devices developed for use in medical diagnosis and treatment. Nanotechnology research focuses on developing efficient and affordable ways to assemble carbon-based molecules into novel nanostructures.

The field of nanotechnology focuses on the creation and manipulation of nano-sized particles, with diameters of 100 nanometers or smaller, to create materials or devices for industrial and medical applications. The upper size limit in nanotechnology is 100 nm, much smaller than bacteria or human cells (Figure 13.15). One nanometer (nm) is the same length as one-billionth of a meter ($10^{-9}$ of a meter), much too small to be seen with a conventional light microscope (see Figure 13.15). To put these tiny dimensions in better perspective, the distances between adjacent atoms in a molecule range between 0.12 nm and 0.15 nm, just over a tenth of a nanometer. In the context of biologically important molecules, the diameter of the DNA double-helix molecule (double-stranded DNA) is 2 nm. Cells range in size from tiny bacteria (about 200 nm long) to the large eggs of birds and dinosaurs and including a typical animal cell that averages about 10 um in length.

## The Amazing Science of Carbon Nanotechnology

The science behind carbon nanotechnology followed from the discovery of the buckminsterfullerene (C60) in 1985 (Figure 13.16). Some chemical elements such as carbon can exist in two or more different structures, or **allotropes**, of that element. The "bucky ball" allotrope is a soccer ball–shaped structure containing 60 carbon atoms with one carbon atom positioned at each corner of the 20 hexagons and 12 pentagons. This specific configuration of ordered carbon atoms makes an extremely stable structure (Figure 13.17). The earliest carbon nanotubes were structurally imperfect, and as a result they did not exhibit particularly interesting properties. In 1990, researchers found that the nanobucky ball C60 structure could be produced in an arc evaporation apparatus, which eventually led to the use of this method to create fullerene-related cylindrical carbon nanotubes. Certain carbon-based nano-sized materials exhibit intriguing physical and chemical properties that change dramatically as nanostructures, such as changing from opaque to transparent, converting inert materials into potent chemical catalysts, and transforming insulating materials into efficient nanoconductors.

To make a carbon nanotube, a graphite sheet is rolled into a nanocylinder, so that the pattern of hexagons arranged around the circumference of the cylinder reflects specific vector indices (n,m) (Figure 13.18). For example, to produce a nanotube with the indices (6,3), the graphite sheet is rolled up so that the atom

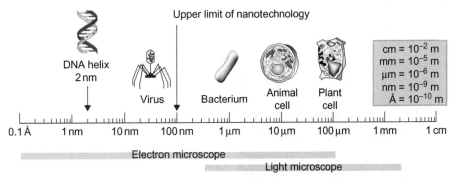

**FIGURE 13.15** Relative sizes of cells and sub-cellular components. The scale shown compares a range of sizes of various cells and cellular components (DNA helix, virus, bacterial cell, animal cell and plant cell). The bar below the ruler indicates the overlapping size ranges of objects that are visible with a light microscope and an electron microscope. The upper size limit for nanotechnology is 100 nm.

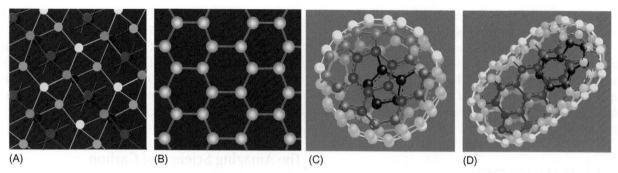

**FIGURE 13.16** Different structures (allotropes) are made by different bonds between carbon atoms. (A) Diamond. (B) Graphite. (C) Fullerene. (D) Nanotube. Colors in (A), (C), and (D) represent depth, with red closest to the viewer, and violet (or blue for [C] and [D]) farthest away.

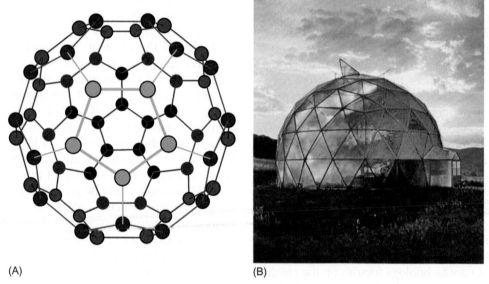

(A)                                                                                                     (B)

**FIGURE 13.17** Geodesic domes resemble buckeyballs. (A) In 1985 a new allotrope of carbon called buckminsterfullerene (C60) was discovered and named after American architect and designer Richard Buckminster "Bucky" Fuller (1895 – 1983). The 60 carbon atoms form the shape of a soccer ball with a carbon atom at each corner of 20 hexagons and 12 pentagons; (B) A geodesic dome home designed by architect Buckminster "Bucky" Fuller.

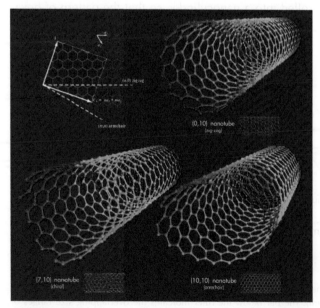

**FIGURE 13.18** Amazing carbon nanotubes often have unusual and unexpected properties.

labeled (0,0) is superimposed in the cylinder on top of the atom labeled (6,3). The extreme light weight and strength characteristics of the tiny carbon nanotube cylinders are a direct result of the physical and chemical properties conferred by the nano-size structure. The nanotube structures are called "zig-zag" (m = 0), "armchair" (n = m), and the chiral tube, which can exist in one of two mirror-image forms (Figure 13.19).

Modern carbon nanotubes have either one wall (single-walled) or more than one wall (multiwalled), but the earliest nanotubes were multiwalled cylinders with an outer diameter measuring from 3 nm to 30 nm (Figure 13.20). In 1993, single-walled carbon nanotubes were created, with narrower diameters (1 to 2 nm) than the multiwalled nanotubes. The ability to construct the new single-walled carbon nanotubes was an important advance in the field because the single-walled nanotubes exhibit novel electric properties that are not exhibited by the multiwalled carbon nanotubes. For example, the single-walled nanotubes make

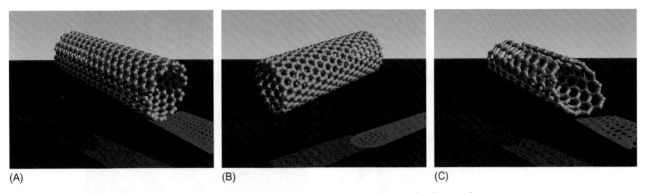

(A)                          (B)                          (C)

**FIGURE 13.19**   Single-walled carbon nanotubes. (A) Armchair wiki. (B) Zig zag wiki. (C) Chiral nanotube.

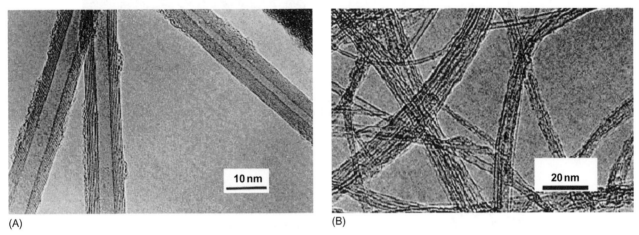

10 nm

20 nm

(A)                                              (B)

**FIGURE 13.20**   Single and multilayered carbon nanotubes visualized using an electron microscope. (A) Multiwalled nanotubes. (B) Single-walled carbon nanotubes. These new fibers exhibit a range of exceptional properties that prompted a surge of research into the new carbon nanotubes.

excellent electrical conductors and play a central role in the first intramolecular field effect transistors. Single-walled nanotubes are also the best candidates for the construction of miniaturized electronic devices that can operate routinely at nanoscale dimensions inside human cells.

The much desired single-walled nanotubes are expensive to produce, but work is under way to find affordable synthetic methods that will allow the routine use of carbon nanotube technology in future applications. Commercial applications of discoveries made in nanotechnology have emerged slowly, partly because of the increased production costs of high-quality nanotubes. Still, the amazing characteristics of carbon nanotubes give this area of nanotechnology great potential for future applications in high-technology fields that require lightweight, extremely strong materials, with nanocharacteristics such as superconductivity.

Interestingly, for the first time two different carbon allotrope structures, a fullerene and a nanotube, were combined to make a single nanobud structure that exhibits some properties from both fullerenes and carbon nanotubes (Figure 13.21). The fullerene-like "bud" is bonded

to the outer wall of the carbon nanotube, which greatly strengthens the mechanical properties of the composite structure and makes an effective molecular anchor that prevents the attached nanotube from moving.

## Clean Energy: Carbon Nanohorns Store Hydrogen

The hydrogen-carbon bonds that form the cone-shaped nanohorn structures are much more stable than the hydrogen-carbon bonds in the nanotube structures, making nanohorns an excellent option for industrial and medical applications of carbon-based nanomaterials (Figure 13.22). Nanohorns not only have remarkable adsorptive and catalytic properties, but French researchers discovered that carbon nanohorns offer an efficient and inexpensive way to store hydrogen in hydrogen-powered fuel cells, a source of clean energy. Hydrogen is an abundant, renewable energy source that could replace fossil fuels, but the use of hydrogen is limited because it is difficult to store. However, the carbon nanohorns trap the hydrogen in an efficient

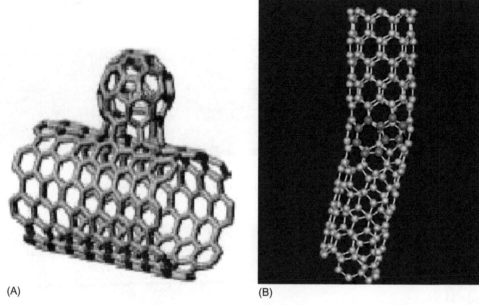

(A)                                                      (B)

**FIGURE 13.21** Hybrid carbon nanostructures. (A) Nanobud. (B) This potential nanodiode was constructed by joining an "even" rolled graphical sheet, which has semiconducting properties, to a "spiral" rolled sheet, which is predicted to have metallic characteristics.

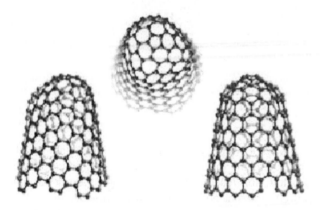

**FIGURE 13.22** Carbon nanohorns. Carbon nanohorns are strong and light, and are much more stable than other carbon-based nanomaterials, including nanotubes.

Nanotechnology produces novel nanomaterials and nanostructures that have a myriad of different applications in diverse areas of science.

## NANOMEDICINE

The applications of nanotechnology to medicine, called nanomedicine, includes advances in drug delivery, imaging and diagnostics, biosensors, targeted drug delivery, tissue engineering, and nanotoxicology. In the future, molecular machines like nanorobots will perform the rapid detection, diagnosis and elimination of pathogens inside the living human body, eventually performing surgery on individual cells. The nanodoctors will be tiny, less than 100 nanometers (nm), the appropriate size needed to operate within a living human cell.

The national importance of nanotechnology research to the federal government was recognized when the National Institutes of Health (NIH) established eight Nanomedicine Development Centers in the United States staffed by teams of biologists, physicians, mathematicians, engineers, and computer scientists. This NIH program provides researchers with an opportunity to combine nanoscale research with the ability to manipulate cellular nanostructures and design novel medical therapies. The scientists compiled extensive information on the chemical and physical properties of

process that permits all of the adsorbed hydrogen to be recovered.

The unique properties exhibited by carbon nanostructures are essential for the development of many novel materials with potential applications in areas such as medicine, electronics, and energy production. As expected for a new scientific field, people have raised questions about nanotechnology, including safety issues, the possible environmental impact of nanomaterials and the need for special federal or state regulation of the rapidly growing nanotechnology industry.

(A)

(B)

**FIGURE 13.23** DNA helix and carbon nanotube act as a biosensor device. (A) Researchers have developed a DNA-nanotube sensor that detects tiny amounts of metal ions in blood, tissue, and cells. (B) When the DNA-nanotube senses metal ions in the environment the DNA wrapped around the nano-tube changes helical conformation, which in turn squeezes an infrared emission from the nanotube sensor.

nanoscale cellular structures and studied the interactions between individual molecules and multiprotein complexes to increase our knowledge of the molecular structures and processes in living cells. This information is necessary for researchers to be able to design strategies to correct the structural defects and functional problems in disease cells. Scientists developed new nanotechnology tools to explore and manipulate nanoscale biological structures.

New nanodevices will be able to perform a wide range of biomedical tasks in the future including tiny nanodoctors that will diagnosis infectious agents, detect metabolic disorders, and repair or replace defective or broken cellular components.

## DNA Nanotube Biosensor

A novel DNA-nanotube sensor developed at the University of Illinois at Urbana-Champaign can detect the presence of metal ions in biological samples such as blood, tissues, and cells. The double-stranded DNA helixes are wrapped around semiconducting, single-walled carbon nanotubes (Figure 13.23). In the presence of ions, the DNA helix changes conformation around the nanotube, which changes the infrared emission from the nanotube sensor.

Researchers have developed nanotubes that are coated with different organic materials that allow the nanotube to act as a diagnostic test for lung cancer by sampling the molecules in a patient's breath. The first such nanodevice contained networks of single-walled carbon nanotubes, each coated with 1 of 10 different

organic materials. Each of the different organic coatings on the nanotubes was designed to emit a unique response when the detector is exposed to more than 200 volatile organic chemicals present in the human breath. The nanotube devices were calibrated using breath from 15 nonsmoking healthy people and 15 individuals with stage 4 lung cancer. The organic compounds in each breath sample were concentrated and identified by gas chromatography and mass spectrometry. The same samples were then tested using the nanotube sensor array and the results compared. The electrical output of the nanotube array varies in response to the exact mixture of organic compounds in the breath. The researchers found they could reliably distinguish between patterns of the healthy and lung cancer patients using the nanotube sensor device.

The amazing products made by carbon nanotechnology have great potential for many applications in diverse fields that require lightweight, extremely strong materials with unusual electrical or magnetic properties.

## Carbon Nanoparticles Make New Delivery Systems

The fight against cancer now includes the field of nanotechnology. Advances in nanotechnology offer a new way to overcome problems limiting the use of platinum-based anti-cancer drugs. Because of delivery problems, these strong drugs often become inactive in the body long before reaching the tumor. Researchers (Massachusetts Institute of Technology, Stanford University) made special carbon nanotubes that function as a "longboat delivery system" designed to protect the platinum-based anticancer drugs as they are carried through the body to the tumor (Figure 13.24).

The researchers attached the platinum drug to a single-walled carbon nanotube and created a longboat delivery system that efficiently transports the "warhead" to the tumor and releases the active platinum drug. Tests in cultured human cells showed that the nanotube longboat delivery system resulted in platinum levels that were significantly higher than when the platinum drug was administered by injection. The new carbon longboats can also transport other types of cargo to other target body sites in addition to the anticancer drugs.

Researchers have developed special nanoparticles designed to target cancer cells after metastasis when the cancer cells have left the primary tumor and spread in the body, making them difficult to remove by surgery. Special nanoparticles with gold shells and carrying mica absorb radiation and generate heat, effectively killing the thermal sensitive tumor cells. The effectiveness of this new system was tested in mice and shown to kill the cancer cells without damaging the surrounding healthy tissues. Researchers in Japan used a similar approach to treat tumor cells in mice by filling the carbon nanohorn structures with a photodynamic agent, zinc phthalocyanine. After exposure to a laser beam for 15 minutes each day for 10 days, the tumors disappeared.

Cells use many approaches to control which genes are transcribed including an RNAi mechanism that blocks the expression of some genes (see Chapter 11). The RNAi control mechanism relies on small interfering RNA molecules (siRNA) that base pair to specific target messenger RNA molecules (mRNA) in the cells, which block synthesis of that protein. Specially designed small interfering RNA molecules can block protein production in mammalian cells and, as a result, stop the growth of cancers and other diseases.

To use siRNA therapy to treat disease, researchers must overcome the dangers that face the RNA molecules in the body, where they are easily destroyed.

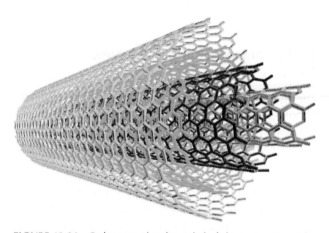

**FIGURE 13.24**  Carbon nanolongboats help fight cancer. Nanotube longboats can carry agents to fight cancer cells.

This complication is avoided by using nanoparticles designed to transport the siRNA molecules directly into the target cells as described for the anticancer drugs. To accomplish this feat, the outer surfaces of the nanoparticles are covered with special proteins that help to deliver the nanoparticles to the appropriate cells. The target cells are covered with proteins that bind to the surface proteins on the nanoparticles. Once inside the cells, the siRNA molecules bind to the target mRNAs and prevent synthesis of the targeted protein.

## Nanotechnology Powered by Biomotor Proteins

Human cells produce a variety of molecular machines that perform essential functions in many complicated biochemical processes such as metabolism, cell division, intracellular transport, and cell signaling. Cell components are often in motion in the cell, powered by special biological motor proteins. Scientists noticed how the kinesin and myosin biomotor proteins move along fibers in the cell, suggesting that the molecular motors could be made to perform similar functions outside the cell (Figure 13.25). Biomotors are the ultimate cellular nanomachines. They do mechanical work by hydrolyzing adenosine triphosphate (ATP) to generate energy. The ATP hydrolysis reaction changes the three-dimensional shapes of the motor proteins, which converts the chemical process into mechanical work that drives the movements of the proteins (see Figure 13.25).

## FOUR-DIMENSIONAL MICROSCOPE REVOLUTIONIZES OUR VIEW

The need to visualize nanoscale events as they occur has increased dramatically with the rapid development of nanotechnology. However, scientists recently

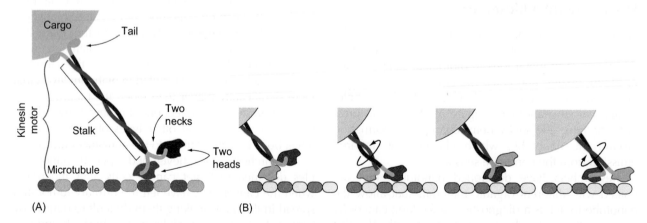

**FIGURE 13.25**  Kinesin motor proteins assemble into a molecular machine that walks along fibers in the cell. (A) The kinesin motor machine hauls cargo in precise 8-nanometer steps along the fibers made up of tubulin proteins (microtubules). (B) Researchers recently discovered that the kinesin motor machine uses its two heads to walk in an asymmetric, hand-over-hand fashion.

visualized nanoevents caught in real time and in real space. The 4D electron microscope produces "movies" of the atomic changes that take place. The new technology is based on the traditional electron microscope, which bounces a stream of electrons off of objects and produces a static image with a resolution of over a billionth of a meter in length. In this case electrons are used to visualize molecules because the wavelength of the radiation source of a microscope must be shorter than the distance between the atoms in the sample.

The research group that developed four-dimensional (4D) electron microscopy at the California Institute of Technology was directed by Ahmed Zewail, who won the 1999 Nobel Prize in chemistry for his work in femtochemistry (one-quadrillionth or 10 to the minus 15th). 4D electron microscopy uses ultrashort flashes of laser to observe the behavior of atoms during fundamental chemical reactions, permiting scientists to watch atoms come together to form new molecules, capturing motion on a femtosecond timescale (one millionth of a billionth of a second). The team used ultrafast "single-electron" imaging, where each electron trajectory is precisely controlled in time and space. The image produced by each electron represents a femtosecond snap shot of events at that moment in time. Much like the still frames in a film, the millions of the sequential images generated over time are assembled into a digital movie of events at an atomic scale. Zewail compared this new high-resolution technology to freeze-frame still photographs taken in the 19th-century, which required use of a strobe light, and proved for the first time that a galloping horse lifts all four hooves off the ground at the same time.

Modern researchers used 4D electron microscope technology to visualize the movement of carbon atoms in material that was heated rapidly, causing the heated carbon atoms to vibrate in a random, nonsynchronized manner. Over time, the vibrations of the individual atoms synchronize and exhibit a heartbeat-like "nanodrumming," a mechanical phenomenon that resonates at a frequency that is much higher than detected by the human eardrum. Zewail used the new 4D electron microscopy to observe atoms in superthin sheets of gold and graphite.

> Researchers are using the 4D microscope to view new images of cellular components in live cells made by following the dynamic changes in complex cellular structures in real space and time, providing a new perspective on the functions of molecules and cells.

## DNA COMPUTERS

In 2003 the world celebrated the 50th anniversary of the discovery of the DNA double-helix structure by James D. Watson, Francis H. Crick, Rosalind Franklin and Maurice Wilkens. Their work explained genetic events in terms of the DNA helix molecule and opened the doors to the next century of biological study of cells. Many thousands of researchers continue working to decipher the diverse ways that genes and DNA control the development and functions of living organisms. The applications involving the extraordinary DNA molecule extend way beyond uses in cell and molecular biology. In part this was due to rapid advances in DNA technology, for example, the modern DNA synthesis machines routinely produce long DNA molecules with predetermined DNA sequences, opening the door to many new applications of DNA in many diverse areas of science.

In 1994, Leonard M. Adleman (University of Southern California) built the first DNA computer that can actually play the game of tic-tac-toe! Since then, scientists have developed special DNA molecules that can perform the logic operations that are typically carried out by a silicon-based computer (Figure 13.26). This DNA computer consists of a series of wells containing DNA molecules called **logic gates**. The logic gate responds to input information in specific ways depending on the secondary structure (internal base-pairing) of the DNA strand. This system, called a molecular array of YES and AND gates (MAYA), can play a restricted game of tic-tac-toe. The next generation of DNA computer, MAYA-II, allows an unrestricted game where the player can make a move in any well (logic gate) after the computer has made the first move in the center square.

The MAYA-II game of tic-tac-toe consists of a three-by-three array of wells containing different DNA gates. The player makes a move by adding a specific single-stranded DNA into each well. The sequence of each input DNA strand corresponds to a specific move in tic-tac-toe such as "middle left square, first round." The player adds the input DNA strand into all the wells, but only one well (the one the player wants to mark) contains the DNA gate that is activated by the specific sequence of that input DNA strand, releasing a signal that represents one move in that round of game play. The hairpin DNA in only one gate responds to the input DNA strands by dramatically changing its hairpin structure, triggering a fluorescent signal (see Figure 13.26). The moves made by the computer are encoded in the DNA sequence of each logic gate (well). The players are identified by fluorescence at different frequencies, so that the game moves made by the human and computer players are easily distinguished by color.

DNA computers have future applications in novel biomedical technologies and in manufacturing and will be essential for the assembly of new nano-sized materials and nanomachines. For example, researchers have used DNA strands to build a nanorobot system containing dozens of flipper-like molecular arms.

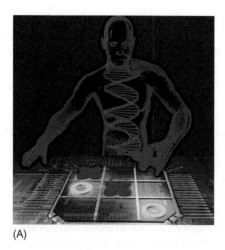
(A)

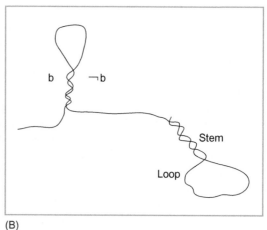
(B)

**FIGURE 13.26** The new age of DNA, the DNA computer. The newest version of the DNA computer can perform the logic operations usually carried out by a silicon-based computer. (A) The MAYA DNA computer can play tic-tac-toe. (B) This single-stranded DNA molecule represents a logical formula with the conflicting assignment *b* and *!b (not b)*. The regions of the single-stranded DNA that are complementary to each other in the same DNA molecule base pair to each other to make short stable double-stranded regions of DNA (stem) connecting to single-stranded loop regions.

This machine relies on nothing more than the specificity of the hydrogen bonds involved in the zipping and unzipping of complementary DNA bases in each gate. In the new DNA computer circuits, these DNA base-pairing interactions perform logic operations such as the "if A and C, then D" (called an A AND C gate), which is usually performed in an electronic computer in a solid-state circuit representing A, C, and D.

## SUMMARY

Much of the research in pharmaceutical biotechnology focuses on the overall goal of improving drug development and improving disease treatment. Success in drug development research requires a detailed understanding of the biological targets of drugs in the cells. Drug design also relies on human genome sequence information and a real appreciation for the intricacies of protein structures and the protein changes responsible for different diseases and disorders. One approach is to design a drug that can mimic the structure of a protein that normally binds to a key protein in the disease pathway. Once the imposter binds to the key protein, that protein can no longer participate in causing the disease. This result would indicate that the imposter compound might make an effective drug to treat the disease in question.

The genomics and proteomics fields both generate important resources for researchers in pharmaceutical biotechnology, because research in these fields focuses on understanding the structure and function of genes and exploring how the gene products interact with each other and with environmental factors. Scientists use microarray technologies to find out which genes are expressed in the healthy compared to diseased cells or to determine how a particular drug affects the expression of large numbers of genes. These questions are central to pharmaceutical biotechnology researchers because genes and proteins have a central role in human disease processes and in drug treatments.

Like many areas of science, pharmaceutical biotechnology continues to benefit from the amazing and intriguing new field of nanotechnology, which created carbon nanoparticles to provide new ways to deliver drugs to specific targets in the body, DNA nanotubes that act as biosensors in the body, and nanohorn structures that offer a novel way to store hydrogen. Recent hybrid nanostructures like the nanobud now provide ways to construct hybrid nanomachines. The amazing characteristics of carbon nanotubes include an extremely strong material vthat is lightweight, and often exhibits unexpected nano-characteristics such as superconductivity. The novel nanoscale properties of these tiny structures have the potential to produce a wide variety of useful nanoscale products with many potential applications in medicine, including nanomachines that will work inside living cells.

Lab on a chip technology has already produced the first handheld bioanalyzer devices that will change the future of standard laboratory testing. Future devices will include special units available for DNA testing at home, and will improve forensic and other testing technologies.

DNA computers are still very new, but the biomedical applications of DNA computers have a bright future. The development of DNA nanocomputers is under way to produce DNA computers that can function inside living human cells. In the future the collaboration of nanotechnology, DNA computers, and lab on a chip technology will produce tiny implanted devices that monitor and control the levels of insulin and sugar in diabetic patients, patrol the body for pathogens or cancer cells, or provide the appropriate doses of analgesic drugs for most effective pain control.

The information available in an individual's DNA genome is rapidly becoming an important tool to improve diagnosis and treatment, and it is the foundation for the practice of personalized medicine. This field is very new but it will continue to grow under the influence of more biotechnology companies that offer the public access to their genes, genotypes, and possible genetic future. The structural approach to designing new drugs is based on understanding the molecular structures of the biological drug targets inside the cell, and the characteristics of the active site on the target enzyme. Some types of new drugs are specifically designed to stop key interactions between a protein enzyme and its target. The fields of genomics, proteomics, personalized medicine, drug discovery, lab on a chip technology, nanotechnology, and nanomedicine, are among the areas making important contributions to the growing field of pharmaceutical biotechnology. Although diverse, most of these scientific specialties focus on research to find and develop new drugs and to design more effective drug treatments. Researchers in the field of pharmaceutical biotechnology are working on developing new and better medicines and therapies to treat human diseases. In the future, pharmaceutical biotechnology will continue to have a significant impact on related research in personalized medicine, targeted drugs, and nanoparticle-based delivery systems for anticancer drugs.

## REVIEW

This chapter describes the field of pharmaceutical biotechnology and updates the status of scientific fields that contribute to drug development and disease treatments, including genomics, proteomics, personalized medicine, drug discovery, lab-on-a-chip technology, nanotechnology, and nanomedicine. To assess your understanding of these areas, answer the following review questions:

1. What is the main focus of research in pharmaceutical biotechnology?
2. Describe the impact of the human genome sequence and genomics on advances in the science of personalized medicine.
3. Explain the role of protein structure in designing new drugs to treat a disease.
4. Explain how changes in protein expression can be used to identify an early diagnosis biomarker for a disease.
5. Describe how microarray and cDNA technologies are used to study differential gene expression in cells and to analyze the genetic results of drug treatments.

6. Give an example of the new physical or chemical characteristics that are exhibited by nanostructures such as nanotubes.
7. Explain which features can be visualized by the new 4D electron microscope that cannot be observed using typical electron microscopy.
8. Describe the important role that 3D protein structure plays in the strategy used to design new drugs.
9. Explain the relationship between a single nucleotide polymorphism (SNP) in an individual's genome and the chance of inheriting a genetic disease.
10. Explain how the use of handheld lab-on-a-chip devices will change routine patient health care in the future, or be an advantage for forensic scientists.

## ADDITIONAL READING

Angier, N., November 11, 2008. Scientists and philosophers find that "Gene" has a multitude of meanings. N. Y. Times page D2 of the New York edition. www.nytimes.com/2008/11/11/science/11angi.html?fta=y&pagewanted=all.

Grady, D., November 6, 2008. A cancer genome: scientists decode set of cancer genes. N. Y. Times page A1 New York edition. www.nytimes.com/2008/11/06/health/research/06cancer.html?pagewanted=2&fta=y. Smart DNA: Programming the Molecule of Life for Work and Play, Joanne Macdonald, Darko Stefanovic and Milan N. Stojanovic, November 10, 2008 Scientific American Magazine www.sciam.com/article.cfm?id=smart-dna.

Harmon, A., 2007. The DNA age: my genome, myself: seeking clues in DNA. N. Y. Times www.nytimes.com/2007/11/17/us/17dna.html.

Kit to spot serious illness early may be just 10 years away. The Guardian (UK). November 2007. www.guardian.co.uk/science2007/nov/14/medicalresearch.healthorhttp://tinyurl.com/27ed2j.

Macur, J., November 30, 2008. Born to run? Little ones get test for sports gene. N. Y. Times page A1 New York Times Edition. www.nytimes.com/2008/11/30/sports/30genetics.html?_r=1&emc=eta1.

Minkel, J.R., May 21, 2007. DNA Computer Works in Human Cells. Sci. Am. www.sciam.com/article.cfm?id=dna-computer-works-in-human-cells.

Nanotube 'longboats' slaughter cancer cells. New Sci. (2672). September 6, 2008. <http://www.newscientist.com/article/mg19926725.100-nanotube-longboats-slaughter-cancer-cells.html>.

Pollack, A., October 6, 2008. Dawn of low-price mapping could broaden DNA uses. N. Y. Times page B5 of the New York edition. www.nytimes.com/2008/10/06/business/06gene.html.

## WEB SITES

Carbon nanotubes (CNTs) are an allotrope of carbon. www.sciencedaily.com/articles/c/carbon_nanotube.htm.

From the Genome to the Proteome, U.S. Department of Energy Office of Science. www.ornl.gov/sci/techresources/Human_Genome/project/info.shtml.

Googling your DNA personal genome science. www.personalgenomes.org.

The science and technology of carbon nanotubes. www.personal.rdg.ac.uk/~scsharip/tubes.htm.

devices that monitor and control the levels of insulin and sugar in diabetic patients, control the body for only organs or cancer cells, or provide the appropriate doses of analgesic drugs for most after-the-pain control.

The information available in an individual's DNA genome is rapidly becoming an important tool to improve diagnosis and treatment, and it is the foundation for the practice of personalized medicine. This field is very new but it will continue to grow under the influence of more biotechnology companies that offer the public access to their genes, genotypes, and possible genetic future. The structural approach to designing new drugs is based on understanding the molecular structures of the biological targets inside the cell, and the characteristics of the active site on the target enzyme. Some types of new drugs are specifically designed to stop key interactions between a protein enzyme and its target. The fields genomics, proteomics, personalized medicine, drug discovery lab-on-a-chip technology, nanotechnology, and nanomedicine are among the areas making important contributions to the growing field of pharmaceutical biotechnology. Although diverse, most of these scientific efforts focus on research to find and develop new drugs and to design more creative drug treatments. Researchers in the field of pharmaceutical biotechnology are working on developing new and better medicines and therapies to treat human diseases in the future. Pharmaceutical biotechnology will continue to have a significant impact on related research in personalized medicine, targeted drugs, and macromolecule-based delivery systems for anticancer drugs.

## REVIEW

This chapter describes the field of pharmaceutical biotechnology and updates the status of scientific fields that contribute to their development and disease treatment, including genomics, proteomics, personalized medicine, drug discovery lab-on-a-chip technology, nanotechnology, and nanomedicine. To assess your understanding of these areas, answer the following review questions:

1. What is the main focus of research in pharmaceutical biotechnology?
2. Describe the impact of the human genome sequence and genomics on advances in the science of personalized medicine.
3. Explain the role of protein structure in designing new drugs to treat a disease.
4. Explain how changes in protein expression can be used to identify an early diagnosis biomarker for a disease.
5. Describe how microarray and DNA technologies are used to study differential gene expression in cells and to analyze the genetic results of drug treatments.

# Animal Biotechnology

before transplanting the genetically modified cells into eggs (Figure 14.1).

**FIGURE 14.1** Fluorescent feline face! The cloned Turkish Angola kitten *(left)*, gives off a red fluorescent glow, whereas an ordinary kitten *(right)* appears green because this picture was taken under ultraviolet light. The genome of the cloned Angola kitten contains a gene that was modified to produce a red fluorescent protein.

## Feline Fluorescence

*Newsday, December 15, 2007*
  By Cho Jin-seo
  Staff Reporter, *Korea Times*
  South Korean scientists have cloned cats that glow red when exposed to ultraviolet rays, an achievement that could help develop cures for human genetic diseases, the Science and Technology Ministry said. Three Turkish Angora cats were born in January and February through cloning with a gene that produces a red fluorescent protein that makes them glow in the dark. One died at birth, but the two others survived, the ministry said. The ministry claimed it was the first time cats with modified genes have been cloned. Scientists from Gyeongsang National University and Sunchon National University took skin cells from a cat and inserted the fluorescent gene into them

## LOOKING AHEAD

This chapter describes the science behind animal biotechnology, which encompasses the strategies and methods used to change the genetic makeup of living animals, including humans. A major goal of this work is to provide animal test models to better understand and treat human diseases. In addition, these methods can be used to create animals to produce medically useful proteins and to develop appropriate animals to provide transplantable organs such as kidney, heart, or lung for human recipients.

On completing the chapter, you should be able to do the following:

- Describe the major methods used to create transgenic animals.
- Explain how a functioning gene can be removed from an animal to create a "knockout" organism.

DNA and Biotechnology

**311**

- Describe the use of animals carrying tissue-specific gene knockouts to understand the role of different genes in human development.
- Explain why scientists might decide to incorporate a transgene or a knockout gene into the germ line cells of an animal, and not the somatic cells.
- Identify five ways in which transgenic, knockout, and conditional knockout animals contribute to understanding human diseases.
- Explain how transgenic genes are expressed in the host transgenic animals created to make useful products, including protein drugs.
- Describe the ethical arguments for and against the xenotransplantation of animal organs to treat human patients.
- Explain the steps used in reproductive and therapeutic cloning and distinguish between the goals and purposes of each approach.
- Describe the concerns raised about animal biotechnology; in particular describe the potential problems associated with recombinant DNA products that may enter the food chain.

**FIGURE 14.2**   Rat growth factor hormone influences the body size of mice and rats. The transgenic mouse on the left has an active rat growth hormone gene inserted into its genome and it is twice the size of the normal mouse on the right.

## INTRODUCTION

The first efforts to alter the genetic information of an animal to make the animal more useful for biomedical research began long ago with the most common laboratory animal, the white mouse. Years of research in molecular genetics revealed that the fundamental rules governing gene expression and protein synthesis in bacteria are very similar to the rules governing the same processes in animal cells but with some important differences. In early experiments, when researchers inserted a ribosomal RNA gene from frogs into bacterial cells, they were surprised to find that the frog gene was expressed and the frog ribosomal RNAs were produced in the bacteria.

The possibility that scientists could move genes from one organism into another organism, and from one species into another species, became the foundation for the modern biotechnology research. This work is aimed at studying the functions of human genes in animal cells with the goal of producing therapeutic proteins and other useful biological products as a result of inserting foreign genes into the genomes of appropriate animals. In 1981, a rabbit β-globin gene, which encodes one of the major proteins in red blood cells, was introduced into mouse egg cells. The resulting offspring were mice that produced the rabbit β-globin protein in their red blood cells. In a different experiment, the rat human growth hormone gene was inserted into mouse oocytes and the offspring grew into especially large adult mice (Figure 14.2).

Modern animal biotechnology focuses on the creation of genetically engineered animals that will aid in the understanding and treatment of human diseases and will make transgenic animals that produce many proteins, drugs, and other products to benefit humans.

## TRANSGENIC ANIMALS ARE GENETICALLY ALTERED

**Transgenic animals** are constructed by inserting one or more foreign genes into the genome of the recipient organism, including bacteria, plants, and animals. Vectors are commonly used to carry a foreign gene into the genome of animal cells. The foreign gene carried by a transgenic animal is a **transgene.** Small transgenic animals such as laboratory mice were initially constructed for research studies. The first successful large transgenic animals were created in 1985, when transgenic rabbits, pigs, and sheep were produced carrying the gene for human growth hormone (HGH). The HGH gene DNA was integrated into the chromosomal DNA in all three species and the HGH protein was successfully produced in transgenic rabbits and pigs. Since then, transgenic cattle expressing human growth hormone have been produced.

### Changing Genes in an Animal Genome

The first challenge to overcome when creating a transgenic animal is to determine the best strategy to safely

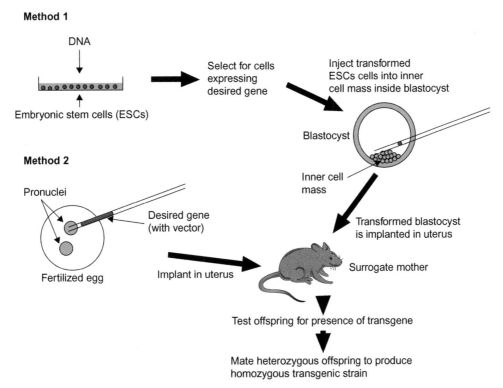

**FIGURE 14.3** How to make transgenic animals. (A) Method 1: The mouse embryonic stem cells are first treated with the vector DNA carrying the transgene. Most cells treated with the DNA do not pick up the DNA and as a result most of the cells in the experiment are killed when they are exposed to the G418 antibiotic. A few cells pick up the DNA vector and the *tk* gene is inserted incorrectly into the genome. These cells are resistant to G418 but are killed by ganciclovir. A very few cells incorporate the transgene and the vector DNA inserted correctly into the genome. Homologous recombination involves the exchange of vector DNA with similar (homologous) DNA sequences in the target animal genome. DNA exchange between homologous regions causes the transgene to be inserted accurately into the target site in the animal genome. The desired population of cells will grow in culture medium containing the antibiotic drugs, G418 and ganciclovir and the cells that survive the selection process become enriched several thousand fold. (B) Method 2: The earliest transgenic animals were created by microinjecting the transgene DNA into one of the two pronuclei in the fertilized oocyte, these are the egg and the sperm nuclei before they fuse. Under a microscope, the scientist uses a glass pipette with an ultrathin tip to deliver the DNA into one pronucleus while the egg cell is held in place with suction from a blunt pipette. The resulting transgenic eggs are introduced into surrogate females to allow the transgenic embryo to grow and develop.

change the genome DNA in every cell in the entire animal. This involves manipulating the germ line cells and embryonic cells of the animals to be altered. Scientists must know the location, structure, and function of the gene in the animal genome that is targeted for change. It is extremely important to use an appropriate DNA vector that will not only to carry the transgene into the cell, but will also dictate the fate of the transgene DNA once it enters the nucleus.

The first transgenic animals were created by microinjecting the transgene DNA into one of the two **pronuclei** in the oocyte, which are the egg and the sperm nuclei before they fuse (Figure 14.3). Using a microscope to visualize the process of microinjection, the scientist uses a glass pipette with an ultrathin tip to deliver a tiny amount of solution containing DNA into one pronucleus while the egg cell is held in place with suction from a blunt pipette. The resulting transgenic embryos are then

introduced into surrogate females to allow the transgenic embryo to grow and develop.

> A transgenic animal is genetically engineered to carry an additional gene (or genes) in the genome of all of its cells, but the transgene protein product might not be expressed in all of the cells.

**Embryonic stem cells (ESCs)** were first isolated from mice and cultured in the lab in 1981 by Martin Evans and Mathew Kaufman of Cambridge University (see Chapter 12). Embryonic stem cells are **undifferentiated cells** that grow inside the hollow **blastocyst** embryo (see Chapter 12). The cells on the outside of the hollow blastocyst will eventually develop into the **placenta**, which physically connects to the embryo to the mother through the umbilical cord. In comparison, the ESCs growing inside the blastocyst embryo have the ability to develop into all of the different types of cells in the human body. This includes **germ line** cells, which produce eggs in the females and sperm in males. The undifferentiated ESCs are **pluripotent** because they have the ability to develop into all of the types of cells in the human body (Chapter 12).

Major progress was made in the approach to creating transgenic animals when scientists began to use genetically altered ESCs to generate transgenic embryos instead of relying on microinjecting DNA into pronuclei (see Figure 14.3). In this new method, the ESCs are removed from inside a blastocyst and grown in the lab and then treated with the vector DNA carrying the transgene. This vector also contains other genes and functional DNA elements including a gene that is expressed only if the vector integrates properly into the cell genome. Vectors typically carry antibiotic-resistance genes that allow only the cells carrying the vector to grow in the presence of the antibiotic.

The scientists then inject the treated ESCs carrying the vector/ transgene DNA into the inner cell mass in blastocyst embryos of the same species. The transgenic blastocyst embryos are then transferred into the uterus of a surrogate female to carry the developing embryos until birth. Transgenic animals created in this way acquire only one copy of the transgene per cell because the transgene was introduced into only one chromosome in the cell. However, the offspring created by mating two transgenic animals produces transgenic offspring that have two copies of the transgene.

> Scientists created the first transgenic animals by injecting the altered gene DNA into the pronuclei immediately after fertilization. Later they used embryonic stem cells to generate lines of transgenic animals for scientific research and the development of biomedical treatments.

Specialized vectors play a significant role in creating custom designed transgenic animals in addition to controlling the delivery and expression of the transferred transgenes. The vectors are engineered not only to carry the transgene DNA into the cell, and into the nucleus, but also to direct the fate of the vector DNA and the transgene inside the recipient cell. The functional DNA elements included on the vector are important components of the genetic selection strategy devised to select for the ESCs carrying the desired genetic alteration(s). This includes whether or not the vector and transgene integrate into the genome and controls the timing of the expression of the transgene product.

Transgenic vectors used to produce transgenic animals often contain the **neomycin resistance gene (neo$^r$)** and **thymidine kinase gene (tk)** as well as the transgene and other DNA elements (Figure 14.4). The neomycin resistance gene codes for an enzyme that inactivates the antibiotics neomycin and G418. The thymidine kinase gene codes for an enzyme that phosphorylates the nucleoside analog, ganciclovir. During DNA replication, the phosphorylated ganciclovir is mistakenly incorporated into the replicating DNA, causing the cells expressing the thymidine kinase gene to be killed by the ganciclovir.

## Targeted Integration of Transgene DNA

Certain DNA elements carried on the vector play important roles determining the fate of the vector DNA once it reaches the nucleus of the recipient cell. Some vectors are designed to **integrate** directly into the chromosome DNA, sometimes at random sites in the genome and sometimes at specific sites in the genome DNA, depending on the type of vector. In a few cells the vector DNA is inserted into the target site in the genome using the rare process of **homologous recombination**. This occurs because the vector contains DNA regions that are similar (homologous) in DNA sequence to the targeted regions of the animal genome (Figure 14.4). The exchange of DNA sequences between homologous DNA regions in the vector and in the genome causes the transgene DNA to be inserted into a known target site in the recipient genome.

There are some drawbacks of integrating the transgene into an animal genome. For example, there would be negative consequences if the vector integrated into the coding region of a gene and disrupted expression of an essential protein. This would be a rare event, however, because the human chromosomes are composed mostly of noncoding DNA sequences, with very little coding DNA (see Chapter 6). It is more likely that the expression of a transgene might be turned off or suppressed if the transgene DNA is inserted into regions of the chromosomes where genes are known to be permanently inactivated and are not expressed.

Vectors that are packaged into virus particles are particularly efficient at **transfecting** cells because these vectors mimic the processes normally used by infectious viruses. Most viral vectors have limitations on the length of the DNA that can be inserted into the vectors, usually because of the constraints imposed by packaging the vector into the viral capsid. To overcome this restriction, scientists constructed artificial chromosome vectors derived from native budding yeast chromosomes, which have no limit on insert size. The P1 plasmid DNA is used to make bacterial artificial chromosome vectors (BAC), which can accommodate an insert of about 300,000 base pairs (300 kb). These vectors also contain the additional DNA sequences needed to regulate transcription of the vector genes including the transgene. Progress has been made to ensure that the transgene will show accurate tissue-specific gene expression, an important feature of transgenic gene expression.

> Specialized DNA vectors play an essential role in inserting genes into specific target sites in the genome by homologous recombination. This natural process involves exchanging DNA on one chromosome for the equivalent region of DNA on the homologous chromosome.

## Transgenic Knockout Animals

A **knockout gene** refers to any gene that has been inactivated or deleted from a genome. A **knockout animal** is a specific type of transgenic animal that has had a specific gene inactivated or deleted from its genome These gene knockout animals have been essential for a variety of studies designed to determine the role of specific genes in cell growth and development in animals. Many genes in diverse organisms such as humans and mice encode similar proteins. Scientists can investigate the functions of the human genes and

proteins by expressing the human protein in different organisms.

Transgenic knockout animals are produced using special vectors engineered to carry out targeted integration of the vector DNA into a specific location in the genome, instead of inserting into random genome locations. The vector carries the mutant (knockout) version of the gene to be deleted, and a gene that encodes resistance to antibiotics such as **neomycin**. In the vector, these two genes are flanked on both sides by regions of DNA that are similar (homologous) to the sequences of the DNA regions that flank the target gene in the genome. The regions of similar DNA sequences in the vector and the genome direct the process of **recombination** (**crossing-over**) so that the vector DNA is inserted into the target site in the genome. In the case of a knockout gene, the vector DNA is inserted in such a way that the wildtype gene is removed from the genome and replaced with the thymidine kinase (tk) gene, which confers cell sensitivity to the antibiotic ganciclovir (see Figure 14.4).

The same approach was used to produce transgenic knockout mice, with each knockout strain lacking a particular mouse gene. After the vector has been introduced into the mouse embryonic stem cells (ESCs), the cells were grown in the presence of antibiotics (neomycin and ganciclovir), which simultaneously imposes positive and negative selections on the growth of the cells. The neomycin resistant cells continue to grow in

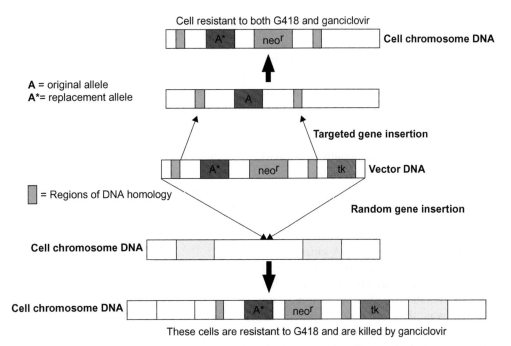

**FIGURE 14.4** Transgenic vectors are used to generate transgenic animals. The *neo*<sup>r</sup> gene encodes an enzyme that inactivates neomycin and G418. The tk gene encodes thymidine kinase, an enzyme that phosphorylates the analog ganciclovir, which is incorporated into newly replicated DNA; as a result, ganciclovir kills cells expressing the tk gene. Shaded sequences are regions of DNA homology (have similar DNA sequences).

the presence of neomycin, whereas those cells killed by the ganciclovir antibiotic lack the tk gene and represent unwanted transformants. The potential ES knock-out cells that continue to grow in culture are injected into three- to five-day-old mouse blastocyst embryos, which are then transferred into the uterus of a surrogate mouse. The knockout mice generated by this process are **heterozygous** for the knockout gene because the knockout vector integrates into only one of the two chromosomes carrying the target gene. The knockout offspring are then bred to produce second-generation transgenic animals with **homozygous knockout** genes. However, if the gene that is inactivated by the knock-out mutation is normally essential for survival, the heterozygous mice will survive but the homozygous knockout alleles will be lethal.

Specialized DNA vectors are used to construct transgenic knockout animals that have had a selected gene(s) completely deleted from their genome DNA. The knockout vector inserts into only one of the two chromosomes, making the first generation mice heterozygous for the knock-out allele (wt/ko). With appropriate breeding, the second generation mice can be homozygous for the knockout mutation and carry knockout alleles on both chromosomes (ko/ko).

## Controlling Transgene Expression in Transgenic Animals

In 2006 the Nobel Prize in medicine was awarded to Andrew Fire of Stanford University and Craig Mello of University of Massachusetts Medical School for discovering RNA interference (RNAi). This novel biological mechanism involves small RNA molecules (siRNAs) that can turn off the expression of selected genes (see Chapter 11). This siRNA technology provides scientists with an extremely powerful tool to explore the functions of specific proteins in the cell, and potentially be used to abolish the function of disease genes in patients.

Fire and Mello first identified the short hairpin shaped RNA molecules as interfering with gene expression in the roundworm *C. elegans*. They determined that the small RNAs were part of a mechanism to protect the cells against invading RNA viruses. The discovery of RNAi opened up a new field of research that identified the many components involved in this complex protection mechanism. Perhaps even more important however was the rapid development of RNAi technologies, which were applied to a very wide range of scientific fields and experimental approaches. Although discovered in worms, scientists have used RNAi technology to control gene expression in many

organisms and during embryo development (Figure 14.5) (see Chapter 11).

The naturally occurring Cre-loxP DNA recombination system was adapted as a molecular tool used to remove specific genes from a genome (Figure 14.6). The *Cre* gene codes for the Cre protein, a site-specific DNA recombination enzyme (recombinase) that acts only on DNA containing a specific 34 base pair sequence called loxP (locus of X-over P) sites. The transgene DNA is inserted into the vector so that the gene is flanked by two loxP DNA sequences. The Cre enzyme stimulates recombination between the loxP sites in the vector and the loxP sites in the genome, causing the DNA region between the two loxP sites to be deleted from the animal genome.

Transgenic animals engineered using the Cre-loxP DNA recombination system have been used in tissue-specific gene expression studies to analyze genes and proteins that are expressed only in certain tissues. In these experiments a set of transgenic animals is constructed containing the transgene of interest in the genome flanked on both sides by loxP DNA sequences. Some of these transgenic animals also carry the Cre

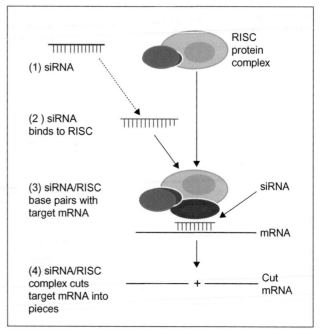

**FIGURE 14.5** Gene expression controlled using siRNA (RNAi). Expression of selected transgenes can be silenced in live cells using siRNA technology. (1) Scientists make their own siRNAs that are complementary to the mRNAs they have targeted for destruction in the experiment. The custom-designed siRNAs are added to the cells *(dotted arrow)*. (2) In the cell, the siRNA binds to special cellular proteins and forms RISC complexes. (3) RISC-siRNA complex searches for mRNAs that can base pair with the siRNA. (4) The siRNA/mRNA hybrid induces enzyme activity, which destroys the target mRNA and prevents it from being translated into protein. (same figure as fig 11.25 (OK to do that??)

gene inserted at a different site in the genome under the control of a tissue-specific promoter. The Cre enzyme is expressed in only one type of cell in the body (green in Figure 14.6). A knockout mutation can be created in the genome by expressing the Cre enzyme in cells carrying a knockout vector when the tissue-specific Cre promoter is active. The targeted gene will be deleted and the knockout mutation created when recombination occurs between the flanking loxP DNA sites (see Figure 14.6). The Cre-loxP system permits researchers to observe what happens when the targeted gene is deleted only from the genomes of specific tissues in the organism, revealing the roles of certain genes and proteins in tissue-specific cellular processes. In order for the genetic changes engineered in the genome by the Cre-loxP system to be inherited by the offspring, the knockout mutation must be introduced into the genomes of the germ line cells, which give rise to egg cells in females and to sperm cells in males.

> The Cre-loxP recombination system has the advantage that it can be used to create transgenic animals carrying controllable knockout genes. This permits scientists to induce a knockout mutation and analyze the consequences of losing a key protein in many different circumstances.

In the years since the first transgenic mice were developed, researchers have created tens of thousands of different strains of transgenic and knockout mice for use as research tools to study specific genes in human development, health and disease. The OncoMouse, for example, is a strain of mice engineered with a specific transgene that makes the mice especially susceptible to tumors, providing an excellent model system to test the effectiveness of potential anticancer drugs. The OncoMouse was developed at Harvard University by Philip Leder and Timothy Stewart and received the first animal patent ever issued in the United States.

Pharmaceutical and biotechnology companies have developed thousands of transgenic and knockout mice strains for the highly competitive and profitable race to discover new drugs. A good example is the company Deltagen, which has generated a collection of more than 500 different knockout mouse strains. Each strain is missing a different gene that codes for a different enzyme or protein and can be licensed by the company as a tool used to test new drugs.

Studies on the G–protein-coupled receptors (GPCRs) are a good example of the use of knockout transgenes applied to members of a large family of important membrane proteins that activate the signal transduction pathways that control various cellular

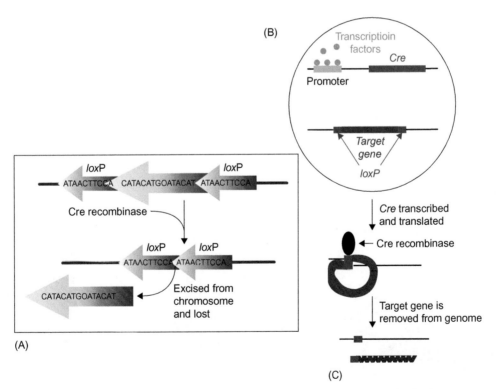

**FIGURE 14.6** Cre-loxP can be used to delete genes from animal genomes. (A) The Cre gene *(red)* codes for the Cre protein, a site-specific DNA recombination enzyme that recognizes and acts on any DNA molecule containing the 34 base pair loxP (locus of X-over P). (B) When the Cre recombinase enzyme is expressed in mammalian cells *(green promoter)* it catalyzes recombination exchange between the loxP sites *(pink)* in the vector and the loxP sites in the genome, causing the DNA *(target gene in blue)* between those two loxP sites in the genome to be deleted. (C) A closeup of the DNA recombination event that releases the target gene from the genome.

responses. GPCRs are involved in cellular responses to hormones, pheromones, sensitivities to light and odor, and the transmission of nerve signals. In different eukaryotic organisms, GPCRs are also called seven trans-membrane domain receptors, hepta-helical receptors, serpentine receptors, and G protein-linked receptors (GPLR) (Figure 14.7). The GPCR researchers created a collection of 236 mouse strains, with each strain carrying a different knockout mutation that

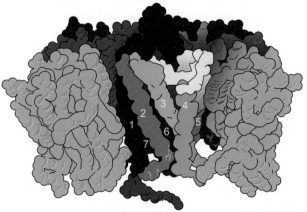

**FIGURE 14.7** The seven transmembrane α-helix structure of a G-protein-coupled receptor.

removes a different G-protein-coupled receptor protein gene from that mouse genome DNA.

Many of the commonly used drugs elicit a cellular response through a biochemical pathway that involves GPCR proteins. For this reason, transgenic GPCR knockout mice have been developed for use in many strategies designed to search for new drugs. Initial searches for new drugs involved cells grown in the laboratory used to identify compounds that are predicted to work by blocking the function of a specific GPCR protein. In subsequent experiments the GPCR knockout animals were treated with a hormone stimulus, which caused measurable effects on the animals tested. The next test for the knockout mice was designed to find out if certain compounds block the effect of the initial hormone-induced stimulus.

Transgenic animals play a key role in drug design and development by pharmaceutical companies. The biochemical pathways involved in a disease can often be blocked in knockout animals, giving researchers clues to types of new drugs that will mimic the knockout allele and alleviate disease symptoms.

## Box 14.1 Will Mutant Mice Help Us to Understand Autism?

In 2007, Tabuchi and colleagues (University of Texas Southwestern Medical Center) reported studies on transgenic mice with mutations in the neuroligin gene, which encodes cell-adhesion molecules in the nervous system. One set of knockout mice have had one copy of the neuroligin gene deleted from the genome. The second set of transgenic mice have a single point mutation in the neuroligin gene (R451C), which changes a cysteine amino acid to an arginine amino acid in the neuroligin-3 protein. The same R451C neuroligin gene mutation was also found in the genomes of some

people with **autism**, a spectrum of disorders that cause significant impairment of mental function and social behavior. The neuroligin R451C mutant mice showed impaired social interactions and they exhibited an increase in inhibitory nerve transmissions (Figure 14.8). This study suggests that the changes in synaptic nerve transmission might contribute to autism in humans. The R451C mice will continue to be a useful model for studying autism behavior and possible treatments in mammals.

**FIGURE 14.8** Transgenic mice offer a new model for studying autism. Normal mice are interested in socializing with other mice *(left)*, but the neuroligin knockout mice ignore the other mice and behave as loners *(right)*.

## PRODUCTS FROM TRANSGENIC ANIMALS: BIOPHARMING

Transgenic animals are used routinely used for research in many agricultural labs and for production by biotechnology companies. Agricultural and biopharmaceutical scientists envision a future with transgenic livestock as the major sources of many therapeutic proteins, and as an approach to generate livestock animals that are resistant to economically important diseases.

The development and marketing of transgenic animals and plants that produce therapeutic proteins and commercial products is often referred to as **pharming** or biopharming. First scientists cloned and expressed recombinant proteins in bacterial cells like *E. coli*. Then scientists discovered that it is essential for many recombinant eukaryotic proteins to be expressed in mammalian cells to ensure that the transgenic recombinant proteins will be correctly posttranslationally modified by the addition of the appropriate carbohydrates (sugars) and other chemical groups. Mammalian cells also provide the appropriate environment for the chains of amino acids to fold properly after synthesis and form the 3D protein structures required so that the protein can function correctly in the cells.

A number of research groups have produced stable lines of transgenic livestock including cows, sheep, and goats, which secrete the transgene products in the milk. In these transgenic animals the expression of the transgene was under the control of the transcriptional promoter for the casein gene that normally drives the expression of proteins that are secreted into milk. The industrial scale of transgenic livestock management and the development of methods to purify the desired transgene proteins presented many challenges. However, since the late 1990s a number of companies have established commercially viable transgenic protein production programs using goats, cattle, pigs, rabbits, and chickens.

The medical applications of transgenic animal products have been slow to be realized. The only transgenic protein made in animals that has been approved for therapeutic use in humans is ATryn (GTC Biotherapeutics), which was given market authorization in August 2006 by the European Commission. In 2007 ATryn received Orphan Drug designation from the Food and Drug Administration (FDA); orphan drugs are used to treat rare conditions that affect 200,000 or fewer individuals at any one time. ATryn is a recombinant form of the human antithrombin protein, which prevents blood clots during surgery on patients with an inherited antithrombin disease. These people lack an important blood protein that normally inhibits blood-clotting. Transgenic animals have been used to produce several clotting factor proteins that are easily harvested from milk, but are very difficult to isolate in sufficient quantities from human sources. Additional biopharming projects that are based on the production of specific proteins in transgenic rabbits and chickens are still in the early research stages or are currently in early clinical trials. Examples include the Granulocyte–Colony Stimulating Factor (G-CSF), which was first discovered because of its ability to stimulate the growth and maturation of granulocyte cells, an important type of blood cell that fights infection. A recombinant form of the Granulocyte–Colony Stimulating Factor (r-hG CSF) protein was produced in bacterial cultures and approved for use in the United States in 2002. In addition, researchers at the French National Center for Scientific Research produced recombinant human C-CSF (r-hG-CSF) protein secreted into the milk of transgenic goats. A Dutch company, Pharming, has used transgenic cows to produce fibrinogen, a key blood-clotting protein, which was released into the cow's milk. This company is also planning to produce a protein sealant in cow's milk that can prevent excessive bleeding of wounds during surgical procedures or after traumatic injury. In 2007 Pharming received the FDA's Orphan Drug designation for this product.

The overall success of the effort to produce drugs in transgenic animals remains uncertain. In 2007, the Council for Agricultural Science and Technology listed 14 companies and research groups in North America and Europe with projects aimed at producing biological products using transgenic livestock. However, recently at least two companies had gone out of business, some were purchased by larger companies, and others had abandoned the research projects that rely on transgenic animal technology. Research groups and biotechnology companies face additional challenges to overcome for the transgenic industry to continue, including the need to establish disease-free transgenic animal herds that stably produce proteins. In addition, better and more cost-effective protein purification methods are needed, which meet the regulatory requirements of the FDA and the comparable agencies in Europe, the United Kingdom, and other developed countries.

Despite these challenges, the potential offered by transgenic biopharming is substantial. Using transgenic cows, GTC Biotherapeutics produced 2 grams of recombinant human antithrombin (AT) protein in every liter of milk. The AT protein was purified from the milk with a loss of only 50%, and the AT protein produced by cows was found to function as well as the antithrombin protein made by human cells. The recombinant proteins produced in bacterial, yeast, or mammalian cell cultures must meet high standards for purification that remove viruses and dangerous bacteria. A big concern is the possibility that the transgenic

**FIGURE 14.9** Mastitus is a common infection in cows caused by dangerous *streptococci* bacteria.

Transgenic animals continue to be used to make important discoveries in basic research and in the search for new drugs. Genetically altered animals can potentially provide cost-effective means to make many products with medical applications. Each transgenic product must be documented as safe before it can be used in humans. As with drugs and recombinant proteins produced by cells in culture, the evidence supporting safety and efficacy of animal transgene products must be developed and evaluated on a case-by-case basis.

Transgenic animals continue to have an important role in producing specialized animal products for human use, ranging from drugs to fibers.

proteins made in animals might become contaminated with the protein-based agent that causes scrapie, a lethal neurological disease or the agent responsible for Mad Cow disease.

In the late 1990s, the increased interest in the use of transgenic animals to produce biopharmaceuticals was driven by the growth of the biotechnology industry and the fact that a facility to produce therapeutic proteins using laboratory cell cultures can cost more than $500 million. In comparison, transgenic dairy goats are much more cost effective—establishing a herd of transgenic goats costs less than a half million dollars. Dairy cows can produce 10,000 liters of milk per year so that tens of kilograms of transgenic protein can be produced by a single transgenic cow.

Transgenic animals are also used to improve the agricultural, commercial, and nutritional value of natural animal products. For example, the composition of animal milk can be altered to improve the growth and survival of offspring, and other approaches are used to improve the nutrition of the transgenic animal products. The fat and cholesterol composition of the meat produced by livestock could be manipulated in transgenic animals to improve the impact on the human diet. Transgenic animals can also help to achieve better resistance to bacterial diseases on the farm. Transgenic cows make milk containing lysostaphin, an enzyme that kills the dangerous bacteria *Staphylococcus aureus*. These transgenic cows are more resistant than normal cows to mastitis, a serious infection in cows caused by the *S. aureus* bacteria (Figure 14.9). Scientists are also working to improve the quality of other animal products such as hair, wool, and fiber, made by transgenic animals. Recently, the protein subunits that make up the strong silk fibers made by spiders were expressed in goats and the spider proteins were secreted in the goat's milk.

## Monoclonal Antibodies Make Disease-Specific Drugs

Transgenic animals have had an impact on **immunology**, an important field that focuses on the study of the physical, chemical, and physiological characteristics of the components of the human immune system in healthy and diseased states. Malfunctions of the immune system cause serious immunological disorders such as autoimmune diseases, hypersensitivities, immune deficiency, and transplant rejection.

Antibodies are key proteins in the immune system with central roles in the prevention and fighting of disease. The human body makes antibodies throughout life in response to natural exposure to infectious agents in the environment. The immune system is responsible for the effectiveness of the vaccines used to immunize children and adults. Vaccines are made using a protein or part of a protein from an infectious bacteria or virus, which triggers the immune system cells to generate antibody proteins that recognize specific bacterial or viral proteins as **antigens** (also called immunogens).

Humans make five types of antibodies known as immunoglobulins, IgG, IgA, IgD, IgE, and IgM. Antibodies are Y-shaped molecules composed of two identical long proteins (Heavy or H chains) and two identical short proteins (Light or L chains) (Figure 14.10). Antibodies function by recognizing and binding to specific antigens in a lock-and-key fashion, forming extremely tight antigen-antibody complexes (see Figure 14.10). Antigens have physical and chemical characteristics that illicit an immune response in the body. Antibody-antigen complex formation triggers one of several processes that lead to the destruction or inactivation of the antigen (bacteria or virus); the antibody binds to the outside of the bacterial cell and acts like a "tag," making the bacteria more easily

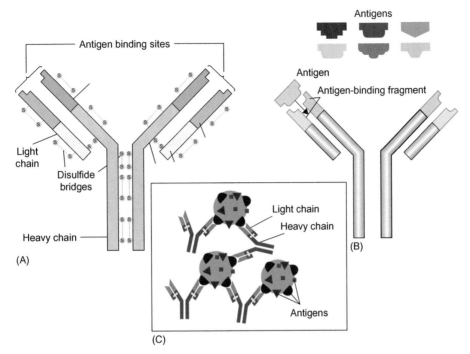

**FIGURE 14.10**   Structure of an antibody. (A) Antibodies are Y-shaped molecules composed of two identical proteins (Heavy or H chains) and two identical short proteins (Light or L chains). (B) Antigens bind to the antigen binding site on the end of the "arms" of the Y. (C) Antibody antigen complexes form to promote a specific immune attack.

detected by phagocytes and disposed of by **phagocytosis**. Antigens can be toxins in the human body, which become inactivated when bound by the antibody, which acts like an **anti-toxin**. By the same mechanism, a virus (antigen) that is bound to an antibody complex can no longer enter the host cells and is no longer infectious or dangerous.

The human immune system produces a complex mixture of different antibodies in response to infection or vaccination. When the body mounts an immune response to an antigen, many different B cell lines are activated to produce a mixture of different **polyclonal** antibodies raised against an array of specific antigens. Antibodies play an important role in basic research as research tools that are used to purify and study proteins of interest. Scientists often make antibodies ("raise" antibodies) against a specific protein antigen and then take advantage of the tight binding interaction between the antibody and the antigen to study the normal function of the antigen protein in the cell. They can find out which cells in the animal express the protein of interest by using immunoprecipitation and protein Western blots (see Chapter 5). Researchers also use antibodies to identify other proteins in the cell that interact with the protein of interest by co-immunoprecipitation assays that can detect physical binding between proteins in the cells.

Antibodies that bind with high specificity to a target molecule (antigen) have the potential to be engineered to make powerful drugs. Antibodies raised against the virus coat protein can neutralize a virus or kill a virus-infected cell. Scientists developed techniques to isolate and grow specific antibody producing cells in the lab, so that each cell line produces only one specific **monoclonal antibody** (**mAb**), which recognizes and binds to only one specific antigen (protein, bacteria, virus, etc.) (Figure 14.11). Monospecific monoclonal antibodies have been developed as drugs to treat cancer. Each monoclonal antibody binds to one specific chemical region on an antigen molecule called an **epitope**. To produce monoclonal antibodies, the mice are first injected with an immunogen (antigen), usually a protein or a part of a protein that will elicit antibody production. The antibody-producing cells are removed from the mouse spleen and fused with tumor cells to make a population of **hybridomas**, which are screened to identify the specific cells that are producing the desired antibodies. A single cell is placed into liquid growth medium in each well of a multiwell plastic culture dish. The cells grow into clones containing many identical cells in each well. The medium in each well is tested for the presence of antibodies that react specifically with the antigen. The cell cultures that produce specific monoclonal antibody proteins are grown in large cultures and the antibody proteins are harvested from the cells and purified. This process is a form of **cloning** cells because a population of genetically identical cell clones is generated from a

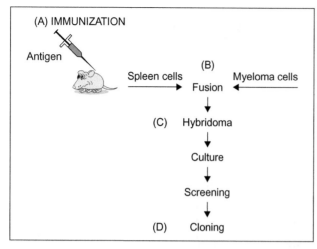

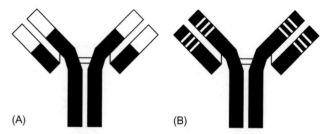

**FIGURE 14.12** Humanized antibodies. (A) Schematic picture shows a chimeric antibody with the rodent heavy chain and light chain *[white]* attached to the human constant regions *[black]*. The fully humanized antibody contains predominantly human protein sequences except in regions required to maintain the specific binding properties of the rodent antibody *[white]*.

**FIGURE 14.11** Monoclonal antibody production. (A) To produce monoclonal antibodies, the mice are first injected with an immunogen (antigen), which is a protein or a part of a protein that will elicit an immune response. (B) The antibody-producing cells are removed from the spleen and fused with tumor cells to make hybridomas. (C) Hybridoma cells are screened to identify cells that are producing antibodies. (D) These cells are placed separately into each well of a multiwell plastic culture dish. In each well the cells divide into clones of cells, and the cell media is tested to detect production of antibodies that react specifically with the antigen (immunogen). The cell cultures that produce the desired monoclonal antibodies are grown in large cultures and the antibodies proteins are harvested and purified. These cells are called clones because they are genetically identical cells, generated from the single cell in each well.

single cell. The antibodies made in this way are termed "monoclonal antibodies".

Monoclonal antibodies have several advantages over polyclonal antibodies in research applications, and for some medical purposes, but polyclonal antibodies continue to have an important role. Monoclonal antibodies are monospecific so they will bind only to the antigen used to raise that particular monoclonal antibody. Polyclonal antibodies bind to an array of different antigens. Monoclonal antibodies form cloned cell lines that can be stored and used at any time to generate more identical cells making the same monoclonal antibodies. Polyclonal antibodies are made by a collection of cells that cannot regenerate and make more immune cells when the initial collection of cells has been depleted.

The key feature of antibody proteins is the ability of antibodies to bind tightly to specific antigens with high specificity. Polyclonal antibodies are a mixture of antibodies that bind to a range of different proteins and antigens. In contrast, each monoclonal antibody binds only to one specific antigen.

Monoclonal antibodies made in mice have proven to be very useful for research applications and diagnostic testing, but the application of these antibodies as drug treatments for human diseases has been limited. In 1986, the first monoclonal antibody drug approved by the FDA for use in humans was a mouse antibody (OKT3) directed against a protein on the surface of immune cells that is involved in organ rejection. Initially the OKT3 antibody was used effectively to prevent or treat tissue rejection in human patients, but repeated treatments were not safe because the mouse monoclonal antibody stimulated the human body to raise an immune response against the mouse antibody proteins. In this response, the antibody proteins made by the patient bound to the mouse monoclonal antibody proteins, causing the monoclonal antibodies to be cleared from the blood and removed from the body through the liver and spleen.

One approach to avoid the problem of tissue rejection was to engineer the mouse antibody proteins to appear more like human antibody proteins. Recombinant DNA techniques were used to swap certain parts of the mouse heavy and light protein chains for the equivalent human protein sequences (Figure 14.12) to make **humanized** antibodies. These hybrid proteins consist of both mouse and human protein sequences, and preserve the immunogen-specific binding capacity of the mouse antibody. Humanized monoclonal antibodies are much less likely to cause an immune reaction in human patients and have been approved for human treatments. The first humanized monoclonal antibody was rituximab, which was raised against a specific protein found in lymph node cancer, and was as effective as a clinical treatment for lymph cancer. Herceptin is a humanized monoclonal antibody that targets the Her2 protein located on the surface of many breast cancer tumors. Herceptin is an effective breast cancer treatment when used together with chemotherapy in patients who have Her2 positive (Her2+) breast cancers (see Chapter 9).

The mouse protein sequences remaining in the structure of the humanized antibodies can still cause an unwanted antibody response. To overcome this problem, scientists are working on different ways to generate human monoclonal antibodies for treatment purposes. In one method the cells that form the human immune system in the bone marrow and thymus were transplanted into mice that are genetically unable to develop their own immune systems. After the human immune system cells are established in the mice, the mice are immunized with the desired protein and they make antibodies that are, in molecular terms, human. These antibody-producing cells can be isolated and cloned as described previously to produce monoclonal antibodies that do not cause tissue or organ rejection in humans.

> Monoclonal antibodies are particularly useful because they represent a pure population of cloned cells producing only one type of antibodies that all bind tightly to the same antigen.

## XENOTRANSPLANTATION

**Xenotransplantation** involves transplanting organs from animals into humans to treat organ failure, an approach that is predicted to reduce the waiting times and increase survival of human patients who need organs. The first successful human kidney transplant was performed in 1954 and worked because the donor and recipient were genetically identical twins. A genetic match between donor and recipient is necessary because certain genetic differences cause the recipient's immune system to attack and kill the foreign transplanted tissue. In 1954 there were no drugs to effectively prevent transplant rejection, but in 1983 the first big advance in immunosuppressive drug therapy came with the discovery of cyclosporine. Other **immunosuppressive drugs** were released in the following years, allowing successful transplants involving liver, heart, heart-lung, single lung, living donor liver, and living donor lungs.

Currently there are more than 100,000 patients on the organ transplant waiting lists that are maintained by the United Network for Organ Sharing (UNOS). This nonprofit organization administers the Organ Procurement and Transplantation Network (OPTN) that was established by the United States Congress in 1984. Worldwide there are not enough organ donors to meet the transplant need, even including living donors who can safely donate a kidney, a part of the liver, or a single lung. Although the numbers and percentages vary from year to year and from organ to organ, approximately 4% of the people waiting for a kidney and 10%

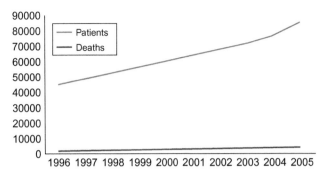

**FIGURE 14.13** Patients on kidney waiting list and reported deaths by year 1996 to 2005.

of those needing a different organ will die while waiting for an organ transplant. This problem continues to get worse with no solution to the organ shortage in sight, while the waiting list of patients in organ failure continues to grow, doubling in the 9 years between 1996 and 2005 (Figure 14.13).

## XenoMouse

In another approach to avoid transplant rejection, the XenoMouse was engineered to produce human antibodies using mouse embryonic stem cell technology. The XenoMouse was developed to generate fully human antibody drugs that can be used to treat a wide range of human diseases (Figure 14.14). The XenoMouse technology was developed by JT America Inc and purchased by the biopharmaceutical company Abgenix, Inc., of Fremont, California. Once immunized, the XenoMouse cells that are producing the desired human antibody can be isolated and cloned to generate cloned cell lines. An example of a monoclonal antibody drug produced using this approach is Panitumumab (marketed by AmGen as Vectibix), which specifically binds to the epidermal growth factor receptor protein (EGFR) and is effective in treating certain cancers including colon cancer.

> Monoclonal antibodies that are partly humanized or have entirely human sequences have proven to be extremely useful in medical therapies, in diagnostic tests, and as highly specific research tools.

## Pigs are the Pick for Human Organs

Faced with a severe lack of human organ donors, and building on the reported success of genetically altered transgenic animals, scientists worldwide have been developing animals with the goal of developing alternative organ donors. Pigs share organ size, anatomy,

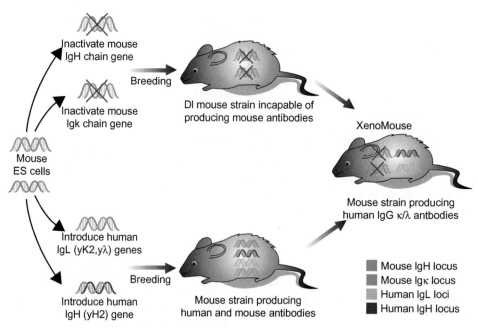

**FIGURE 14.14** Constructing the XenoMouse. The transgenic XenoMouse (Abgenix, Inc.) was developed to generate fully human antibodies for use as drugs to treat a range of human diseases.

and basic physiology with humans and pigs have been genetically altered to reduce the potential for rejection of pig organs by the human transplant recipients. Unlike human cells, pigs have a specific sugar (1,3-alpha-galactose) displayed on the surfaces of their cells that will generate the production of antibodies if transplanted into a human patient. An organ grown in a pig and transplanted into a human would be attacked and destroyed by these antibodies. In the human body, antibodies bind to the 1,3-alpha-galactose sugars on the pig cells, which triggers the complement cascade and eventually kills the foreign pig cells. When triggered by antibodies, a series of more than 30 complement proteins launch a stepwise process that leads to the assembly of protein complexes that make holes in the cell membrane, killing the pig cells. This process, called **complement activation**, can lead to the death of the pig cells transplanted into the human body. However, organs transplanted from engineered transgene pigs carrying the one of three different genes for human proteins that inhibit complement activation survived for three weeks after transplantation.

Different strategies have been developed to avoid rejection of the organs transplanted from pigs. In one approach, knockout pigs were created in which the gene coding for the protein that catalyzes the synthesis of the 1,3-alpha-galactose sugar was deleted from the pig genome. This knockout mutation prevented the production of the 1,3-alpha-galactose sugar in the knockout pig cells, and avoided the rejection mechanism, suggesting that these animals might be appropriate

hosts to grow human organs for human transplantation. The biopharmaceutical company Living Cell Technologies found that coating the insulin-producing pancreatic islet cells from pigs with a complex sugar acted to protect the pig cells from attack and rejection by the human immune system.

In a different approach, transgenic pigs were created, which express human proteins that inhibit the complement cascade. The complement system consists of a number of small proteins called zymogens or proenzymes, which circulate in the blood as inactive enzyme precursors. A **zymogen** is a protein that requires a biochemical change for it to become an active enzyme. When stimulated by one of several trigger signals, **protease enzymes** in the complement cascade cleave specific proteins to release peptide **cytokines** (Greek *cyto-*, cell, and *-kinos*, movement). Cytokines are a large and diverse family of polypeptide regulators that are made throughout the body, and function to amplify a cascade of protein cleavages. This massive signal amplification increases the assembly of attack complexes used by the complement cascade to kill cells. The outcome of the complement response is cell lysis, and phagocytosis of the antigens, which are engulfed by immune system cells. The immune complexes are cleared from the immune system and deposited in the spleen and liver.

Tissue rejection can also be caused by the T-cell immune response. Immunosuppressive drugs can blunt the T cell–mediated responses, but not without a significant risk of drug toxicity. An alternate strategy

## Box 14.2 Goodbye, Dolly

### It's So Sad to See You Go

Dolly is gone, but in the decade since she was cloned in 1997, many other animals have been cloned successfully. The birth and short life of Dolly, the lamb cloned from an adult cell, not only sparked worldwide debate about the ethics of cloning humans from adult cells, but it also represented the answer to a scientific debate that had been going on since 1938. Embryologists wanted to know more about the process of cell development to find out if as cells specialized (differentiated) into all the different cells of the body, the genetic information in the DNA of these cells had changed irreversibly, so that the DNA could no longer provide the genetic information needed to develop an entire new organism from the start. The scientists replaced the nucleus of an egg cell with the nucleus of an adult somatic cell, not a germ line cell that will generate a sperm or egg, using a process called somatic cell nuclear transfer (SCNT) (see Chapter 12). This process was used successfully in frogs in the 1950s and in fish in 1984 when scientists cloned frogs from tadpole cells and from adult frog intestinal cells, and the fish (carp) were cloned using cultured fish kidney cells. SCNT was not successful in mammals until 1996 when researchers at the Roslin Institute in Scotland reported that they had cloned two lambs from embryonic sheep cells. Technically, the development and birth of the two lambs was a major advance, but questions remained unanswered about the state of the genetic information in the sheep somatic cells.

The birth of Dolly the sheep in February 27, 1997—who was cloned from adult sheep skin cells by Ian Wilmut and colleagues at the Roslin Institute—generated news coverage and debate around the world. The media questions generally focused on human cloning, especially on the ethics and risks of generating a clone from the cells of an adult human. In subsequent years, the debate about cloning humans focused on the low success rate of mammalian cloning and on the high rate of fetal death and developmental abnormalities. Dolly was the only live lamb born from 29 embryos transferred to surrogate mothers, and the 29 were the only embryos that were suitable for transfer from 277 nuclear transfers, a success rate of 0.3%. The success rates with fetal or embryo nuclei were higher, about 5% to 6%, but the death of lambs at around the time of birth was high (12%) regardless of the source of the nuclei. Similar low efficiencies and high rates of fetal and neonatal deaths are also common in SCNT experiments conducted using other mammalian species.

Most people, including many scientists and bioethicists, conclude that the high risk makes it unethical to use SCNT in human patients.

---

involves establishing immunological tolerance through the use of a pig thymus tissue transplant. Baboons with transplanted pig organs and with modified immune systems fail to react to the pig organs as foreign tissue, allowing the baboons to survive using only the pig kidney for up to 90 days.

Despite the contributions of animal research to progress made in the area of organ transplantation, the barriers to transferring pig organs into humans still remain. The risk of infection from the porcine endogenous retrovirus (PERV), an RNA virus that can cause immune system damage, is unlikely because tests on hundreds of patients who were exposed to pig tissues such as pancreatic islet cells, skin, and whole livers and spleens for extracorporeal blood perfusion showed no evidence of PERV.

Given all the barriers to the use of pig organs to treat human patients by xenotransplantation, it is not surprising that research into the potential of pig organs to meet the urgent need for transplantable organs has declined.

## ANIMAL CLONING

Many animals have now been cloned using somatic cell nuclear transfer (SCNT) including goats, cattle, mice, pigs, cats, rabbits, horses, and dogs. These animals were cloned for various reasons. Some were personal, such as wanting to replace a beloved pet, or in commercial agriculture, to promote the inheritance of desired traits, such as high levels of milk or meat production by a champion farm animal, for the development of an animal model for research, the recovery of species facing extinction, or to prolong the genetic lifetime of a champion racehorse.

## Regulation of Animal Biotechnology

The regulations that control the development and commerce of animal biotechnology products in the United States are based on the 1986 guidelines produced by the White House Office of Science and Technology Policy (OSTP), incorporating policies from the U.S. Department of Agriculture, the Environmental Protection Agency, the Food and Drug Administration, and the Occupational Safety and Health Agency (OSHA). These guidelines direct that the regulation of animal biotechnology should be based on the products made, not on the process needed to make the products.

Regulations that control the use of transgenic animals in the United States are based on the Animal Welfare Act, which is administered by the U.S. Department of Agriculture. The act, first passed in

1966 and amended in 1970, 1976, 1985, and 1990, established regulations to control the handling and treatment of animals by commercial and research facilities. All facilities must be licensed, and research laboratories must request approval from the Institutional Animal Care and Use Committee (IACUC) to comply with all the standards and procedures; the IACUC will review all proposed research using animals for compliance with the rules. Each research protocol is reviewed by the committee to determine that the proposed study is necessary and that the proposed procedures would minimize the pain and discomfort to the animals. The Animal and Plant Health Inspection Service (APHIS) of the USDA Veterinary Services' National Center for Import and Export regulates the import, export, and interstate transport of all animals and animal products (e.g., tissues, blood, and semen), including transgenic animals altered by genetic engineering. The Food and Drug Administration, in keeping with the 1986 OSTP guidelines, regulates a genetically altered animal as a new drug, because the genetic changes affect the structure and function of the transgenic animal.

In 2003, the FDA and the National Academy of Sciences (NAS) evaluated the available scientific evidence and concluded that the food products derived from cloned animals and their offspring are as safe to eat as food derived from animals that were not cloned. Research showed that healthy adult cloned animals were virtually indistinguishable from their conventional counterparts. In January of 2008 a subsequent risk assessment performed by the FDA supported the scientific findings that cloned animals are a safe source of food. The potential use of cloned animals as food sources for general consumption required clearly written labels on food to indicate that the product comes from cloned animals or from **transgenic plants** (see Chapter 15). Also included are products from animals exposed to biotechnology products. A good example is recombinant bovine growth hormone (somatotropin) (bST), which is present in the milk and dairy products derived from cows who received treatments with this recombinant hormone to increase milk production. Bovine growth hormone is normally made in the pituitary gland in cows and other mammals, but many people want the recombinant products to be labeled for the consumer's information. For the commercial production of bovine growth hormone protein, the bST gene was isolated from the cow genome, inserted into a DNA vector and introduced into bacterial cells using recombinant cloning methods (see Chapter 4). The engineered bacteria produce large amounts of the bovine growth hormone, which is purified for use as a drug to treat the cows to increase milk production.

Approved by the FDA in 1994 after 14 years of review, the bovine growth hormone case became the subject of state laws and legal cases in Vermont and Illinois. The Ben and Jerry's Ice Cream company publically supported clear labeling as a right-to-know issue and probably a marketing issue as well. Vermont enacted a state law requiring appropriate labeling. However, the U.S. Appeals Court judged that the Vermont law requiring the labeling of milk products from bST-treated cows was unconstitutional based on a First Amendment argument that the plaintiffs would do themselves irreparable harm by using product labels that failed to distinguish milk derived from bST-treated cows from milk derived from non-bST-treated cows. The state of Illinois prohibited labeling milk and milk products as bST-free despite campaigns by the Pure Food Campaign (arguing that consumers had the right to know) and the Humane Farming Association (arguing that mastitis requiring antibiotic treatment is more frequent in bST-treated cows). In fact, dairy cows are evaluated regularly for infections and sick cows are no longer used to produce transgenic milk products.

The effort to label consumer products as "bST-free" could represent a marketing strategy to attract consumers who are environmentally conscious and choose to buy organic foods, as well as people with a gut-level fear of biotechnology. The label currently allowed by federal law reads "No Artificial Growth Hormone," which is defined as "No significant differences have been shown between milk derived from rBST-treated and non-rBST treated cows."

> Much controversy and public debate worldwide surrounds the sale of products made from transgenic animals, including basic food staples like milk. The public favors labeling these products for the benefit of the consumer.

## Human Therapeutic Cloning

Human reproductive cloning, the creation of an identical genetic copy of an individual human achieved by the transfer of a nucleus from an adult cell into an "empty" egg cell, is widely viewed by scientists and most of the public as dangerous and unethical. Although the procedure has been successful in a number of different animal species, the process is very inefficient, the frequency of live births is generally low, and many of the animals born using the procedure have developmental challenges. These problems led most scientists, including Ian Wilmut, who successfully cloned the sheep Dolly, to conclude that the risk of creating a human child with developmental problems was so high that cloning is unethical for use in humans.

Attempts at reproductive cloning of other, nonhuman primates such as monkeys have been very difficult. Gerald Schatten at the University of Pittsburgh worked for years to clone monkeys using known somatic cell nuclear transfer methods. Eventually his research team produced two cloned rhesus monkeys but they were generated by embryo splitting and not by nuclear transfer. The monkey embryos were made in the laboratory by combining eggs and sperm and culturing the cells in the lab to form eight-cell-embryos. The embryo cells were physically separated into four embryos containing two cells each. These two-cell embryos are genetically identical twins. In initial studies, 107 embryos were used to generate 368 split two-cell embryos, of which only two embryos survived to the blastocyst embryo stage and were implanted into separate female monkeys. Only one of the monkeys was born and survived, emphasizing the difficulty and inefficiency of working with primate embryos. Later, the same research group produced a second split-embryo monkey.

The process of **reproductive cloning** involves the production of individual animals with identical nuclear genomes (see Chapter 12). The nucleus is removed from an adult cell and transferred into an empty oocyte (an **enucleated egg cell**) which no longer contains a nucleus. The oocyte begins to undergo cell division when activated with a pulse of electricity (or similar trigger). After a period of growth in culture, the embryos were transferred into the uterus of an appropriate female animal treated with hormones to support pregnancy. Reproductive cloning can also be accomplished using embryonic stem cells, as well as oocytes.

**Therapeutic cloning** allows scientists to produce genetically matched embryonic stem cells by somatic cell nuclear transfer. Until recently, SCNT was accomplished most reliably in mice, and in 2007, scientists produced embryonic stem cells from rhesus macaque (primate) embryos using SCNT and the nuclei from adult male skin cells. As expected, the process was quite inefficient; only 16% formed blastocysts (35 out of 213). Given the risk of harvesting and manipulating human oocytes, and a myriad of ethical reasons, this procedure is unlikely to be applied to humans.

## iPS Cells Cure Sickle Cell in Mice

In 2007, the Yamanaka group (in Japan) and James Thomson and colleagues (at the University of Wisconsin) announced that they had induced the development of pluripotent human cells that exhibit the properties of human embryonic stem cells (see Chapter 12). They published their research in two highly reputable peer-reviewed scientific journals (*Cell* and *Science*). The amazing feat of producing pluripotent cells with unlimited developmental potential from adult skin cells was achieved by introducing a specific set of transcription factor genes into the skin cells. Yamanaka's team inserted genes encoding Oct3/4, Sox2, Kif4, and c-Myc into the skin cells. Later they found that they could generate pluripotent mouse stem cells without using the c-Myc gene, which was good since c-Myc was implicated in converting iPS cells into cancer cells. Thomson's group took a similar approach and created iPS cells from adult skin cells using DNA encoding the transcription factor proteins Oct4, Sox2, Nanog, and Lin28. These transcription factor proteins work together as master regulator proteins that keep cells in an embryonic-stem-cell-like state (see Chapter 12).

The potential of iPS cells to treat a genetic disease has been realized in the research lab. In 2007, researchers at MIT and the University of Alabama showed that they could successfully treat mice with sickle-cell anemia by reprogramming the mouse cells to make induced pluripotent stem (iPS) cells (see Chapter 12). To try to treat sickle cell disease in mice, the scientists induced the mouse iPS cells to develop into the precursor stem cells in adult bone marrow that generate mature blood cells (Figure 14.15). Sickle cell anemia, the most common inherited blood disorder in the United States, is caused by a single base pair change in the β-globin gene (see Chapter 10). As a result, the mutant β-globin protein forms defective

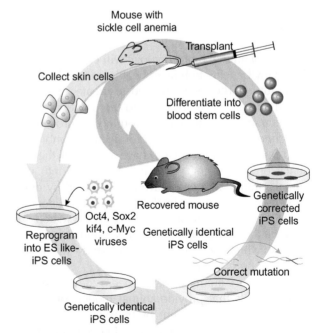

**FIGURE 14.15** Sickle cell anemia successfully treated in mice by reprogramming the mouse genome. The approach used by scientists (MIT/Whitehead) to successfully treat mice carrying the human beta-globin gene with the sickle-cell mutation begins by directly reprogramming the mice's own cells to an embryonic-stem-cell-like state.

hemoglobin complexes called hemoglobin S (HbS), which distort the red blood cells into sickle shapes that clog the blood vessels and cause severe pain and other symptoms (see Chapter 10).

The scientists replaced the mutant β-globin gene in the iPS mouse cells (see Chapter 12) with the wildtype β-globin gene DNA to try to correct the sickle cell defect using a gene therapy strategy. The genetically treated iPS cells were injected into the mice with sickle cell disease. The treated mice expressed the wildtype β-globin proteins and produced functional hemoglobin HbA complexes, effectively curing sickle cell anemia using the iPS cells.

Scientists are encouraged by the generation of "embryo free" iPS cells that could eventually provide genetically matched, differentiated cells for the personalized treatment of many human diseases. Of course additional research is needed to establish conditions for the isolation and differentiation of the iPS cells to accommodate all the different specialized cells needed to treat most human genetic diseases.

> It was a major development in animal biotechnology as well as in human health and medicine when scientists successfully induced adult human skin cells to become embryonic stem cell-like. Like other pluripotent cells, the iPS cells have the ability to develop into a large number of different types of cells in the human body.

## SUMMARY

The field of animal biotechnology will continue to provide many benefits to human health and medical care in the future. The methods used to create genetically altered animals have advanced to the point where the use of transgenic animals is routine. Improvements in technologies involving microinjection, embryonic stem cells, and new vectors have increased the efficiency of introducing genes into cells and into new animals. Important genes have been studied by deleting gene copies from animal genomes using knockout technology. Transgenic animals have been created and manipulated using interference RNA and the Cre/lox systems, which permit scientists to study the tissue-specific expression of genes.

Transgenic animals have been developed as research tools that provide animal models of genetic diseases. The development of transgenic animals to produce biopharmaceuticals has become an active area of research and currently involves engineering animals so that the drug, usually a protein, is secreted in the milk or other product such as hair. Although a great deal of progress has been made in this research,

the efforts to move this research into human clinical trials has been understandably slowed because of the regulatory agencies' concerns about the purity, production, and safety of biopharmaceuticals made for humans by transgenic animals.

There is a recognized but as yet unrealized potential to genetically engineer agriculturally important animals to improve the reproduction and health of the animals and to increase the quality of products such as meat and wool. The FDA has concluded that milk or meat from cloned animals do not present a significant risk even if they did enter the food chain. The resistance of animal rights activists and the uncertainty about consumer acceptance of transgenic animal products have slowed corporate sponsored research and development.

The transplantation of organs from animals into humans (xenotransplantation) has the advantage that it could treat organ failure and reduce the waiting times for human organs for transplant, but concerns about safety and the recent progress made in therapeutic cloning technology have reduced commercial interest in this approach.

The safety and ethics issues surrounding animal reproductive cloning research have essentially halted all human reproductive cloning supported by federal funds in the United States. Therapeutic cloning holds a very promising approach for the future treatment of human diseases. An exciting major step forward occurred when scientists showed that a small number of transcription factor genes are capable of converting human skin cells into iPS cells, pluripotent cells that have the ability to develop into a large number of specialized human cells. This discovery shifted the focus away from human embryos as the major source of human embryonic stem cells for research and for many applications of stem cell technology (see Chapter 12). Experiments with mice carrying the mutant human β-globin gene that causes sickle cell anemia disease showed that the induced pluripotent stem cells (iPS) have the potential to cure human genetic disease.

## REVIEW

To test your understanding of the concepts in this chapter, answer the following questions:

1. How were the earliest transgenic animals produced?
2. What is a blastocyst, and what role does it play in the generation of embryonic stem cells?
3. What are knockout transgenic animals, and how are they produced?
4. Why is it necessary to make certain genetic alterations in the germ line cells?

5. What arguments favor the production of biopharmaceuticals in the milk of transgenic animals?

6. What are some of the challenges scientists face when trying to produce drugs and other products in transgenic animals?

7. What social and ethical issues surround the field of xenotransplantation?

8. What is human reproductive cloning, and why is it considered to be unethical?

9. What are some of the biopharmaceutical products that have been successfully produced in transgenic animals?

10. How does reproductive cloning differ from therapeutic cloning?

## ADDITIONAL READING

Bioengineering; not just humans benefit from animal biotechnology. November 2007. Biotech Business Week. 5:1842.

Cloned rabbit fine five months after birth. July 24 2007. *Chinadaily. com.cn.*

Pollack, A., July 30, 2007. Without U.S. rules, biotech food lacks investors. New York Times Section Ahttp://www.nytimes.com/2007/07/30/washington/30animal.html?_r=1&pagewanted=2&fta=y.

Safe as milk? January 6 2007. New York Times. Sect. A; Editorial Desk (14). http://www.nytimes.com/2007/01/06/opinion/06sat4.html?pagewanted=print.

Saletan, W., January 7, 2007. The fruit of our sirloins. Washington Post page B03 http://www.washingtonpost.com/wp-dyn/content/article/2007/01/05/AR2007010501912.html.

Sparshott, J., August 27, 2006. Diving into the gene pool; biotechnology changing farms, pharmaceuticals. Washington Times http://www.questia.com/library/encyclopedia/guelph-university-of.jsp.

Tabuchi, K., Blundell, J., Etherton, M.R., Hammer, R.E., Liu, X., Powell, C.M., Südhof, T.C., October 2007. neuroligin-3 mutation implicated in autism increases inhibitory synaptic transmission in mice. Science 318 (5847), 71–76 DOI: 10.1126/science.1146221.

The duplicate dog. August 5 2005. New York Times. pg 14 New York edition.

Wade, N., June 7, 2007. Biologists make skin cells work like stem cells. New York Times Section A. http://www.nytimes.com/2007/06/07/science/07cell.html.

Washington Times Staff reporters. August 26, 2007 Regulating biotech food animals. Washington Times. http://goliath.ecnext.com/coms2/browse_R_T009-200708_226_300.

Weiss, R., October 17, 2006. FDA is set to approve milk, meat from clones. Washington Post, A01. http://www.washingtonpost.com/wp-dyn/content/article/2006/10/16/AR2006101601337.html.

## WEB SITES

Biotech Chronicles—from Access Excellence. www.accessexcellence.org/RC/AB/BC/index.html#Anchor-6296. A useful introduction to biotechnology.

Church of Scotland site on stem cells and cloning. www.srtp.org.uk/cloning.shtml. Has thoughtful informed essays.

FDA Center for Food Safety and Applied Nutrition. Biotechnology Notices, <www.cfsan.fda.gov/~lrd/biotechm.html/>.

Food and Drug Agency. http://fda.gov. The FDA regulates prescription and over-the-counter drugs as well as food.

GTC Biotherapeutics. www.transgenics.com/science/howitworks.html. A company developing biopharmaceutics in goat milk.

Hastings Center on Bioethics. www.thehastingscenter.org/default.asp. A well-respected center for the discussion of the ethical issues of medicine and biotechnology.

International Society for Stem Cell Research. www.isscr.org/public/basics.htm. Has descriptions that are very accessible.

Online Mendelian Inheritance in Man, entry for Williams syndrome. www.ncbi.nlm.nih.gov/entrez/dispomim.cgi?id=194050. Entry to extensive resources on human disease genes.

Origen Therapeutics. www.origentherapeutics.com/main-technology.php. Company is developing technology for producing biopharmaceutics in chicken eggs.

Pew Initiative on Food and Biotechnology. http://pewagbiotech.org/research. Pew is a neutral think-tank.

President's Council on Bioethics, white paper on alternative sources of pluripotent stem cells. www.bioethics.gov/reports/white_paper/index.html.

Primer on Ethics and Human Cloning, Glenn McGee. www.action-bioscience.org/biotech/mcgee.html. An ActionBioscience.org original article. McGee is in the bioethics group at the University of Pennsylvania and has written widely on this and other bioethical issue. Has links to other articles.

Report on the biology of nuclear cloning and the potential of embryonic stem cells for transplantation therapy by Rudolf Jaenisch, www.bioethics.gov/background/jaenisch.html. Jaenisch is a pioneer in the field and co-author of the recent paper describing the efficacy of induced pluripotent stem cells in treating sickle cell disease in mice.

Risk-Based Approach to Evaluate Animal Clones and Their Progeny (draft), http://www.fda.gov/cvm/CloneRiskAssessment.htm.

Stem Cells in the Spotlight, http://learn.genetics.utah.edu/units/stemcells. From the University of Utah, this site has animations.

Transgenic Animal Web at University of Michigan, www.med.umich.edu/tamc/links.html#Tg%20Web%20Sites.

United States Regulatory Agencies Unified Biotechnology web site, http://usbiotechreg.nbii.gov.Brings together positions and decisions from FDA, EPA, and USDA.

U.S. Department of Agriculture, Animal and Plant Health Inspection Service. www.aphis.usda.gov/biotechnology/index.shtml. Control importing or exporting animals, including transgenic animals and transgenic products.

# Agricultural Biotechnology

## Scientists Create "No Tears" Onions

*Agence France-Presse*, February 1, 2008

Scientists in New Zealand and Japan have created a "tear-free" onion using biotechnology to switch off the gene behind the enzyme that makes us cry, one of the leading researchers said Friday. The discovery could signal an end to one of cooking's eternal puzzles, why does cutting up a simple onion sting the eyes and trigger teardrops? The research institute in New Zealand, Crop and Food, used gene-silencing to make the breakthrough which it hopes could lead to a prototype onion hitting the market in a decade's time.

The engineering of a no-tears onion represents the application of one of the most advanced methods of genetic engineering to a common consumer problem. Researchers discovered that human tears are a response to the sulfur compounds that are released by the onion when the onion cells are sliced. The no-tears onion was created using an RNAi method where small interfering RNAs (siRNAs) were used to prevent the expression of a specific gene and the encoded protein product. The Nobel Prize for medicine in 2006 was awarded to Andrew Fire of Stanford University and Craig Mello of University of Massachusetts Medical School for their development and use of siRNA method in their research (see Chapter 11). RNA interference using siRNA gene suppression has been used to silence the expression of different genes in many animal and plant cells in addition to the onion plant.

## LOOKING AHEAD

This chapter describes how the tools of molecular biology and recombinant DNA technology have been used to analyze and modify the DNA found in agriculturally important plants. Like transgenic animals, genetically modified (GM) plants are useful but also have potential dangers and raise important controversies. On completing the chapter, you should be able to do the following:

- Describe the different approaches used to introduce new genes into plants.
- Explain why and how a gun is sometimes used in changing the genes of a plant.
- List some major food crops that have been genetically altered.
- Describe the mechanisms used by pest- and weed-killer-resistant genes that are advantageous to farmers who plant genetically altered crops.
- Describe the roles of the Environmental Protection Agency (EPA), Food and Drug Administration (FDA), and the U.S. Department of Agriculture in the regulation of GM plants.
- Explain the potential advantages and disadvantages of an application of agricultural biotechnology.
- Describe the efforts to produce protein drugs in plants.
- Describe the environmental and safety issues raised concerning agricultural biotechnology.

- Describe how the development of plant or agricultural sources of biofuels will reduce the use of fossil fuels in the United States.
- Consider and discuss the impact of agricultural biotechnology in developing countries.

## INTRODUCTION

Scientists working in the field of agricultural biotechnology use the tools of molecular biology to analyze the genetic information in plants, to detect bacterial and viral infections in plants, and to modify the genetic information in plants, which alters the proteins made by the plant cells. Genetically modified soybean and corn plants were first tested in the 1980s to determine how the modified plants would respond to herbicides (chemical weed-killers). GM plants with the genetically enhanced ability to repel invasion by worm pests were first commercially available in the United States in 1995–1996.

The technical successes of **genetically modified organisms** (GMOs) used for agricultural purposes are numerous, but agricultural biotechnology and GMOs have not been widely accepted by the international

### Box 15.1   Ice-Minus Bacteria Prevent Frozen Strawberries

The first crisis in the new field of agricultural biotechnology did not involve a genetically altered plant but a genetically altered bacterium. Steve Lindow, a researcher in California, discovered the "ice-minus protein" gene in *Pseudomonas syringe,* a bacteria commonly found growing on the outside of strawberry fruit in the field. The ice-minus protein offered the ability to prevent damage to strawberries from ice crystals that form on the surfaces of the strawberries when the temperature drops below freezing. To test this idea, Lindow inserted the gene coding for the ice-minus protein into a strain of *Pseudomonas* bacteria commonly found on strawberries. He wanted to test whether the engineered ice-minus *Pseudomonas*, called Frostban, could outgrow the resident population of *Pseudomonas* and protect

the strawberries in the field from the cold temperatures. At the time, the proposed test was approved by the National Institutes of Health Recombinant DNA Advisory Committee (RAC). However, Lindow's proposal to test Frostban in an open field raised serious concerns among local residents and national opponents of genetic engineering. Eventually the Frostban test was performed with technicians and observers in the field outfitted in HazMat jumpsuits (Figure 15.1). Although the test was technically successful—the Frostban bacteria grew more rapidly than the resident *Pseudomonas* bacteria and protected the strawberries from low temperatures—the controversy discouraged any further development of Frostban as a commercially available biotech product.

(A)

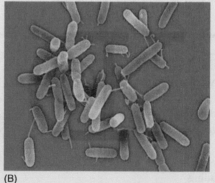
(B)

**FIGURE 15.1**   Frostban ice-minus bacteria are used to treat strawberry plants. (A) Technician in biohazard suit treats strawberries with Frostban; (B) Frostban is a genetically modified strain of *Pseudomonas* bacteria.

community. The public is particularly concerned about the use of unfamiliar gene-altering technologies to change the properties of the foods they eat, especially familiar staples like milk (see Chapter 14). In the United States and Europe, agricultural biotechnology has been portrayed by opponents as an unnatural process that has unknown effects on the crops, the environment, and the health of humans and animals. To try to reassure the public, the companies developing GMOs sought government regulations to oversee advances in agricultural biotechnology. The Environmental Protection Agency (EPA), the U.S. Department of Agriculture (USDA), and the Food and Drug Agency (FDA) developed a regulatory system to review research advances in the field. The successful use of GMOs in agricultural applications is reflected in the number of global acres devoted to growing crops for biotechnology applications, which increased from 4.3 million acres in 1996 to 252 million acres in 2006. The United States leads the world in the number of acres planted with genetically modified crops, with 54.6 million acres consisting of soybeans, corn, cotton, canola, squash, papaya, and alfalfa.

## Flavr Savr Tomatoes

A second crisis in agricultural biotechnology occurred after the federal agencies responsible for regulating agricultural biotechnology were established. GM plants that had been extensively tested in closed greenhouses were granted a nonregulated status by the Animal Plant Health Inspection Service of the USDA and grown in open fields. Calgene, a California company, was given nonregulated status for Flavr Savr tomatoes, which were engineered to ripen but not soften while growing on the vine (Figure 15.2). This tomato was made more resistant to rotting by using RNAi technology (see Chapter 11) to silence the gene encoding the enzyme polygalacturonase (PG). Normally the unmodified tomatoes

FIGURE 15.2 Calgene created Flavr Savr tomatoes, which were engineered to ripen but not to get soft on the vine. The tomato was made more resistant to rotting using RNAi antisense technology to silence a key enzyme in the biochemical process of decomposition.

are harvested before they are ripe, which permits ease of handling and shipping, but then the unmodified tomatoes must be artificially ripened after harvest using ethylene gas. The Flavr Savr tomatoes could be allowed to ripen on the vine without compromising shelf life.

The genetically engineered Flavr Savr tomatoes were approved after the FDA concluded that the Flavr Savr tomatoes were as safe as conventional tomatoes and first sold to consumers in 1994. However sales were low as many people, including restaurant chefs, consumers, and anti-GMO activists, boycotted the Flavr Savr tomato. In the end, the Flavr Savr tomatoes were a disappointment because although the silenced PG gene had a positive effect on shelf life, it did not alter the firmness of the GM tomatoes. This meant that the Flavr Savr tomatoes had to be harvested like any other unmodified vine-ripe tomatoes, negating one of the advantages of making the genetically altered Flavr Savr tomatoes. However, after the Flavr Savr tomato plants were approved, scientists tested other GMOs, including corn that is resistant to the European corn borer, virus-resistant squash, herbicide tolerant cotton, soybeans resistant to an herbicide, and canola (rapeseed) oil with commercial and industrial uses.

> A primary focus of agricultural biotechnology research involves the development of transgenic plants designed to, for example, improve the taste of the fruit or increase the disease resistance of the transgenic plants.

## AGRICULTURAL BIOTECHNOLOGIES

### Genetically Modified Organisms (GMOs)

The goal of transferring new genes into plants is to alter the plant genome with the intention of expressing transgenic proteins in the plant. Several methods have been tested since the early 1980s to find the best way to introduce DNA into plant cells. For the DNA to transform the cells, the DNA must enter the cell nucleus where the genes carried on the vector are expressed as mRNAs and translated into transgenic proteins in the plant cells.

### A Complete Plant Can Be Grown from a Single Cell

Since the 1920s, scientists have known that it is possible to grow an entire new plant from a single root cell that is nurtured in tissue culture media and exposed to appropriate plant hormones (Figure 15.3). The fact that an entire plant can be grown from a single cell indicates that the genetic changes introduced into the single cells will be transmitted to the whole plant and can give rise to many genetically identical plants (clones).

In the United States, GM plants can be patented, which temporarily protects the company's invention or discovery. However, the Plant Patent Act, passed in 1930, gave protection only to **asexually propagated** plants (from bulbs, tubers, or cuttings) because at that time plant scientists doubted that **sexually propagated** plants (from seeds) would have offspring that inherit the traits from their parents. Later, contrary genetic evidence led to the Plant Variety Protection legislation in 1970, which provided patent protection for sexually propagated plants. The demonstration of successful recombinant DNA techniques in microorganisms led researchers to experiment with mutations in plant genes (see Chapter 4). Success with GM plants convinced companies to invest in research to generate new transgenic plants that could be patented by the companies and sold to growers.

## INSERTING GENES INTO PLANTS

### Agrobacterium Tumefaciens

A common method used to introduce genes into plants involves the soil bacterium, *Agrobacterium tumefaciens*. Outside the lab, this bacterium causes crown gall disease, which produces large tumor-like growths on the trunks and branches of certain trees (Figure 15.4). These tumors grow because the *A. tumefaciens* bacterium can take over the plant cell and alter the cellular metabolism. *A. tumefaciens* bacteria contain a DNA plasmid that carries a segment of bacterial DNA called the Ti complex. When the bacteria adhere to a plant cell, the Ti DNA is transferred into the plant cell and then into the nucleus where the Ti DNA becomes inserted into the plant cell genome. As a result the plant cell expresses Ti bacterial genes and produces the *A. tumefaciens* proteins. The genes carried on the Ti DNA plasmid alter the normal growth of the plant cells, causing the crown gall tumors.

To introduce foreign DNA into a plant genome, the scientists insert the Ti DNA into a bacterial plasmid that is grown in large amounts in *E. coli* cells in the laboratory. The plasmid DNA contains many sites that are recognized and cut by restriction enzymes. These enzymes have the distinguishing feature that they will cleave a DNA double-helix molecule at only specific DNA sequences (see Chapter 4). Any change in a restriction site, even a single base pair change, will prevent the restriction enzyme from cutting the DNA

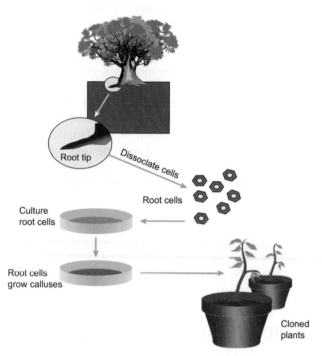

**FIGURE 15.3** A complete plant can be generated from a single plant cell. An entire new plant can be generated from a single cell that is grown in tissue culture media. The cell is treated with appropriate plant hormones that induce the expression of certain genes during development. The root cells develop into calluses in culture, which then grow into whole plants.

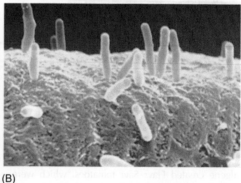

**FIGURE 15.4** *Agrobacterium tumefaciens* can move DNA into plant cells. (A) *Agrobacterium tumefaciens* Crown Gall tumor. (B) *Agrobacterium tumefaciens* bacteria adhere to the surface of the plant cell, an essential step in the transfer of DNA from the bacterium into the plant cell.

at the altered site. For the *A. tumefaciens* plasmid to be used as a DNA vector to carry genes into plant cells, the scientists used recombinant DNA methods to remove the tumor-causing genes from the Ti plasmid and to add an antibiotic resistance gene into the Ti plasmid DNA. This gene permits bacterial cells containing the Ti DNA (and the antibiotic resistance gene) to grow in the presence of the antibiotic, whereas bacteria lacking the Ti DNA cannot grow. The modified Ti plasmid DNA was introduced into the *A. tumefaciens* bacteria, which were then used to infect the plant cells (see Figure 15.4).

Over the years, as scientists learned more about the molecular mechanisms of the Ti plasmid DNA, they devised powerful strategies to apply *A. tumefasciens* technology to many areas of agriculture. However, the increasing use of vectors containing antibiotic resistant genes has raised concerns about potential environmental and health dangers. The widespread planting of GM crops carrying vectors with antibiotic resistance genes has the potential to spread the antibiotic resistance genes to pathogens living in the soil. These organisms can acquire antibiotic resistance genes, making the pathogens much more dangerous and reducing the overall effectiveness of the handful of antibiotics available to treat diseases in humans and animals.

To eliminate this potential danger, scientists have developed methods to remove the antibiotic resistance gene from the plant genome after the gene has served its purpose. For example, researchers used the Cre-loxP site-specific recombination system to delete antibiotic resistance gene DNA from the plant genome (see Chapter 14). Briefly, the Cre enzyme catalyzes the recombination events between two loxP DNA sequences, which deletes the gene that was flanked by the loxP sites in the cell genome.

> Transgenic animals and transgenic plants can be produced in more than one way depending on the characteristics of the animal or plant, the transgene, and the expected outcome of transgene expression.

## Gene Transfer into Protoplasts

Agrobacteria can naturally infect only certain plants, so new methods have been developed to introduce new genes into plants that are resistant to agrobacterium infection. Successful approaches include transferring DNA into **plant protoplast cells**, or shooting the DNA into cells using a **gene gun** (see Chapter 11). All plant cells have an external **cell wall** made of a carbohydrate polymer of **cellulose**, which is part of the physical barrier that protects the cell and the nucleus from the

outside environment. Enzymes that degrade cellulose are often used to remove the plant cell wall to generate **protoplasts** (Figure 15.5). The uptake of DNA by protoplasts can be increased by treatment with polyethylene glycol or by the use of a mild electric pulse called **electroporation**. Both methods gently perturb the **phospholipid cell membrane** and promote DNA uptake into the cell (see Chapter 10 and Chapter 11).

## SHOOTING GENES INTO CELLS

Another method used to make plant cells take up foreign DNA is the **biolistic gene gun** (Figure 15.6). The particles of tungsten or gold are coated with the DNA to be delivered into the cells and loaded into a shell in the gene gun. The shell is rapidly accelerated out of the gene gun, using either a pneumatic or a mechanical propulsion method. A perforated plate stops the shell to avoid damage to the cells, while the metal particles continue into the cells and deliver the DNA into the chloroplast, nucleus, or mitochondria.

### Insertion of DNA into Chloroplasts

Green plant cells contain specialized membrane bound organelles called **chloroplasts**, which are the site of **photosynthesis**, the process by which plants use sunlight to produce carbohydrates from carbon dioxide ($CO_2$) (Figure 15.7). The gene gun can deliver the DNA vector (and genes) into the chloroplasts within a plant cell. In these experiments the DNA vector contains the gene(s) to be transferred, as well as a chloroplast marker gene, and an antibiotic resistance gene that can be expressed in the chloroplast. The DNA vector also contains chloroplast-specific sequences that direct integration of the DNA vector into a precise location in the chloroplast genome. Chloroplasts can express

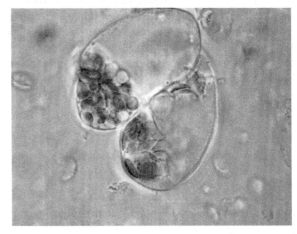

**FIGURE 15.5** Protoplasts are plant cells with the cell walls removed. These protoplasts were made from spinach cells.

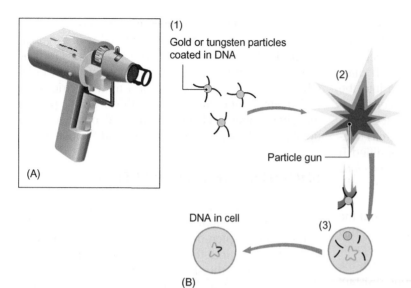

(1)

Gold or tungsten particles coated in DNA

(2)

Particle gun

DNA in cell

(3)

(B)

**FIGURE 15.6** A gene gun successfully delivers DNA into plant cells. (A) The gene gun. (B) (1) The gene gun fires 'bullets' made of tiny particles of gold or tungsten coated with vector and gene DNA. (2) The gene gun is fired at the plant cells, which forces the DNA particles through the plant cell wall without damaging the cell. (3) The gene gun can deliver the DNA into the nucleus, mitochondria or chloroplast organelles depending on the type of vector used.

(A)

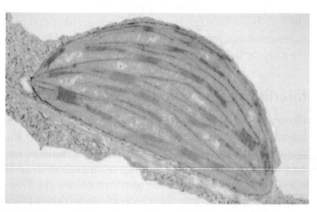

**FIGURE 15.7** Chloroplasts are the site of photosynthesis in green plant cells. A false-color transmission electron micrograph of a chloroplast from a tobacco leaf shows the many stacked membranes in the chloroplast.

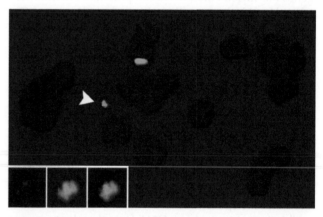

**FIGURE 15.8** Minichromosomes are small versions of native chromosomes. The minichromosome produced by telomere-mediated truncation is indicated by the arrowhead. Each minichromosome contains centromere DNA *(green)*. Telomeres are added to the chromosome by transformation. The chromosome DNA breaks at that site and adds genes at the tip of the chromosome *(red)*. *(Inset)* Transgenes *(red)*, centromere region *(green)*, and the merged image *(red/green)*. Chromosomes are stained blue.

high levels of foreign proteins that fold properly and are biologically active inside the chloroplasts.

Since the early 1990s, numerous foreign genes have been expressed in the chloroplasts of higher (multicellular) plants, including herbicide resistant genes, insect-resistant genes, a drought resistance gene, and genes that can degrade or metabolize mercury. A number of biopharmaceutical compounds and a vaccine against cholera toxin have been produced using chloroplasts, a possible source of vaccines against serious diseases.

> Scientists routinely introduce DNA into the nuclei of different types of cells, but now it is possible to introduce DNA into the genomes of mitochondria and chloroplasts, which are organelles located in the cytoplasm of eukaryotic cells.

## Improving Vectors Chromosomes

The production of transgenic plants has some of the same drawbacks that limit options when constructing transgenic animals. Anytime a gene is added to a plant or animal genome, the new gene can physically disrupt an essential gene in the genome or alter gene expression by inserting into a control region of the gene such as a promoter or enhancer. Recently Daphne Preuss (University of Chicago and Chromatin, Inc.) and her team constructed **minichromosomes**, circular segments of the corn genome that replicate as circles in the corn cells (Figure 15.8). The corn minichromosomes contain repeated DNA sequences that are typically found in the centromere of each plant chromosome, the region of the chromosome that attaches to spindle fibers and moves the chromosome during cell division (see Chapter 9). The centromere DNA on the minichromosome allows the circular chromosomes to be inherited correctly through several generations of corn cells. In comparison, minichromosomes without centromeres are very unstable and are physically lost when the cell divides. Without a functional centromere region, the minichromosome cannot attach to the spindle fibers when cells divide.

Minichromosomes are effective vectors for use in complex plant cells because they contain the DNA elements needed to function as a native chromosomes including centromeres for stable inheritance and replication origins to ensure chromosome duplication. The corn minichromosomes are circular but they can be double-stranded linear DNA molecules.

(A)                              (B)

**FIGURE 15.9** Comparison of teosinte and maize (corn). (A) Teosinte and reconstructed primitive maize were created by crossing teosinte with Argentine popcorn and selecting the smallest offspring. (B) Ear of teosinte *(left, Zea mays ssp. mexicana),* maize *(right),* and the F1 hybrid *(center)* made by crossing the teosinte and maize.

## HOW AGRICULTURAL BIOTECHNOLOGY IS REGULATED IN THE UNITED STATES

The agencies of the United States federal government, including the Environmental Protection Agency, the Department of Agriculture, and the Food and Drug Administration, are responsible for reviewing proposals for the field testing and commercialization of genetically modified crops. Much of the regulatory review focused on the products made for human consumption and not on the scientific methods used to create the product. Safety issues include the possibility that insertion of a gene into a genome might disrupt the function of an essential gene or cause the production of new proteins that might cause an allergic reaction in people or be otherwise harmful to people or to the environment.

Farmers, universities, and seed companies have been altering the properties of plants for hundreds of years through conventional breeding programs. The domestication of wild plants, which began 10,000 years ago, changed the genetic makeup of plants through conventional breeding. For example, Native Americans derived what we know as modern corn from **teosinte**, a related plant with small kernels and ears (Figure 15.9). Genetic manipulation using recombinant DNA methods to produce transgenic plants and animals is often more precise than the genetic changes that occur by conventional breeding methods.

(corn, oilseed canola, soybeans, and cotton) that are protected from insects. Herbicide-resistant sugar beets and potatoes with altered starch composition are under review. The resistance to GMOs that is evident in most EU countries is reflected in the different approaches to uncertainty and risk taken by regulators in the United States and the European Union. The U.S. regulators make their decisions on GMOs based on a systematic risk assessment and application process devised by each agency. The EU relies on the precautionary principle, formulated at a United Nations meeting on biological diversity, which says that if the negative impacts of a proposed new technology (such as genetic engineering) may be severe and are unknown, it is prudent to wait until these negative consequences can be shown not to occur before the technology is allowed. Using this approach, the scientific difficulty in proving a negative essentially prevents the development of new technologies such as agriculture biotechnology.

Public opinion about biotechnology and GM plants are shaped in large part by the extreme views reported in the popular press instead of relying on scientific facts. Many people with different motives are invested in the controversy over transgenic plants and animals.

## WHAT HAS CAUSED RESISTANCE TO AGBIOTECH IN EUROPE?

The European Union (EU) limits the cultivation of genetically engineered crops and requires stringent testing. In 2006, Spain, France, Germany, the Czech Republic, and Slovakia each reported growing about 1.23 million acres of GM corn. They authorized the importation of GM food and animal feed produced from plants

## NEW GENES IN THE FIELDS

Most people in the United States eat a diet that has included steadily increasing amounts of products derived from genetically engineered crops since 1992. These products include soy food, food made with corn including corn flour, corn oil and corn syrup, canola oil, cottonseed oil, sugar from beets, potatoes, squash, papaya, and beets. Recently the Agriculture Research

(A)

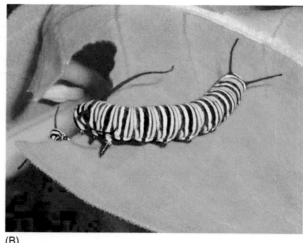

(B)

**FIGURE 15.10** Bt powder threatens Monarch butterfly larvae. (A) Adult monarch butterfly. (B) Monarch butterfly larvae.

Service (ARS), the research division of the USDA, developed and received "non-regulated" status for a genetically modified plum tree that is resistant to the plum pox virus. The ARS accomplished this feat using RNAi technology to silence the production of the plum pox virus coat protein, which halted the virus lifecycle in the plum tree.

Many transgenic plants were engineered to carry genes that confer resistance to an insect, a virus, or an herbicide. Herbicides kill crop plants as well as weeds. A crop that is resistant to an herbicide allows the farmers to use substantially fewer herbicide treatments to control weeds without harming the crop. Many types of transgenic plants are engineered to be resistant to the herbicide glyphosate, which is called Roundup by Monsanto. Glyphosate is toxic to plants because it inhibits the action of an enzyme called 5-enolpyruvylshikimate-3 phosphate synthetase (EPSPS), which is required for the plant to synthesize some essential amino acids. In this case scientists used a gene from *E. coli* that encodes the bacterial form of the EPSPS enzyme, and confers resistance to glyphosate. When inserted into the plant genome, the bacterial gene makes the recombinant plants resistant to the glyphosate herbicide.

The soil bacterium *Bacillus thuringiensis* (Bt) has been marketed as a **biopesticide** because recombinant Bt cells have been engineered to produce many different protein products, including pesticides. But even without carrying a foreign gene, Bt bacteria are toxic to many kinds of insects.

Wild-type Bt cells normally express genes for six different crystalline (Cry) toxin proteins, which dissolve in the **alkaline** environment of the insect's stomach (Figure 15.10). The Cry proteins bind to Cry protein-specific **receptors** on the surfaces of the cells that make up the intestine wall in the insect. This causes

pores to form that allow lethal fluid uptake by the insect's gut. Bt cells have limitations as a biopesticide because the powdered preparation of Bt cells must be applied frequently since the Bt powder is washed away by rain. Also the levels of Bt toxicity are unpredictable in prolonged sunlight.

Recombinant Bt cells offer important advantages for the expression and preparation of large amounts of eukaryotic proteins in the lab. A DNA vector carrying a promoter that normally drives transcription of the crystalline protein genes is used to clone foreign protein genes. The foreign protein coding gene is inserted into the vector under the transcriptional control of the promoter for the crystalline protein gene. Like the normal crystalline proteins, the recombinant proteins are made at high levels and can be harvested in large quantities from insect cells grown in culture in the lab.

In 1999, the results of a small research study by John Losey (Cornell) published in *Nature* reported that Monarch butterfly larvae were harmed by eating pollen from corn plants treated with Bt powder (Figure 15.11). In the study the Monarch butterfly larvae were allowed to feed on milkweed leaves covered with pollen from the Bt corn. Monarch butterfly larvae typically feed on the leaves of milkweed plants near cornfields, raising the possibility that the pollen from the Bt corn would harm the monarchs. However, new research was published that disagreed with the initial conclusions of Losey's research. Scientists from the USDA-ARS, U.S. and Canadian universities, and industry and environmental groups coordinated the development and review of grant applications to test Losey's conclusion. This subsequent research disproved the original conclusions and was published in a series of papers in the *Proceedings of the National Academy of Sciences, USA*.

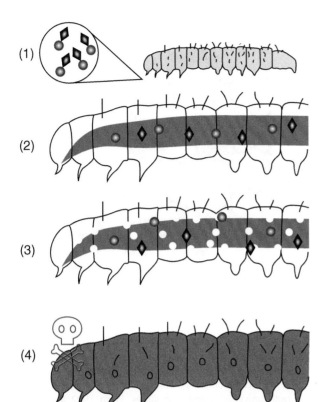

**FIGURE 15.11** Mechanism of Bt toxicity in Monarch caterpillars. The proposed mode of action of Bt includes the following: (1) Bt spores are ingested by the insect; (2) the spores are dissolved in the midgut and the toxins are activated by proteases (enzymes that cleave proteins); (3) toxins bind to specific receptors in the midgut cells causing pores to form in the epithelial cells of the gut; (4) the larvae dies and the Bt spores germinate and are spread.

In 2000, a type of Bt corn called Starlink was developed by the biotechnology company Aventis, which attracted negative publicity when it became known that the DNA from the engineered Cry9C corn had been detected in tacos for sale in stores in the United States and Mexico. The detection of Cry9C in a human food source aroused concern because Starlink had received approval for use in animal feed but had not been approved for human consumption. The level of contamination of the taco with Cry9C was tiny and was only detected using DNA probes that base pair to the Cry9C DNA and the polymerase chain reaction (PCR). Even without more substantial evidence, public concern was raised, and Aventis collected and destroyed all the corn products containing and contaminated with Starlink. Eventually 34 people reported to the FDA that they had suffered allergic reactions to the tacos contaminated with Cry9C. However, the actual impact is unclear as none of the people tested had raised antibodies against the Cry9C protein as would be expected for an allergic reaction to Cry9C.

This experience shed light on the potential problems if products are approved for use in animal feed but not approved for human consumption. In many cases, the farmer, trucker, or processor do not keep the two crops separate along the way from field to taco factory, or the trucks are not fully cleaned between loads. In the case of Starlink, all it took was a kernel remaining in a truck when the next load of corn was loaded, the tiniest contamination can be detected using the highly sensitive polymerase chain reaction.

> Biocontrol involves the use of one kind of organism to control the population of a different type of organism, for example, insects used to control bacteria.

## BENEFITS OF GM CROPS

The International Service for the Acquisition of Agri-Biotech Applications (ISAAA) is a nonprofit group sponsored by philanthropic organizations and agricultural companies. The ISAAA compiles information on GMO crops worldwide and provides information that supports the economic and environmental benefits of adopting GMOs. These include the environmental and economic benefits of switching to no-till or low-till farming. Tillage refers to the need to turn the soil over before a new crop is planted, a process that helps with weed control and seed planting but that is also associated with the loss of carbon in the soil, which adds carbon dioxide to the atmosphere.

The ISAAA calculated the benefits from the reduced use of herbicides and pesticides for the 10 years between 1996–2005 and found that the largest global benefit came from GM soy crops. The economic benefits were higher for all GM crops used in developing countries that have adopted agricultural biotechnology (South Africa, Paraguay, India, and Mexico). In fact, developing countries using GMOs accounted for 55% of the economic benefits realized in 2005. In addition, as a result of switching to no/low till practices, GM crops have decreased the use of herbicides and pesticides by 7% and reduced the environmental impact by over 15%. Not assessed was the expected reduction in greenhouse gas emissions that accompany the adoption of no/low till methods and the need for less frequent applications of pesticides and herbicides.

> Many farmers in developed and under developed countries have adopted the use of GM crops. Although activists object to their use, especially in food crops, the economic benefits to farmers and the environment have led to the increased use of GMOs.

## Pharming-Protein Drugs from Genetically Altered Plants

Applications of biotechnology have produced a number of novel protein drugs that have been approved for human use by the FDA and by drug regulatory agencies in the United Kingdom and Europe. These biotechnology products were produced by transgenic bacteria, yeast cells, or mammalian cells grown in bulk culture in large fermentation vats. The use of transgenic plants to produce recombinant proteins was pioneered by a number of biotechnology companies who see transgenic plants as relatively inexpensive options compared to the bulk growth of cells in tissue culture. Also, transgenic plants can perform the appropriate **posttranslational modifications** of proteins that are required for function, including the addition of carbohydrates, phosphates, or other chemical groups to proteins. The failure to properly modify the eukaryotic proteins can prevent proteins from interacting with receptor proteins or can cause the protein to be metabolized and cleared from the blood too quickly or too slowly.

Research into the use of plants to produce transgenic proteins is at an early stage, and most plant-produced proteins are sold as research reagents and not for use in humans. Dow Agro Sciences gained FDA approval to market a poultry vaccine that was produced in plant cell culture. Scientists are also working on the production of a vaccine in the edible portion of a plant, the fruit. This would be a cost-effective way to deliver the vaccine in developing countries with limited refrigeration for vaccine storage and few skilled medical workers to administer conventional vaccines by injection. The plant is engineered with a transgene that expresses a specific part of a bacterial protein known to cause antibody production in an animal.

Researchers at the University of Arizona developed and tested an oral vaccine for humans against hepatitis B (HepB), which they produced by transgene expression in potatoes. The transgene expressed in potatoes was the same hepatitis virus DNA sequence that was incorporated to make the standard injectable HepB vaccine. The volunteers who ate the transgenic potatoes (uncooked to avoid destroying the antibodies with heat) developed antibodies to HepB, but those who ate the conventional potatoes lacking the transgene (also uncooked) did not. The potato vaccine was not pursued further in part because of the requirement to eat the potatoes raw. This and another study on potatoes engineered to express a Norwalk virus protein showed surprisingly that these proteins survive in the stomach and can stimulate an immune response in the human body. Additional edible vaccines are under development.

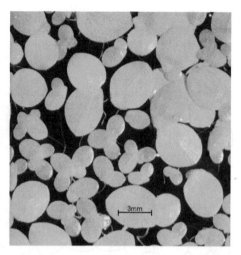

**FIGURE 15.12** Transgenic duckweed makes antiviral interferon. Interferon blocks the production of viruses by infected cells and was produced in duckweed, a simple plant that grows in water.

Transgenic plants producing various biopharmaceuticals have been constructed and analyzed in research performed at both universities and companies. From 1992 to 2002, more than 20 different recombinant vaccines directed against disease-causing viruses or bacteria were expressed in transgenic fruits or vegetables.

In both the United States and Europe, many companies are developing programs to produce biopharmaceuticals in transgenic plants. In some cases, the transgenic plant products have advanced to trials in humans, including a cancer vaccine produced in tobacco plants and a stomach enzyme designed to treat patients with cystic fibrosis and produced in corn. Transgenic interferon, which blocks the production of viruses in infected cells, was produced in duckweed, a simple plant that grows in water (Figure 15.12). The barriers to making biopharmaceuticals from transgenic plants include not only the dosage, production quality, and efficacy of the process but also the standard challenges of cost-effective expression and purification of the protein. Regulators and activists raised environmental and safety concerns about the inadvertent exposure of humans to the biopharmaceutical products from transgenic plants, for instance, workers involved in growing and processing transgenic plants risk exposure to the transgenic products.

Also of concern is the possibility that a transgenic plant expressing a biopharmaceutical product could accidentally enter the food chain. This almost happened in 2002 when soy for human consumption was found to be contaminated with corn engineered to produce an enzyme to treat cystic fibrosis. The contamination was caused when the company ProdiGene planted the soy crops in fields that had been previously used to grow the transgenic corn. The contamination was detected before the soy beans entered the food chain, the crop was destroyed, and ProdiGene paid a fine. The issues of containment and the need for buffer zones between crops were raised for all genetically modified crops. Understanding these concerns is critically important for the safe production and use of biopharmaceuticals produced by plants.

## BIOTECHNOLOGY TOOLS HELP DIAGNOSE PLANT DISEASES AND DETECT TRANSGENES

### Polymerase Chain Reaction (PCR) Can Detect a Single DNA Molecule

The molecular evidence that a plant has a disease or carries a transgene can be detected using the **polymerase chain reaction** (PCR) to amplify specific DNA sequences. PCR amplification is used to copy a single DNA molecule into millions of identical DNA copies, providing scientists with large quantities of specific amplified DNA for study. Using PCR, scientists can specifically detect the presence of the transgene DNA in plant cell genomes. PCR can also be used to detect the DNA or RNA molecules made by a disease-causing organism in an infected plant, even if these molecules are present in very, very small amounts. Kary Mullis was awarded the Nobel Prize in 1993 for his discovery of PCR technology. PCR uses thermostable DNA polymerase enzymes isolated from organisms that grow at high temperatures. Through multiple repeated cycles of heating, cooling, and DNA replication, the PCR method generates millions of copies of a specific DNA sequence (see Chapter 5).

> PCR is an extremely powerful research tool used in many scientific fields, including agriculture. PCR can detect viroids in plant cells, which are tightly folded RNA strands that do not encode proteins. PCR was used in 1993 to identify 4 different viroids that infect hops and fruit, 16 viruses that infect cabbage, beans, corn, tomato, peas, and gladiolus, and 7 bacteria, 11 fungi, and 3 nematodes that infect a wide variety of plants.

### Microarray Technology

Scientists use microarray technology to study large scale gene expression, for example, to compare the genes expressed in healthy cells compared to the genes expressed in diseased cells (see Chapter 13). Other microarray assays include genes expressed in diseased cells compared to the genes expressed in the diseased cells treated with a specific drug. In agricultural biotechnology, the microarray technology has many applications including analysis of the expression of transgenes in plant cells and tests to detect and diagnose infectious pathogens including plant viruses. Microarray technology can be used to screen infected cells to determine which type of plant virus might be causing a particular plant infection.

The candidate viral DNA strands are located in different known positions on the microarray grid pattern. The scientists harvested total RNAs from the infected plants, and copied the RNAs into cDNA probes containing a radioactive or florescent tag for subsequent detection. The microarray grid is incubated in buffer containing the single-stranded cDNA probe that will base pair to the DNA on the microarray slide corresponding to the virus that expressed mRNAs in the infected cells. This technique has been used to identify plants infected by potato viruses and cucumber viruses. A group of British and European academic, government, and corporate scientists (Diag Chip) is developing microarray grids that can be used to detect a large number of the important plant pests and pathogens. The USDA supports research on methods of pathogen identification, and the National Institutes of Health (NIH) and National Science Foundation (NSF) also provide support for basic research on plant pathogens.

> Microarray technology allows scientists to distinguish between the genes expressed by cells under different environmental and genetic conditions.

### Breeding Better Plants by Analyzing Genetic Variation

Every year, plant breeders and farmers try to improve their crop yield by selecting plants that are known to respond well to environmental challenges such as drought or flood, and to be resistant to microbial or insect pathogens. Advances in automated DNA sequence analysis have allowed scientists to determine the sequences of many plant genomes. As with human genomes, the DNA sequences of different plant genomes are compared to each other to search for genes associated with important traits such as pathogen resistance. Resistance to a specific pathogen might be controlled by a single gene, but a complex trait such as crop yield is unlikely to be controlled by a single gene. In fact, plant geneticists have identified **quantitative trait loci (QTL)**, which are specific chromosome regions containing several genes that influence a complex trait such as crop yield. Scientists use this information to identify genes that are involved in specifying certain traits. Once identified, a gene can be isolated from the genome, sequenced, and studied further to determine the specific function of the protein product. Then scientists can make mutant versions of the gene and test the impact of the mutant transgene on the plant.

The new field of **bioinformatics** (see Chapter 7) provides scientists with powerful research tools for the management and analysis of computer databases containing complex genetic information from genome studies. Once it became routine to sequence entire genomes, scientists compared the sequences of individual human genomes and found sequence variations that correlate with specific genetic traits. Similar genetic variations were found when the genome sequences of different lines of the same plant species were compared. DNA variations act as **genetic markers** in plant genomes, and can be correlated with specific traits or phenotypes in the genome. This information is essential for plant breeding programs to obtain the desired progeny. Commonly used genetic markers based on genetic variation include **restriction fragment length polymorphisms** (RFLP) and **single nucleotide polymorphisms** (SNPs) (see Chapters 8 and 10).

The RFLP and SNP markers are powerful tools used to study gene function, even though these DNA markers are not usually associated with biological function. RFLPs are simply DNA sequences that differ between genomes of the same species. RFLPs involve a DNA change that alters a restriction enzyme cleavage site in the genome sequence. For example, if the DNA genome sequences from two individual plants of the same species are digested with the same restriction enzymes, comparison of the resulting DNA fragments will reveal differences in the DNA fragment lengths that indicate the position of the RFLP in the genome. RFLP analysis performed on the genomes of plants from the same species but with different phenotypes will provide information that helps the geneticist to decide which plant crosses (matings) will result in the desired progeny. The inheritance patterns of the SNPs and RFLPs are important to search for genes that contribute to the ability of plants to grow and tolerate extreme environmental conditions. The search is made more difficult by the likelihood that multiple interacting genes are responsible for many plant traits.

## BIOCONTROL ALTERNATIVES TO PESTICIDES AND FERTILIZERS

After six years of laboratory testing scientists at the University of Minnesota are field-testing a strategy using the Chinese wasp (*Binodoxys communis*) to control soybean aphids. In 2000, these aphids appeared in the soybean fields in Minnesota, which cost state farmers an estimated $200 million a year in lost crops, including the added expense of spraying pesticides. The Chinese wasps only sting and kill soybean aphids but are not harmful to humans or pets. Other wasp species from the same region of China are also under evaluation.

**Biological control (biocontrol)** involves approaches that use parasites, predators, or pathogens to control an unwanted organism; in other words the use of bugs, infectious bacteria, or viruses to control pests. The application of biocontrol methods for plants includes the conservation of naturally occurring biological control agents, the release of additional native biological control agents, or the classic introduction of an exotic, non-native, biological control agent to control an exotic pest.

The first documented example of biocontrol occurred in 1883 in Maryland, where the wasp *Apanteles glomeratus* was introduced to control the accidentally imported cabbageworm *Pieris rapae*. The *A. glomeratus* wasp parasitizes the cabbageworm larvae, killing the insect and saving the cabbage crop. Early attempts at biocontrol methods were effective so further tests involving biocontrol took place in Florida and California. Many states have active programs to monitor pests and to identify potential biocontrol agents, often using exotic organisms that originate where the pest is native. The USDA Agriculture Research Service provides a database of invertebrate and microbial biological agents that are available to control invertebrate, weed, and microbial pests. The USDA regulates the release of these organisms, to guard against the use of imported biocontrol agents that can become pests.

> Biocontrol strategies are less successful in controlling weeds than in controlling insects and microbial pests. In the past some biocontrol agents became pests when they outgrew the native species and interfered with the growth of native plants.

## TERMINATOR TECHNOLOGY: ARE THE GM SEED COMPANIES EVIL OR PRUDENT?

In 1990, scientists in Belgium constructed transgenic plants carrying a gene with an another-specific promoter (from tobacco) located in front of the barnase gene from the bacterium *Bacillus amyloliquefaciens*. The barnase gene codes for a **ribonuclease**, an enzyme that degrades RNA in the cell and is toxic to the tapetal

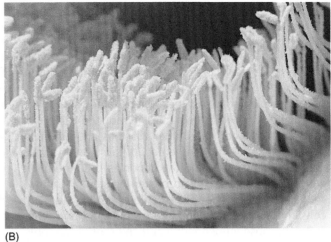

(A)                                    (B)

**FIGURE 15.13** Pollen grains carry the male sperm cells. (A) This scanning electron microscope (SEM) image of pollen grains from a variety of common plants including sunflower (*Helianthus annuus*), morning glory (*Ipomoea purpurea*), prairie hollyhock (*Sidalcea malviflora*), oriental lily (*Lilium auratum*), evening primrose (*Oenothera fruticosa*), and castor bean (*Ricinus communis*). (B) Pollen is deposited on the ends of the anthers.

cells, which are required for pollen development in the plant. The results of the tests were dramatic. The transferred gene prevented normal pollen development and caused **male sterility** in the transgenic plants. These observations made it possible to use the transgenic barnase gene to induce male sterility in other crops. Called terminator technology by the companies, male sterility is a genetic condition that prevents plants from producing functional **anthers**, **pollen**, or male **gametes** (sperm cells) (Figure 15.13). Pollen grains protect the sperm cells while they move from the stamens of one flower to the pistil of the next flower, causing the formation of seeds.

In 2003, news headlines announced that the terminator technology developed by the large seed companies was a new advance in agricultural biotechnology. What raised considerable public controversy was the idea that the seed companies planned to use terminator technology to engineer the male sterility trait into the genetically modified seeds of commercially important crops. Perhaps this response was not surprising considering the earlier negative public reaction to the perceived threat of genetically engineered plants and other GMOs, seen as a threat to agriculture, the environment, and human and animal health in developed and developing countries.

Some farmers worried that if they grew crops using the seed company's sterile "terminator" plants that do not make seeds, they would be unable to save seeds from one harvest to use for the next planting. They were very concerned that they would be forced to buy seeds each year from the seed companies using the terminator technology. People were also concerned that the plant gene involved in male sterility would

be accidentally transferred into other plant genomes and spread the male sterility trait to other crops. As is often the case with new technologies, at least some of the resistance to terminator technology was caused by a general lack of understanding about male sterility in plants in part because of the failure of the companies to clearly explain the basic science behind male sterility. The molecular mechanisms responsible for male sterility vary in different plants, providing the seed companies with a choice of paths to achieve the goal of male sterility. For example, the Pioneer seed company created corn with the male sterility trait by inserting the *E. coli dam* gene into the plant cells under the control of a promoter that turns on expression of the *dam* gene only in the anther cells. The *dam* gene encodes a DNA methylase enzyme (Dam methylase) that transfers a methyl group (–CH3) onto the G base in the GATC sites in the plant genome DNA. Methylation interferes with normal expression of the plant genes, causing the anthers to die, and conferring male sterility.

Traditionally, farmers would save seeds from one harvest for planting in the next, a process called brown bagging, but this practice has some genetic risk and has become less popular. The hybrid F1 generation created by crossing two parental lines (P1 X P2) often exhibits desirable traits that exceed those of both parents, which is called **hybrid vigor** or **heterosis**. When farmers save seeds from a crop one year to plant the crop the next year, the crop planted with the saved seeds is the F2 generation. Unfortunately the F2 generation is genetically unpredictable and has characteristically lower yield. Corn crops have been bred for thousands of years by picking and breeding the plants with the most desirable traits. The practice of brown bagging corn in the United States is essentially unknown now because almost all modern corn crops are hybrids. Farmers buy hybrid corn seeds each year based on the results of field-testing the hybrid strains in their locations. There are test plots of hybrids and now GM crops planted across the United States.

Initially the seed companies introduced male sterility traits into corn and other crops as an answer to the criticism that the transgenic DNA in the GM crops could be transmitted into the genomes of unmodified crops. Other approaches to conferring sterility involve changing the mitochondria or chloroplasts in the plant cells and are being tested in corn, soybeans, and other crops. Because mitochondria and chloroplasts are only inherited from the mother, the foreign transgene that confers sterility in these cases would not be present in the male sperm cells in pollen.

The possibility that genetically engineered plants may create environmental or health risks is real, and techniques such as male sterility and transgenic chloroplasts or mitochondria are being used to reduce the risk of contamination by GM crops.

## BIOFUELS

There are many serious problems associated with the use of fossil fuels including the high cost of oil and gasoline and the risk of global warming, which triggered an international race to develop alternative sources of fuel to run cars, heat homes, and power factories. But the pressure to quickly solve this worldwide problem has already raised many dilemmas. A good example was the large political and public pressure to produce ethanol (EtOH) from corn plants, arguably at the expense of the human food supply. Until the ethanol boom in the 2000s, more than 60% of the corn harvested annually

in the United States was fed to domestic cattle, hogs, and chickens for human consumption in food or beverages. Many thousands of food items contain corn or corn byproducts and in Mexico, where corn is a staple food; the price of tortillas skyrocketed when corn grown in the United States and Mexico was diverted into ethanol production. Of course, the corn farmers were happy with this development, since the ethanol processors bought more corn for much higher prices. As the demand for ethanol increased, so did the demand for corn. As a result, however, the cost of corn used for animal feed and as a source of human food both rose dramatically. By 2006 the price of corn had doubled. Consumers in the United States felt the upward pressure on food prices. In developing countries the high price of corn motivated many farmers to clear more land to grow additional corn crops by burning down the jungle and rain forests.

In 2007, The President of the United States made it a national goal to reduce American gasoline consumption by 20 % in the next decade. The new mandatory fuel standard requires a capacity of 35 billion gallons of renewable and alternate fuels by 2017, a challenging goal given that the current oil consumption in the United States is more than 70 billion barrels each year.

Concerns about global warming due to greenhouse gas emissions due to fossil fuels combined with the high price and uncertain availability of oil from the Mideast have generated interest and investment in **biofuels**, sustainable fuels that come from plants. In 2008, the United States consumed 12 million barrels of oil from fossil fuels. Considerable political pressure encouraged the federal government to focus on the production of ethanol derived from corn. To support this effort, ethanol (15%) was added to gasoline as an alternate oxygenate to methyl tertiary-butyl ether (MBTE), which was widely used as an additive from the 1970s to the 1990s to reduce carbon monoxide air pollution. However, the EPA later concluded the MBTE is a potential human **carcinogen** at high doses, and when MBTE was detected in water wells in a number of states as a result of leaking underground tanks at gas stations, it became essential to find an alternate oxygenate for gasoline.

The industry and the federal government turned to ethanol and more recently biological sources of diesel as alternative additives for fossil fuels. Diesel fuel, which is ignited by compression and not a spark like gasoline, was patented as a fuel in 1893. The shift in the industry to ethanol had a huge impact on corn farmers. Corn acreage rose from 70 million acres planted and 78 million acres harvested in 2006 to 83 million acres planted and 93.6 million acres harvested in 2007.

Corn grain was chosen as the first source of ethanol for a number of reasons. First the process used to ferment the corn and distill the ethanol was well established. Also, despite the evidence that ethanol can damage engines, the automotive technology companies and consumers have long accepted gasoline with 15% ethanol, and gasoline containing 85% ethanol (E85) is available in gas stations. U.S. car manufacturers sell models that will run on E85, but these vehicles suffer a 15% drop in fuel economy. Research is also under way to identify different efficient sources of ethanol that might replace corn as the main source of ethanol for fuel, including cellulose stalks (from corn and other crop plants), rice hulls and bagasse, the sugarcane stalks that remain after the sugar juice has been extracted (Figure 15.14).

**FIGURE 15.14** Sugarcane stalks—source of biofuel? The sugarcane stalks that remain after the sugar juice has been extracted are possible sources of a new biofuel.

**Biodiesel** is a fuel can be derived from soybeans, canola, sunflower, soybean oil, palm oil, and other crops as well as used vegetable oil from restaurants. The interest in biodiesels grew in the United States because biodiesel offers the potential to reduce dependence on imported petroleum and to reduce the negative impacts of global climate change by lowering the net carbon dioxide emissions from transportation. Advocates of biodiesel have gained publicity and legislative tax incentives. New biodiesel fuels must meet EPA requirements and the standards of the American Society of Testing and Materials. How easily a diesel fuel will ignite under pressure, is reflected by the **cetane number** of the fuel; a higher cetane number indicates the fuel ignites more quickly. The cetane number for petroleum based diesels ranges from 40 to 42, and the biodiesel made from soybean has a cetane number of about 51.

> Ethanol made from corn quickly became a popular idea and a uniformly acceptable component of gasoline for automobile fuel in the United States. But ethanol from corn has disadvantages that have inspired scientists to look for alternatives to ethanol from corn.

Three factors are considered when deciding how beneficial different alternative fuel sources are to the environment:

- The net energy balance (NEB) between the two fuels (diesel compared to biodiesel for example)
- The levels of greenhouse gas emissions (GHG)
- The production of useful co-products (such as glycerol made from soybeans)

These factors are used to compare the costs and benefits of developing different alternate fuels: ethanol, biofuel, and biodiesel. To calculate the net energy balance ratio, the energy output is subtracted from the energy input needed to produce the fuel, which is a measure of whether or not the fuel requires more energy to produce than it yields. Production of soy biodiesel would cause a decrease in the release of the pollutants such as phosphorus and pesticides. Currently, the use of corn grain ethanol results in 87% of the greenhouse gases released, compared to fossil fuels that generate equivalent amounts of energy; soybean biodiesel production and combustion release only 59% of the greenhouse gases.

In 2005, the production of corn grain ethanol accounted for 1.7% of gasoline in the United States and soy derived biodiesel accounted for 0.9% of the diesel usage. It was estimated that if all of the corn and

(A)                                                     (B)

**FIGURE 15.15** Fungus as a source of biofuels. (A) Microscope image of *T. reesei* cells with hyphae extensions containing vesicle membranes *(red)* and chitin that makes up the cell wall *(blue)*; (B) Studies on fungi focus on reducing the high cost of converting lignocellulose (shown here) into fermentable sugars that can be used to generate biofuels.

soybean production in 2005 were used to make biofuels, it would provide only 12% of the gasoline usage and 6% of the diesel used in the United States. If we depend on current technologies, it is not possible for the United States to sufficiently decrease dependence on foreign oil or to reduce greenhouse gas emissions enough to make a difference in climate warming. In addition, there is a risk of food shortages and negative environmental consequences. With the rising demand for biofuels made from food crops, the price of food will also continue to increase.

Other sources of ethanol in addition to or instead of corn remain experimental. The Department of Energy has committed $385 million to six companies to help support research on developing the transgenic plants and associated technologies needed to produce ethanol from cellulose and lignocellulose. During the process of biofuel production, **lignin** physically blocks the access of enzymes that are needed to break down the polysaccharides in the cellulose. To solve this problem, scientists are altering the genes involved in lignin production. The first strategy involved inserting a transgene into plants to reduce the lignin amounts, but the transgenic corn and sorghum plants in this experiment grew very slowly. In another approach researchers engineered transgenic corn with genes that produce bacterial enzymes to break down the cellulose at times in the life cycle of the plants that avoid negative effects on growth. So far, high expression levels of this transgene have not yet been achieved. The productivity and net energy balance of this new biofuel will not be known until the technologies are tested in both pilot plants as well as in manufacturing facilities.

Studies are underway to identify new potential sources of biofuels using strains of fungus such as *T. reesei* (Figure 15.15). This research focuses on reducing the prohibitively high cost of converting lignocellulose to fermentable sugars to produce the next generation of biofuels. The plan is to use fungi to generate industrial enzyme "cocktails", combinations of enzyme proteins that will enable the more economical conversion of biomass into biofuel. Different feedstock includes the perennial grasses *Miscanthus* and switch grass, wood from fast-growing trees like popular, agricultural crop residues, and municipal waste.

The long range goal of this research is to replace the gasoline-dependent transportation sector of the United States economy with a carbon-neutral source of fuel from plants or microorganisms.

The use of algae to produce biofuels received recent support from the Department of Energy, which had previously funded research on renewable transportation fuels from algae (1978 to 1996). Numerous algae strains were collected and characterized and a number of algae genes involved in oil production were identified and cloned (Figure 15.16). At the time the program was canceled, the biofuel produced from algae would have been able to compete against the price of oil at that time. Of course the recent surge in oil prices and recent advances in biotechnology have refueled the race to develop commercially competitive and affordable algae biofuels. In 2007, the Chevron corporation entered into a research and development (R&D) agreement with the Department of Energy designed to identify algal strains and biochemical processes that can be used to produce algae-based biodiesel fuel. Whereas

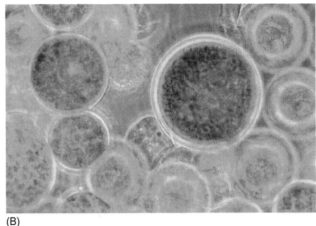

(A)

(B)

FIGURE 15.16 Algae are a potentially rich source of future biofuel. (A) The LiveFuels Alliance is tapping into the oil-producing potential of algae with an ambitious initiative to replace millions of gallons of fossil fuels with algae-based biofuel by 2010; (B) Algae cells growing in liquid culture.

several companies like Algae BioFuels, Greenfuels, and Solix Biofuels are independently working toward this goal, the LiveFuels Alliance differs in that it is a collaboration led by Sandia National Laboratories, and the U.S. Department of Energy National Laboratory and will sponsor dozens of research programs and hundreds of scientists. This is the largest national effort focused on producing **commercial biocrude** fuels.

The scientists at the LiveFuels Company plan to refine the processes needed to increase algae oil production at competitive prices and focus on the specialized aspects of the process involving the algae cells, including breeding algae to find the best high-fat strains, refining the fat and oil extraction process, and developing cost-effective harvesting techniques. Extracting oil from algae cells is basically the same as any other biofuel extraction technology, but cultivating the algae has other advantages. The algae cells do not require prime agricultural land and the potential biofuel yield from algae far exceeds other renewable sources. Algae are predicted to yield between 1000 to 20,000 gallons of oil per acre, depending on the specific algae strain, which is significantly more oil per acre than soy, which produces about 117 gallons of oil per acre. The potential yield from 20 million to 40 million acres of agriculturally marginal land in the United States could produce enough algae to replace imported oil and preserve 450 million acres of current farmland for use growing food crops. Under the right conditions, algae cells divide rapidly in fresh, brackish, or even wastewater. The amazing algae cells are nontoxic, biodegradable, and have the potential to supply a large fraction of the domestically grown biofuels available in the future.

Green algae are good source of a new biofuel as they can be grown in large cultures and do not compete with animals or people as a food source. Scientists predict that in the future algae-based biofuel will replace millions of gallons of fossil fuels used in the United States.

## DEVELOPING COUNTRIES

The use of transgenic crops in developing countries might not make much sense, particularly since most farms are small and must support a number of crops without the aid of pesticides or fertilizer. Land in tropical areas is often nutrient poor, acidic, and has high aluminum and low phosphorous content. Drought is a problem in many developing countries where farmers depend on rain because there is no widespread irrigation. Although GM crops can offer marginal improvement in yield, which might be substantial for farmers with small farms in developing countries, the economic and infrastructure resources that make GM crops attractive to large-scale farms are generally not available in developing countries. The international seed companies have been slow to respond to the needs of farmers in developing countries.

The 2001 Human Development Report from the United Nations supported the use of transgenic crops by farmers in developing nations and urged the nations to identify and develop the transgenic crops that are well suited to growing conditions in their local regions such as drought and nutrient-poor soil. The World Bank reviewed the Consultative Group on International Agriculture Research (CGIAR), which supports and connects the scientists at 15 international research centers, and concluded that the CGIAR group had not responded sufficiently to the biotechnology revolution, underscoring the continuing problems in developing countries. Just as issues of intellectual property (patents) and profit have slowed the use of medicines and vaccines to treat people in the developing world, these problems have slowed the development of transgenic crops suited for use in developing countries. Recently, more attention is being paid to diseases of developing countries with the efforts to publicize the astounding advances that agricultural biotechnology has made that should be available to benefit the world's poor.

## SUSTAINABLE AGRICULTURE

**Sustainable agriculture** is a system of farming that takes into account the environmental, social, and economic issues relating to the growth and distribution of crops. Modern, large-scale (factory) farms that usually grow one (monoculture) or just a few crops and rely completely on the use of fertilizers, pesticides, and herbicides are not conducting sustainable agriculture, even if they use genetically modified plants. Yet these farming practices have successfully provided people in the developed world with abundant, inexpensive food.

Sustainable agriculture is often used synonymously with **organic farming** and **organic food**, but this is not accurate. It does not take much experience food shopping to realize that organic products are significantly more expensive than non-organic food. This is because organic products are raised, grown, or processed without the use of synthetic hormones, pesticides, or fertilizers, which is overall a less efficient process and therefore more expensive. But sustainable agriculture is not an option if humans are to become better stewards of the earth. Scientists have a responsibility to continue to focus on developing the best tools, modern or traditional, that will effectively protect the security of the world's food supply, the environment of the entire planet, and provide for the well-being of people in resource-rich and resource poor lands. Biofuels play a large role in achieving these goals. Success will require collaborations between developed and underdeveloped countries, involving governments, universities, and representatives of the companies and the public sector.

Future success in the equitable use of the Earth's resources will require people to better balance the use of modern biotechnology and increased productivity with ethical concerns about environmental impact, profitability, and social equality worldwide.

## SUMMARY

This chapter covered ways that advances in the concepts, tools, and methods of molecular biology and recombinant DNA technology have influenced the rapidly growing field of agricultural biotechnology. Researchers routinely identify, isolate, and characterize genes and their protein products from microorganisms and plants as well as animals and humans. DNA cloning methods have been developed to insert genes into DNA vectors used to introduce new genes into the genomes of plants to generate transgenic plants including corn, soybean, sugar beets, cotton, canola, squash, and many others. Transgenic plants are designed to improve the crops used in farming because the transgenic plants are engineered to be resistant to infections by some viruses, insects, and bacteria and can grow well in the presence of herbicides. Scientists have developed different ways to introduce the vector DNA carrying a transgene into the plant genome including transfection by an *Agrobacterium* plasmid and physical methods such as a gene gun that shoots DNA projectiles into the plant cells.

The introduction of genetically modified plants into the environment raised concerns from some environmentalists about the unknown impact of transgenic plants on the environment and human, animal, and plant health. The USDA, EPA, and the FDA agreed on rules and procedures to reduce the potential risks posed by transgenic plants. Despite the continuing controversy in both the United States and in Europe, the use of GM plants has increased in the United States and in other developed countries. Researchers have genetically engineered plants to produce proteins for medical use in humans and animals, and several are now in clinical trials to assess safety and effectiveness. Oral vaccines have been developed using transgenic potatoes and bananas that can be eaten instead of injected. A vaccine against the bacteria that cause tooth decay is being developed in tobacco plants.

The use of transgenic plants to produce protein drugs for use in humans intensified public concerns about the accidental exposure to transgenic plants, especially if the genetically altered plant is a food or if the GM pollen could unintentionally spread the altered genomes to human food crops. Genetically altered chloroplasts may be a preferable option because chloroplasts are only inherited through the maternal germ line. It is not possible for a chloroplast transgene to be transmitted through the male sperm cells in the pollen. Like human and animal genes, plant genes and associated diseases are linked to known RFLP and SNP genetic markers in plant chromosomes that can be visualized through methods involving PCR amplification.

The sudden increase in gas prices in 2008, the shortage of fossil fuels, and the critical issue of global warming have led to new investment in plant-source fuel alternatives. Ethanol derived from corn grain and biodiesel derived from soy and other plants are currently available to the consumer, but this biofuel has put additional pressure on the human food supply, especially corn-based products. However, possible alternatives such as ethanol derived from corn stalks instead of corn and oil made from algae, are mostly experimental at this time.

The United Nations has urged the international community of scientists to participate in the transfer of appropriate agricultural biotechnology to developing countries. The use of transgenic crops and other aspects of agricultural biotechnology in developing countries will improve the yield of subsistence farms, but to develop an agricultural industry that will substantially improve the quality of life will require the collaboration of governments, companies, academics, private-sector stakeholders, and people who are often involved in tribal warfare and suffering from widespread diseases.

## REVIEW

To test your knowledge of the concepts in this chapter, answer the following questions:

1. What methods are used to deliver DNA into plant cells to create transgenic plants?
2. Why are plant cells more difficult to transform with DNA than animal cells?
3. How do the Cry proteins kill insects?
4. What federal government agencies must approve plans to plant a transgenic crop in an open field?
5. What are the concerns expressed by people opposed to use of the genetically modified crops?
6. How are DNA microarrays used to detect the pathogen responsible for an infected plant?
7. Explain the basic mechanism behind PCR, and describe the basic characteristics of the final PCR products after many PCR cycles.
8. What special concerns focus on the transgenic plants that produce protein drugs for human use?
9. What are the limitations of ethanol made from corn as a fuel to replace fossil fuels and reduce greenhouse gases?
10. How do scientists explain the mechanism of genetically engineered male sterility, and why are some people opposed to the concept of terminator technology?

## ADDITIONAL READING

Another big trade worry for the U.S. June 27, 2003. The Kiplinger Agriculture Letter, 74 (13). <http://findarticles.com/p/articles/mi_hb5973/is_200306/ai_n24134839/>.

Biofuels could be a 'green' disaster. September 14, 2007. Farmers Weekly. <http://www.accessmylibrary.com/coms2/summary_0286-32892773_ITM/>

Bauers, S., August 25, 2007. Scientists hope a weevil works wonders. Philadelphia Inquirer (Pennsylvania) <www.highbeam.com/The+Philadelphia+Inquirer+(Philadelphia,+PA)/publications.aspx?date=200708-/>.

Boscariol, J., Silva, O. May 3, 2004. Canada: GMO dispute raises novel issues. Legal Week Global. <www.lawyers.com/Ontario/Toronto/McCarthy-TAtrault-LLP-1333090-f.html-796k-/>.

Carter, C.A., Miller, H.I., May 17, 2007. Why ethanol backfires; politicians love the idea of fuel from corn, but it carries the seeds of serious problems. Los Angeles Times Sect. MAIN NEWS; Editorial Pages Desk; Part A (23).

$CO_2$ loving algae is our green friend. May 22, 2007. Modern Power System. Define precautionary principle to avoid biotechnology clashes under world trade rules. September 18 2006. Space Daily. <http://www.accessmylibrary.com/coms2/summary_0286-17696800_ITM/>

Eichenwald, K., Kolata, G., January 25, 2001. Redesigning Nature: Hard lessons learned; biotechnology food: From the lab to a debacle. The New York Times Section A page 1 of the New York edition. <www.nytimes.com/2001/01/25/business/redesigning-nature-hard-lessons-learned-biotechnology-food-lab-debacle.html/>.

Geiselman, B., June 11, 2007. USDA sells agencies on biobased benefits. Waste News 15 http://www.accessmylibrary.com/coms2/summary_0286-31279211_ITM.

Sissell, K., January 28, 2004. USDA to overhaul GM regulations. Chem Week, 10.

Sussmann, P., April 2, 2007. CNN Biofuel: green savior or red herring? CNN.com http://edition.cnn.com/2007/TECH/science/04/02/biofuel.debate/.

Uchtmann, D.L. 2002. Starlink TM -- A case study of agricultural biotechnology regulation. Drake University, Drake Journal of Agricultural Law. 159-211.

Weintraub, A., Gogoi, P., July 2003. The outcry over 'terminator' genes. Bus Week. <http://www.businessweek.com/magazine/content/03_28/b3841091.htm/>.

## WEB SITES

AgBioForum: an online journal of Agbiotechnology Management and Economics, <www.agbioforum.org/index.htm/>.

Agbiotech Infonet. www.biotech-info.net/index.html. Sponsored by some environmental groups skeptical of the safety of agbiotechnology; a good source for articles supporting their positions.

Agricultural Biotechnology and Its Regulation, A Harvard Business School Case Study (2000). <http://harvardbusinessonline.hbsp.harvard.edu/b02/en/common/item_detail.jhtml?id=701004&referral=8636&_requestid=51941/>.

Animal and Plant Health Inspection Service of the USDA. www.aphis.usda.gov/biotechnology/status.shtml. This site has a lot of information on environmental releases

Biotechnology Industry Organization. http://bio.org/foodag. The lobbying group for the industry.

Bourne, J.K. Jr. 2007. Green Dreams: Making fuel from crops could be good for the planet—after a breakthrough or two, by Joel K. Bourne, Jr., National Geographic Staff, 2007, <http://ngm.nationalgeographic.com/2007/10/biofuels/biofuels-text/2/>.

*Electronic Journal of Biotechnology* (from Chile) technically challenging but gives a look at what's going on elsewhere <www.ejbiotechnology.info/index.html/>.

Environmental and Energy Study Institute, www.eesi.org/publications/Briefing%20Summaries/briefingsummaries.htm. A lobby group with some interesting position papers.

Environmental Protection Agency (EPA). www.epa.gov. The page on biopesticides provides a good view of how this bureaucracy addresses the topic; <www.epa.gov/oppbppd1/biopesticides/regtools/index.htm/>.

Food and Drug Agency (FDA). www.fda.gov. The Biotechnology page has a number of good articles and provides a list of recent decisions and policy statements; <www.cfsan.fda.gov/~lrd/biotechm.html/>.

Information Systems for Biotechnology. www.isb.vt.edu. Hosted by Virginia Tech, which has the database of the USDA field releases and nonregulated transgenic organisms and a very good list of agbiotech websites, technical, political, and economic

Power Your Car with Algae: Algae Biocrude by LiveFuels, by Ali Kriscenski, 2007. <www.inhabitat.com, http://images.google.com/imgres?imgurl=http://www.inhabitat.com/wp-content/uploads/algae5.jpg&imgrefurl=http://www.inhabitat.com/2007/10/22/power-your-car-with-algae-algae-biocrude-by-livefuels/&usg=__6GASudGOiqSpcEy77A310afM-FE=&h=358&w=537&sz=58&hl=en&start=3&tbnid=u3vWNoD6wEcs8M:&tbnh=88&tbnw=132&prev=/images%3Fq%3Dalgae%2Bbiofuel%26hl%3Den%26sa%3DG/>.

United States Department of Agriculture (USDA), www.usda.gov. The page on Biotechnology has a lot of information on projects and regulation; <www.usda.gov/wps/portal/!ut/p/_s.7_0_A/7_0_1OB?navid=BIOTECH&parentnav=AGRICULTURE&navtype=RT/>.

# Genes and Race

## Eliminating Racial and Ethnic Health Disparities

### Centers for Disease Control

African-American, American Indian, and Puerto Rican infants have higher death rates than white infants. In 2000, the black-to-white ratio in infant mortality was 2.5 (up from 2.4 in 1998). This widening disparity between black and white infants is a trend that has persisted over the last two decades.

African-American (black) children are two and a half times as likely to die as infants than Caucasian (white) children in the United States. Could this disparity possibly be caused by a difference in genetics due to race? Some people argue that the differences in health statistics result from the different genetics that define different racial populations. Others maintain that this idea is nonsense, and attribute the differences in heath statistics not to genetics, but to the environmental, social, and economic conditions under which many black people live in the United States.

This chapter will examine the scientific evidence available to determine whether race is a genetic or biological concept. In cases where there are clear differences in health issues between black and white populations, can these differences be explained by human genetics?

## LOOKING AHEAD

In this chapter, we will examine the idea that different human races can be defined in terms of genetics or biology. On completing the chapter, you should be able to do the following:

- Explain why it is problematic to define race in terms of human genetics.
- Describe how modern scholars have revised the depiction of race in ancient texts.
- Name some human diseases that have been labeled as 'racial' or 'ethnic' diseases, and explain whether or not it is accurate to use these labels to describe these diseases.

- Describe whether there is more genetic variation occurring within groups called races or between these groups.
- Explain how the geographical origins of human populations are represented by differences in human genome DNA sequences.
- Explain some of the complexities involved in establishing accurate relationships between intelligence, IQ and race.

## INTRODUCTION

"We are living through an era of the ascendance of biology, and we have to be very careful," said Henry Louis Gates Jr., director of the W. E. B. Du Bois Institute for African and African American Research at Harvard University. "We will all be walking a fine line between using biology and allowing it to be abused."

(In DNA Era, New Worries About Prejudice, Amy Harmon 2007, *New York Times*)

Tremendous advances in molecular biology and genetics, including the complete human genome DNA sequence, have recently shed additional light on the subject of race and have greatly illuminated the immensely complex issues involved. The goal here is to understand what is meant in biological terms by "**race**" and to understand the implications of 'race' on advances in medicine and biotechnology. For example, are there important biological differences between people that can be ascribed totally to race? Should a patient's race be considered when prescribing drugs to treat diseases? Is there any situation in which the consideration of race is necessary for the delivery of appropriate medical care? These types of questions may be difficult to discuss but race discrimination in the United States has had a tremendous impact on medicine and healthcare. In this chapter we address race from a biological standpoint to better understand the connections between race, healthcare, and medicine and to help promote equal access to the amazing advances in biotechnology and medicine that are generated by research in modern genome science.

In 1972 Richard Lewontin, a leading evolutionary geneticist, published a study comparing the amino acid sequences of different human proteins (Figure 16.1). Lewontin found that at least 85% of the genetic differences between human individuals have nothing to do with race, as this genetic variation is present even between individuals of the same race. Lewontin's results on protein sequence variation were later confirmed by research on DNA genomes, including the human genome (see Chapter 6). In 2001 when the first DNA sequence of the entire human genome was

**FIGURE 16.1** Richard C. Lewontin. One of the most brilliant evolutionary biologists of his time, Lewontin is also a leading critic of the misuse of scientific discoveries. He exposes common misconceptions that interfere with the ability of people to understand biology and evolution. Lewontin's work contributed to our understanding of the functional connection between genes and the environment and the role of genes and DNA in evolution.

released, scientists initially concluded that humans are remarkably similar to each other, and that the DNA genomes of any two people are at least 99% identical. Since that time, researchers have updated the sequence of the 3.2 billion base pairs in the human genome. They found that the difference in DNA sequences among individual human genomes is 99.5%, much greater than previously thought (Figure 16.2A). About 10 million single nucleotide polymorphisms (SNPs) (DNA differences) have been identified in human genomes and are used as extremely important tools in many applications of DNA technology (see Chapters 6, 8, 10). The sequence variations between human genomes are used in many applications such as DNA forensics to identify potential criminals (including innocent people behind bars), paternity testing to identify biological family relationships, DNA used as a 'barcode' to track poached meat from endangered species, and in medical research to study the genes associated with human diseases. A single base pair mutation in a human gene can cause a life threatening disease such as sickle cell anemia.

The discussion of genes and race in this chapter will include three parts. First we will explore the history of race to find out if race has always been a concept in human society. Second we will examine the genetic definition of 'race' and look for biological evidence to support the concept of racial diseases. The current and future research on human genomes from people from all over the world has revealed important information

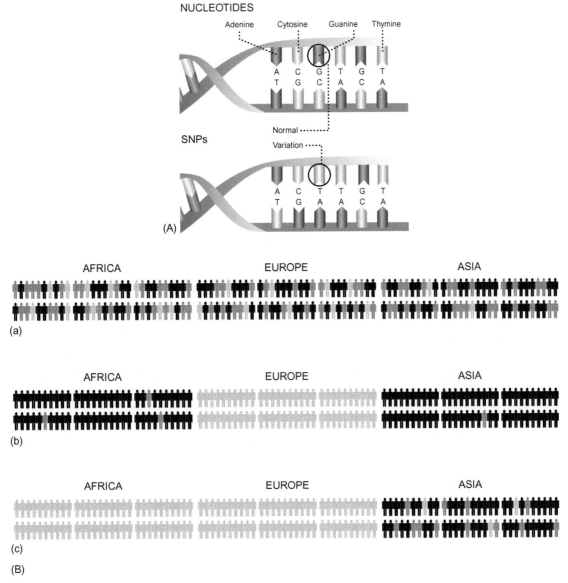

**FIGURE 16.2** Genetic variation: Small sequence differences between human genomes. (A) Comparison of the sequences of different human genomes shows that most of the sequences are identical. There are also about 10 million sites in the genome that differ at a particular base; one person's genome might have a "C" at that position while another person might have a "T" in that position. These single base differences are called SNPs (snips) and are distributed across the genome sequence. (B) About 10% of the small genome sequence differences (SNPs) are distributed unevenly across the genome DNA and are correlated with different traits. (a) Most SNPs such as the ones shown in different shades here occur at about the same frequency in people from Africa, Europe and Asia (genetic variation is shown in different shades). (b) The pale skin color common in Europeans is caused by an SNP that is never found in populations in Africa or Asia (people with this SNP are shown in a lighter shade). (c) A DNA change that is found exclusively in East Asian populations correlates with a reduced ability to sweat and includes almost all people of Chinese ancestry (darker shades).

about the frequency of genetic variation among the individual human genomes (see Figure 16.2B).

In the last section of this chapter we will discuss the controversies surrounding the impact of race on genes, IQ, healthcare, and lifestyle and discuss the genetic explanations for the apparent differences between African-Americans (black people) and European Americans (white people).

## THE HISTORY OF RACE

### Race is a Modern Concept Created by Humans

During the eighteenth and nineteenth centuries and much of the twentieth, the notion of "race" was presented as a concept that had always been part of human culture. At the time, people tried to justify the idea that black people are fundamentally inferior, some scholars turned to the writings of ancient cultures and used the texts of the ancient Greeks and Romans as examples of the early origins of race as a human concept. Franz Boas (1858–1942) was a pioneering anthropologist who was among the first to call into question the traditional notion of race at the time. He initially encountered resistance within his own scientific community, but later those scientists were among the first to denounce the use of race as a way to define and classify human differences.

In the late twentieth century, scholars began to re-examine the ancient texts more closely. Frank Snowden and Ivan Hannaford, (Woodrow Wilson International Center for Scholars) re-visited the origins of race. Hannaford's work places the beginning of racial theory in 1684 with the publication of a book by Francois Bernier, which divided people according to their observable characteristics in terms of race and ethnicity. For much of the seventeenth, eighteenth, and even into the twentieth century, people believed that race had been used by humans to classify humans since the start of civilization. Part of the ideology behind the concept of different races included the idea that the different human races evolved separately. Whereas blacks evolved in Africa, it was said, whites evolved separately in the Aryan Plain. This was the origin of the name Aryans for whites, a term adopted by the Nazis. In the twenty first century we have clear evidence from genomic studies and archaeological research showing that all modern humans evolved at one time in Africa, about 200,000 years ago (see Chapter 8).

> The concept of "race" is a relatively recent invention of society and was not always part of human culture. The constant characteristics attributed to different races led to the incorrect idea that different races of humans evolved separately.

## THE GENETICS OF PHYSICAL CHARACTERISTICS

If we define "race" using biological traits such as skin color, facial characteristics, height, weight, then in that specific context, race is genetically determined. These physical traits, or phenotypes, are determined by the particular alleles of genes that are inherited by the individual. Alleles are different versions of the same gene, such as a wild type (normal) allele and a mutant (altered) allele of the same gene.

The term "race" is sometimes used interchangeably with terms such as "variety" (as in plants) and "subspecies." However, biologists categorize a plant or animal group as part of a separate race (subspecies, or variety), if it satisfies one of the following two criteria.

First, the group must have its own separate genetic lineage, which means that individuals never, or rarely, would mate outside the group. Clearly, the human groups currently called races do not satisfy this criterion, and according to historical and archaeological evidence they never have.

Second, the genetic differences between one group and another group would have to be significantly greater than the genetic variation within members of either group. Human races do not fulfill this requirement either. In fact modern genome studies have proven that there is actually more genetic variation (DNA differences) within races than exists between people from different races (see Figure 16.2). For example, the genetic variation among Africans, including African Americans, is greater than the genetic variation in genomes from the rest of the human population.

Genetic studies show that different people will sometimes have alleles in common if they have lived in the same geographic region for many generations. Studies also show that there is no single gene that is found in one race but is not found in a different race, once again questioning the genetic basis of the term, race. As expected, researchers did identify differences in **allelic frequency** among the genomes of people belonging to different races. The differences between these groups represent from 4% to10% of the total amount of genetic variation in the human genome, but do not shed light on the biological concept of race. Scientists might never agree on a clear genetic definition of 'race' because so far there are no clear genetic differences that distinguish between different 'human races'.

## Race is Not a Genetic Concept

To date (2009) there is no reproducible evidence from genome research that supports the existence of separate races (subspecies) of modern humans. Different alleles have been found that are responsible for physical traits such as pale skin and hair color, but there are no consistent patterns of human genes or alleles that can distinguish humans in one race from those in another. Scientists have found no evidence to support the classification of humans based on ethnicity. In fact, genome studies have proven that there is actually more genetic variation (DNA differences) within 'races' than exists between people of different races.

Most scientists agree that evolution of the modern human (homo sapiens) (Figure 16.4) involved waves of migration of the first hominids (homo erectus) out

**FIGURE 16.3** Small differences in the DNA genome of an animal can cause dramatic differences in physical characteristics such as fur color. (A) This jaguar has the familiar coat color that is common to most of the South American jaguar population. (B) This dark-colored jaguar represents about 6% of the South American population and is caused by a polymorphism in the jaguar genome.

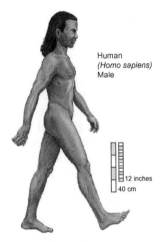

**FIGURE 16.4** *Homo sapiens* : the modern human. Most scientists agree that the modern human (*homo sapiens*) evolved from *homo erectus*, a human-like primate who walked upright on two feet (see Chapter 8).

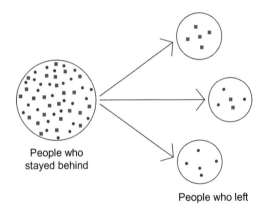

**FIGURE 16.5** Genetic variation and the founder effect. The migration of hominids from Africa to Europe involved a small number of individuals. As a result the level of DNA sequence variation present in the chromosomes of those who migrated was much less than the sequence variation present in the genomes of the individuals who stayed behind, which is called the founder effect. (Genetic variation is indicated by the red and blue shapes).

of Africa and into Europe and Asia about 1 million to 2 million years ago (1.0-2.0 mya) and a second wave of migration followed about 100,000 to 200,000 years ago (see Chapter 8). The theories proposed to explain the evolutionary origin of modern human *homo sapiens* predict different types of encounters between the most recent African migrants and other hominid groups. Some evidence indicates that the groups co-existed with possible interbreeding between the recent African migrants and other hominids. Scientists are using DNA analysis to determine if the human *homo sapiens* genome is derived solely from the DNA of the most recent wave of African migrants and to find out whether the DNA record includes evidence of interbreeding with the other hominids along the route to Europe and Asia.

The evolutionary pressures from natural selection and genetic drift (random fluctuations in the frequencies of DNA sequence variants) act on genetic variation but do not cause the actual changes in the DNA sequence. In the natural environment outside the lab, the genetic variation in genome DNA sequences are caused by mistakes in DNA replication (see Figure 16.2) and by DNA damage due to environmental toxins and radiation like the UV rays in sunlight (see Chapter 9). The genetic variation present across the current population of human genomes reflects the differences among the genomes of the human ancestors. Because a comparatively small number of humans made the migratory trek out of Africa, the total amount of DNA variation present in their collective DNA pool was much less than that of the rest of the *homo sapiens* who stayed behind, known as the "founder" effect (Figure 16.5). The human concept of race once again fails to meet the previously stated criteria used to identify a species or a subspecies.

The socially defined concept of race meets neither of the criteria required to establish a genetically defined species and subspecies. There are no alleles or genes present in one 'race' and not in another 'race'. There is no evidence to support a biological concept of human races that justify prejudice and discrimination based on the artificial concept of race.

to find any evidence to support the evolution of separate human species or races.

Races (species) evolve over an extended period of physical separation and many thousands of years. It is very unlikely that there was geographical separation for different human races to evolve.

## Geographical Origins Reflected in DNA Differences

Could the different human races have evolved during the hominid migration out of Africa? This is possible, because we know from evolution that one species evolves from other species. Clearly evolution occurred over this time period; after all, people living in Norway do not look like Africans. On the other hand, having ancestors from one geographical area as opposed to another does not mean that these individuals will have fundamental genetic differences.

To understand what is necessary for species (or races) to evolve, consider the basic steps during the evolution of a new species. Two animals are said to be of the same species if they can mate and produce fertile offspring. However, a mating that produces live but sterile offspring indicates that the parents belong to species that are evolutionarily very closely related but are two separate species. In order for a new species to evolve as a distinct species, the species must be physically (geographically) isolated for enough time for genetic changes to occur that prevents the two species from successful breeding. Darwin's finches are two species of birds living on the **Galapagos Islands**, which diverged into two species because they were geographically separated into two different environments over many thousands of years, with no crossover of birds from one group to the other. The idea that geographical separation could cause different species to evolve was one of Darwin's great insights.

For a subspecies or race of humans to evolve, the two human populations would need to be geographically (physically) separated for a time period much longer than it took for humans to migrate out of Africa. Physical separation of the migrating populations was also unlikely. Genetic and archaeological data support a gradual migration, with constant interactions among the different villages as the groups of people migrated along the same or parallel paths. Research is underway to sequence human genomes from all over the world, which will substantially increase the genome sequence information available for comparison analyses (see Chapter 6). Researchers have searched the human genome sequence databases but were unable

## Blood Type Genes Used To Classify Humans

In 1900 and 1901, Karl Landsteiner showed that humans have only three blood type alleles A, B, and O (Table 16.1), which he classified according to the types of antigens on the surfaces of red blood cells and the kinds of antibodies that bind to those antigens (Figure 16.6). For this work, Landsteiner won the 1930 Nobel Prize in physiology or medicine. Scientists now know that there are a number of other, rare, antigens that complicate actual blood typing, but none of the methods found a correlation between blood type and race.

Landsteiner's research was one of the first to show that different human biochemical characteristics are not divided along racial lines. This work was used as a powerful weapon to argue against racial prejudice and segregation, and provided important evidence to support the idea that heritable traits are not be related to the human concept of race.

Two protein antigens, A and B, occur on the surface of red blood cells (see Figure 16.6). An antigen is usually a protein that when introduced into the body can stimulate the production of an antibody response. This means that antibodies are made in response to a specific antigen. Antigens can be toxins, bacteria, pollen, animal dander, and foreign blood cells. An antibody is a protein complex made by the immune cells that are secreted into the blood or lymph in response to an antigen. This is one way that the immune system protects the body from disease. Each antibody recognizes a specific antigen and binds extremely tightly to the antigen. This is part of the mechanism allowing the body to recognize its own cells and tissues as "self".

Most everyone knows that blood type is important in case you need a blood transfusion at a hospital, but few people know why. Using blood with the wrong blood type can be fatal, which is why healthcare workers carefully cross-check the patient's blood type and the blood products before treating the patient. The A and B protein antigens on the surface of the blood cells interact with antibodies A and B. If treated with the wrong blood type, antibody A binds to antigen A, causing the blood cells to agglutinate (clump together). This is dangerous because the agglutinated blood cells

**TABLE 16.1** Human blood type groups

| Phenotype | Genotype |
| --- | --- |
| A | AA or AO |
| B | BB or BO |
| AB | AB |
| O | OO |

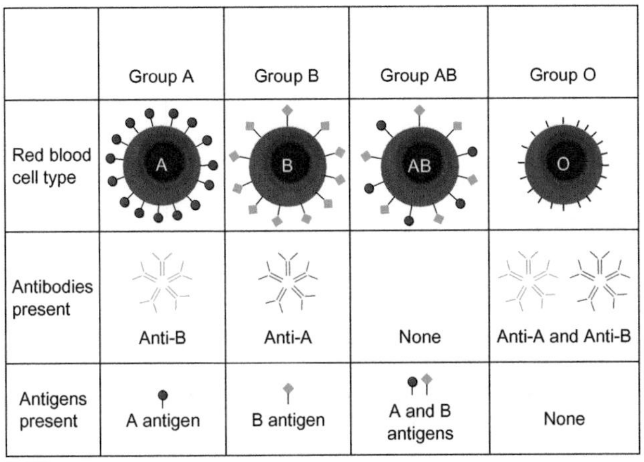

**TABLE 16.2** Blood type alleles predicted from parental alleles (ABO) × (ABO)

| Parent Alleles | A | B | O |
| --- | --- | --- | --- |
| A | AA (**A**) | AB (**AB**) | AO (**A**) |
| B | AB (**AB**) | BB (**B**) | BO (**B**) |
| O | AO (**A**) | BO (**B**) | OO (**O**) |

|  | Group A | Group B | Group AB | Group O |
| --- | --- | --- | --- | --- |
| Red blood cell type | A | B | AB | O |
| Antibodies present | Anti-B | Anti-A | None | Anti-A and Anti-B |
| Antigens present | A antigen | B antigen | A and B antigens | None |

**FIGURE 16.6** Blood type (or blood group) is determined by the ABO blood group antigens on the surfaces of red blood cells. The four different human blood groups involve the following components: Blood group A has A antigens and B antibodies. Blood group B has B antigens and A antibodies. Blood group O has no antigens but has both A and B antibodies. Blood group AB has both A and B antigens, and no A or B antibodies.

leak and are toxic (see Figure 16.6). The genetic inheritance patterns of the A, B, O alleles that determine the blood groups are shown in Table 16.2.

The four different human blood groups A, B, and O have the following components:

- Blood group A has A antigens and B antibodies.
- Blood group B has B antigens and A antibodies.
- Blood group O has no antigens but has both A and B antibodies.
- Blood group AB has both A and B antigens, and no A or B antibodies.

The Punnett square method can be used to predict all the possible combinations of the maternal and paternal alleles for each gene studied. The Punnett square diagram is used to organize the parental alleles in a genetic cross so that they can be used in a logical way to predict the genes inherited by the potential offspring of that particular mating. The Punnett square provides a summary of every possible combination of maternal and paternal alleles of each gene involved in the cross. In the case of human blood types, the parental alleles (ABO) and (ABO) are placed in the top row and

the left column of the grid with the genotype shown in regular font, and the phenotype indicated in bold italics and parentheses (Table 16.2).

The blood typing process depends on genetics and biochemistry, with little or no contribution from environmental factors. However, the relative frequencies of the blood type alleles and the distribution of the different blood types show considerable variation in different populations and geographical locations (Figure 16.7) (Table 16.3).

Due to the extensive mixing of human populations over time, no single blood type currently predominates in the human population, including blacks and whites in the United States. The evolutionary significance of the geographical distributions of human blood types is not yet known, although some preliminary results suggest that blood type O might make a person slightly less susceptible to bubonic plague.

When scientists and others attempt to define different populations, either culturally, geographically, or both, within these populations they find many genetic variations, with an uneven distribution of genetic variations throughout different populations, as is observed for blood type.

> Most human populations have all the possible blood types, with no one blood type predominating. This is partly due to the extensive mixing of human populations that helped to distribute the blood types worldwide.

## CONTROVERSIES IN HEALTH, MEDICINE, AND THE IQ TEST

There are well-known disparities in health and medical care among the designated human races in the United States (Table 16.4). That African Americans experience illness more often than European Americans and that blacks in the United States have a shorter life span are scientific facts. But the reasons for these differences are under dispute and require additional information. What are the relevant scientific facts?

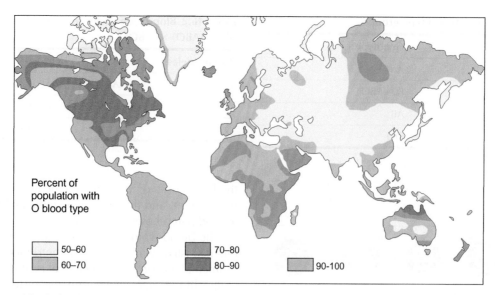

**FIGURE 16.7** Worldwide distributions of human blood type groups.

**TABLE 16.3** Human blood type distribution varies in different populations and geographical locations (%)

| Population | O | A | B | AB |
|---|---|---|---|---|
| Aborigines | 61 | 39 | 0 | 0 |
| Arabs | 34 | 31 | 29 | 6 |
| Bororo (South American Indian) | 100 | 0 | 0 | 0 |
| Eskimos (Alaska) | 38 | 44 | 13 | 5 |
| Eskimos (Greenland) | 54 | 36 | 23 | 8 |
| Jews (Germany) | 42 | 41 | 12 | 5 |
| Jews (Poland) | 33 | 41 | 18 | 8 |
| Kikuyu (Kenya) | 60 | 19 | 20 | 1 |
| United States (blacks) | 49 | 27 | 20 | 4 |
| United States (whites) | 45 | 40 | 11 | 4 |

Three factors affecting cancer risk were found to impact mortality in the United States African American population: (1) blacks are also more likely to develop cancer, (2) blacks often wait longer than whites to get medical care, and (3) blacks have lower survival rates after cancer diagnosis, compared to whites. To what extent are these factors responsible for the differences in cancer deaths between blacks and whites and is there a genetic basis for these differences?

A new study by the University of California, Los Angeles (UCLA) published in the Journal of General Internal Medicine (2009), found that for most types of cancer, the difference in mortality is almost entirely due to the fact that African Americans are more likely to get cancer in the first place. The stage at diagnosis and survival time played a much smaller role in mortality.

This was the first time that research clarified the role that these differences play in increasing cancer or decreasing the life expectancy of African Americans.

Breast cancer is a notable exception identified by this study. Whereas white women were more likely to get breast cancer than African American women, the difference in mortality between white and black women was mostly due to the gap in breast cancer screening and treatment. This study was significant because it involved many people; the researchers analyzed data sets from the Surveillance and Epidemiology End Result (SEER) cancer registry and the National Health Interview Survey (NHIS) involving about 2.7 million white and 291,000 African American cancer patients from 12 geographic regions in the United States.

**TABLE 16.4**  Health status measures in racial/ethnic groups in the United States, 1998

| Cause of death | White | Black | Hispanic | Asian |
|---|---|---|---|---|
| | | Age-adjusted death rates* | | |
| All causes | 450.4 | 690.9 | 432.8 | 264.6 |
| Heart disease | 121.9 | 183.3 | 84.2 | 67.4 |
| Coronary disease | 79.2 | 92.5 | 54.7 | 42.9 |
| Stroke | 23.3 | 41.4 | 19.0 | 22.7 |
| Cancer | 121.0 | 161.2 | 76.1 | 74.8 |
| Coronary obstructive pulmonary disease | 21.9 | 17.7 | 8.5 | 7.4 |
| Pneumonia/influenza | 12.7 | 17.4 | 9.8 | 10.3 |
| Liver disease | 7.1 | 8.0 | 11.7 | 2.4 |
| Diabetes mellitus | 12.0 | 28.8 | 18.4 | 8.7 |
| HIV infection | 2.6 | 20.6 | 6.2 | 0.8 |
| External causes | 46.7 | 68.8 | 44.7 | 24.4 |
| *Infant mortality (*)* | 6.0 | 13.6 | 5.8 | 5.5 |
| Life expectancy (yrs) | 77.3 | 71.3 | >80 ? | >80 ? |

The researchers concluded that continuing to improve cancer treatment and screening methods are undoubtedly important to improve the life expectancy for all adults, but substantial disparities in cancer mortality will probably persist unless we can find ways to address "the enormous impact of racial differences in cancer incidence."

## The New Era of Human Genomics

When the Human Genome Project (HGP) was first announced, its goals were articulated in terms of the anticipated medical benefits (see Chapter 6). This international public project was funded in the United States through the National Institutes of Health and the Department of Energy. Great Britain funded almost a third of the overall project through the Wellcome Trust, and other countries, such as Japan, also contributed funding and personnel. Several private corporations also worked on the human genome sequence including J. Craig Venter who founded Celera Genomics (see Chapter 6).

The Human Genome Project (public and private groups together) not only gave us the complete DNA sequence of the human genome, but also provided a technical foundation for research on many other genomes in addition to the human genome. Research on the complexity of human genes revealed the ubiquitous presence of introns and exons that often code for overlapping genetic information. Scientists have refined the traditional definition of a gene to more appropriately reflect the new characteristics of human genes based on extensive research on the human genome (see Chapter 6).

Among the benefits promised by the human genome research was the ability to develop effective treatments for devastating human genetic diseases such as Huntington disease (HD), muscular dystrophy and cystic fibrosis (see Chapter 10). In some cases scientists know the precise DNA mutation that causes a genetic disease. Using this precise information, scientists can locate the wild type (normal) alleles, which can be used in gene therapy methods to correct mutant genes by treating the patient with the wild type genes. This approach was used to successfully treat blindness in humans and dogs and to alleviate the symptoms of Parkinson's disease and other genetically inherited human diseases (see Chapters 10 and 11).

## Personalized Medicine: Genetic Risk and the Environment

People in the United States are afraid of getting cancer, but in terms of actual risk, people should be more concerned about developing heart disease or diabetes. These diseases, and many more like depression and bipolar disorder, result from problems involving a complicated interplay of multiple mutant genes that are further complicated by the influence of environmental factors.

Medicines seldom work the same way in all patients because the specific genetic and biochemical makeup of each person is a little bit different when compared to another person. A person's makeup depends on the influence of many factors in addition to genetics, including lifestyle and diet. Both can increase the risk of heart disease, but the impact of these factors is often small compared to the influence of genes that cause heart disease to 'run in a family' (are inherited).

The new genetic approach to future healthcare is based on patients who suffer side effects from medications and might help to improve the efficiency, price, and effectiveness of specialized drugs and therapies. Drug dosages based on genetic information, in addition to other patient characteristics, would allow doctors to avoid testing a large number of potential medicines by trial-and-error to decide how to treat each individual patient. In the future, advocates claim that the doctor will compare each patient's genome with sequences in a large DNA database to select an appropriate medicine judged by the patient's genome to be most effective drug available, with the least risk of side effects. In recent years important progress has been made in areas of medicine where human genetics has overlapped with behavior, depression, and personality traits. For example, scientists found out that the rate that drugs are metabolized in the human body is strongly influenced by the genetics of the liver, especially for drugs used to treat social anxiety, depression, and similar disorders. Some DNA mutations slow the body's drug metabolism, possibly causing serious side effects as a result of buildup of drugs and their breakdown products (metabolites). The widespread use of pharmacogenomics and personalized medicine could have a large impact on how physicians routinely prescribe medications if it results in a more accurate way to administer doses of medicine for individuals.

There is much research to do in order to better understand the details of the interactions between human genes and the environment. Sometimes researchers use "surrogates" or arbitrary characteristics to define specific groups of people who react better to one medicine or another medicine in clinical trials. Increasingly, researchers have used "race" as a surrogate characteristic, which raised ethical and moral dilemmas for many people. Currently there are some drugs available that are marketed solely to members of specific racial groups, which claim to be more effective for use in one "race" just as some medications are marketed primarily to members of one gender. To provide the correct medical information when advising patients, doctors and other healthcare providers must keep up to date with new developments in science and medicine in their specialty and they must be aware of the economic, social, cultural, and ethical issues associated

> **Box 16.1   The BiDil Story**
>
> On June 23, 2005, the first so-called ethnic or racial drug, BiDil (pronounced bye-dill), was approved for use by one racial-ethnic group, African Americans. Hailed at the time as a preview of the many future drugs that would also target specific ethnic and racial groups, it was applauded as an example of the medical establishment finally paying attention to African Americans, a group very long neglected by the healthcare industry. Critics claim that the difficulties encountered by BiDil in the FDA approval process had mostly to do with commercialization and patent expiration dates.
>
> Sold as "for blacks only", BiDil is a fixed-dose combination of isosorbide and hydralazine, designed to treat congestive heart failure, which is common in blacks. BiDil became controversial as the first drug approved by the FDA for use in only one racial-ethnic group.
>
> Critics note that the research supporting the attributes of BiDil involved testing BiDil in African American people only. The efficacy of BiDil to treat congestive heart failure in other human populations remains unknown.
>
> After the original patent covering the production of BiDil was not renewed, the issue of marketing a drug to one race became more controversial. The research to test the efficacy of BiDil involved small numbers, only 49 African Americans in one group and just 1050 self-identified African Americans in another, both small sample sizes even by government testing standards. On the basis of this questionable research a new patent was filed, andthe FDA approved BiDil with a race-specific label, for use in treating black patients only.
>
> Many questions concerning this race-targeted drug remain unanswered, including the efficacy of treating patients of one race compared to patients belonging to another race. Even if, as advertised, the BiDil drug works better in African Americans than in other races, it is still questionable policy to restrict its use to African Americans. This limitation potentially poses a real risk that some people who might benefit from taking the drug, will not have access to the medication because they cannot prove they belong to the correct race. How will doctors define race with regard to access to the drug? Must a patient be 100% black or Asian American? What about people who are one half or one quarter African American, are they black enough to have access to this drug?

with each new application. One good example is a drug marketed "for blacks only", which is very popular even without appropriate drug safety testing.

## GENES AND THE IMPACT OF ENVIRONMENT

Some of the differences in health status observed among people of different 'races' might be due to the influence of the environment. In the past it was common to try to quantify the contributions from nature (i.e., genes) and nurture (i.e., environment) to an individual, but now we know that it is not possible to accurately quantify the contributions of genetics and environment to a particular trait and phenotype of an individual.

All the cells in an individual's body have identical genome DNA sequences and carry the exact same genes. Even though all the cells carry identical DNA genomes, the genes are expressed differently in different cell types. For example, muscle cells express different genes than nerve cells, which is why muscle cells contract and extend and nerve cells transmit electrical nerve signals (see Chapter 6). It is the expression of different genes in different cell types that allow cells to carry out specialized biochemical functions in the body. This is a very important concept because it helps people to grasp the idea that all of our cells have the same genes, it is the differential gene expression in different cells that generates different types of cells and tissues. The cells in an individual human body all contain identical copies of the human genome, but humans often inherit different versions of the genes, the alleles. Humans inherit two copies of each gene one from the mother and one from the father (see Chapters 9 and 10). The two copies of the gene can be identical. Either both alleles are normal (wildtype), both alleles are mutant, or the inherited alleles are different from each other—one allele is wildtype and the other allele contains the gene mutation.

The environment has a large impact on gene expression and on the functions of the gene product. However, in the case of a genetic disease such as sickle cell anemia, inheriting two copies of the mutant globin gene has an overwhelming impact on the individual, and causes a serious disease. Living in a healthy environment and following a healthy lifestyle is good but will not prevent or reverse sickle cell disease in people who inherit two mutant globin alleles. The impact of the environment tends to be larger when the disease is caused by multiple genes, such as heart disease and diabetes (type II) (see Chapter 10). Like most cancers, the development of prostate cancer involves contributions from both genes and environment. The prostate glands of Japanese men develop different forms of prostate cancer depending on whether they are living in the United States or Japan. In this case it is a combination of environment and genetics that causes cancers to develop (see Chapter 9).

## Heart Disease is Caused by Many Genes and the Environment

Cardiac (heart) disease is a good example of a human disease that is impacted by environmental factors such as lifestyle and diet as well as by inherited genes.

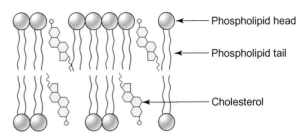

**FIGURE 16.8** Cholesterol functions in the cell membrane. Cholesterol has important functions in the cell plasma membrane. The membrane is a double layer of phospholipids (blue) with their charged 'head units' exposed on the outside of the membrane and the fatty acid tails located inside the membrane. Cholesterol is transported by the lipoprotein carriers, LDL and HDL, to different cells in the body.

Cardiac disease can be caused by defects in cardiac genes and proteins, but heart disease is often caused by processes that are not directly involved with the structure of the heart. Scientists discovered a clear correlation between the incidence of heart disease and high levels of certain forms of cholesterol in the blood. Cholesterol is important because it is one risk factor that people can control in order to reduce the risk of heart disease.

Cholesterol is a necessary chemical needed to build new cell membranes and make non-protein hormones. However, because of the chemical structure of cholesterol the molecule cannot dissolve in water or in the blood (Figure 16.8). The body solves this problem by providing two types of lipoprotein carriers (HDL, LDL) to transport the cholesterol in the blood to and from the cells.

- Low-density lipoprotein (LDL), known as "bad" cholesterol
- High-density lipoprotein (HDL), known as "good" cholesterol
- Triglycerides
- Lp(a) cholesterol

## Total Cholesterol Count

The total cholesterol count (LDL, HDL, triglycerides and Lp(a) cholesterol) is determined by a blood test. The LDL (bad) cholesterol carries 60% to 75% of the cholesterol in the blood. If the LDL levels are too high, the LDL slowly builds up on the inner walls of the vessels, clogging the arteries to the heart and brain. LDL promotes the formation of plaque, a thick, hard deposit that narrows the arteries and causes atherosclerosis. The plaque can form a clot and cause a heart attack or stroke.

About 25% to 40% of the blood cholesterol is carried by high-density lipoprotein (HDL), which is healthy because high HDL levels protect against having a heart attack. HDL helps to remove excess cholesterol from arterial plaques, which slows the accumulation of plaque. HDL also carries cholesterol back to the liver, where it is removed from the body. Triglycerides are a type of fat that is also measured by a blood test. High triglyceride levels can be caused by obesity, physical inactivity, cigarette smoking, excess alcohol and a carbohydrate rich diet, and are common in people with heart disease and/or diabetes. Lp(a) is a genetic variant of the bad LDL. High levels of Lp(a) indicate a significant risk factor for premature fatty deposits in the arteries.

The blood levels of the LDL and HDL forms of cholesterol reflect contributions from diet, lifestyle (amount of exercise), and genetics. Some people inherit a combination of genetic alleles that permit them to consume high levels of fatty cholesterol-rich foods without disrupting the relatively low levels of the bad LDL cholesterol in the blood. Other people inherit gene alleles that act to maintain high levels of cholesterol despite eating a very low-fat diet and getting plenty of exercise. Recent medications called statins have been very successful at reducing the level of LDL cholesterol and increasing the HDL levels in the blood, reducing the impact from environment, diet, and lifestyle.

## Hypertension: A Silent Killer

High blood pressure (hypertension) is also influenced by genetic and environmental factors. High blood pressure is often called the "silent killer," because there are no warning symptoms and untreated hypertension can lead to stroke and heart disease at a young age. Health studies show that compared to other Americans, African Americans typically have a much higher incidence of uncontrolled high blood pressure. However, it is not yet clear the extent to which this is caused by genetic or environmental factors. It is interesting to note a recent study showing that native Africans living in Africa, are genetically very similar to African Americans, but do not have the uncontrolled high blood pressure found in black Americans. This study emphasizes that there is a great deal of research yet to be done to understand the genetic relationship between genes, hypertension, and the distribution of this silent killer in various population groups.

## WHAT CAUSES GENETIC DISEASES TO PREDOMINATE IN CERTAIN HUMAN POPULATIONS?

If we accept the idea that race is not a genetic concept, then why do certain genetic diseases predominate in certain racial groups (Table 16.4)? Here we will examine sickle cell anemia, a genetic disease that affects many

more African Americans than Caucasian Americans (see Chapter 10).

The obvious risk in using race as a surrogate concept is that this could hide a genetic reason that one drug might work better for certain people than other drugs and might interfere with the necessary genetic analysis. Race might be a surrogate for some genetic differences between people, but even if a drug works better for African Americans than Caucasians, there is no sharp distinction (and no genetic difference) between the two populations. All humans have the same genes regardless of race; in other words there is no gene, no allele, and no DNA sequence that is present only in the genomes of people in one race but not in another race. Furthermore, an increasing number of Americans are of mixed race and do not fit into the existing categories of race based on physical appearance.

Scientist studied the comparative death rates for people of different races who have succumbed to cancer or other chronic diseases (Table 16.4 and Table 16.5). To overcome the age bias inherent in different populations, most comparative death studies use age-adjusted rates. This is because a population with a higher average age is going to have a higher overall death rate due to chronic diseases such as heart disease and cancer, which predominantly affect an older population. On the other hand, a younger population will have a larger number of deaths from accidental causes, such as automobile accidents. To make sure that the comparison actually addresses differences among races, it is important to adjust the death rates for the ages of the population under study.

## Sickle Cell Anemia is a Genetic Disease, Not a Racial Disease

Sickle cell anemia, which is caused when a person inherits two copies of a mutant beta-globin gene, is much more common in African Americans in the U.S. than in other groups. The normal (wildtype) beta-globin gene encodes beta-globin protein, one of the two proteins that make up the hemoglobin in red blood cells. Hemoglobin normally contains four proteins, two beta-globin proteins and two alpha-hemoglobin proteins, which assemble together with heme and iron (Fe) to produce the hemoglobin molecule that carries oxygen (see Chapter 10). The severe symptoms of sickle cell disease occur in people who inherit two mutant beta-globin genes. Their blood cells cannot make normal beta-globin proteins so the body must survive on hemoglobin containing mutant beta-globin proteins.

In diseased red blood cells, the mutant beta-globin proteins (HbS) assemble with the alpha-globin proteins into mutant hemoglobin molecules. Because

**TABLE 16.5** Death rates from malignant neoplasms (cancers) in racial/ethnic groups in the United States, 1998

| Breast Racial/ethnic group | Total (per 100,000) | | Lung | | |
|---|---|---|---|---|---|
| | Men | Women | Men | Women | Women |
| White | 146 | 106 | 49.4 | 27.4 | 19.0 |
| Black | 208 | 129 | 70.8 | 27.2 | 26.2 |
| American Indian/ Alaskan Native | 96 | 74 | 33.9 | 16.5 | 10.8 |
| Asian/Pacific Islander | 91 | 63 | 24.6 | 11.2 | 9.3 |
| Hispanic | 93 | 64 | 21.4 | 8.3 | 12.5 |

*Data source: National Center for Health Statistics. Health, United States, 2000 with Adolescent Health Chartbook Hyattsville, Maryland: 2000.*

of the amino acid change in the beta-globin protein, inside the cells the abnormal hemoglobin molecules aggregate together into stiff protein rods that grow long enough to distort the physical shapes of the red blood cells from normal Frisbee-like disks to banana-shaped sickle cells (see Chapter 10). Unfortunately the sickle red blood cells cannot pass easily through the thin capillaries in the body they block the blood vessels and cause sudden attacks of severe pain, fever, swelling and possibly organ damage. Attacks can be triggered by any physical activity that increases the body's requirement for oxygen, which creates a low oxygen environment that causes the mutant red blood cells to sickle.

## A Crucial Link Between Sickle Cell Trait and Malaria Resistance

HbS has been at the center of a medical and scientific puzzle since the mid-1940s when doctors in Africa first noticed that patients with sickle cell anemia were much more likely to survive malaria than the European patients who do not carry sickle cell trait (Figure 16.9). The *Plasmodium falciparum* parasite that causes malaria begins the human part of its life cycle when a mosquito carrying the sporozoite form of the parasite bites a person (Figure 16.10). The sporozoites enter the bloodstream and migrate to the liver where they infect the liver cells and continue to develop and multiply into the merozoite form of the parasite, which rupture the liver cells and enter the bloodstream (Figure 16.11). The merozoites infect the red blood cells where they continue to develop and eventually produce parasite gametocytes, which infect a biting mosquito during a blood meal and continue the next parasite life cycle.

In many parts of the world, malaria is a serious disease that kills over 1 million people every year. Scientists

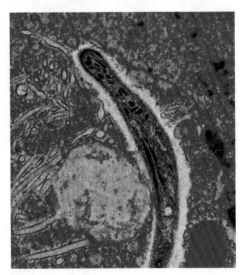

**FIGURE 16.9** Malaria is caused by a parasite. The *Plasmodium falciparum* parasite that causes malaria has a complicated life cycle. The sporozoite form of the parasite is shown here in the cytoplasm of a mosquito midgut epithelial cell (false-color electron micrograph).

found that the sickle cell gene (the mutant beta-globin gene) is especially prevalent in the genomes of people from the parts of Africa that are typically hardest hit by malaria. Evolutionary biologists have proposed that the sickle cell mutation in the beta-globin (HbS) gene became permanently established in the human population after especially serious outbreaks of malaria took place in Asia, the Middle East, and Africa.

People (carriers) who inherited one mutant beta-globin allele and one normal (wildtype) beta-globin allele do not typically suffer sickle cell symptoms unless they over-exercise in an environment of low oxygen. These people have red blood cells that make both the wildtype and mutant beta-globin proteins in the same cell and as a result the cells retain their disk-like shape under normal oxygen conditions. Even though the red blood cells in sickle cell carriers make only about half the normal number of wildtype beta-globin proteins, apparently there is sufficient beta-globin available to assemble enough functional hemoglobin molecules to significantly decrease the formation of sickle-shaped cells and prevent disease symptoms.

## Selective Advantage for Sickle Cell Trait and Malaria Resistance

People who are carriers of sickle cell disease are the result of natural selection, not because the mutant HbS

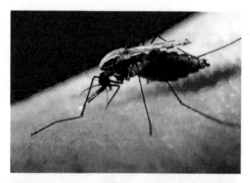

**FIGURE 16.10**  The mosquito is a vector for malaria. This mosquito (Anopheles *albimanus*) is feeding on a human arm. The mosquito carries the parasite that causes malaria, making mosquito control an effective way of reducing the risk of the malaria disease.

beta-globin gene causes sickle cell disease but because the mutant beta-globin gene is closely linked genetically to a nearby gene that confers resistance to malaria. The genes encoding sickle cell and the resistance to malaria are almost always inherited together by the same individual, indicating that the mutant beta-globin gene is located adjacent to the malaria resistance gene on the human chromosome DNA. This means that if a person inherits a mutant beta-globin gene (sickle cell trait), there is a very high probability that the person will also inherit a gene that confers resistance to malaria. How did this tight genetic connection affect the distribution of the sickle cell mutation in the human population?

Scientists were surprised to learn that in some parts of Africa as much as 40% of the population carries one copy of the mutant beta-globin gene (sickle cell trait) and suffer little or no serious symptoms from malaria. When the mutant sickle cell allele is frequent in the population, the resistance to malaria conferred by the sickle cell trait gives carriers a significant selective advantage during frequent outbreaks of malaria. But people who are sickle cell carriers also face the disadvantage of passing a potentially lethal disease gene to a biological child. A Punnett square can be used to predict the inheritance of the sickle cell genes (Figure 16.12).

Two parents who are carriers of sickle cell each have one copy of the mutant beta-globin allele (heterozygous)

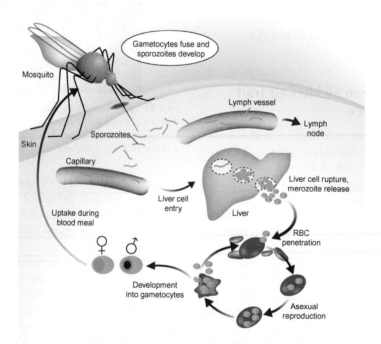

**FIGURE 16.11**  The life cycle of the malaria parasite in the human body. The *Plasmodium falciparum* parasite enters a human when an infected mosquito bites a person and draws blood. A form of the parasite called the sporozoite enters the bloodstream and travels to the liver. The parasite infects the liver cells, multiplies into merozoites and causes the liver cells to rupture. Once in the bloodstream, the merozoites infect red blood cells, where they continue to develop into ring forms, then trophozoites (a feeding stage), then schizonts (a reproduction stage), then back into merozoites. When the parasite gametocytes are produced in the bloodstream they are taken up by a feeding mosquito, where they are ready to continue the life cycle in the next bitten human.

(and one wildtype copy), which gives them a 25% chance that their biological child will inherit two mutant beta-globin genes and have sickle cell disease and resistance to malaria (see Figure 16.12). Carriers also have a 25% chance that their child will inherit two wildtype beta-globin genes, which means normal hemoglobin, but no protection against malaria. The carrier parents have a 50% chance of having a heterozygous child who will inherit one mutant and one wildtype beta-globin allele, be a genetic carrier of the sickle cell trait, and exhibit some protection from malaria.

## A Molecular Mechanism to Explain Malaria Resistance

The intriguing connection between sickle cell trait and malaria stems from observations that individuals who inherit the sickle cell trait (allele) are also protected from malaria. Exploring the genetics and biology of sickle cell disease has helped scientists to interpret the complex influence of evolutionary pressures on the inheritance of these disease genes. Researchers investigated possible mechanisms to explain how resistance to malaria might be connected to the sickle cell trait. Scientists finally connected the dots leading from sickle cell, to African-American blacks, to malaria resistance. Scientists have now described and characterized the malaria resistance gene that is often inherited along with sickle cell trait, and answered some long standing questions.

Scientists discovered a strong genetic link between a null (knockout) mutation in the Duffy gene and the mutant beta-globin gene (sickle cell trait) located nearby on the same chromosome. The co-inheritance of the Duffy knockout allele and the mutant beta-globin gene (sickle cell trait) suggests that the Duffy protein is part of the biological mechanism that protects carriers against malaria. In addition, the Duffy null (knockout) phenotype is most common in people whose ancestors came from populations in regions of Africa where malaria is endemic (native to the region).

How might the Duffy protein be involved in malaria? Interestingly, the Duffy protein is normally located on the outside surface of red blood cells, where it is used by the malaria parasite to attach to and enter the red blood cells. The mutant Duffy alleles cannot make Duffy proteins so the sickle cell carriers have far fewer Duffy receptor proteins on the red blood cell surfaces, dramatically reducing the ability of the parasite to attach to the red blood cells, and greatly reducing the chances of a malaria infection in sickle cell carriers (see Figure 16.9).

African Americans in the United States have a higher frequency of the sickle cell allele, and a higher incidence of sickle cell disease, compared to European Americans. But there are many people who are not black but carry the sickle cell allele or have sickle cell disease and do not live in the United States. These people often originate from countries such as Greece or Saudi Arabia, where malaria is also common. Of course, most African-Americans now in the United States brought their sickle cell and malaria genetic heritage with them when they came to the New World as slaves, and most came from West Africa where malaria is endemic. Genetics and

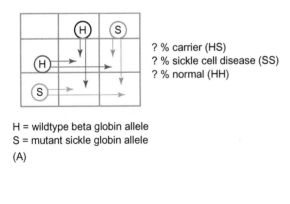

? % carrier (HS)
? % sickle cell disease (SS)
? % normal (HH)

H = wildtype beta globin allele
S = mutant sickle globin allele

(A)

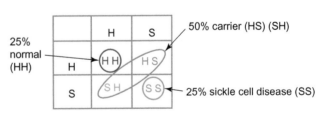

25% normal (HH)

50% carrier (HS) (SH)

25% sickle cell disease (SS)

H = wildtype beta globin allele
S = mutant sickle globin allele
HS = heterozygous alleles (sickle cell trait)
SS = sickle cell disease (homozygous alleles)
HH = normal beta globin genes (and proteins)

(B)

**FIGURE 16.12**  What are the chances that a child will inherit sickle cell disease? (A) The Punnett square is a simple grid diagram that geneticists use to organize the parental alleles (egg and sperm) so that the alleles can be used to predict the genotypes of the offspring. The Punnett square provides a summary of the possible combinations of one maternal allele with one paternal allele for each gene. In this case both parents are heterozygous for the beta-globin sickle cell allele: (H, S) and (S, H). The H and S alleles are placed across the top row and down the left column of the grid. To predict the genotypes of the offspring, recall that each fertilization event involves a sperm and an egg, carrying one copy of each chromosome, gene and allele. The colored arrows indicate how each allele forms the genotypes of the fertilized zygotes. (B) The genotypes of the possible offspring from two carriers of sickle cell disease are shown in the center of the Punnett square: Homozygous (HH) (normal or wildtype) inherited two normal beta globin alleles (25%). Homozygous (SS) (sickle cell disease) inherited two mutant beta globin alleles (25%). Heterozygous (HS) (sickle cell carriers) inherited both a wildtype beta globin allele (H) and a mutant sickle cell beta globin allele (S) (50%).

environment provide an ideal setting for evolutionary pressure to select for carriers of sickle cell who are protected from malaria.

> It is clear why scientists think it is misleading to label sickle cell anemia (or any disease) a racial disease (or any drug as a racial drug). African Americans who are sickle cell carriers are also protected from malaria because they carry the Duffy knockout allele located near the mutant globin gene in the genome. Over time, sickle cell carriers resistant to malaria have a selective survival advantage because they do not succumb to malaria.

## THE CONCEPTS OF RACE AND INTELLIGENCE

Among all the possible issues of racial disparity, those involving race, intelligence (IQ) and genes are perhaps the most difficult to interpret without bias. From an evolutionary perspective it is unlikely that any modern human populations have been in isolation for a sufficient period of time to detect any differences. As humans migrated out of Africa and around the world, there must have been sufficient contact between the villagers moving on and those staying behind to make sufficient isolation impossible (see Chapter 8). Earlier studies traced the migration of humans out of Africa and onto the other continents, and now the science of molecular genetics has confirmed these conclusions.

Scientists analyzed the lineage of DNA sequences from the 22 autosomal (non-sex) human chromosomes, and the X and Y chromosomes to study human evolution. The Y chromosome is used to trace the male lineage, and the mitochondrial DNA genome is used to trace the maternal female lineage from mothers to sons and from mothers to daughters. The results from studies using different methods have been remarkably consistent and support the idea that the human population has not been subjected to sufficient periods of geographical isolation to allow different human races to evolve naturally. For this study the scientists defined the different races in the context of a subspecies, but others point to the undeniable differences in average IQ scores, about one standard deviation or 15 points, between whites and blacks in America.

### Limitations of the Standard IQ Test

The IQ test was designed to measure intelligence but it is difficult to believe that a human characteristic as complicated and subjective as intelligence can be represented by a simple test. Nonetheless, IQ testing is commonly used in the United States and is ranked highly as an indicator of intelligence.

The history of IQ testing gives interesting insights into the usefulness of this test to assess this valued human quality. When Francis Galton invented the pseudo-science of eugenics, based on the erroneous belief that every human trait is inborn, he legitimized a concept that has often been misused for the benefit of prejudiced people. In 1892, Galton published his book, *Hereditary Genius*, which was widely viewed as the first scientific investigation of intelligence. He originated the idea that intelligence can be measured by a test. One of Galton's students, James McKeen Cattell brought intelligence testing to America for the first time in the 1890s. Ironically, particularly given the extensive use of testing today in U.S. schools, Cattell's work soon fell out of favor because the results did not correlate well with the success of the students in school.

At about the same time in France, Francis Binet was asked by school authorities to devise a test that could accurately predict the good and poor French students. Binet worked with average and disabled students, and decided which activities a "normal" child should be doing at specific ages, leading to the notion popular at the time that it was possible to determine a child's "mental" age. German psychologist Wilhelm Stern proposed that the ratio between a child's mental and chronological ages was an indication of the child's intelligence and American Lewis Terman coined the term *intelligence quotient (IQ)*. Despite extreme reservations on the part of the inventors of the IQ test, Stern and Binet, who always doubted that intelligence could be measured by a simple test, the IQ test soon gained wide acceptance across America. The Army adopted the IQ test in 1917 as a way to rapidly determine which draftees were suited for which jobs in the military. Despite the fact that the original IQ tests were designed to be given individually in an interview setting, the Army developed rapid IQ tests for potential soldiers. Soon after, the IQ test found its way into almost every school system, public and private in the country.

The controversy over IQ testing continued to grow and in the 1960s and 1970s, the IQ test fell out of favor, in part because the test contained a number of culturally biased questions easily understood by white suburban middle-class children but unfamiliar to black kids from the city. In time the IQ tests were revised to be more culturally neutral and are once again used widely in public schools. Critics include Howard Gardner, a Harvard professor who proposed the concept of "multiple intelligences" in 1983. In a more modern context it is correct to say that intelligence is very complex with many components and interacting variables, rather than something tangible that can be measured by a single number or even by a battery of tests.

## Stereotype Threat could Explain Racial Bias

Claude Steele (Stanford University) statistically analyzed the Standard Achievement Test (SAT) verbal scores of black and white students and found no difference between these two groups even on the difficult verbal test from the Graduate Record Exams. Surprisingly, there was a difference between the test scores of the two groups when the task was presented as a way that measures cognitive ability, but the difference *disappeared* when the test was presented as a problem solving task. Similar results were obtained when males were compared to females and when whites were compared to Asian math students. In situations where students expect to do poorly, because that is what they were told to expect, then the students follow through with the perceived expectation and tend to *do* poorly. On the other hand, in cases where the students feel confident about their expected abilities, they tend to do well. These test taking tendencies are known as the "stereotype threat."

Stereotype threat is an important factor used to help explain diverse abilities among many different groups. Before we ascribe these differences totally to innate abilities (i.e., genes), it is important to determine that the differences are not affected by the external factors in the environments in which the different groups live.

> It is fair to say that human intelligence is far too complex to be measured or quantified by a single number, like IQ; possibly the IQ discrepancies between African Americans and European Americans are more likely to be explained by stereotype threat, rather than by genetics.

## SUMMARY

History tells us that since about 1684 the idea of race was created by people and promoted by people as a social and cultural concept. For whatever reason, most modern societies classify people according to the prevailing concept of race, although most societies do not adhere to the rigid apartheid rules used by the former South African government to control black people. It is common for people to categorize people by physical characteristics such as short and tall, left-handed and right-handed, plump and skinny, and using talents such as athletic and nonathletic, and so forth. People are also categorized, often in a negative way, using skin color and other physical attributes such as slanted eyes or a wide nose that are commonly recognized as "racial traits".

The concept of race was a human invention of the seventeenth century, used as a mechanism to perpetrate racism, the idea that one group of people is inherently (genetically), inferior to another group. There is no scientific evidence to support the idea that humans evolved into different races, so the current practice of categorizing people according to race using labels such as African Americans, Chinese American, etc., primarily for the sake of clarity must be accompanied by a statement about the terrible impact of discrimination due to racism and prejudice.

## REVIEW QUESTIONS

1. What are the arguments for and against considering race in humans to be a biological concept versus a social construct?
2. How has the term race viewed from the perspective of history?
3. Why is it possible to say that race is not well defined from a genetic point of view?
4. Is there more variation within groups called races or between them? Explain the significance of the distribution.
5. Explain how the geographical origins of humans are reflected in the differences between human genome DNA sequences.
6. What is stereotype threat, and what does it have to do with the issue of the difference in intelligence among the races?
7. Explain what cancer tells us about the relationship between genes and the environment.
8. Define what is meant by the new field of pharmacogenomics.
9. What factors can explain disparities in the health of different races?
10. Explain some of the complexities in the relationships among race, intelligence, and IQ.

## ADDITIONAL READING

Bamshad, M.J., Olson, S.E., December 2003. Does Race Exist?. Sci. Am.

Cooper, R.S., 2004. Genetic Factors in Ethnic Disparities in Health: Critical Perspectives on Racial and Ethnic Disparities in Late Life. National Academy of Sciences/National Research Council Press, Washington, DC 267–309.

Gould, S.J., 1996. The Mismeasure of Man Ed.. Norton, New York.

Graves Jr., J.L., 2004. The Race Myth: Why We Pretend Race Exists in America. Dutton, New York.

Harmon, A., November 11, 2007 In the DNA era, new worries about prejudice NY Times. <http://www.nytimes.com/2007/11/11/us/11dna.html?_r=1&pagewanted=1&ei=5087&em&en=654c92ed2a9ed7fe&ex=1195016400/>

Kelley, R., Lewis, E. (Eds.), 2000. To Make Our World Anew: a History of African Americans. Oxford University Press, New York.

Khan, J., 2007. Race in a bottle. Sci. Am. http://www.scientificamerican.com/article.cfm?id=race-in-a-bottle.

Lehrman, S., February 2003. Profile: The reality of race. Sci. Am.

Lewontin, R.C., 2003. The Triple Helix, Gene, Organism, and Environment. Harvard University Press.

Racial disparities in cancer mortality rates between blacks and whites quantified Science Daily. March 22, 2009. <http://www. sciencedaily.com/releases/2009/03/090321103811.htm/>.

Smedley, B.D., Stith, A.Y., Nelson, A.R. (Eds.), 2003. Unequal Treatment: Confronting Racial and Ethnic Disparities in Health Care. National Academy Press, Washington, DC.www.nap.edu/ catalog.php?record_id=10260#tocr

Zakai, N. A., McClure, L. A., Prineas, R., Howard, G., McClellan, W., Holmes, C. E., Newsome, B. B., Warnock, D. G., Audhya, P., Cushman, M., December 9, 2008. Correlates of anemia in American blacks and whites, The REGARDS Renal Ancillary Study (REasons for Geographic And Racial Differences in Stroke) American Journal of Epidemiology DOI:10.1093/aje/kwn355

Published by the Johns Hopkins Bloomberg School of Public Health. http://aje.oxfordjournals.org/cgi/content/abstract/kwn355v1

## WEB SITES

Genetic Anthropology, Ancestry, and Ancient Human Migration. <http://www.ornl.gov/sci/techresources/Human_Genome/elsi/ humanmigration.shtml/>.

Minorities, Race, and Genomics <http://www.ornl.gov/sci/ techresources/Human_Genome/elsi/minorities.shtml/>.

How sickle cell protects against malaria. <http://sickle.bwh.harvard. edu/malaria_sickle.html/>.

A Mutation Story: sickle cell and malaria. <http://www.pbs. org/wgbh/evolution/library/01/2/l_012_02.html/>below.

**3′ cohesive end**  Single-stranded complementary DNA end containing a 3′ OH (hydroxyl group)

**5′ cohesive end**  Single-stranded complementary DNA end containing a 5′ phosphate group

**5′ to 3′ direction**  Starting at the 5′ end of a DNA strand and reading toward the 3′ end of the same DNA strand

**adenine (A)**  A purine base found in the DNA and RNA nucleic acids of cells in plant and animal cells

**adult stem cells**  Undifferentiated cells growing in a specialized tissue

**affinity chromatography**  Method to separate molecules based on a highly specific chemical interaction between molecules and the affinity matrix

**alkaline**  Having a pH greater than 7.0; also called basic

**alleles**  One particular version of a gene, or more broadly, a particular sequence of any location (locus) on a molecule of DNA

**allelic frequency**  The percentage of any allele in a population's gene pool

**allotropes**  Two or more different structures of a chemical element

**amino acids**  Any of a class of organic compounds containing amino ($NH_2$) and the carboxyl (COOH) groups; forming the main constituents of protein

**aminoacyl tRNA synthetase**  Enzyme that attaches an amino acid to a tRNA molecule

**aneuploidy**  A condition in which a cell inherits the wrong number of chromosomes (too many or too few)

**angiogenesis**  The development and growth of new blood vessels

**anthers**  Plant structures that contain pollen

**anthropologists**  Scientists who study the origin, behavior, and the physical, social, and cultural development of humans

**anticodon**  Group of three consecutive bases in tRNA that are complementary to a three base codon in the mRNA

**antigens**  Molecules that cause an immune response in the body and are recognized and bound by antibodies

**antiparallel**  DNA strands with polarity (5′ to 3′) are arranged in opposite directions in the DNA helix molecule

**anti-toxin**  An antibody or other protein produced in the body in response to the presence of a toxin or poison

**apoptosis**  Programmed cell death; cell suicide

**archea**  One of three major branches of life (the other two being bacteria and eukaryota); it comprises halophiles and thermophiles

**asexually propagated**  The deliberate reproduction of whole plants using vegetative cells, tissues, or organs

**assimilation model**  One of three theories of the evolution of *homo erectus* into the anatomically modern human (*homo sapiens*) which proposes that the recent African migrants did interbreed with some other hominid groups, but that the degree of interbreeding varied greatly from one geographic region to another and from one time period to another

**attenuation**  The step-wise decrease in expression of a particular gene

**autism**  A spectrum of disorders that cause significant impairment of mental function and social behavior

**autologous stem cell transplants**  Treatment in which the patient's stem cells are used to replace damaged or diseased tissues in the patient's body

**bacterial artificial chromosome**  One type of vector used to clone DNA fragments

**bacterial chromosome**  Structure within bacterial cells that contain the bacteria DNA

**bacterial restriction and modification system**  A system that protects bacteria against bacteriophage infection by destroying the bacteriophage DNA but not the bacterial chromosome

**bacterial strains**  Related but genetically distinct bacteria

**bacteriophages**  Viruses that infect bacterial cells

**baculovirus**  DNA virus that infects insects and related invertebrates and is widely used as a vector for animal genes

**bases**  Alkaline chemical substances; in molecular biology refers to the cyclic nitrogen compounds found in DNA and RNA

**benign tumors** Tumors that only grow in a single location in the body and do not metastasize to other tissues and organs

**biodiesel** A fuel that is derived from soybeans, canola, sunflower, soybean oil, palm oil, and other agricultural crops as well as vegetable oil recycled from restaurants

**biofuels** Sustainable fuels that come from plants

**bioinformatics** The computerized analysis and manipulation of large amounts of biological sequence data including DNA, RNA, and amino acid sequences

**biolistic gene gun** Gene delivery system designed to shoot DNA projectiles directly into sub-cellular compartments in eukaryotic cells

**biological control (biocontrol)** Approaches that use parasites, predators, or pathogens to control an unwanted organism

**biological database** A collection of biological information including DNA, RNA and protein sequences organized for efficient access to data

**biomarkers** Specific proteins that visibly change with the onset of a disease

**biopesticide** A pesticide composed of a biological control agent

**blastocyst** A very early stage in embryo development

**blunt ends** Ends of a double-stranded DNA molecule that are fully base paired without unpaired single-stranded overhang

**bottleneck** Evolutionary event in which most of the individuals in a population are killed or are otherwise unable to reproduce

**cancer genome** A set of DNA mutations that act together to convert a normal cell into a cancer cell

**capsid** Protective protein layer that surrounds the DNA or RNA genome in a virus particle

**carcinogen** An agent that causes a normal human cell to become a cancer cell

**cell division cycle** Series of stages that a cell goes through between cell division events

**cell-free extract** Subcellular fraction created *in vitro* that retains biological activity

**cellulose** Structural polymer of β-1,4-linked glucose that is a major component of plant cell walls

**cell wall** The layer covering a plant cell outside the plasma membrane

**central dogma** Basic flow of genetic information in living cells starting with genes (DNA), then messenger (mRNA), and finally, proteins

**centromere** Region on each eukaryotic chromosome where microtubules attach to move chromosomes during mitosis and meiosis

**cetane number** A measurement of the combustion quality of diesel fuel during compression ignition

**chaperones** Proteins that help newly synthesized proteins fold into proper three-dimensional structures

**chimera** A hybrid created by fusing together proteins (or genes) from two species

**chloroplasts** Specialized membrane-bound organelles that are the sites of photosynthesis in green plants

**chromatid** One of the paired daughter DNA strands produced by DNA replication

**chromatin** DNA-protein complexes that are major components of eukaryotic chromosomes

**chromosomes** Structures in the cell nucleus that each contain one linear double-stranded DNA molecule

**clone** A genetically identical population of cells derived from a single parent cell

**cloned gene** A recombinant DNA gene carried in a DNA vector that is propagated in pure populations of cells

**cloning** The production of identical copies of a specific DNA fragment or grow genetically identical cells from a single parent cell

**codominance** A situation in which gene products from both alleles of a gene are made in the cell and both proteins influence phenotype

**codons** A group of three consecutive RNA or DNA bases that encode a single amino acid

**cohesive end sites (cos)** Complementary single-stranded cohesive ends of the lambda (λ) viral genome DNA

**coincidental match** A questionable DNA match scored between a suspect and the evidence

**column** Vertical tube used in chromatography methods

**combinatorial control** Control of gene expression involving the presence or absence of a particular combination of regulatory proteins in different cells

**Combined DNA Index System (CODIS)** The Federal Bureau of Investigation computer system that enables local, state, and federal law enforcement officers to search the forensic DNA databanks of law enforcement agencies throughout the United States

**commercial biocrude** Biofuels that are used commercially

**conjugation** Process in which genes are transferred between bacteria by cell-to-cell contact

**complement activation** A process in which a series of complement proteins launch a stepwise cascade that leads to the assembly of protein complexes that make holes in the membrane of the target cells

**complementary DNA (cDNA) library** A DNA collection containing the DNA copies of RNA transcripts produced in the particular cell type used to generate the DNA clones in the library

**consensus sequence** A sequence created by comparing many promoter sequences and choosing the bases that appear most frequently at each position

**conserved** Refers to an amino acid sequence in a protein that has remained essentially unchanged throughout evolution, or a DNA or RNA sequence that is common in the genomes of different organisms

**constitutive** A type of gene promoter that almost always expresses high levels of RNA and the encoded protein

**continuous replication** Refers to the mechanism of DNA replication used to synthesize the leading strand of DNA at each replication fork

**control region** The part of a gene containing the promoter and regulatory DNA sequences that control transcription or replication

**cosmid vectors** Vectors that can carry very long DNA fragments, enter the cells with high efficiency, and replicate as independent plasmids in the cells

**covalent modification** Altering the structure of a DNA, RNA or protein molecule by enzymatic means, changing the chemical properties of that macromolecule; for example: phosphorylatyion and methylation

**cyclin-dependent kinases (Cdk)** Special protein kinase enzyme that is activated by a cyclin protein and controls eukaryotic cell division

**cyclins** Family of proteins that fluctuate with the cell cycle; at high levels cyclins bind to CdK to make MPF (mitosis promoting factor) and control the cell cycle

**cytokines** Short peptides that stimulate the growth of immune cells

**cytokinesis** The final stage of cell division that physically separates the two daughter cells

**cytological staining** Method of staining cells so that specific subcellular structures are visible when the stained cells are magnified in a light microscope

**cytoplasm** The protoplasm surrounding the nucleus of a eukaryotic cell

**database mining** Searching the vast amount of information in a database to find the few pieces of data of interest to the researcher

**diploid** The genome complement of cells that inherit two copies of each chromosome and two copies of each gene

**discontinuous replication** Refers to the mechanism of DNA replication used to synthesize the lagging strand of DNA at each replication fork

**disease-specific** Pertaining to a particular disease

**divergence** A process in which differences in nucleic acid and protein sequences accumulate over time

**DNA (deoxyribonucleic acid)** Nucleic acid polymer of bases which make up the genes

**DNA hybridization probes** Tools used to identify specific target DNA sequences using a library or genome screen

**DNA ligase** Enzyme that catalyzes the formation of a covalent chemical bond between the 3' and 5' ends of two DNA strands, sealing the DNA backbone and joining DNA fragments end to end

**DNA polymerase** An enzyme that copies DNA templates into new DNA strands when chromosomes are being replicated

**DNA profile** The genotypes and other chromosome markers that identify the unique genome sequence of each individual

**DNA replication origin** Site on the DNA helix where DNA replication begins

**dominant** A gene that directs a phenotype even when present at only one copy per cell

**dominant allele** A gene allele that determines a phenotype even when present at only one copy per cell

**E. coli DNA polymerase I** One of three DNA polymerase enzymes that replicate DNA in E. coli cells

**electroporation** Technique that uses an electric field to make cells take up DNA

**elongation factor** Proteins that are required for the elongation of a growing polypeptide during translation (protein synthesis)

**embryonic stem cell lines** Cell lines generated from embryonic stem cells (ESCs)

**embryonic stem cells (ESCs)** Stem cells derived from the inner cell mass within the blastocyst embryo

**endonuclease** An enzyme that cleaves internal covalent bonds and degrades DNA molecules

**enhancers** Regulatory sequences in the genome DNA that are often located at long distance from the promoter region of eukaryotic genes

**enol** A specific chemical structural conformation of a DNA base

**enucleated egg cell** An egg cell (oocyte) from which the nucleus has been removed

**enzyme** A protein that acts as a catalyst, increasing the rate at which a chemical reaction occurs, without itself changing in molecular structure

**epigenetic** Refers to inherited phenotypes that are not due to changes or mutations in DNA sequence

**epigenetic memory** A process of protein modification that allows cells to maintain an undifferentiated state

**epigenetic modification** Changes that impact the entire genome, altering gene expression and changing the state of the cell

**epigenome**   The newly reprogrammed genome in the developing embryo

**epitope**   Localized structure in an antigen to which an antibody binds

*Escherichia coli (E. coli)*   A species of bacterium that normally lives in the human gut and is commonly used in research in genetics and molecular biology

**eukaryotic**   The type of biological cell containing a nucleus and chromosomes

**exonucleases**   Enzymes that cleave single mononucleotides from the 5′ or 3′ end of the DNA strand

**Expect value (E)**   For a BLAST search, this value measures the number of hits (matching sequences) that can be expected by random chance using a particular query (the lower the E-value, the more likely that the sequences are related to each other)

**expressed**   The process of copying a DNA region (gene) into RNA (transcription); the process of protein synthesis (translation) used by living cells

**expressed sequence tags (ESTs)**   A special type of cloned DNA derived from DNA sequences that were transcribed into RNA strands in the cell

*ex vivo*   When diseased cells are removed from the body, treated and returned to the patient's body

**FASTA**   Online program that compares the sequences of proteins or nucleotide sequences

**forensic DNA databanks**   Databases that contain the DNA profiles (DNA fingerprints) of people who have been convicted of crimes as well as DNA profiles from evidence samples collected at unsolved crimes

**free radicals**   Very reactive ions that carry an unpaired electron and can damage and mutate DNA genes and proteins

**Galapagos Islands**   An archipelago of volcanic islands distributed around the equator in the Pacific Ocean, where Darwin studied evolution

**gametes**   Haploid egg or sperm cells produced for sexual reproduction

**gel electrophoresis**   An electric field that moves charged biomolecules through a gel matrix in order to sort DNA, RNA and protein molecules by size

**gene copy number**   The number of copies of each gene present in one genome

**gene expression**   The process by which genes send instructions to cells through transcription and translation

**gene guns**   Gene delivery system that shoots DNA projectiles directly into the subcellular organelles in the target cells (nucleus, mitochondrion, chloroplast)

**gene knockout**   A DNA mutation made when a specific gene is removed from a genome or disabled by an insertion mutation that disrupts the gene

**genes**   Biological genetic units of heredity

**gene-specific DNA probes**   Short single-stranded DNAs used to test for a specific mutant or wildtype gene in an individual's genome or library

**genetically linked**   When a gene or other chromosome marker and a particular gene are inherited together through many generations

**genetically modified organisms**   Plants and animals that have had their genetic makeup altered to exhibit traits or produce proteins that are not typical

**genetic code**   The sequence of tandem nucleotide triplets (codons) in DNA or RNA that specifies the amino acid sequence of a protein

**genetics**   The branch of biology dealing with heredity and the laws governing genetic inheritance

**genetic screen**   A method to search for rare cells using the genetic characteristics of the cells

**genome reprogramming**   The cell's genome is reprogrammed at fertilization, essentially initiating a pluripotent state and permitting embryo development to occur

**genomic DNA library**   A DNA library containing 2 to 3 copies of all the DNA fragments from an entire genome

**genomic map**   A diagram showing the relative positions of every gene and other locus indicated along the linear DNA molecule contained in each chromosome

**genomics**   The study of the structure and function of the genomes of all organisms

**genotype**   The DNA sequence characteristics of the alleles for all the genes in a specific individual genome

**germ line**   Reproductive cells that produce egg or sperm cells required to generate the next generation (in eukaryotic organisms)

**glycosylation**   Posttranslational modification reaction in which enzymes catalyze the addition of sugar units (carbohydrates) to specific amino acids in certain proteins

**guanine (G)**   One of the five fundamental bases that make up DNA and RNA sequences

**haploid**   Having inherited a single set of chromosomes

**helicase**   An enzyme that unwinds the DNA double helix at each replication fork

**hematopoietic stem cells (HSCs)**   Stem cells that generate all of the different types of specialized cells in the mammalian blood and immune systems

**herpesvirus**   A DNA virus that causes a variety of diseases including tumors; the virion contains a double-stranded DNA genome and an outer envelope surrounds the nucleocapsid

**heteroduplex**   A molecule in which an RNA strand is base paired to its complementary DNA strand

**heterosis** When a hybrid F1 generation created by crossing two parental lines (P1 × P2) exhibits desirable traits that exceed the traits of both parents

**heterozygous** Having inherited two different alleles of the same gene on two homologous chromosomes

**histone** Special positively charged proteins that bind to DNA and create and alter the dynamic structure of eukaryotic chromosomes

**hominids** Human-like primates that walked upright on two feet

**homologous** Similar in DNA, RNA or amino acid sequence

**homologous recombination** Recombination or genetic exchange between two DNA regions that are identical, or very similar in DNA sequence

**homozygous** Having inherited the same form (allele) of a gene on two homologous chromosomes

**homozygous knockout** Second-generation transgenic animals that carry knockout alleles of the same gene on both chromosomes

**hormone** Molecules, often short proteins, which carry signals to different cells and tissues inside multicellular organisms

**host cells** Living cells invaded by an infectious agent or target cells carrying foreign DNA usually introduced into cells on a vector

**human artificial chromosomes (HACs)** Recombinant minichromosomes that replicate and segregate just like native chromosomes when introduced into human cells

**human immunodeficiency virus (HIV)** A retrovirus that causes AIDS

**humanized** Replacing parts of a foreign protein with the equivalent human amino acid sequences

**hybridization** Base pairing of single strands of DNA or RNA to each other via hydrogen bonding between the complementary bases

**hybridomas** Hybrid cells made by researchers in which an antibody-producing B cell is fused with a myeloma cell to form a self-proliferating cell that produces only one specific monoclonal antibody

**hybrid vigor** When a hybrid F1 generation created by crossing two parental lines (P1 × P2) exhibits desirable traits that exceed those of both parents

**immunology** Scientific field that focuses on the study of the physical, chemical, and physiological characteristics of the components of the human immune system in healthy and diseased states

**immunosuppressive drugs** Agents capable of suppressing immune responses in the human body

**induced pluripotent stem cells (iPS cells)** Type of pluripotent stem cell artificially derived by introducing transcription factor genes into highly differentiated adult cells, which triggers genome reprogramming and a pluripotent state

**inducer** Molecule with a regulatory influence on gene expression

**inducible** Describes a type of promoter or operon that turns on gene expression in response to an inducer signal

**informed consent** Agreement by a patient to participate in an experimental treatment only after the patient understands the risks involved

**inheritance** The acquisition of characteristics, qualities or traits by the transmission of genes from parent to offspring

**initiation factors** Proteins that are required to initiate synthesis of a new polypeptide

**insulin** Small protein hormone made by the pancreas cells that controls the level of sugar in the blood and does not function properly in diabetes

**integrate** To insert a DNA strand into a DNA chromosome (or plasmid)

**intelligence quotient (IQ)** The ratio of a person's mental age to his or her chronological age

**interphase** The part of the cell cycle between two successive cell divisions, during which cellular metabolism and DNA synthesis occur

**interrupted genes** DNA genes that contain both intron and exon sequences

*in vivo* Refers to processes in living cells, including the treatment of diseased cells inside the body

**karyotype** The full set of mitotic chromosomes in the nucleus that is characteristic for each eukaryotic species; human nuclei contain 23 or 46 chromosomes

**keto** A specific chemical structural conformation of a DNA base

**kilobase pair (kB)** A unit of 1000 consecutive DNA base pairs

**knockout animal** A transgenic animal that has had a specific gene inactivated or deleted from its genome

**knockout gene** A gene that has been inactivated or deleted from the genome of an organism

*lac* **operon** An inducible operon encompassing three genetic loci involved in the uptake and breakdown of lactose in *E. coli*

**lagging strand** The DNA synthesized in short single-strands called Okazaki fragments during discontinuous DNA replication

**lambda (λ)** A bacteriophage that infects *E. coli* cells

**leading strand** The strand of DNA that is synthesized continuously during DNA replication and does not contain Okazaki fragments

**liposome fusion**   A vesicle used to transfer DNA genes into cells by fusing the vesicle with the cell plasma membrane

**locus**   A specific position or location on a chromosome DNA

**logic gates**   In DNA computers, a series of wells containing DNA molecules that respond to input information in specific ways depending on the secondary structure of the DNA strand

**lysogenic**   A viral pathway in which the virus genome DNA integrates directly into the host chromosome, called a prophage, which can be transmitted to daughter cells at each cell division, causing the production of new phages through the lytic cycle

**macrophage**   Large, mononuclear, highly phagocytic cells derived from monocytes

**major groove**   The wide groove in each DNA helix (compared to the narrower, minor groove)

**male sterility**   In plants, the failure to produce functional anthers, preventing sexual reproduction

**malignant tumors**   Cancer cells that spread from a primary tumor to other body location

**markers**   Polymorphisms and other genetic loci and landmarks identified on human chromosome maps

**meiosis**   The process of cell division by which reproductive cells are formed and chromosome number is reduced by half

**mesenchymal stem cells (MSCs)**   Multipotent adult stem cells that can differentiate into a variety of muscle-related cell types

**messenger RNA (mRNA)**   RNA that encodes proteins and is made by transcribing genes

**metastasis**   Process in which cancer cells migrate away from a primary tumor and move around the body to form secondary cancers at other locations

**methylation**   Posttranslational modification in which methyl groups are added to specific amino acids in proteins to influence protein functions

**microarray technology**   A technique that permits the simultaneous detection of all of the mRNAs transcribed from thousands of genes in a genome at any one point in time

**micro RNAs (miRNAs)**   Small regulatory RNA molecules made in eukaryotic cells

**mini-chromosomes**   Small chromosome-like DNA-protein structures that sometimes form in different types of eukaryotic cells

**minor groove**   The narrower groove in each DNA helix molecule

**mitochondrial DNA**   A double-stranded circular DNA molecule that contains only 37 genes and is located within the mitochondria in the cytoplasm of eukaryotic cells

**mitosis**   The process of cell division that results in the formation of two daughter cells, without changing the chromosome numbers of the cells involved

**mitotic chromosomes**   Highly compact structures composed of histone proteins bound to long linear double-stranded chromosome DNA

**model**   Certain organisms such as the mouse, fruit fly and bacteria are the focus of active research and are now very well characterized organisms

**monoclonal antibody (mAb)**   A pure population of identical antibody proteins with a unique sequence that recognizes only one specific antigen (made by a cell line cloned from a single B cell)

**monogenic diseases**   Diseases caused by mutations in single human genes

**monomer**   A protein molecule of relatively low molecular weight that can bind to itself and other proteins to form dimers, trimers, or other protein complexes, sometimes bound to DNA or RNA

**motifs**   Amino acid sequences that are conserved among proteins from many different organisms

**MPF (maturation promoting factor or M-phase promoting factor)**   The cyclin-Cdk protein kinase is assembled when the cyclin proteins build up in the cell and bind to CdK to make the active protein kinase enzyme (MPF)

**multigenic**   Diseases and disorders caused by interactions among the products of more than one mutant gene and often several mutant genes

**multiple sequence alignment**   Comparing one sequence in parallel with several other sequences to look for similarities and differences at each position in the linear sequence

**multipotent**   Cells that are capable of differentiating into more than one cell type but cannot generate all of the cell types needed in the human body

**multiprotein**   Consisting of several proteins that interact together in a complex

**multiregional model**   One of three theories of the evolution of *homo erectus* into the anatomically modern human (*homo sapiens*) which proposes that the original *homo erectus* population migrated into Europe and Asia and then *homo erectus* evolved into *homo sapiens* simultaneously in several different geographic regions

**mutagen**   An agent that induces genetic mutation by changing DNA sequence (for example, a "G" changed to a "T")

**mutagenesis**   The creation of a genetic mutation by changing the DNA sequence

**mutagenic**   Having the ability to induce genetic mutation by changing the DNA sequence

**mutation**   Change in a DNA (or RNA) sequence

**myc**   One of the proto-oncogenes that encodes a transcription factor protein that normally regulates the expression of several different genes

**nanotechnology**  Technology dealing with structures smaller than 100 nanometers in size, and involves developing materials or devices with novel properties; the technology of molecular manufacturing outside the cell, including molecular machinery inside the cell

**neomycin**  A type of antibiotic used in genetic screens

**neomycin resistance gene (neo$^r$)**  A gene often contained in transgenic vectors used to produce transgenic animals; the neo$^r$ enzyme inactivates neomycin and G418 and renders the cells resistant to these antibiotics

**neural progenitor cells**  Another name for neural stem cells

**neural stem cells (NSCs)**  Multipotent adult stem cells that generate the specialized cells required for the function of the nervous system and brain

**neurospheres**  Multicellular structures generated by neural stem cells growing in tissue culture plates

**neurotransmitter**  Molecule that carries signals across the synapses, the gaps between the ends of the nerve cells

**noncoding RNA (ncRNA)**  RNAs made in cells that do not encode a protein

**non-protein-coding DNA (non-coding DNA)**  The strand of DNA that does not carry the information necessary to make proteins

**nuclear DNA**  Found in the nuclei of eukaryotic cells; for example, each human cell nucleus contains 23 or 46 chromosomes

**nucleoid**  A DNA-containing region lacking a surrounding membrane, found in prokaryotes

**nucleotides**  The monomer component of a nucleic acid (DNA and RNA), consisting of a pentose sugar plus a base and a phosphate group

**nucleus**  This spherical compartment in eukaryotic cells is enclosed within a double membrane (nuclear envelope) and contains all the chromosomes and one or more nucleoli (site of ribosome assembly)

**obesity**  Excessive accumulation of fat in the body, partly in response to gene expression

**Okazaki fragments**  The short single-stranded DNA synthesized on the lagging strand during DNA replication

**oncogene**  The mutant form of a gene that promotes cancer development

**one gene–one enzyme hypothesis**  The early idea that each gene in a cell can produce only one specific protein or cellular enzyme

**operator**  Site on DNA where a repressor protein recognizes and binds to a specific DNA sequence

**operons**  Groups of prokaryotic genes that are transcribed together into a single polycistronic mRNA that is translated into two or more proteins

**organelle**  Membrane-bound compartments with distinctive morphology and function present in the cytoplasm of all eukaryotic cells

**organic farming**  The process of producing food naturally, without the use of manufactured chemicals for pest and weed control and fertilizers

**organic food**  Foods grown without the use of chemicals, including pesticides and fertilizers; also meat from animals raised without hormones and other drugs

**origin of replication (ori)**  Site on a DNA molecule where DNA replication begins (initiation)

**pair-wise sequence alignment**  Comparing one DNA, RNA or amino acid sequence directly with one other DNA, RNA or amino acid sequence

**palindrome**  A double-stranded DNA sequence that reads the same on the top strand (5′ to 3′) as the complementary bottom strand sequence read backwards (5′ to 3′). For example, ACCTAGGT and its complement, TGGATCCA, represent a palindrome

**patient-specific**  Having to do with the needs of an individual patient

**phagocytosis**  The engulfing of microorganisms, other cells and foreign particles by specialized cells called phagocytes

**pharming**  The development and marketing of transgenic animals and plants to produce novel therapeutic proteins and commercial products

**phenotype**  Observable characteristics of an individual as directed by the expression of the individual's genes

**phosphodiester bond**  The covalent chemical bond that links nucleotides in a nucleic acid polymer and consists of a central phosphate group esterified to flanking sugar hydroxyl groups

**phospholipid cell membrane**  All biological membranes contain phospholipids

**phosphorylation**  Posttranslational modification reaction in which special enzymes add phosphate groups to selected proteins

**photosynthesis**  The process by which plants use the energy of sunlight to produce carbohydrates from carbon dioxide

**phylogenetics**  The study of evolutionary relationships among and between organisms and the involvement of genetic variation in these changes

**phylogenetic trees**  Diagrams showing the evolutionary relationships among different organisms over time

**pilus**  A thin tube made in certain bacterial cells and used to pass the bacterial chromosome DNA from the donor cell to the recipient cell during conjugation (mating)

**placenta**  The tissue that joins the mother and embryo or fetus; allowing diffusion of nutrients from the mother's blood into the fetus's blood and diffusion of waste products from the fetus back to the mother

**plant protoplast cells**   Plant cells that have had their cell wall completely or partially removed by either mechanical or enzymatic means

**plaque**   A clear circular zone that forms in a lawn of bacteria growing on a plate when a virus destroys the lawn of host cells

**plasmids**   Self-replicating double-stranded circular DNA molecules that are most often found in prokaryotic cells and also in some in eukaryotic cells

**ploidy**   Chromosome number specific to each type of cell and influenced by the stage of the cell cycle

**pluripotent**   The ability to develop into all of the types of cells in the human body

**pluripotent stem cells**   These cells exist in an undifferentiated state and have the potential to develop into the many types of specialized cells needed in the human body

**polarity**   In DNA this refers to the two chemically different ends of each DNA strand called 3′ and 5′; in the helix, the two DNA strands are arranged in opposite directions: 5′ to 3′ and 3′ to 5′. In cell development, polarity can refer to the different biological activities taking place at the different ends (poles) of the cells

**pollen**   Tiny fine spores containing male gametes; for example, pollen carried in the anthers of a flowering plant

**polycistronic transcript**   An mRNA transcript that encodes multiple genes and is translated into multiple proteins in bacterial cells

**polyclonal**   A population of different antibody proteins that specifically recognize different epitopes of the same antigen

**polymerase chain reaction (PCR)**   Method used to amplify a DNA sequence by repeated cycles of DNA strand separation and replication

**polymorphisms**   Locations in the DNA genome where the DNA sequences differ between different individuals

**posttranscriptional processing**   The processing of precursor (or primary) RNA transcripts to generate mature messenger RNAs (mRNAs) required for translation in eukaryotic cells

**posttranslational modifications**   Chemical modifications (phosphorylation, methylation, glycosylation) added by enzymes to selected proteins in eukaryotic cells

**precursor mRNAs**   Primary RNA transcripts that are copied from a gene that are processed into mature mRNAs

**primase**   Enzyme that participates in DNA replication by making RNA primers needed to initiate DNA replication

**primer**   A short RNA (or DNA) strand that is base paired to the template DNA strand at the position on the helix where DNA replication will start

**product rule**   States that the probability a series of independent events will happen is equal to the product of the probabilities of the individual events separately

**prokaryotic**   A type of biological cell that lacks a nucleus and other organelles (bacteria)

**promoter**   Region of DNA in front of a gene that interactions with the RNA polymerase when gene expression is turned on

**promoter strength**   Transcriptional control can be measured by the rate of transcription initiation in the absence of additional gene regulation; strong and weak promoters

**pronuclei**   The parental male and female nuclei in a fertilized egg just before nuclear fusion

**protease enzymes**   Various enzymes that catalyze the hydrolytic breakdown of proteins into peptides or single amino acids

**protein-coding sequence**   The specific DNA sequence of a gene that codes for a specific amino acid sequence

**protein modeling**   Computer models of protein structures are used to predict the shapes of proteins and study function

**proteins**   Unbranched chains of amino acid units that fold into functional 3D structures and perform countless jobs in cells

**protein synthesis**   The production of proteins by the process of translation in cells

**proteome**   The total set of proteins encoded by a specific cell's genome or the total protein complement of an entire organism

**proteomics**   Study of the complete protein complement of each organism

**proto-oncogenes**   Unmutated (wildtype) form of oncogenes that encode proteins with important cellular functions

**protoplasts**   Plant cells without cell walls

**provirus**   Virus genome copy that is integrated into the chromosome DNA of the host cell

**pyrimidines**   Type of nitrogenous base with a single chemical ring found in DNA and RNA

**quantitative trait loci (QTL)**   Specific chromosome regions containing several genes that influence a complex trait such as crop yield

**query**   A request for the analysis of a specific DNA, RNA, or amino acid sequence submitted to a database

**random match probability (RMP)**   The probability that a DNA profile in question would be found in a person who was randomly selected from the same racial/ethnic group to which the defendant in a criminal trial belongs

**receptor**   Protein that binds to another molecule, such as a hormone or a nutrient, and participates in cell to cell signaling; receptors are located on the outer surface of the cell membrane

**recessive**   Phenotypic expression of a genetic allele only in homozygous cells that lack wildtype forms of the gene

**recessive allele** Usually encodes a nonfunctional protein product

**reciprocal translocation** Translocation involving an equal exchange of DNA sequences between two chromosomes

**recognition sequence** A specific DNA sequence that is recognized by a specific protein such as a restriction endonuclease that will cleave a DNA molecule at a specific DNA sequence

**recombinant DNA** A molecule containing foreign DNA sequences covalently linked to vector or host DNA sequences

**recombinant DNA cloning** A double-stranded DNA vector carrying a foreign gene that is cloned and propagated in bacteria

**recombinant DNA plasmids** Circular double-stranded molecules containing both vector DNA and a foreign gene(s)

**recombination (crossing-over)** DNA exchange creates new combinations of genes and genetic information

**reference databases** Collections of DNA profiles of many people who were not involved in crimes but who volunteered to give their DNA to law enforcement agencies to determine the frequencies of common marker alleles and genotypes in the general population

**regeneration** This healing process involves the processes of inflammation, proliferation, and tissue remodeling

**rejection** A serious complication that often occurs whenever a patient receives a transplant of cells or tissues from a person who is a genetically unrelated unmatched donor

**release factors** Proteins that recognize a stop codon during translation and cause the finished polypeptide to be released from the ribosome

**repeated DNA sequences** Stretches of DNA bases that are repeated in many locations throughout a genome

**replica plating** A technique in which one or more secondary petri plates containing different types of solid (agar-based) selective growth media are inoculated with the same colonies of microorganisms from a primary plate (or master dish); a sterile cloth is pressed gently onto the colonies on the first plate and then the cloth is pressed onto an empty agar plate to reproduce the original spatial pattern of colonies

**replication fork** The region of replicating DNA that encompasses half of a replication bubble; the replicating fork contains replicating DNA and the enzymes and other proteins involved in DNA synthesis

**replication origin** A specific DNA sequence can be involved in the initiation of DNA replication

**repressor proteins** Regulatory proteins that repress expression and prevent a gene from being transcribed into RNA

**reproductive cloning** The production of genetically identical animals using somatic cell nuclear transfer methods

**restriction fragment length polymorphisms (RFLPs)** Differences in the DNA sequences of restriction enzyme cleavage sites in an individual's genome can produce DNA fragments of different lengths when the individual's genome is cut by that restriction enzyme; these differences are used to map genes and identify people

**result** The information obtained from a query to a database or the data obtained from a scientific experiment

**retrovirus RNA genomes** The viruses infect cells by reversing the usual flow of genetic information in the cell; the viral RNA is copied into DNA which is integrated into the chromosome DNA (provirus)

**reverse transcriptase** A protein enzyme that copies single-stranded RNA into double-stranded DNA

**ribonuclease** An enzyme that degrades RNA molecules in the cell

**ribosomal RNA (rRNA)** RNA molecules that bind to ribosome proteins and have an essential part of the structure and function of a ribosome

**ribosomes** Very large multiprotein-rRNA complexes that perform protein synthesis (translation) in the cell

**ribozyme** An enzyme that is composed of protein and RNA components in an RNA-protein complex

**RNAi gene silencing** A research technique in which the expression of a gene in live cells can be turned off when desired

**RNA polymerase** Enzyme that copies DNA templates into RNA strands (transcription)

**RNA polymerase I** Eukaryotic RNA polymerase that transcribes the DNA genes encoding the large ribosomal RNAs

**RNA polymerase II** Eukaryotic RNA polymerase that transcribes the DNA genes encoding proteins in the cell (also called structural genes/proteins)

**RNA polymerase III** Eukaryotic RNA polymerase that transcribes the genes for 5S ribosomal RNA and the transfer RNA genes

**Score (S)** The numerical value assigned to each match resulting from a BLAST database search (the higher the score, the better the match)

**segregated** When chromosomes move to opposite poles during cell division

**selectable marker gene** A gene that kills the cells that have not taken up a plasmid and allows only cells carrying the plasmid to grow (a selection method)

**semiconservative** Mode of DNA replication in which each daughter DNA molecule contains one of the two original DNA strands and one new complementary DNA strand

**senescence** The process of growing old for organisms and individual cells

**sequence alignment** A way of arranging DNA, RNA, or protein sequences to identify regions of sequence similarity that might be of functional, structural, or evolutionary significance

**serum**  Blood plasma containing clotting factor proteins

**sexually propagated**  The reproduction of plants using seeds or spores

**Shine-Dalgarno sequence**  A conserved sequence near the start of mRNA that is recognized and bound by a rRNA in the prokaryotic ribosomes

**short tandem repeat (STR)**  Subset of VNTR (variable number tandem repeats) made up of short repeated DNA sequences

**shuttle vectors**  Vectors that can replicate in more than one type of cell and can transfer or shuttle between the two organisms as desired

**signal sequence**  Hydrophobic amino acid residues at the amino terminus of secretory or integrated membrane proteins that function to direct the protein to the appropriate cell membranes

**single nucleotide polymorphisms (SNPs)**  Single base pair differences between two individual genomes

**small interfering RNAs (siRNAs)**  Short RNA molecules involved in controlling gene expression

**spliced**  To form new genetic combinations by intron removal from precursor RNA to connect the exons and produce the mature mRNA

**spliceosome**  A ribonucleoprotein complex that removes the introns from precursor mRNAs and joins the exons together to form functional mRNA

**spontaneous mutation**  A change in DNA sequence usually due to errors in the normal functioning of cellular enzymes

**Src**  The first oncogene to be identified and studied in the laboratory

**stem cell niche**  The immediate environment surrounding the locations of stem cells in the body

**stringency**  Conditions that influence the base pairing (hybridization) interactions between complementary nucleotide sequences

**structural genes**  DNA sequences that code for structural proteins in the cell

**sustainable agriculture**  A system of farming that takes into account the environmental, social, and economic issues relating to the growth and distribution of crops

**synapse**  The gap in the junction between the ends of two nerve cells

**syndactyly**  The most common congenital anomaly of the hand, marked by the persistence of webbing between fingers and toes after birth

**TATA box**  DNA binding site for a transcription factor that guides RNA polymerase II to the promoter region of eukaryotic genes

**telomere**  Special repeated sequences located at the ends of linear eukaryotic chromosomes that are required to replicate the DNA at the ends of the chromosomes

**teosinte**  A tall annual grass that is closely related to and possibly the ancestor of Indian corn

**termination signal**  A specific DNA sequence that signals the release of both RNA polymerase and the newly made RNA transcript from the DNA template

**tetranucleotide repeat**  A type of short tandem repeat in which a four-base pair sequence is repeated 5 to 50 times in the human genome; repeats are used for most forensic DNA testing methods

**the most recent common ancestor (TMRCA)**  The amount of time or number of generations since individuals have shared a common ancestor

**therapeutic cloning**  Cloning of an embryo for the purpose of deriving embryonic stem cells for therapeutic treatments

**thymine (T)**  A pyrimidine base found in DNA but not in RNA

**thymidine kinase gene (tk)**  A gene often carried in transgenic vectors used in mammalian cells coding for an enzyme that phosphorylates and inactivates the nucleoside analog ganciclovir

**traits**  Genetically determined characteristics and phenotypes exhibited by an individual

**transcription**  Process by which genetic information in DNA is copied into an RNA transcript

**transcriptionally silent**  Region of a genome in which no transcription occurs (gene expression is off)

**transcription factor**  Protein that regulates gene expression by binding to DNA in the control regions of the gene and interacting with RNA polymerase enzyme

**transcription initiation**  The initiation of transcription events by interactions between proteins, promoters, and RNA polymerase

**transduction**  The transfer of genetic information from one bacterium to another as a result of a bacteriophage carrying bacterial chromosome DNA

**transfecting**  The process of transferring DNA (usually of a gene) into a cultured cell using a virus-based vector

**transfection**  A process in which virus-based vectors are used to transfer recombinant DNA into cells

**transfer RNA (tRNA)**  Short RNA molecules that carry amino acids to the ribosome and help to translate the genetic code during protein synthesis

**transformation**  Process in which genes are transferred into bacterial or mammalian cells as foreign DNA carried in a plasmid-based vector

**transforming substance**  In early experiments, DNA was found to change (transform) the biochemistry of bacterial cells

**transgene**  The foreign gene that is inserted into a plant or animal of interest using genetic engineering and recombinant DNA methods

**transgenic animals** Animals carrying foreign DNA transgene integrated into their genomes

**transgenic plants** Plants containing an integrated foreign gene (transgene)

**translocations** Breaking the DNA helix in one chromosome and attaching it to a different DNA helix in a completely different chromosome

**trisomy** A condition in which an extra copy of an entire chromosome is inherited by an individual's cells (for example, Trisomy 21 is Down Syndrome)

**trophectoderm** The outer layer of a developing embryo after the differentiation of the ectoderm, endoderm, and mesoderm layers

**tumor suppressor genes** Genes that act to prevent unwanted cell division and as a result they suppress the development of cancer cells

**undifferentiated cells** Cells that have not yet become specialized or exhibit the morphological and functional characteristics they will acquire upon differentiation

**uniregional model** One of three theories of the evolution of *homo erectus* into the anatomically modern human (*homo sapiens*) which proposes that recent African migrants did not interbreed with hominids they encountered in Europe and Asia but, rather, these other hominid groups became extinct and the African migrants replaced them

**upstream** Regions of the DNA genome located to the 5′ side of each gene (before the start of the transcribed region

**uracil (U)** A pyrimidine base found in RNA but not in DNA

**vectors** DNA molecules modified for use as carriers to transport foreign genes

**virus packaging extract** An extract from infected cells that rapidly packages lambda (λ) vector DNA, along with any inserted foreign DNA, into virus particles that transfer the DNA into the bacterial cells

**xenotransplantation** Transplanting organs between species, such as organs transplanted from animals into humans to treat organ failure

**X-linked gene** A gene located on the X chromosome

**x-ray diffraction** A process in which crystallized molecules are rotated and bombarded with x-rays to determine structural information about the molecule

**yeast artificial chromosome (YAC)** A synthetic chromosome vector made from yeast genome DNA that can carry large inserts of foreign DNA and replicate in yeast cells as normal linear chromosomes

**zygote** The cell resulting from the union of a male and a female gamete during fertilization

**zymogen** A protein that requires a chemical change for the molecule to become an active enzyme

## A

Acridine compounds 50
Actin filaments 236f
Active site proteins 295f
ADA *See* Adenosine deaminase deficiency
Adenine 17
Adenosine deaminase deficiency (ADA) 222, 252–254, 261
Adenoviruses 246–247
Adleman, Leonard M. 307
Adult cells 274, 275–278
Adult stem cells 266, 271
Affinity chromatography 116–117
Agbiotech 337–339
Age-related macular degeneration (AMD) 260
Agilent Bioanalyzer 299f
Agricultural biotechnology 331–332, 333–334
*Agrobacterium tumefaciens* 334–335, 334f
Air pollution 209–210
Alkaline 338
Alleles 175, 218, 362
   dominant 219, 233
   parental 357t, 364–365
   recessive 219, 233
Alpha protein 47f
ALS *See* Amyotrophic lateral sclerosis
Alternative RNA splicing 74–75, 63f
Alzheimer, Alois 223–224
AMD *See* Age-related macular degeneration
Amerinds 188f
AMH *See* Anatomical modern human
Amino acids 44–46, 45f
   chains 44, 296
   chromosomes and 50–53
   first 50–51
   phenylalanine 52f
   sequence 151–154, 160–165
Aminoacyl tRNA synthetases 65–66, 66f
Amyotrophic lateral sclerosis (ALS) 224
Anatomical modern human (AMH) 186
Aneuploidy 197
Angiogenesis 200–201, 260
Animal biotechnology
   introduction to 312
   looking ahead 311–312
   regulation of 328
Animal cloning 327–328
Animal genomes 312–314
Anthers 342
Antibiotic resistance genes 335
Antibodies
   humanized 317–318, 322f

monoclonal 316–317, 323–326, 322f
polyclonal 316
proteins 317
structure of 321f
Anticancer drugs 297f
Anticodon 55–56
Antigens 316
Antithrombin protein (AT) 314–315
Anti-toxins 316
Antiviral interferon 340f
*Apanteles glomeratus* 342
Apoptosis 192, 213f, 214f *See also* Programmed cell death
   embryos with 10 figures/10 toes 214
   genetic cell death and 213–214
Applied bioinformatics 165–170
*Arabidopsis thaliana* 129–130, 153–154, 168
*Archaea* 130, 153
Artificial bacterial genome 102
Artificial chromosomes 109
Asexually propagated plants 333–334
Assimilation models 186, 187
Asymmetric cell division 265, 269f
AT *See* Antithrombin protein
Attenuation 70, 71
Autism 237, 318
Autologous stem cell transplants 266
Automated DNA sequence 138f
Avery, Oswald 11f

## B

*Bacillus amyloliquefaciens* 342
*Bacillus subtilis* 102, 110–111
*Bacillus thuringiensis* (Bt) 338
Bacterial cells 106, 64f
Bacterial chromosome 79
Bacterial reproduction 29
Bacterial restriction 84
Bacterial RNA polymerase 57f
Bacterial strains 111
Bacterial viruses 12f
Bacteriophage lambda 107f
Bacteriophages 81–82, 51f
Baculovirus 123, 123f
BamHI enzyme dimer 88f
BamHI recognition source 88f
Base-paired secondary structure 142f
Basic Local Alignment Search Tool (BLAST)
   from comparing nucleotide sequences 162–165, 164f
   home pages online 163f
   protein sequences with 165

searches for 166
Beano 262
Benign tumors 201
Beta globin protein 47f
BiDil 360
Binet, Francis 366
Biocontrol alternatives 342–344
Biodiversity 140
Biofuels 344, 346, 347–348
Bioinformatics 149, 245
   applied 165–170
   computer science and 170
   explosion of data fueled 150–151
   field of 341–342
   introduction to 150
   looking ahead 150
   using 160–165
Biolistic gene gun 335
Biological control 342
Biological data 154–160
Biomarkers 289–290
Biomedical research 238–239
Biomotor proteins 306
Biopesticide 338
BLAST *See* Basic Local Alignment Search Tool
BLAST searches (blastx) 162–164
blastx *See* BLAST searches
Blind dogs 255–256
Blindness 132
Blood cell diseases 219f
Blood pressure 226
Blood types *See also* Human blood types
   to classify humans 356–357
   from parental alleles 357t
   relative frequencies of 358t
Blood vessels 201f
Boas, Franz 354
Bone marrow transplant therapy 254f
Bottlenecks 187–188
Boyer, Herbert 88–89
Brain health 300f
Brain plaques 224f
Breast cancer 208–209, 209f
Breathing problems 226
Bt *See* Bacillus thuringiensis
Bucky ball allotrope 301
Budding yeast cell cycle 199f

## C

Caenorhabditis elegans 129–130, 131f
Calgene 333f
Caliper Technologies Corp. 298